Springer Series in Statistics

Springer Series in Statistics

L. A. Goodman and W. H. Kruskal, Measures of Association for Cross Classifications. x, 146 pages, 1979.

J. O. Berger, Statistical Decision Theory: Foundations, Concepts, and Methods. xiv, 425 pages, 1980.

R. G. Miller, Jr., Simultaneous Statistical Inference, 2nd edition. xvi, 299 pages, 1981.

P. Brémaud, Point Processes and Queues: Martingale Dynamics. xviii, 354 pages, 1981.

E. Seneta, Non-Negative Matrices and Markov Chains. xv, 279 pages, 1981.

F. J. Anscombe, Computing in Statistical Science through APL. xvi, 426 pages, 1981.

J. W. Pratt and J. D. Gibbons, Concepts of Nonparametric Theory. xvi, 462 pages, 1981.

V. Vapnik, Estimation of Dependences based on Empirical Data. xvi, 399 pages, 1982.

H. Heyer, Theory of Statistical Experiments. x, 289 pages, 1982.

L. Sachs, Applied Statistics: A Handbook of Techniques, 2nd edition. xxviii, 707 pages, 1984.

M. R. Leadbetter, G. Lindgren and H. Rootzen, Extremes and Related Properties of Random Sequences and Processes. xii, 336 pages, 1983.

H. Kres, Statistical Tables for Multivariate Analysis. xxii, 504 pages, 1983.

J. A. Hartigan, Bayes Theory. xii, 145 pages, 1983.

F. Mosteller, D.S. Wallace, Applied Bayesian and Classical Inference: The Case of *The Federalist* Papers. xxxv, 301 pages, 1984.

D. Pollard, Convergence of Stochastic Processes. xiv, 215 pages, 1984.

R. H. Farrell, Multivariate Calculation. xvi, 367 pages, 1985.

Roger H. Farrell

Multivariate Calculation

Use of the Continuous Groups

Springer-Verlag
New York Berlin Heidelberg Tokyo

Roger H. Farrell
Mathematics Department
Cornell University
Ithaca, NY 14853

AMS Classification: 62HXX

Library of Congress Cataloging in Publication Data
Farrell, Roger H.
Multivariate calculation.
(Springer series in statistics)
Bibliography: p.
Includes index.
1. Multivariate analysis. 2. Groups, Continuous.
I. Title. II. Series.
QA278.F37 1985 519.5′35 84-13971

Typeset by Asco Trade Typesetting Ltd., Hong Kong.
Printed and bound by R. R. Donnelley & Sons, Harrisonburg, Virginia.
Printed in the United States of America.

9 8 7 6 5 4 3 2 1

ISBN 0-387-96049-X Springer-Verlag New York Berlin Heidelberg Tokyo
ISBN 3-540-96049-X Springer-Verlag Berlin Heidelberg New York Tokyo

Preface

Like some of my colleagues, in my earlier years I found the multivariate Jacobian calculations horrible and unbelievable. As I listened and read during the years 1956 to 1974 I continually saw alternatives to the Jacobian and variable change method of computing probability density functions. Further, it was made clear by the work of A. T. James that computation of the density functions of the sets of roots of determinental equations required a method other than Jacobian calculations and that the densities could be calculated using differential forms on manifolds. It had become clear from the work of C. S. Herz and A. T. James that the expression of the noncentral multivariate density functions required integration with respect to Haar measures on locally compact groups. Material on manifolds and locally compact groups had not yet reached the pages of multivariate books of the time and also much material about multivariate computations existed only in the journal literature or in unpublished sets of lecture notes. In spirit, being more a mathematician than a statistician, the urge to write a book giving an integrated treatment of these topics found expression in 1974–1975 when I took a one year medical leave of absence from Cornell University. During this period I wrote *Techniques of Multivariate Calculation.*

Writing a coherent treatment of the various methods made obvious required background material. In *Techniques* I tried to include some of this material on null sets, exterior algebra, symmetric functions, and consequences of the uniqueness theorem for Haar measures, especially the implications about factorization of measures.

Techniques received favorable commentary. S. Karlin asked me to include material of his that utilized a direct application of the uniqueness theorem for Haar measure. I had helpful conversations with among others James Bondar, Sam Karlin, Jack Kiefer, Avi Mandelbaum, James Malley, Akimichi

Takemura, and Brenda Taylor. Further, with major revisions, the Editorial Board of the Springer Series in Statistics was receptive to the idea of a hardcover book.

In the intervening years I have not found any mathematical errors in *Techniques*, and the material included there seems as relevant today as it did then. Consequently the purpose of this new book is to include new material. The new material includes exposition of Karlin's ideas, more material on null sets, further development of the use of multiplicative functionals to compute Jacobians, material on Gaussian processes, recent developments about zonal polynomials, and an exposition of topics about multivariate inequalities. Time limitations did not allow exploration of papers by A. P. Dawid which use random variable methods, S. A. Andersson on maximal invariants and eigenvalue problems, and R. A. Wijsman on cross-sections and maximal invariants. Had time allowed I would have used some of this material.

Techniques included a discussion of the matrix algebra needed to study the coefficient matrices of homogeneous polynomials, this leading to an algebraic analysis of zonal polynomials. There has now emerged four distinctly different approaches to the discussion of zonal polynomials: combinatorial-group representations, which is the method used by A. T. James; the algebra of coefficient matrices, which is the method of *Techniques*; zonal polynomials as the eigenfunctions of a commutative family of linear transformations, which is the development of Constantine, Saw, Kates, and Takemura; and zonal polynomials as the solutions of partial differential equations, which is the development of R. Muirhead. Each of these approaches has its own validity and each method yields results seemingly not obtainable by the other methods. An adequate discussion of all these approaches would require a separate book. Consequently the discussion of *Techniques* has been kept with but few changes, which are in Section 12.2 to reflect the more elegant results that are available, Section 12.7 to reflect a deeper understanding of the Schur functions, and Sections 12.13 and 12.14 which I have added to reflect the new results due to A. Takemura and L. K. Kates.

I do not believe that this is a definitive book. I can only hope that the book will help others to understand and that the book will provide a background from which further development can proceed.

Ithaca, New York ROGER H. FARRELL

Contents

Important Examples and Definitions*

* Numbered by section number in order of occurrence.

CHAPTER 1

Introduction and Brief Survey

1.1. Aspects of Multivariate Analysis

Multivariate analysis originated with problems of statistical inference in the work of Pearson and Fisher, men with thorough grounding in applied statistics. The first important book on the subject, Anderson (1958) gives a balanced view of the subject by treating, in each case, first the question of inference, and then, the calculation of the multivariate density function of the resulting multivariate statistic, most often for a likelihood ratio test. At that time the multivariate statistics considered had density functions. In the past decade study of estimators of multivariate parameters of Poisson and negative binomial random variables has made discrete problems an important part of multivariate inference. Material about discrete problems was outside the scope of Anderson (1958) and remains outside the scope of recent books such as Eaton (1983) and Muirhead (1982).

Many of the hard mathematical problems arise in connection with the problem of the calculation of density functions. Anderson (1958) was part of a general development of the 1950's of techniques for finding the multivariate density functions. Random variable techniques found expression in Wijsman (1957, 1958), the use of Jacobians and change of variable found expression in Anderson's book, the use of differential forms on manifolds was developed by James (1954), and the use of invariance, matrix decompositions and maximal invariants found expression in Stein (1956c, 1959) and Karlin (1960). The use of Fourier transforms was widely known. In the past twenty years no really new methods have appeared. Instead there has been a consolidation of results in books like those by Eaton (1983), Farrell (1984) and Muirhead (1982).

The techniques of the 1950's produced answers to many previously unanswered noncentral problems. See particularly James (1954, 1955a, 1960, 1961a, 1964), Constantine (1960, 1963, 1966), Constantine and James (1958), Karlin (1960), Schwartz (1966a, 1967a), and Wijsman (1966). It was found that the expression of the answer in many of these problems involved integrals of functions of a matrix argument, integrals that apparently cannot be evaluated in closed form in terms of the usual elementary functions. This was noted, especially by James (1955a, 1955b).

The development of the 1960's has centered about special functions and their use as an alternative to integrals as a means of representing the answers to these problems. One example of the use of special functions occurs in Mathai and Saxena (1969, 1978) where H-functions and Mellin transforms are used to study the distribution of a product of noncentral chi-square random variables, needed for inference about the generalized variance. But more in fashion today is the use of hypergeometric series which are sums of zonal polynomials multiplied by hypergeometric coefficients. This latter approach originated with Herz (1955) who defined the hypergeometric functions by use of Laplace and inverse Laplace transforms. Constantine (1963) showed that the functions defined by Herz were representable as weighted sums of the zonal polynomials that were being developed by James (1960, 1961a, 1961b, 1964, 1968). Also, Herz (1955) and Constantine (1966) dcfine Laguerre polynomials of a matrix argument and these polynomials are finding use in numerical analysis. More recently the book by Muirhead (1982), especially, emphasizes the expression of results by formulas with hypergeometric coefficients and hypergeometric functions.

An alternative to the use of special functions is the use of approximate answers, i.e., asymptotic methods. A frequently cited early reference is Box (1949) who used inversion of Fourier transforms to obtain asymptotic expansions. Since the middle 1950's a vast literature about asymptotic expansions has appeared. A typical literature item gives several terms of an asymptotic expansion without provision for any error bound on the remainder. An exception to this pattern is Korin (1968) who obtains a complete asymptotic series which he then uses to check previous approximations by others. The statistical problem examined by Korin is that of making tests about the covariance matrix. Sometimes asymptotic series are obtained accidentally, as for example Dynkin and Mandelbaum (1983). Today a primary source about asymptotic series is the set of research papers of Muirhead, and also, his book, op. cit. Many references may be located in the bibliography of this book and in Muirhead (1982).

In the past decade high-speed computation has become very available, and actual computation of probabilities from multivariate density functions has become a reality. Likewise more complex Monte-Carlo calculations are being undertaken using multivariate quantities such as random orthogonal matrices. The expression of answers, as done in this book, as integrals by a Haar measure, or, as a hypergeometric function, or, as an infinite series of zonal polynomials, does not face up to the difficulties of computation. The author has seen a number of comments by users of zonal polynomials that indicate the series representing hypergeometric functions as weighted sums of zonal polynomials converge very slowly, hence are not really computable. A major area of concern today, then, should be computation. The asymptotoic results such as those of Muirhead seem central to the computation question.

As noted earlier the subject of inference underlies all of multivariate analysis. The problems of inference may be classified in several ways which we mention here. Estimation and tests of hypotheses; invariance; maximum likelihood and Bayes; admissible, minimax, locally minimax, unbiased, asymptotically best; the analysis of variance, regression, linear hypotheses and the general linear hypothesis; design and combinatorial questions; ranking problems; confidence sets and their optimality; improving on point and set estimators; robustness; and so on.

In much of the literature determination of a suitable statistic and its distribution or density function is a far as the author carries a given development. Anderson (1958) follows the pattern: problem, likelihood ratio or maximum likelihood, distribution, without consideration of the "good" properties of the procedure. Muirhead (1982) places primary emphasis on multivariate statistics that arise in tests of hypotheses and in this context mostly restricts consideration to likelihood ratio tests. But he does try to lay groundwork for consideration of noncentral problems, power function computation, and consideration of robustness through the consideration of eliptical as well as spherical distributions.

The optimality literature is thin due to the difficulty of the subject. See Kiefer (1966, 1974). In recent years a series of papers, not referenced in the bibliography, but see Marden (1982), by Marden and Perlman, have carried the methodology of decision theory to the point of establishing admissibility and inadmissibility of many classical invariant tests within the class of invariant tests. The methodology has depended on the null hypothesis being a single orbit in the parameter space so that the null hypothesis is simple within the class of invariant tests. The earlier literature on unbiasedness and monotone properties of power functions seems to have consisted of Anderson (1955), Anderson and Das Gupta (1964), Cohen and Strawderman (1971), Gleser (1966), Lehmann (1959), Mudholkar (1965, 1966a, 1966b), Sugiura and Nagao (1968). With the development of the FKG-inequality, see Kemperman (1977), a theory of log-concave measures, see Rinott (1976), and a theory of majorization, see Marshall and Olkin (1979), further progress has been made in, for example, Olkin and Perlman (1980).

The relationship between being minimax or most stringent and being invariant is discussed in Lehmann (1959). From an example due to Stein published in Lehmann (1959) it follows that the interesting Hunt–Stein theory fails to apply in many interesting statistical examples in which the transformation group is "too big." Such groups include the full linear group and nonparametric examples in which random variables are transformed by groups of monotone functions. An outstanding problem in the subject of most stringent tests is to show the UMP tests of $a = 0$ vs $a \neq 0$ using the Hotelling T^2 statistic are most stringent tests on orbits of the parameter space. Giri, Kiefer and Stein (1963) reduced this problem to that of solving an integral equation in the simplest case thus showing the Hotelling T^2 test to be minimax. Salaevskii (1971) has completely solved the integral equation in question for two dimensional random variables (cf. *Math. Reviews* **42** 5380). The English translation is referenced in the bibliography. It has been rumored that the complete problem was solved by Salaevskii in an enormous calculation but we know of no published source. This is one of the simplest cases of an invariant multivariate statistic and the minimax question in harder cases is untouched. Marden, op. cit., does not seem to feel his work will give results about the minimaxity of invariant tests. If the group is amenable then the Hunt–Stein theory does apply. See Bondar and Milnes (1981).

The first complete class theorem for a multivariate problem was proven by Birnbaum (1955). Some years later Birnbaum's result was generalized by Farrell (1968). The sufficiency part of Birnbaum's proof was greatly simplified by Stein (1956b) who restated the form of the result and then used the restated form to prove admissibility of the Hotelling T^2 test. The ideas of Birnbaum, Farrell and Stein have in part provided the basis for the work of Marden and Perlman. Stein's idea was used by Schwartz (1966a, 1967a). Roughly at the same time Kiefer and Schwartz showed many classical multivariate tests are Bayes and thus admissible and this was further explored by Schwartz (1966a, 1966b). Stein's method of obtaining the density function of a maximal invariant, Stein (1956c), was developed by Schwartz (1966a) in order to get a complete class theorem for the general linear hypothesis. Related methodology has been interesting in its own right, see Andersson (1982), Bondar (1976), Koehn (1970), and Wijsman (1966, 1984).

A result following from the results of Kiefer and Schwartz, op. cit., is that many invariant multivariate tests are admissible tests. That the corresponding result for multivariate estimation might be false and is in fact false was first shown by Stein (1956a). For tests of hypotheses the admissibility proofs have depended in an essential way on the actual distribution of the multivariate statistic in question whereas the inadmissibility proofs for estimators come from the construction of improved estimators. The constructions seem to have an almost nonparametric character in that the computations seem to depend only on the "moment structure" of the problem, to borrow a term from Berger.

In recent years much work has been done on improving multivariate estimators. Most of the key literature is not referenced here, but it is generally recognized that major contributions have been made by J. Berger, L. D. Brown, C. Stein, and J. Zidek, to name a few. Lehmann (1983) is now a starting point for learning about inference in multivariate estimation. There are few research papers using zonal polynomials for the discussion of inference. Olkin and Perlman (1980), and Zidek (1978) are included in the bibliography as examples of a test of hypothesis problem and of an estimation problem in which computations with zonal polynomials are made. In testing, discrete density functions are inconvenient, since randomization may be necessary. This difficulty does not arise in the theory of estimation and contemporary work on improving estimators also includes results on discrete random variables. Those working with estimation problems have not always been so lucky as to obtain optimality results. Instead the subject has been one of multivariate inequalities, solutions to multivariate difference and differential inequalities, and numerical computation to evaluate magnitudes of improvement. Whereas it is known that very large improvements do result, it is not generally known how to obtain improved estimators that are admissible.

Complete class results for estimators are in general more difficult than those obtained for tests. The literature is not referenced in the bibliography. The original paper, L. D. Brown (1971), on estimation of a multivariate normal mean, now has a sequel in I. Johnstone (1981), Ph.D. Thesis, Cornell University, regarding estimation of a multivariate Poisson mean vector, and a partial sequel in Mandelbaum (1983) regarding estimation of the drift of Brownian motion. Mandelbaum identifies all admissible linear estimators of the drift. Brown and Farrell (1983) have obtained a complete class theorem for linear estimators of a Poisson mean vector.

Literature on ranking problems is now vast and we only mention the earlier book by Bechhofer, Kiefer and Sobel (1966), together with the names, Bechhofer, Gupta, regarding inference about best subset selection. Design problems have been studied extensively by Kiefer and Wolfowitz and subsequently by Kiefer's students. The great theory paper is Kiefer (1974).

Various of the material in the chapters of this book bear on these problems. The author has viewed this book as trying to lay an adequate mathematical background for reading the literature on inference. To actually try and discuss questions of inference is beyond the scope of this book. Overall the subject of multivariate analysis in the continuous case is too vast for one book to make an inclusive treatment. It was decided to write a methods book that would provide necessary mathematical background and that would illustrate methods of density function computation and forms of density function expression that have been used in the literature. In 1976 the background and methods were not treated in books on the subject and only now with Farrell (1976), Muirhead (1982), and Eaton (1983) is the synthesis of the subject into books beginning to catch up.

1.2. On the Organization of the Book

The discussion of topics in inference makes it clear that the literature on inference might today warrant several books. In this book the discussion has been limited to a discussion of the calculations and side issues related to inference have been skipped. Chapters 2, 4, 5, 9, 10, and 11 are directly concerned with distinct techniques of computing or otherwise determining density functions while Chapters 3, 6, 7, and 8 give the development of needed mathematical background. Because of the importance and mathematical complexity of the subject, Chapters 12 and 13 on zonal polynomials have been included but the content of these chapters really belongs to the subject of special functions and the evaluation of multivariate integrals. Chapter 14 contains isolated but important material needed for inference and as background for looking at the eigenvalue function $\lambda(A)$ of symmetric matrices A.

We assume the reader knows about Jacobians and changes of variables and has access to Anderson (1958) and the recent books by Eaton and Muirhead. We use Jacobians sparingly and do not go out of our way to discuss them in this book. Usually the multipliers that result from change of variable can be determined by algebraic manipulation of multiplicative functionals, which we emphasize, as being more elegant. Representation of density functions as integrals by Haar measures often are more revealing of the structure that decision theory methods of inference use to try and decide the structure of good tests. This consideration has in part dictated that in this book answers be represented as integrals rather than as hypergeometric functions. The answers obtained may require special functions for numerical evaluation. If so that is beyond the scope of this book. Likewise asymptotic series are not discussed. Inference is not discussed except in Section 1.1, nor are the statistical problems that underlay various statistics discussed.

As a methods book the book is very long. Yet it is sparsely written and assumes a great deal of its readers. Standard complex variable theory is needed in Chapter 2 and for the references to Chapter 12. Measure theory equivalent to most of Halmos (1974) is assumed. If the reader has not seen a development of regular measures in metric spaces, locally compact Hausdorff spaces, and locally compact groups, then it is assumed the reader will take the sketch presented in this book and make good use of references like Halmos (1974), Loomis (1953) and Nachbin (1965). The theory of analytic manifolds cannot be done completely here and Dieudonne (1972) can be used for reference. On the algebraic side, a good understanding of quadratic forms, positive definite matrices, and canonical forms is essential. Chapter 8 can help here. Often graduate students today do not learn about determinants and forms, in which case the reader should find a suitable source and read. The algebra of this book is basically easy except for Chapter 12. In Chapter 12 we develop a theory of the mathematics of zonal polynomials and extensive reference will be made to Loomis (1953) and Weyl (1946) for material on algebras and the symmetric group, to Littlewood (1940, 1950) for material on group characters and symmetric functions, and Helgason (1962) on group representations and spherical functions. There are a number of more contemporary books on group representations referenced in the bibliography not used as references in the text. Chapter 12 is nearly self contained. I have tried to take the necessary material from these sources and make a coherent development, without use of group representations, of the Constantine–James theory of zonal polynomials. For the most part, the use of zonal polynomials is beyond the scope of this book, although a small amount of material is included in Chapter 13, which consists almost totally of problems on the use of zonal polynomials together with some connective material and references to source.

Many of the results of Chapter 2 are only sketched and are essentially problems but are stated as results in the overall context of the discussion. Other chapters have some problems, except Chapter 3 and Chapter 12. Because of its length, Chapter 13 consists of the problem set for Chapter 12. In Chapters 4 through 11 and in Chapter 14 the problems are collected together in a section at the end of the chapter. A number of items of theory needed at later stages of the book are stated as problems and in the sequel are referenced by their problem number. Thus the problems contain partly theory, partly calculations of intermediate results needed later, and partly illustrative examples of the distributions of statistics used in the literature.

There is a considerable amount of relevant commentary that is not formally stated as a lemma, theorem, or problem. Usually this material is in a numbered remark which makes subsequent reference possible.

The contents of the auxillary chapters are as follows. Chapter 3 on Haar measures stresses the uniqueness theorems which explain why various invariant measures factor to allow integration out of variables. This concept is at the heart of Karlin's idea developed in Chapter 5 and at the heart of Stein's idea developed in Chapter 10. If differential forms are used then the same factorizations occur and are seen in the way differential forms for invariant measures occur in the expressions, thus allowing identification of measures constructed from the integration of multivariate differential forms in Chapters 6, 7, and 9 as well as in examples that occur in Chapters 3, 5, and 10.

Since the book is really about multivariate normal density functions, Chapter 2 introduces this topic using Laplace transforms. Chapter 4 starts with a geometric concept, obtains the Wishart density, and develops a method whereby, after a central distribution problem has been solved, the noncentral problem is also solved free of much extra work. The ideas explored in Chapter 4 are really a motivation for parts of Chapters 9, 10, 12, and 13. Chapters 6 and 7 develop elementary manifold techniques sufficient for the computations with differential forms that occur throughout the book. However, Chapter 6 is not a complete development in that the existence of local coordinates is not shown, merely suggested in some of the examples. The actual factorizations of manifolds result from various matrix decompositions and these are discussed in Sections 8.1, 8.2, and 8.5. Section 8.4 is on sets of measure zero and represents results generally needed but not widely known. Section 8.3 on Hilbert space valued Gaussian processes is mostly about operators on Hilbert space, i.e., infinite dimensional matrices.

The main techniques of computation discussed are as follows. Laplace and Fourier *transforms* are discussed in Chapter 2 and are illustrated by development of the multivariate normal distribution theory and by obtaining the density functions of the noncentral chi-squares. *Geometrical* reasoning is explained by example in Chapter 4. Wishart (1928) has been used to obtain a first derivation of the Wishart density function. Then James (1955a) has been used to show how, once extra variables in the central problem have been integrated out, the noncentral problem can be solved. Implicit in the development of Chapter 4 is the factoring of the space of matrices into invariants. In cases in which a *transitive group* of transformations acts on the space of matrices Karlin's methods are applicable and are developed by examples in Chapter 5. Implicit again is the factorization of the group and space of matrices. The invariants are not always "nice" and determination of the actual integrating measures being induced on invariants such as sets of eigenvalues requires methods such as the *decomposition of manifolds* and the factorization of differential forms. Examples are in Section 5.3 and Chapter 9. The concept here results from the realization that the variables to be integrated out naturally take their values in homogeneous spaces, hence the description in terms of manifolds. As in Section 3.4 and Chapter 5, in Chapter 10 uniqueness of Haar measures on homogeneous spaces is used to *factor invariant measures* and to obtain the density functions of maximal invariants. This is illustrated by the example of the general linear hypothesis. And, *random variable* techniques, Chapter 11, are developed mostly by example. An important example, the analysis of variance and best linear unbiased estimation, is treated in detail. Section 8.3 on Gaussian processes is random variable and linear in character.

This book is an expansion, revision and correction of Farrell (1976). The author has been accumulating the material included for approximately twenty-eight years. The original concept, that of presenting several methods in isolation, which were to be transforms, geometrical arguments, differential forms, random variable techniques, has long since proven inadequate since there are far too many interrelations between steps of one method and those of another. And, it is only as one tries to work with a method that the necessary mathematical preliminaries are discovered. Since the author continues to find new material relevant to the original concept, various parts of the book have grown disproportionately. Whereas in Farrell (1976) it might have been possible to read the parts of the book in sequence, that is much less true of this book. In terms of the cross references made, an approximately linear ordering of material is now:

Chapter 8 excepting Section 8.3. Matrices and null sets.
Chapter 6. Exterior algebra and differential forms.
Chapter 2. Transforms. Multivariate normal density functions.
Chapter 11, Section 8.3. Random variable techniques.
Chapter 14. Multivariate inequalities.

Chapter 3. Invariant measures.
Chapter 7. Invariant measures on specific manifolds.
Chapter 4. Geometric reasoning and noncentral problems.
Chapter 5. Use of uniqueness to compute density functions.
Chapter 10. Cross-sections, uniqueness, factorization of measures.
Chapter 9. More examples using differential forms.
Chapter 12. Some of the theories of zonal polynomials.
Chapter 13. Problems related to Chapter 12.

If the matrix algebra of Sections 8.1 and 8.5 is familiar, and, if the reader knows the theory of differential forms, or, if the reader is prepared to skip the references to differential forms that occur in Chapters 2, 3, and 5, remembering to later reread these parts, then the present ordering of material gives a sequential presentation.

1.3. Sources and the Literature

The part of the book original to the author is the synthesis of many different sources into a whole. Much material has been passed to the author through lectures, personal communications, and through notes used to supplement courses. All such material remains outside the open literature. The author therefore acknowledges the following sources: Chapter 2, lectures by J. L. Doob (1958); Chapter 3, lectures by P. R. Halmos, I. Segal, and conversations with J. Bondar; Chapter 5, lecture notes by S. Karlin (1960); Chapter 6, lectures by G. Hunt; Chapters 6, 7, 9 and parts of Chapters 3 and 4, the writings of A. T. James; Chapter 8, T. W. Anderson (1958) and conversations with J. Malley and A. Mandelbaum; Chapter 10, conversations with J. Bondar, R. Schwartz, and R. A. Wijsman; Chapter 11, lectures by R. A. Wijsman (1958); Chapter 12, the writings of A. A. Albert, A. T. James, J. Saw, A. Takemura, H. Weyl; Chapter 13, the writings of A. J. Constantine, A. T. James, J. Saw; Chapter 14, lectures by, conversations with, the writings of, J. Bondar, C.-S. Cheng, J. H. B. Kemperman, J. Kiefer, J. Rinott.

In the eight years following Farrell (1976) the topics that have appeared to be active, as demonstrated by conversations, correspondence, and publication, are, results on sets of measure zero, on maximal invariants and their density functions, further development of the theory of zonal polynomials, multivariate inequalities, and Hilbert space valued Gaussian processes. The form of the book, particularly in these areas, has been influenced by the ideas of the people mentioned above. And this interaction has resulted in Chapters 5, 8, 10, 14 and Sections 12.2, 12.13 and 12.14 being in their present form.

No effort has been made to fully reference the current research literature. Instead, some effort has been made to include mention of key sources to contemporary thinking on the topics that seem to be evolving. The bibliography is therefore heavy on the side of contemporary books, as it is here that comprehensive overviews of the subject may be found. In the area of inequalities one should thus look at Eaton (1982) and Marshall and Olkin (1979). For zonal polynomials, Takemura (1982) to appear in the I.M.S. monograph series, and Muirhead (1982). Asymptotic expansions and special functions are not discussed in this book, but, because of their importance, a number of papers and books are referenced and starting points can be located from the bibliography. See also Muirhead (1982). Schur functions and symmetric polynomials are still subjects of great interest and a number of recent books mentioned to me have been listed in the bibliography. Infinite dimensional Gaussian random variables (see Section 8.3) is an emerging subject. The references included reflect many conversations with A. Mandelbaum and also Mandelbaum (1983).

Earlier important bibliographies may be found in Anderson (1958), Gupta (1963) and Anderson, Gupta and Styan (1972). Other general sources on multivariate problems are Anderson (1958), Anderson, Gupta and Styan (1972), Bechhofer, Kiefer and Sobel (1968), Bondar and Milnes (1979), Dempster (1969), Eaton (1972, 1983), Gupta (1963), Karlin (1960), Kiefer (1966, 1974), Kullback (1959), Lehmann (1959, 1983), Marshall and Olkin (1979), Miller (1964), Muirhead (1982), Olkin (1966), Scheffe (1959), Stein (1956a, 1956b, 1956c, 1959, 1966), Wijsman (1957, 1966, 1984), and Wilks (1962).

1.4. Notations

An effort has been made to make the notations used global rather than local. If one reads Weyl (1946) it is clear that the notations were the creation of his typewriter, as for example on page 107 we find the display

$$\#\sigma \subset \sigma, \quad \Sigma \subset \#\Sigma. \tag{1.4.1}$$

To a large extent the notations of this book have been determined by the author's typewriter, in this case, it has been an I.B.M. personal computer. The ability to program one's own characters has increased the available notations somewhat.

The book is almost entirely about matrices. Usually capital letters A, B, C, D, E, H, I, P, Q, R, S, T, U, V, W, X, Y, and Z are notations for matrices. The availability of bold face such as **H** has allowed notations for sets and random variables to be set in bold face. It must be determined from context whether the reference is to a set, random matrix, or random vector, or, random variable. Usually the notations for vectors and numbers will be small case, x for the value of x. There is a tendency to use H, U, and V for orthogonal matrices, P, Q for idempotents or for permutation matrices, and

W, X, Y, and Z for the values of random variables. The ordinary matrix sum of X and Y is $X + Y$ and the product is XY. The transpose of X is X^t. Many matrices considered are in fact functions, so that random matrices are functions on some probability space, and in the discussion of manifolds the matrix entries are functions of local coordinates on the manifolds. A frequently used notation is to form the ij-entry function of a matrix as $(Z)_{ij}$, this being particularly useful when writing a differential form. dY would mean, compute the differential of each element of Y and form the matrix of these differentials, and $(dY)_{ij}$ is the ij-entry of this matrix. Because of the frequent use of integrals and differentials, the letter d is used only in the traditional meanings of differential or of derivative or of derivative of the measure. Traditionally in random matrices the row vectors represent independently distributed random vectors, and where relevant this tradition is kept in this book. Nonsquare matrices are usually $n \times k$ or $n \times h$ representing n observations on k-dimensional or h-dimensional row vectors. All vectors in this book are column vectors and for this reason we write rows as, $a^t = (a_1, \ldots, a_n)$. Where it is necessary to partition, $n = p + q$ or $k = p + q$. Square matrices arise with multiplicative functions that are either Jacobians, modular type functionals, or are the multiplier obtained from the formation of differential forms of maximal degree. Invariants obtained from nonsquare matrices usually are in the form of square matrices. Usually these will be $k \times k$ or $h \times h$ in size. In general we will write $\mathbf{E}\mathbf{X} = M$, and will use the mean vector a only in the rank one case. No formal notation is introduced for the outer product since in this book the notation $x^t y$ is sufficient. Use of **vec** and expression of covariances by Kronecker products is avoided.

Whereas sets are designated by bold face, some sets with structure are designated by German script. Used in various contexts are $\mathfrak{A}$, $\mathfrak{B}$, $\mathfrak{C}$, $\mathfrak{D}$, $\mathfrak{F}$, $\mathfrak{G}$, $\mathfrak{H}$, $\mathfrak{M}$, and $\mathfrak{N}$. The letter ∂ is reserved for partial derivatives.

Much of the book is concerned with multilinear forms. In the abstract discussions $\mathbf{E}$ becomes an n-dimensional module over a commutative ring with multiplicative unit, whereas in the applications $\mathbf{E}$ becomes either a vector space over the reals $\mathbb{R}$ or the complex numbers $\mathbb{C}$. In all discussions a fixed free basis $e_1, \ldots, e_n$ of $\mathbf{E}$ is assumed and the canonical basis of the dual space to $\mathbf{E}$ is $u_1, \ldots, u_n$. This is the space of multilinear 1-forms. In Chapter 6 $\mathbf{M}(\mathbf{E}^q, \mathfrak{C})$ refers to the space of multilinear q-forms on $\mathbf{E}^q$ with coefficients in the ring $\mathfrak{C}$, and this notation is carried over to the development of multilinear algebra in Chapter 12. But in Chapter 12 the ring $\mathfrak{C}$ becomes the complex numbers $\mathbb{C}$. In Chapters 6 through Chapter 9 the ring is always a ring of functions such as a ring of globally defined $\mathbf{C}_2$ functions. In Chapter 12 the emphasis is on endomorphisms of $\mathbf{M}(\mathbf{E}^k, \mathbb{C})$ and the main new notations used are $\bigotimes_{i=1}^k X_i$ for the tensor product of matrices $X_1, \ldots, X_k$ and $\bigotimes_{i=1}^k X$ for the tensor product of X with itself k times.

The Greek letters μ, ν, and λ are reserved for measures. In Chapter 6 and Chapter 12 the permutations of $1, \ldots, k$ play a special role. Greek letters

σ, τ, π are used for permutations. Γ is used for the gamma function. In Section 12.3 the letters a, d, g, h, k are used for permutations as a local usage. In Chapter 12 associated with the cyclic decomposition of permutations of $1, \ldots, k$ are the partitions κ of k. Sometimes γ of g. Here we have followed the parallelism used by James (1964). The letter φ has a special role in the development of Chapter 6. χ for the chi-square statistics and $1_{\mathbf{A}}$ for the indicator function of the set $\mathbf{A}$. If $\mathbf{X}$ is $n \times h$ with independently and identically distributed rows, each normal (a, Σ) then Σ is the covariance matrix in this context and is the notation for summation in other contexts. Likewise, depending, π can be a permutation or a number, while Π is the product sign.

Often integrals are with respect to a differential form, in which case the integration notation of Section 6.5 is used; if with respect to a Haar measure on a group like the orthogonal matrices $\mathbf{O}(n)$ then dU is written; if with respect to an invariant measure induced on the homogeneous space $\mathbf{S}(k)$ of symmetric matrices then dS is used.

Except for Chapters 3, 12, and 13, matrices are with real number entries. The special sets of matrices discussed are $\mathbf{GL}(n)$, the general linear group of real $n \times n$ matrices, $\mathbf{CGL}(n)$ for the complex matrices, $\mathbf{S}(n)$ and $\mathbf{O}(n)$ as above, $\mathbf{T}(n)$ for the group of lower triangular matrices with real entries an positive diagonal entries, $\mathbf{CT}(n)$ for the lower triangular matrices with complex number entries below the diagonal, positive real number entries on the diagonal, and $\mathbf{D}(n)$ the set of diagonal $n \times n$ matrices with positive diagonal elements. These notations are used throughout and are another context in which bold face notation is used. Overlining such as $\bar{a}$ may mean complex conjugate, to be determined from context. a^* may mean the conjugate transpose of a, and $f * g$ the convolution of functions f and g. The use of primes, X', does not mean transpose but instead means X' is a matrix to be distinguished from X. In Chapter 14 the notation $\mathbf{A}^*$ is used to distinguish this set from $\mathbf{A}$. Throughout the complement of $\mathbf{A}$ is $\mathbf{A}^c$.

In Chapter 7 the Grassman and Stiefel manifolds are defined and notations established. These together with $\mathbf{S}(k)$ are the homogeneous spaces discussed. $\mathbf{T}(k)$ is somewhat special in that it is both a locally compact group and a homogeneous space of $\mathbf{GL}(k)/\mathbf{O}(k)$.

CHAPTER 2

Transforms

2.0. Introduction

In this book we do not need to use the Levy continuity theorem, cf. Feller (1966). For the uses illustrated below inversion theorems for the various transforms are sufficient. Inversion of Fourier transforms has been rendered almost trivial by the elegant calculation that appears in Feller, op. cit. Multidimensional uniqueness theorems for the (complex) Laplace transform are proven easily by induction on the dimension and these proofs are stated in the sequel. For more detail than is presented in this chapter suitable references are Feller (1966), Widder (1941) and Wiener (1933).

Aside from the Laplace transform and its special case, the Fourier transform, we discuss briefly the Mellin transform. Several authors of statistical literature have used inversions of Mellin transforms to determine multivariate density functions. Generally this last approach is used in cases where the moments have nice expressions but the Laplace transform appears to be intractable. Nonetheless the Mellin transform of a random variable $\mathbf{X}$ is $\mathbf{E}\mathbf{X}^{t-1}$ and so long as $\mathbf{X}$ is positive an obvious change of variable reduces this to a Laplace transform. Inversion theorems for the Mellin transform may be found in Widder (1941).

Section 2.1 develops the necessary uniqueness theory. Section 2.2 gives a development of the distribution theory of multivariate normally distributed random variables. The author obtained this presentation from lectures of Doob (1958). In Section 2.3 transforms are used to derive the noncentral chi-square density function. In this section the noncentral F- and t-density functions are derived. In Section 2.4 we have included a brief discussion of other inversion theorems and Hermite polynomials. Section 2.5 relates inversion of Laplace and Mellin transforms to inversion of the Fourier transform. Section 2.6 references a few pieces of significant literature which illustrate use inverse transforms in the statistical literature.

The chapter is sparsely written. Many of the proofs are only sketches intended to let the reader think through the details. The chapter does assume familiarity with Lebesgue integration in $\mathbb{R}^n$, with the computation of Jacobians of linear transformations $y = Ax$, and with the canonical form for symmetric matrices with real entries. As regards the latter, see Theorem 8.1.1 for a discussion. The Jacobian computation is related to the uniqueness theorem for Haar measures. Inasmuch as the uniqueness theorem, Theorem 3.3.2, underlies much of the material of this book we discuss the specific application here.

Theorem 2.0.1. *If μ is Lebesgue measure on $\mathbb{R}^n$ then for every nonsingular $n \times n$ matrix A and Borel subset* $\mathbf{C}$

$$\mu(A\mathbf{C}) = |\det A|\mu(\mathbf{C}). \tag{2.0.1}$$

PROOF. We may define a measure ν by $\nu(\mathbf{C}) = \mu(A\mathbf{C})$. Then ν is invariant, that is, $\nu(a + \mathbf{C}) = \nu(\mathbf{C})$. By the uniqueness theorem, there exists a function $m: \mathbf{GL}(n) \to \mathbb{R}$ such that

$$\mu(A\mathbf{C}) = m(A)\mu(\mathbf{C}), \tag{2.0.2}$$

valid for all $A \in \mathbf{GL}(n)$ and Borel subsets $\mathbf{C}$. Clearly m is a homogeneous function of degree n, $m(AB) = m(A)m(B)$, and m is nonnegative. If U is an orthogonal matrix and $\mathbf{C}$ is the unit sphere then $U\mathbf{C} = C$ so that $m(U) = 1$ follows. By Lemma 6.6.2 it follows that $m(A) = |\det A|$, the absolute value of the determinant. □

2.1. Definitions and Uniqueness

If μ is a finite signed measure defined on the Borel subsets of $(-\infty, \infty)$ then the Fourier transform of μ is

$$\int \exp(itx)\mu(dx), \qquad i = \sqrt{-1}, \qquad t \quad \text{real}, \tag{2.1.1}$$

and the Laplace transform of μ is

$$\int \exp(tx)\mu(dx), \qquad t \quad \text{any complex number}. \tag{2.1.2}$$

In this book we consider only those values of t for which the integrals are absolutely convergent. There is an extensive literature on conditional convergence, some of which may be located using Widder (1941). If the support of μ is $[0, \infty)$ then the Mellin transform of μ is

$$\int_0^\infty x^{t-1}\mu(dx), \qquad t \quad \text{a complex number.} \tag{2.1.3}$$

In this book integration is Lebesgue integration in the sense of Halmos (1974). Thus as noted above we consider only values of t for which the integrals are absolutely convergent. This is automatic for (2.1.1). In terms of the positive part μ_+ and negative part $\mu_- = \mu_+ - \mu$, cf. Halmos (1950), page 121, we require for (2.1.2) $\int \exp(tx)\mu_+(dx) < \infty$ and $\int \exp(tx)\mu_-(dx) < \infty$. The Mellin transform will be computed only for positive measures μ and we require $\int x^{t-1}\mu(dx) < \infty$.

Lemma 2.1.1. *If the Laplace transform of a signed measure μ is absolutely convergent at $t = a_1 + ib_1$ and $t = a_2 + ib_2$ then the integral* (2.1.2) *is absolutely convergent for all complex numbers t in the strip $a_1 < \operatorname{Re} t < a_2$ and is an analytic function in this strip.*

Proof. Show that the line integrals $\int_{\mathbf{C}} dt \int \exp(tx)\mu(dx) = 0$ for every smooth enough closed curve $\mathbf{C}$ in the strip. The order of integration can be changed because of absolute convergence. □

Lemma 2.1.2. *If $\varepsilon > 0$ and the Laplace transform $\int \exp(tx)\mu(dx) = \phi(t)$ of a signed measure μ is absolutely convergent in the strip $|\operatorname{Re} t| < \varepsilon$ then the Fourier transform $\int_{-\infty}^{\infty} \exp(itx)\mu(dx)$ uniquely determines ϕ.*

Proof. Two analytic functions which agree on a countable set with limit point are identical on simply connected domains which contain the limit point. □

Lemma 2.1.3. $(2\pi\sigma^2)^{-1/2} \int_{-\infty}^{\infty} \exp(itx)\exp(-\frac{1}{2}x^2/\sigma^2)\,dx = \exp(-\frac{1}{2}\sigma^2 t^2)$.

Proof. Complete the square in the exponent. Show that the result is equivalent to showing $\sqrt{2\pi} = \int_{-\infty}^{\infty} \exp((x - it)^2/2)\,dx$, $t \in (-\infty, \infty)$. This may be shown using contour integration. □

Lemma 2.1.4. *Let $\{a_n, n \geq 1\}$ be a real number sequence, and $\lim_{n\to\infty} a_n = 0$. Let $\{\mathbf{X}_n, n \geq 1\}$ be a real valued random variable sequence such that $P(|\mathbf{X}_n| > a_n) \leq a_n$, $n \geq 1$. If $\mathbf{Y}$ is a random variable and y is a continuity point of the distribution of $\mathbf{Y}$, then $\lim_{n\to\infty} P(\mathbf{Y} + \mathbf{X}_n \leq y) = P(\mathbf{Y} \leq y)$.*

Proof. Write this as a convolution and use bounded convergence. □

The integral formula (2.1.4) stated next is contained in Feller (1966), pages 480–481. Application of Lemma 2.1.4 to the situation $\sigma \to 0$ shows that the right side of (2.1.4) converges to $\nu((-\infty, s])$ at all continuity points s of ν. The left side of (2.1.4) is clearly a linear functional of the Fourier transform of ν.

Lemma 2.1.5. *Let ν be a probability measure. Then*

$$\int_{-\infty}^{s} dt \int_{-\infty}^{\infty} d\zeta \int_{-\infty}^{\infty} (2\pi)^{-1/2} \exp(-\zeta^2/2\sigma^2) \exp(-i\zeta t) \exp(i\zeta x) \nu(dx)$$
$$= \int_{-\infty}^{s} \int_{-\infty}^{\infty} \sigma(2\pi)^{-1/2} \exp(-\sigma^2(x-t)^2/2) \nu(dx)\, dt. \tag{2.1.4}$$

The right side of (2.1.4) is the distribution function of the sum of two independent random variables.

Lemma 2.1.6. *Let $f\colon(-\infty, \infty) \to (0, \infty)$ be a Borel measurable function. Suppose for all Borel sets* **A**, *that*

$$\int_{\mathbf{A}} f(x)\mu(dx) = \int_{\mathbf{A}} f(x)\nu(dx), \tag{2.1.5}$$

where μ and ν are σ-finite Borel measures. Then $\mu = \nu$.

PROOF. Obvious. □

Theorem 2.1.7 (Uniqueness of the Laplace Transform). *Let μ and ν be two positive σ-finite Borel measures. Assume that the Laplace transforms of μ and ν are absolutely convergent on the strip $a < \operatorname{Re} t < b$ and that on this strip the Laplace transforms are equal. Then $\mu = \nu$.*

PROOF. Let $a < t' < b$, t' a real number, and define finite positive measures μ' and ν' by, if **A** is a Borel set, then

$$\mu'(\mathbf{A}) = \int_{\mathbf{A}} \exp(t'x)\mu(dx) \quad \text{and} \quad \nu'(\mathbf{A}) = \int_{\mathbf{A}} \exp(t'x)\nu(dx).$$

By Lemma 2.1.6 it is sufficient to show that $\mu' = \nu'$. To show this, there exists $\varepsilon > 0$ such that the Laplace transforms of μ' and ν' are absolutely convergent on the strip $|\operatorname{Re} t| < \varepsilon$. In particular μ' and ν' are finite positive measures with the same Fourier transforms (i.e., $\operatorname{Re} t = 0$). From Lemma 2.1.5 and the equation (2.1.4) it follows that the Fourier transform uniquely determines the measure. Applied here it follows that μ' and ν' must be the same. As noted, Lemma 2.1.6 then implies that $\mu = \nu$. □

Lemma 2.1.8. *If $\mathbf{A} \subset \mathbb{R}^n$ is an open set, if $1 \le m < n$ and $t \in \mathbb{R}^m$, then the t-section $\mathbf{A}_t = \{y \mid (y, t) \in \mathbf{A}\}$ is an open subset of $\mathbb{R}^{n-m}$.*

Theorem 2.1.9 (n-dimensional Uniqueness Theorem). *Let μ and ν be positive n-dimensional Borel measures. Let $\mathbf{A} \subset \mathbb{R}^n$ be an open set such that if $t \in \mathbf{A}$ then $\int \exp(t \cdot x)\mu(dx) = \int \exp(t \cdot x)\nu(dx) < \infty$. Then $\mu = \nu$. ($t \cdot x$ is the dot product of t and x.)*

PROOF. We assume $n > 1$ as the theorem holds for the case $n = 1$. The argument is by induction on the dimension n.

Choose a real number t such that for some $y \in \mathbf{A}$, $y^t = (y_1, \ldots, y_n)$ and $t = y_n$. Then there exist numbers $t_1 < t < t_2$ such that for some y' and y'', $y_n' = t_1$ and $y_n'' = t_2$. Let $x^t = (x_1, \ldots, x_n)$ and define measures μ' and ν' by

$$\mu'(\mathbf{B}, t) = \int 1_{\mathbf{B}}(x_1, \ldots, x_{n-1}) \exp(x_n t) \mu(dx), \tag{2.1.6}$$

and

$$\nu'(\mathbf{B}, t) = \int 1_{\mathbf{B}}(x_1, \ldots, x_{n-1}) \exp(x_n t) \nu(dx).$$

In (2.1.6) $1_{\mathbf{B}}$ is the indicator function of the $(n-1)$-dimensional Borel subset $\mathbf{B}$. The Laplace transforms of μ' and ν' clearly satisfy, if $(s_1, \ldots, s_{n-1}, t) \in \mathbf{A}$ then

$$\begin{aligned} &\int \exp(x_1 s_1 + \cdots + x_{n-1} s_{n-1}) \mu'(dx, t) \\ &= \int \exp(x_1 s_1 + \cdots + x_{n-1} s_{n-1} + x_n t) \mu(dx) \\ &= \int \exp(x_1 s_1 + \cdots + x_{n-1} s_{n-1} + x_n t) \nu(dx) \\ &= \int \exp(x_1 s_1 + \cdots + x_{n-1} s_{n-1}) \nu'(dx, t). \end{aligned} \tag{2.1.7}$$

By the inductive hypothesis it follows that if $t_1 < t < t_2$ then $\mu'(\ , t) = \nu'(\ , t)$. Define measures $\mu''(\ , \mathbf{B})$ and $\nu''(\ , \mathbf{B})$ by

$$\mu''(\mathbf{C}, \mathbf{B}) = \int 1_{\mathbf{B}}(x_1, \ldots, x_{n-1}) 1_{\mathbf{C}}(x_n) \mu(dx),$$

and (2.1.8)

$$\nu''(\mathbf{C}, \mathbf{B}) = \int 1_{\mathbf{B}}(x_1, \ldots, x_{n-1}) 1_{\mathbf{C}}(x_n) \nu(dx).$$

If $\mathbf{B}$ is a $(n-1)$-dimensional Borel set and if $t_1 < t < t_2$ then

$$\begin{aligned} \int \exp(zt) \mu''(dz, \mathbf{B}) &= \mu'(\mathbf{B}, t) = \nu'(\mathbf{B}, t) \\ &= \int \exp(zt) \nu''(dz, \mathbf{B}). \end{aligned} \tag{2.1.9}$$

By the uniqueness theorem for dimension $n = 1$ it follows that $\mu''(\mathbf{C}, \mathbf{B}) = \nu''(\mathbf{C}, \mathbf{B})$ for all Borel subsets $\mathbf{C} \subset \mathbb{R}$ and $\mathbf{B} \subset \mathbb{R}^{n-1}$. By Fubini's Theorem, $\mu = \nu$ now follows. □

We state a second form of an n-dimensional uniqueness theorem as Theorem 2.1.10. The proof proceeds by a similar induction on the dimension.

Theorem 2.1.10. *Let λ be a nonatomic positive Borel measure on the Borel subsets of $\mathbb{R}$. Let $\mathbf{A} \subset \mathbb{R}^n$ be a Borel subset such that $(\lambda x \cdots x\lambda)(\mathbf{A}) > 0$. Let μ and ν be positive n-dimensional Borel measures such that if $t \in \mathbf{A}$ then $\int \exp(t \cdot x)\mu(dx) = \int \exp(t \cdot x)\nu(dx)$. Then $\mu = \nu$.*

Analogous results hold for signed measures $\mu = \mu_+ - \mu_-$, and $\nu = \nu_+ - \nu_-$. For if μ and ν have equal absolutely convergent Laplace transforms then

$$\begin{aligned} &\int \exp(t \cdot x)\mu_+(dx) + \int \exp(t \cdot x)\nu_-(dx) \\ &= \int \exp(t \cdot x)\mu_-(dx) + \int \exp(t \cdot x)\nu_+(dx). \end{aligned} \tag{2.1.10}$$

By Theorem 2.1.9 it follows that $\mu_+ + \nu_- = \mu_- + \nu_+$ and hence that $\mu = \nu$.

Since by change of variable the Mellin transform becomes a Laplace transform, corresponding uniqueness theorems hold. We have not stated a n-dimensional uniqueness theorem for the Fourier transform. Such a result may be proven either by induction following the proof of Theorem 2.1.9 or by using the multivariate analogue of (2.1.4). However see Theorem 2.4.7.

2.2. The Multivariate Normal Density Functions

Functions $K\exp(-\frac{1}{2}x^tAx)$, $K > 0$ a real number, $x \in \mathbb{R}^n$, A a $n \times n$ symmetric matrix, are considered here. In this and the remaining chapters of this book the transpose of a matrix A is the matrix A^t. We write dx for n-dimensional Lebesgue measure.

Lemma 2.2.1. *$\int_{-\infty}^{\infty} \cdots \int_{-\infty}^{\infty} \exp(-\frac{1}{2}x^tAx)\,dx$ is an absolutely convergent n-dimensional integral if and only if the $n \times n$ symmetric matrix A is positive definite, denoted by $A > 0$. In case $A > 0$ the value of the integral is $(2\pi)^{n/2}(\det A)^{-1/2}$.*

Proof. Let U be a $n \times n$ orthogonal matrix such that UAU^t is a diagonal matrix. Make the change of variable $y = Ux$ having Jacobian $= \pm 1$. Integrate over spheres centered at 0 so the region of integration is invariant under the change of variable. In this proof we use Theorem 2.0.1. □

Lemma 2.2.2. *The multivariate normal density function*

$$(\det A)^{1/2}(2\pi)^{-n/2}\exp(-\tfrac{1}{2}x^tAx) \tag{2.2.1}$$

has Laplace transform

$$\exp\left(\tfrac{1}{2}s^tA^{-1}s\right). \tag{2.2.2}$$

PROOF. Complete the square in the exponent. □

Lemma 2.2.3. *If the random n-vector* $\mathbf{X}$ *has multivariate normal density function* (2.2.1) *then* $\mathbf{E}\,\mathbf{X} = 0$ *and* $\mathbf{Cov}\,\mathbf{X} = A^{-1}$.

PROOF. Compute the first and second order partial derivatives of the Laplace transform (2.2.2). □

If $\mathbf{E}\,\mathbf{X} = 0$, $\mathbf{Cov}\,\mathbf{X} = A^{-1}$, and $\mathbf{X}$ has a multivariate normal density function, then the random vector $\mathbf{Y} = \mathbf{X} + a$, $a \in \mathbb{R}^n$, has as its Laplace transform

$$\begin{aligned} \mathbf{E}\exp(\mathbf{Y}\cdot s) &= \mathbf{E}\exp((\mathbf{X}+a)\cdot s) \\ &= \exp\left(s\cdot a + \tfrac{1}{2}s^tA^{-1}s\right). \end{aligned} \tag{2.2.3}$$

Clearly

$$\mathbf{E}\,\mathbf{Y} = a \quad \text{and} \quad \mathbf{Cov}\,\mathbf{Y} = A^{-1} \tag{2.2.4}$$

so that the multivariate density function of $\mathbf{Y}$ is

$$(\det A)^{1/2}(2\pi)^{-n/2}\exp(y-a)^tA(y-a). \tag{2.2.5}$$

In the sequel we will say that a random n-vector $\mathbf{Y}$ which has multivariate density function (2.2.5) is normal (a, A^{-1}).

Lemma 2.2.4. *If* $\mathbf{X}^t = (\mathbf{X}_1, \ldots, \mathbf{X}_n)$ *has a multivariate normal density function then* $\mathbf{Y}$ *defined by* $\mathbf{Y}^t = (\mathbf{X}_1, \ldots, \mathbf{X}_{n-1})$ *has a multivariate normal density function.*

PROOF. Compute $\mathbf{E}\exp(\Sigma_{i=1}^n s_i\mathbf{X}_i)$ and then set $s_n = 0$. If $\mathbf{Cov}\,\mathbf{X} = A^{-1}$ with $A^{-1} = \begin{vmatrix} B^{-1} & b \\ b^t & c \end{vmatrix}$ then since $A^{-1} > 0$, so is $B^{-1} > 0$. □

Lemma 2.2.5. *Suppose* $\mathbf{X}^t = (\mathbf{X}_1, \ldots, \mathbf{X}_n)$ *has a multivariate normal density function. Suppose that* $\mathbf{E}\,\mathbf{X}_n = 0$ *and* $\mathbf{E}\,\mathbf{X}_i\mathbf{X}_n = 0$, $1 \le i \le n-1$. *Then* $\mathbf{X}_n$ *is stochastically independent of* $(\mathbf{X}_1, \ldots, \mathbf{X}_{n-1})$.

PROOF. Show that the relevant Laplace transform factors. □

Theorem 2.2.6. *Let* $\mathbf{X}$ *be normal* (a, A^{-1}). *Let* $b_i^t = (b_{i1}, \ldots, b_{in})$, $1 \le i \le k$, *be k linearly independent vectors. Then the random k-vector* $(b_1^t\mathbf{X}, \ldots, b_k^t\mathbf{X})^t$ *has a multivariate normal density function with mean vector* $(a^tb_1, \ldots, a^tb_k)^t$ *and covariance matrix with ij entry* $b_i^tA^{-1}b_j$.

PROOF. The Laplace transform is

$$\begin{aligned}\mathbf{E}\exp\left(\sum_{i=1}^{k} s_i(b_i^t\mathbf{X})\right) &= \mathbf{E}\exp\left(\left(\sum_{i=1}^{k} s_ib_i\right)^t\mathbf{X}\right)\\ &= \exp\left(\frac{1}{2}\left(\sum_{i=1}^{k} s_ib_i\right)^t A^{-1}\left(\sum_{i=1}^{k} s_ib_i\right)\right)\exp\left(\left(\sum_{i=1}^{k} s_ib_i\right)^t a\right).\end{aligned} \tag{2.2.6}$$

Since $b_1, \ldots, b_k$ are linearly independent, the exponent of the covariance part of the transform vanishes if and only if $s_1 = \cdots = s_k = 0$. From (2.2.3) the desired conclusion now follows. □

Theorem 2.2.7. *If the random vector* $(\mathbf{Y}, \mathbf{X}_1, \ldots, \mathbf{X}_n)^t$ *has a multivariate normal density function with zero means then there exist constants* $c_1, \ldots, c_n$ *such that* $\mathbf{Y} - (c_1\mathbf{X}_1 + \cdots + c_n\mathbf{X}_n)$ *is stochastically independent of* $\mathbf{X}_1, \ldots, \mathbf{X}_n$.

PROOF. In view of Lemma 2.2.5 it is sufficient to find constants $c_1, \ldots, c_n$ such that $\mathbf{E}\mathbf{X}_i(\mathbf{Y} - c_1\mathbf{X}_1 - \cdots - c_n\mathbf{X}_n) = 0$, $i = 1, \ldots, n$. By Lemma 2.2.4 the random vector $(\mathbf{X}_1, \ldots, \mathbf{X}_n)^t$ has a nonsingular covariance matrix A^{-1}. Thus the system of equations for $c_1, \ldots, c_n$ has as matrix of coefficients the nonsingular matrix A^{-1} and the equations have a unique solution. By Theorem 2.2.6, the random $(n+1)$-vector $(\mathbf{Y} - \Sigma_{i=1}^n c_i\mathbf{X}_i, \mathbf{X}_1, \ldots, \mathbf{X}_n)^t$ has a multivariate normal density function with covariance matrix of the form $\begin{vmatrix} a^{-1} & 0 \\ 0^t & A^{-1}\end{vmatrix}$. By Lemma 2.2.5 or directly from the Laplace transform which is $\exp(\frac{1}{2}ra^{-1}r)\exp(\frac{1}{2}s^tA^{-1}s)$, independence follows. □

Lemma 2.2.8. *Let* $(\mathbf{Y}, \mathbf{X}_1, \ldots, \mathbf{X}_n)$ *have a joint normal probability density function with zero means and let constants* $c_1, \ldots, c_n$ *be such that* $\mathbf{Y} - \Sigma_{i=1}^n c_i\mathbf{X}_i$ *and* $(\mathbf{X}_1, \ldots, \mathbf{X}_n)^t$ *are stochastically independent. Then the conditional expectation is*

$$\mathbf{E}(\mathbf{Y}|\mathbf{X}_1, \ldots, \mathbf{X}_n) = \sum_{i=1}^{n} c_i\mathbf{X}_i. \tag{2.2.7}$$

Lemma 2.2.9. *If* $\mathbf{X}_1, \ldots, \mathbf{X}_n$ *are mutually independent random variables each normally distributed, then the random n-vector* $(\mathbf{X}_1, \ldots, \mathbf{X}_n)^t$ *has a multivariate normal density function.*

PROOF. Write the product of the Laplace transforms. □

Lemma 2.2.10. *If* $\mathbf{X}_1, \ldots, \mathbf{X}_n$ *are independently distributed random* $h \times 1$ *vectors, and if* $\mathbf{X}_i$ *is normal* (a_i, A^{-1}), $i = 1, \ldots, n$, *then the random* $n \times h$ *matrix* $\mathbf{X}$ *with* i-th *row* $\mathbf{X}_i^t$, $i = 1, \ldots, n$, *has a multivariate normal density function*

$$(2\pi)^{-nh/2}(\det A)^{n/2}\exp(\operatorname{tr} A(X-M)^t(X-M)), \tag{2.2.8}$$

where $M = \mathbf{E}\mathbf{X}$ and "tr" *means trace of the matrix.*

2.3. Noncentral Chi-Square, F-, and t-Density Functions

Although these are not multivariate random variables these random variables and their density functions play a central role not only in the analysis of variance but in parts of distribution theory. It will appear from the Bartlett decomposition of the sample covariance matrix, see Sections 11.0 and 11.2, and use of this decomposition, that determination of many normalizations can be reduced to computation of moments of normal and chi-square random variables. This method will be used in Chapters 5, 12, and 13. The discussion here is about the following problem. If the random $n \times 1$ vector $\mathbf{X}$ is normal (a, A^{-1}) and A^{-1} is the identity matrix, then one wants to write the density function of $\mathbf{X}^t B\mathbf{X}$, where B is a $n \times n$ symmetric positive definite or positive semidefinite matrix. In the case of idempotents, $B = B^2$, the density function is the density function of a noncentral chi-square random variable. Any choice of B nonidempotent leads to a problem without a neat answer about which there is a growing literature. See Good (1969), Graybill and Milliken (1969), Press (1966), Shah (1970), and Shanbhag (1970) for some starters.

In the following the basic argument is the same as in Section 2.2. The Laplace transform of a gamma density function is readily computed and this class of density functions includes all the central chi-square density functions. The noncentral chi-square (random variable) is defined as a sum of squares of independently distributed normal random variables and the corresponding product of Laplace transforms is readily inverted to obtain the density function of a noncentral chi-square to be a weighted infinite sum of central chi-square density functions.

Definition 2.3.1. If $\mathbf{X}_1, \ldots, \mathbf{X}_n$ are independently distributed real valued random variables such that $\mathbf{X}_i$ is normal $(a_i, 1)$, then the density function of $\mathbf{Y} = \mathbf{X}_1^2 + \cdots + \mathbf{X}_n^2$ is the noncentral chi-square density function with noncentrality parameter $a = \frac{1}{2}(a_1^2 + \cdots + a_n^2)$ and n degrees of freedom. (It is shown below that the density function of $\mathbf{Y}$ depends only on a and not individually on $a_1, \ldots, a_n$.) In speaking of a central chi-square with n degrees of freedom, we speak of the case $a = 0$ and will write χ_n^2.

Definition 2.3.2. The two parameter family of density functions

$$\begin{aligned} f_{a,b}(x) &= (\Gamma(a)b^a)^{-1}x^{a-1}\exp(-x/b), && x > 0, \\ &= 0, && x \le 0, \end{aligned} \tag{2.3.1}$$

is called the family of gamma density functions, named after the gamma function.

Lemma 2.3.3. *The gamma density function* (2.3.1) *has Laplace transform*

$$(1 - bs)^{-a}, \qquad \textit{convergent if} \quad bs < 1. \tag{2.3.2}$$

PROOF. Combine exponents in $\int_0^\infty (\Gamma(a)b^a)^{-1}x^{a-1}\exp(-x/b)\exp(sx)\,dx$ and determine the normalization required to make the integral equal one. □

Lemma 2.3.4. *If* $\mathbf{X}_1$ *is normal* $(0, 1)$ *and* $\mathbf{Y} = \mathbf{X}_1^2$ *then* $\mathbf{Y}$ *has a gamma density function with parameters* $a = \frac{1}{2}$ *and* $b = 2$. *On the other hand, by change of variable one may calculate that*

$$P(\mathbf{Y} \le y) = \int_0^y (2\pi x)^{-1/2}\exp(-\tfrac{1}{2}x)\,dx, \tag{2.3.3}$$

and therefore

$$\Gamma(\tfrac{1}{2}) = \sqrt{\pi}. \tag{2.3.4}$$

Corollary 2.3.5. *The Laplace transform of* χ_1^2 *is*

$$(1 - 2t)^{-1/2} \tag{2.3.5}$$

and the Laplace transform of χ_n^2 *is*

$$(1 - 2t)^{-n/2}. \tag{2.3.6}$$

Hence the density function of the central chi-square with n degrees of freedom is

$$\begin{aligned} &(\Gamma(\tfrac{1}{2}n)2^{n/2})^{-1}x^{(n/2)-1}e^{-x/2}, \qquad && x > 0,\\ &0, && x \le 0. \end{aligned} \tag{2.3.7}$$

Lemma 2.3.6. *Let* $\mathbf{X}_1, \ldots, \mathbf{X}_n$ *be mutually independent random variables such that if* $1 \le i \le n$ *then* $\mathbf{X}_i$ *is normal* $(a_i, 1)$. *Let* $a = \frac{1}{2}(a_1^2 + \cdots + a_n^2)$. *Then the Laplace transform of* $\mathbf{X}_1^2 + \cdots + \mathbf{X}_n^2$ *is*

$$(\exp(-a))\sum_{j=0}^\infty (1 - 2t)^{-(n/2)+j}a^j/(j!). \tag{2.3.8}$$

PROOF. Write $\mathbf{E}\exp(t(\mathbf{X}_1^2 + \cdots + \mathbf{X}_n^2))$ as a n-fold integral and complete the square in the exponent. After integration obtain

$$(1 - 2t)^{-n/2}\exp\left(-\frac{1}{2}\sum_{i=1}^n a_i^2\right)\exp\left[\left(\frac{1}{2}\sum_{i=1}^n a_i^2\right)/(1 - 2t)\right]. \tag{2.3.9}$$

Substitute a in (2.3.9) and expand $\exp(a/(1 - 2t))$ in a power series in the variable $x = a/(1 - 2t)$ to obtain (2.3.8). □

Theorem 2.3.7. *The noncentral chi-square density function is*

$$\begin{aligned} &\sum_{j=0}^{\infty} \frac{x^{(n+2j)/2-1}e^{-x/2}}{2^{(n+2j)/2}\Gamma(\frac{1}{2}(n+2j))} \frac{e^{-a}a^j}{j!}, \qquad x > 0, \\ &= 0, \qquad x \le 0. \end{aligned} \tag{2.3.10}$$

PROOF. The numbers $(a^j/j!)\exp(-a)$ are the Poisson probabilities. Thus the Laplace transform (2.3.8) is a mixture of transforms of chi-square density functions. One may invert the transform at once and read off the noncentral chi-square density function to be (2.3.10). □

Definition 2.3.8. The parameter $a = \frac{1}{2}\Sigma_{i=1}^n a_i^2$ of the density function (2.3.10) is called the *noncentrality parameter*.

Lemma 2.3.9. *Let* **X** *and* **Y** *be positive random variables, independently distributed, such that* **X** *has density function* f *and* **Y** *has density function* g *relative to Lebesgue measure. Then the density function of* **X**/**Y** *is*

$$h(s) = \int_0^{\infty} yf(sy)g(y)\,dy, \qquad s > 0. \tag{2.3.11}$$

Definition 2.3.10. Let **X** and **Y** be independently distributed random variables such that **X** is a noncentral χ_n^2 with noncentrality parameter a, and **Y** is a central χ_m^2. Then the ratio $(m/n)(\mathbf{X}/\mathbf{Y})$ has the density function of the noncentral $F_{n,m}$-statistic.

Theorem 2.3.11. *Let* **X** *and* **Y** *be as in Definition* 2.3.10. *The random variable* $\mathbf{Z} = \mathbf{X}/\mathbf{Y}$ (*i.e., omit the normalization* m/m) *has the following density function:*

$$\sum_{j=0}^{\infty} \frac{e^{-a}a^j}{j!} \frac{\Gamma(\frac{1}{2}(m+n+2j))}{\Gamma(\frac{1}{2}(n+2j))\Gamma(\frac{1}{2}m)} \frac{z^{(n+2j-2)/2}}{(1+z)^{(m+n+2j)/2}}, \qquad z > 0. \tag{2.3.12}$$

Theorem 2.3.14 states the form of the noncentral t-statistic. The suggested method here is contained in Lemma 2.3.9. An alternative method is used in Chapter 5. See Section 5.1. A number of different ways of expressing this density function are available. Cf. Kruskal (1954).

Lemma 2.3.12. *If* **Y** *is a central* χ_m^2 *random variable then* $\sqrt{\mathbf{Y}}$ *has the density function*

$$2y^{m-1}\exp(-\tfrac{1}{2}y^2)/2^{m/2}\Gamma(\tfrac{1}{2}m), \qquad y > 0. \tag{2.3.13}$$

Definition 2.3.13. Let **X** an **Y** be independently distributed random variables such that **X** is normal $(a, 1)$ and **Y** is a central χ_m^2 random variable. Then the random variable $Z = \mathbf{X}/(\mathbf{Y}/m)^{1/2}$ has the density function of a noncentral t-statistic with noncentrality parameter a and m degrees of freedom.

Theorem 2.3.14. *One form of the density function of* $Z = \mathbf{X}/\mathbf{Y}^{1/2}$, *that is, an unnormalized noncentral t-statistic, is*

$$(2^{m-1}\pi)^{-1/2}(\Gamma(\tfrac{1}{2}m))^{-1}e^{-a^2/2}\int_0^\infty y^m e^{-(1+z^2)y^2/2}\, e^{azy}\,dy, \qquad -\infty < t < \infty. \tag{2.3.14}$$

2.4. Inversion of Transforms and Hermite Polynomials

Theorems on inversion of transforms are readily available and this topic is not treated in much detail here. Standard references for such material are Widder (1941), Wiener (1933) together with more modern books like Gelfond (1971) and texts on complex variables which are directed towards physics and electrical engineering.

The result of Lemma 2.1.3 can be rephrased as saying

$$\int_{-\infty}^{\infty} (2\pi)^{-1/2}\exp(isx)(2\pi)^{-1/2}\exp(-\tfrac{1}{2}x^2)\,dx = (2\pi)^{-1/2}\exp(-\tfrac{1}{2}s^2). \tag{2.4.1}$$

Or in words, the (normalized) Fourier transform of a normal (0, 1) density function is the normal (0, 1) density function. Relative to the weight function $\exp(-\frac{1}{2}x^2)$ we seek the sequence of polynomials $\{h_n, n \ge 0\}$ such that if $n \ge 0$ then h_n is a polynomial of degree n, and if $m, n \ge 0$ then

$$\begin{aligned}(2\pi)^{-1}\int_{-\infty}^{\infty} h_m(x)h_n(x)\exp(-x^2)\,dx &= 0, \qquad m \ne n,\\ &= 1, \qquad m = n.\end{aligned} \tag{2.4.2}$$

This makes the sequence $\{h_n(x)(2\pi)^{-1/2}\exp(-\frac{1}{2}x^2), n \ge 0\}$ an orthonormal sequence of functions relative to Lebesgue measure. Basic properties established by Wiener (1933), Chapter I, are

Theorem 2.4.1. *The sequence* $\{h_n(x)(2\pi)^{-1/2}\exp(-\frac{1}{2}x^2), n \ge 0\}$ *is a complete orthonormal system in* $\mathbf{L}_2$ *of the real line. Each of these functions is its own Fourier transform, except for a change of normalization by a constant having absolute value one.*

Theorem 2.4.2 (Plancherel's Theorem). *If f belongs to* $\mathbf{L}_2$ *of the real line define the* $\mathbf{L}_2$ *function g by* $g(y) = \lim_{a\to\infty}\int_{-a}^{a}(2\pi)^{-1/2}\exp(iyx)f(x)\,dx$. *This limit exists in* $\mathbf{L}_2$ *norm, and, in* $\mathbf{L}_2$ *norm,* $f(x) = \lim_{a\to\infty}\int_{-a}^{a}(2\pi)^{-1/2}\exp(-iyx)g(y)\,dy$.

The function $\hat{f} = g$ is called the Fourier transform of f. The map $f \to \hat{f}$ is an isometry of period four of $\mathbf{L}_2$ of the real line.

We now consider in detail the Lévy inversion theorems for functions of bounded variation of a real variable. For this purpose we need several lemmas.

Lemma 2.4.3. *Let a contour* $\mathbf{C}$ *in the complex plane be given by* $z = r\exp(is)$, $0 \le s \le \pi$. *Then*

$$\left| \int_{\mathbf{C}} z^{-1} \exp(iz)\, dz \right| < \pi/r. \tag{2.4.3}$$

PROOF. In parametric form the line integral is

$$\begin{aligned} a = \left| \int_0^\pi (\exp ir(\exp is))\, ds \right| &\le \int_0^\pi \exp(-r\sin s)\, ds \\ &= 2\int_0^{\pi/2} \exp(-r\sin s)\, ds. \end{aligned} \tag{2.4.4}$$

By convexity, if $0 \le s \le \frac{1}{2}\pi$, then $\sin s \ge 2s/\pi$ so that

$$\begin{aligned} a \le 2\int_0^{\pi/2} \exp(-2rs/\pi)\, ds &\le (\pi/r)\int_0^r \exp(-s)\, ds \\ &< \pi/r. \end{aligned} \tag{2.4.5}$$

□

Lemma 2.4.4. $\left|\pi - \int_{-r}^{r} x^{-1} \sin x\, dx\right| < \pi/r$.

PROOF. Taken over a closed contour $\mathbf{C}'$ consisting of the line segment $[-r, r]$ together with the semicircle $\mathbf{C}$ (see Lemma 2.4.3), we have

$$\begin{aligned} 0 = \int_{\mathbf{C}'} z^{-1}(\exp(iz) - 1)\, dz &= \int_{-r}^{r} s^{-1}(\exp(is) - 1)\, ds \\ &+ \int_{\mathbf{C}} z^{-1}(\exp(iz) - 1)\, dz. \end{aligned} \tag{2.4.6}$$

By transposition we obtain

$$\begin{aligned} i\int_{-r}^{r} x^{-1}\sin x\, dx &= \int_{-r}^{r} x^{-1}(\exp(ix) - 1)\, dx \\ &= \int_{\mathbf{C}} z^{-1}\, dz - \int_{\mathbf{C}} z^{-1}\exp(iz)\, dz. \end{aligned} \tag{2.4.7}$$

Thus from (2.4.3) we obtain

$$\left| \int_{-r}^{r} x^{-1}\sin x\, dx - \pi \right| \le \left| \int_{\mathbf{C}} z^{-1}\exp(iz)\, dz \right| < \pi/r. \tag{2.4.8}$$

□

In the sequel we need to consider the function

$$\lim_{r\to\infty}(2\pi)^{-1}\int_{-r}^{r}\frac{(\exp(is(b-x))-(\exp(is(a-x))}{is}\,ds,\ a<b. \tag{2.4.9}$$

When expressed in terms of sines and cosines, the cosine terms, being even functions, integrate to zero. Thus (2.4.9) is equal to

$$g(x)=(2\pi)^{-1}\left[\int_{-\infty}^{\infty}\frac{\sin s(b-x)}{s}\,ds-\int_{-\infty}^{\infty}\frac{\sin s(a-x)}{s}\,ds\right]. \tag{2.4.10}$$

Using Lemma 2.4.4 it follows at once that

$$\begin{aligned}&\text{if}\quad a<x<b\quad\text{then}\quad g(x)=1,\\ &\text{if}\quad x=a\quad\text{or}\quad x=b\quad\text{then}\quad g(x)=\tfrac{1}{2},\\ &\text{if}\quad x<a\quad\text{or}\quad x>b\quad\text{then}\quad g(x)=0.\end{aligned} \tag{2.4.11}$$

Theorem 2.4.5 (Lévy). *Suppose F is monotonic, bounded, with normalized Fourier transform $\hat{F}$. Define F_1 by $F_1(s)=\frac{1}{2}(F(s+)+F(s-))$. Then*

$$F_1(b)-F_1(a)=\lim_{r\to\infty}(2\pi)^{1/2}\int_{-r}^{r}\hat{F}(s)\left[\frac{e^{-isb}-e^{-isa}}{is}\right]ds. \tag{2.4.12}$$

Proof. The double integral implied by (2.4.12) is absolutely convergent for each fixed value of r. Let μ_F be the Borel measure determined by F. By Fubini's Theorem,

$$\begin{aligned}&\lim_{r\to\infty}(2\pi)^{-1}\int_{-r}^{r}\left[\frac{e^{-isb}-e^{-isa}}{is}\right]\int_{-\infty}^{\infty}e^{isx}\mu_F(dx)\,ds\\ &\quad=\lim_{r\to\infty}(2\pi)^{-1}\int_{-\infty}^{\infty}\int_{-r}^{r}\frac{\sin s(b-x)-\sin s(a-x)}{s}\,ds\,\mu_F(dx).\end{aligned} \tag{2.4.13}$$

By Lemma 2.4.4 the inner integral is a bounded function of x and r so that by the bounded convergence theorem, passing to the limit under the first integral sign, with g defined in (2.4.10), obtain

$$\int_{-\infty}^{\infty}g(x)\mu_F(dx)=F_1(b)-F_1(a). \tag{2.4.14}$$

□

In case F is absolutely continuous with density function f of bounded variation, then defining f_1 by analogy to F_1, we have from (2.4.12) that if f is absolutely continuous with derivative f' and if $\lim_{|x|\to\infty}f(x)=0$ then an integration by parts yields

$$f_1(b)-f_1(a)=\lim_{r\to\infty}(2\pi)^{-1/2}\int_{-r}^{r}\frac{e^{-ixb}-e^{-ixa}}{-ix}\,dx\int_{-\infty}^{\infty}e^{ixy}f'(y)\,dy$$

$$= \lim_{r\to\infty} (2\pi)^{-1/2} \int_{-r}^{r} dx \int_{-\infty}^{\infty} (e^{-ixb} - e^{-ixa}) e^{ixy} f(y)\, dy$$
$$= \lim_{r\to\infty} (2\pi)^{-1/2} \int_{-r}^{r} (e^{-ixb} - e^{-ixa}) \hat{f}(x)\, dx. \tag{2.4.15}$$

We summarize this in a slightly different form.

Theorem 2.4.6. *Suppose f is absolutely integrable, and f has a continuous absolutely integrable first derivative f'. Let $\hat{f}$ be the unnormalized Fourier transform of f. Then*

$$\lim_{r\to\infty} (2\pi)^{-1} \int_{-r}^{r} e^{-isx} \hat{f}(s)\, ds = f(x), \qquad -\infty < x < \infty. \tag{2.4.16}$$

PROOF. The hypotheses imply that $\lim_{r\to\infty} f(x) = 0$. An integration by parts is required. At this step the constant of integration may be chosen in a helpful fashion. □

Without trying to develop a general n-dimensional theory we note the obvious generalization of Theorem 2.4.6.

Lemma 2.4.7. *Suppose $h\colon \mathbb{R}^n \to \mathbb{R}$ is in $\mathbf{L}_1$ of Lebesgue measure, has zero integral, and that*

$$f(x_1, \ldots, x_n) = \int_{-\infty}^{x_1} \cdots \int_{-\infty}^{x_n} h(y_1, \ldots, y_n)\, dy_1 \cdots dy_n; \quad \hat{f}(s_1, \ldots, s_n)$$
$$= (2\pi)^{-n/2} \int_{-\infty}^{\infty} \cdots \int_{-\infty}^{\infty} \exp i(s_1 x_1 + \cdots + s_n x_n) f(x_1, \ldots, x_n)\, dx_1 \cdots dx_n, \tag{2.4.17}$$

assumed to be absolutely convergent. Then

$$f(x_1, \ldots, x_n) = \lim_{r_1\to\infty} \cdots \lim_{r_n\to\infty} (2\pi)^{-n/2}$$
$$\times \int_{-r}^{r} \cdots \int_{-r_n}^{r_n} \exp -i(s_1 x_1 + \cdots + s_n x_n) \hat{f}(s_1, \ldots, s_n)\, ds_1 \cdots ds_n. \tag{2.4.18}$$

2.5. Inversion of the Laplace and Mellin Transforms

We discuss here the univariate case. The multivariate generalizations follow the argument of Lemma 2.4.7.

We suppose that if $s_1 < s < s_2$ then $\int_{-\infty}^{\infty} \exp(sx) f(x)\, dx$ is an absolutely convergent integral. We will require in addition that f be everywhere differentiable with derivative f' which is Riemann integrable and that $\int_{-\infty}^{\infty} \exp(sx) \times |f'(x)|\, ds < \infty$. It then follows that if $s_1 < s_0 < s_2$ then the function $\exp(s_0 x) f(x)$ is of bounded variation. By Theorem 2.4.6

$$\lim_{r\to\infty}(2\pi)^{-1}\int_{-r}^{r}\exp(-(is+s_0)y)\,ds\int_{-\infty}^{\infty}\exp((is+s_0)x)f(x)\,dx$$
$$=\exp(-s_0y)[\exp(s_0y)f(y)]=f(y). \tag{2.5.1}$$

Theorem 2.5.1. *Given the hypotheses of the preceding paragraph, then* (2.5.1) *holds.*

The Mellin transform $\int_0^\infty x^{s-1}f(x)\,dx$ *becomes a Laplace transform* $\int_{-\infty}^{\infty}\exp(sy)f(\exp y)\,dy$ *under the change of variable* $x=\exp y$. *It then follows that*

$$\lim_{r\to\infty}(2\pi)^{-1}\int_{-r}^{r}\exp((-is-s_0)y)\,ds\int_0^\infty x^{is+s_0-1}f(x)\,dx$$
$$=f(e^y), \tag{2.5.2}$$

so that with the choice $w=\log y$, *we obtain*

$$\lim_{r\to\infty}(2\pi)^{-1}\int_{-r}^{r}y^{-is-s_0}\int_0^\infty x^{is+s_0-1}f(x)\,dx\,ds=f(y). \tag{2.5.3}$$

The limit (2.5.3) *exists if enough smoothness holds. We will not make a formal statement of smoothness assumptions.*

The inversion theorems stated here are adequate for most multivariate calculations. Less restricted assumptions may be found in Widder (1941).

2.6. Examples in the Literature

Kullback (1934) proved inversion theorems for Fourier transforms and used this method to calculate the probability density functions of products of independently distributed chi-square random variables. In his paper it is shown that the distribution of Wilks' generalized variance (central case) is the same as the distribution of a product of chi-squares. The answers are expressed in terms of residues of products of gamma functions without an explicit calculation of these residues being given.

Herz (1955) extended ideas of Bochner and defined a doubly infinite sequence of hypergeometric functions of complex symmetric matrix arguments. These functions were defined by using the Laplace transform and inverse Laplace transform to generate new functions. The hypergeometric functions of Herz have been given infinite series representations by Constantine (1963) in which the individual terms of the series are zonal polynomials, as defined by James (1961). It has been implied but never stated by James (1964, 1968), that the zonal polynomials (of a matrix argument) are spherical functions in the sense of Helgason (1962). We prove this fact in Section 12.11.

Box (1949) used inversion of Fourier transforms as a method of obtaining asymptotic series as the sample size tends to ∞. This is one of several methods currently in use in the literature of asymptotic approximations to the distributions of random variables.

Meijer's functions (c.f. Erdélyi, et al., (1953)) have been generalized by Braaksma (1964). Meijer functions and inversion of Mellin transforms have been used by Consul (1969). Mathai and Rathie (1971) have extended the work of Consul, op. cit., and Mellin transforms, in the study of the distribution of products, claiming H-functions are the most general type of special function. See also Mathai and Saxena (1978) for a more recent treatment of the use of H-functions.

CHAPTER 3

Locally Compact Groups and Haar Measure

3.0. Introduction

This chapter is intended to summarize some results needed later. We state an existence and uniqueness theorem for Haar measure but rather than copy a proof we leave this result unproven. See for example Halmos (1950) or Loomis (1953). The applications made in these notes are to matrix groups in their usual metric topology. Hence all topologies used in applications are Hausdorff topologies with countable base for open sets. More generality will be found in Loomis, op. cit., or Nachbin (1965). The manifolds discussed later are analytic manifolds for which the invariant measures can be given explicit representations using differential forms. Since the existence of invariant measures will usually be shown by explicit construction the part of the theory important to this book is usually the uniqueness part.

Nonetheless the existence and uniqueness theorems of Section 3.3 are stated in full generality. This requires use of the terms *Borel set* and *Baire set*. Full descriptions may be found in Halmos (1950) in the discussion of measures on locally compact spaces. In this book, the set of Borel sets is the least σ-algebra of subsets containing all the open, hence closed, sets. The set of Baire sets is the least σ-algebra of subsets in which all the real valued continuous functions are measurable. For the metric topologies the distance of a point x to a closed set $\mathbf{C}$ defines a continuous function d such that $\mathbf{C} = \{x | d(x) = 0\}$. Hence the Borel sets and the Baire sets are the same in the case of metric topologies.

Some arguments require the use of the regularity of measures. A countably additive nonnegative measure ν which is defined for the Baire sets and such that the measure of compact Baire sets is finite is necessarily a regular measure. That is, given $\varepsilon > 0$ and a Baire set $\mathbf{C}$ of finite measure there exists a compact Baire set $\mathbf{C}' \subset \mathbf{C}$ such that $\nu(\mathbf{C}') + \varepsilon \geq \nu(\mathbf{C})$. Again, see Halmos, op. cit.

At various places in this chapter matrix groups of interest are defined and used to illustrate points of the discussion. It was decided to use differential forms in these examples in spite of the fact that the discussion of differential forms comes later. The matrices discussed in this chapter have real numbers for entries. Later, in Chapter 12, use of complex numbers becomes important. The manifolds discussed include $\mathbf{GL}(n)$, the full linear group of $n \times n$ matrices; $\mathbf{O}(n)$, the group of $n \times n$ orthogonal matrices; $\mathbf{S}(n)$, the homogeneous space of $n \times n$ positive definite symmetric matrices; $\mathbf{T}(n)$ the group of lower triangular matrices with positive diagonal; $\mathbf{D}(n) \subset \mathbf{T}(n)$ the diagonal matrices. In each case matrix multiplication is the group operation so that the group identity is the identity matrix, and the group inverse is the same as the matrix inverse. These notations will be used throughout this book.

In the computation of differential forms it is convenient to compute dX, meaning, compute the differential of each entry of the matrix X and form the matrix of corresponding differentials. Then $(dX)_{ij}$ is the ij-entry of dX and $\bigwedge_{i \leq j} (dX)_{ij}$ a wedge product of the indicated differentials. As will appear in the examples which follow, this short notation leads at once to the differential forms for the Haar measures of the matrix groups.

Section 3.1 gives a summary of basic point set topology for locally compact groups. Section 3.2 discusses quotient spaces. Section 3.3 gives the uniqueness theorems for invariant measures on locally compact groups and quotient spaces. This material provides the basis for a discussion of the factorization of invariant measures and the factorization of manifolds discussed in Section 3.4, and again in Chapter 10. The uniqueness theorems are also the basic tool used in Chapter 5. Section 3.5 discusses the modular function, needed for Chapter 10. Section 3.6 discusses in the abstract the construction of differential forms for Haar measures on matrix groups. Section 3.7 discusses briefly the problem of cross-sections, which provides one way of doing the theory of Chapter 10. Last Section 3.8 briefly discusses material related to the Hunt–Stein theory of minimax invariant statistical procedures and related to material about amenability of groups. No problems were written for this chapter.

3.1. Basic Point Set Topology

The set of group elements $\mathfrak{G}$ is assumed to have a locally compact Hausdorff topology. In this topology the map of $\mathfrak{G} \times \mathfrak{G} \to \mathfrak{G}$ given by $(x, y) \to xy$ is to be jointly continuous and the map $x \to x^{-1}$ is to be continuous. For each fixed y, the maps $x \to xy$ and $x \to yx$ are to be continuous. Thus these latter three maps are homeomorphisms of $\mathfrak{G}$.

If $\mathbf{V} \subset \mathfrak{G}$, $\mathbf{V}$ is an open set, and the unit e of $\mathfrak{G}$ is in $\mathbf{V}$, then $\mathbf{V}^{-1} = \{v | v^{-1} \in$

$\mathbf{V}\}$ is an open set and $e = e^{-1} \in \mathbf{V}^{-1}$. Given y, as noted, the map $x \to yx$ is a homeomorphism. Thus $y(\mathbf{V} \cap \mathbf{V}^{-1})$ is an open set containing y and is said to be a *symmetrical neighborhood* of y. Further, if $\mathbf{U}$ is an open set, then $\mathbf{U} \cdot \mathbf{U} = \{z \mid \text{exist } x, y \in \mathbf{U}, z = xy\}$ is an open set since $\mathbf{U} \cdot \mathbf{U} = \bigcup_{x \in U} (x\mathbf{U})$. Thus the map $(x, y) \to xy$ is an open mapping.

Let $\mathbf{U}$ be an open set and $e \in \mathbf{U}$. Then the inverse image of $\mathbf{U}$ under the map $(x, y) \to xy$ is an open set of $\mathfrak{G} \times \mathfrak{G}$ and hence there exists a set $\mathbf{V} \times \mathbf{V}$, $e \in \mathbf{V}$, $\mathbf{V}$ open, such that $\mathbf{V} \times \mathbf{V}$ is contained in the inverse image. Thus $e \in \mathbf{V} \cdot \mathbf{V} \subset \mathbf{U}$. Similarly there exists $\mathbf{V}_1$ an open set, $e \in \mathbf{V}_1$, with $\mathbf{V}_1^{-1} \cdot \mathbf{V}_1 \subset \mathbf{U}$, and there exists $\mathbf{V}_2$ an open set, $e \in \mathbf{V}_2$, with $\mathbf{V}_2 \cdot \mathbf{V}_2^{-1} \subset \mathbf{U}$.

If $\mathbf{U}$ and $\mathbf{V}$ are compact subsets of $\mathfrak{G}$ then $\mathbf{U} \times \mathbf{U}$ is a compact subset of $\mathfrak{G} \times \mathfrak{G}$ and hence $\mathbf{U} \cdot \mathbf{V}$ is a compact subset of $\mathfrak{G}$. It follows that there exist compact symmetric neighborhoods of e.

In general, if $\mathbf{V}$ is an open set and $y \in \mathbf{V}$ then $e \in y^{-1}\mathbf{V}$, an open set. Thus every neighborhood of y has the form $y\mathbf{U}$, $\mathbf{U}$ an open set, $e \in \mathbf{U}$.

Lemma 3.1.1. *If* $\mathbf{W} \subset \mathfrak{G}$ *then the topological closure* $\mathbf{W}^c$ *of* $\mathbf{W}$ *is* $\mathbf{W}^c = \bigcap_v \mathbf{W} \cdot \mathbf{V} = \bigcap_v \mathbf{V} \cdot \mathbf{W}$ *taken over all neighborhoods of e.*

Proof. If $e \in \mathbf{V}$ then $\mathbf{W} \subset \mathbf{W} \cdot \mathbf{V}$. Let $x \in \mathbf{W}^c$ and $\mathbf{V}$ be an open set, $e \in \mathbf{V}$. Then $(x\mathbf{V}) \cap \mathbf{W}$ is nonempty and contains w_0, say. Then for some $v_0 \in \mathbf{V}$, $xv_0 = w_0$ so that $x = w_0 v_0^{-1} \in \mathbf{W} \cdot \mathbf{V}^{-1}$. Thus $x \in \mathbf{W} \cdot \mathbf{V}^{-1}$ for all open neighborhoods $\mathbf{V}$ of e. That is, $\mathbf{W}^c \subset \bigcap_v \mathbf{W} \cdot \mathbf{V}$. Conversely, suppose $x \in \mathbf{W} \cdot \mathbf{V}$ for all open neighborhoods $\mathbf{V}$ of e. If $\mathbf{U}$ is a neighborhood of x then $x^{-1}\mathbf{U} = \mathbf{V}$ is a neighborhood of e, so that $x \in \mathbf{W} \cdot \mathbf{V}^{-1}$, $x = w_0 v_0^{-1}$, and $w_0 = xv_0 \in \mathbf{U}$. Hence $\mathbf{U} \cap \mathbf{W}$ is not empty. Thus $x \in \mathbf{W}^c$. Use of the map $x \to x^{-1}$ gives the second identity $\mathbf{W}^c = \bigcap_v \mathbf{V} \cdot \mathbf{W}$. □

3.2. Quotient Spaces

We let $\mathbf{H} \subset \mathfrak{G}$ be a subgroup and the points of $\mathfrak{G}/\mathbf{H}$ be the left cosets of $\mathbf{H}$. The projection map π is defined by

$$\pi(x) = x \cdot \mathbf{H}. \tag{3.2.1}$$

Topologize $\mathfrak{G}/\mathbf{H}$ with the finest topology such that the projection map is continuous. Thus $\mathbf{V} \subset \mathfrak{G}/\mathbf{H}$ is an open set if and only if $\pi^{-1}(\mathbf{V})$ is an open set of $\mathfrak{G}$.

$$\pi^{-1}(\mathbf{V}) = \{x \mid \pi(x) \in \mathbf{V}\} = \{x \mid x \cdot \mathbf{H} \in \mathbf{V}\} = \bigcup_{x \in \pi^{-1}(v)} x \cdot \mathbf{H}. \tag{3.2.2}$$

π is an open mapping since if $\mathbf{U}$ is an open subset of $\mathfrak{G}$ then

$$\pi^{-1}(\pi(\mathbf{U})) = \bigcup_{x \in u} x \cdot \mathbf{H} = \mathbf{U} \cdot \mathbf{H}, \tag{3.2.3}$$

which is a union of open sets. If $\mathbf{H}$ is a closed subgroup of $\mathfrak{G}$ then each coset $x \cdot \mathbf{H}$ is a closed set in $\mathfrak{G}$, so that points of $\mathfrak{G}/\mathbf{H}$ are closed (i.e., complements of points are open).

As shown, $\mathbf{H}$ a closed subgroup implies $\mathfrak{G}/\mathbf{H}$ is a T_1-space. We now show that if $\mathbf{H}$ is a compact subgroup then $\mathfrak{G}/\mathbf{H}$ is a Hausdorff space. Let $\mathbf{H}$ be a compact set and $x \cdot \mathbf{H}$ and $y \cdot \mathbf{H}$ be disjoint sets. Then $y \cdot \mathbf{H} = \bigcap_v \mathbf{V} \cdot (y \cdot \mathbf{H})$, by Lemma 3.1.1, and for some neighborhood $\mathbf{V}$ of e, and $h \in \mathbf{H}$, $x \cdot h$ is not in $\mathbf{V} \cdot y \cdot \mathbf{H}$. We may choose a neighborhood $\mathbf{V}_1$ of e such that $\mathbf{V}_1^c$ is compact and $\mathbf{V}_1 \subset \mathbf{V}_1^c \subset \mathbf{V}$. Then $\mathfrak{G} - \mathbf{V}_1^c \cdot y \cdot \mathbf{H}$ is a neighborhood of $x \cdot h$ disjoint from $\mathbf{V}_1 \cdot y \cdot \mathbf{H}$. The compact set $x \cdot \mathbf{H}$ may thus be covered by a finite number of open sets $\mathfrak{G} - \mathbf{V}^c \cdot y \cdot \mathbf{H}$ and each is the inverse image under π of an open subset of $\mathfrak{G}/\mathbf{H}$.

Lemma 3.2.1. *The projection map* (3.2.1) *under the topology described for* $\mathfrak{G}/\mathbf{H}$ *is an open continuous map. If* $\mathbf{H}$ *is a closed subgroup then* $\mathfrak{G}/\mathbf{H}$ *is a* T_1*-space and if* $\mathbf{H}$ *is a compact subgroup then* $\mathfrak{G}/\mathbf{H}$ *is a Hausdorff space.*

Lemma 3.2.2. *If the group* $\mathfrak{G}$ *is a locally compact Hausdorff space and if* $\mathbf{W} \subset \mathfrak{G}/\mathbf{H}$ *is a compact subset then there exists a subset* $\mathbf{W}_1 \subset \mathfrak{G}$ *which is compact and such that* $\pi(\mathbf{W}_1) = \mathbf{W}$.

Proof. Choose an open set $\mathbf{V}$ such that $e \in \mathbf{V}$ and $\mathbf{V}$ has compact closure. Choose points $w_1, \ldots, w_k$ of $\mathbf{W}$ and $g_1, \ldots, g_k$ of $\mathfrak{G}$ such that $w_i = g_i \cdot \mathbf{H}$, $1 \le i \le k$, and such that $\bigcup_{i=1}^k \pi(g_i \cdot \mathbf{V}^c)$ covers $\mathbf{W}$. Then $\mathbf{W}_1 = \pi^{-1}(\mathbf{W}) \cap \bigcup_{i=1}^k (g_i \cdot \mathbf{V}^c)$ is a compact subset of $\mathfrak{G}$ and $\pi(\mathbf{W}_1) = \mathbf{W}$. □

3.3. Haar Measure

As noted in the introductory section the basic results are stated without proof. We use as a source for our definitions Halmos (1950). It is necessary in this section to state results in terms of regular Borel and Baire measures which have been defined and discussed briefly in Section 3.0. The reader should refer to and reread this section.

Theorem 3.3.1 (Existence). *If* $\mathfrak{G}$ *is a locally compact group which is a Hausdorff space then there exist regular Borel measures* μ, ν *such that* μ *and* ν *are nonzero, and,*

$$\text{If } g \in \mathfrak{G} \text{ then } \mu(g \cdot \mathbf{W}) = \mu(\mathbf{W}) \text{ and } \nu(\mathbf{W} \cdot g) = \nu(\mathbf{W}) \text{ for all Borel subsets } \mathbf{W} \text{ of } \mathfrak{G}. \tag{3.3.1}$$

Regularity, being nonzero, together with (3.3.1) *imply,*

If $\mathbf{V} \subset \mathfrak{G}$ *is an open Borel subset then* $\mu(\mathbf{V}) > 0$
and $\nu(\mathbf{V}) > 0$; (3.3.2)

compact Borel subsets have finite measure. (3.3.3)

A regular Borel measure with the left invariance (3.3.1) is called a *left invariant Haar measure*. A regular Borel measure with the right invariance (3.3.1) is called a *right invariant Haar measure*.

Theorem 3.3.2 (Uniqueness). *If* μ_1 *and* μ_2 *are left invariant Baire measures then there exists constants* c_1 *and* c_2 *such that* $c_1\mu_1 = c_2\mu_2$. *If* ν_1 *and* ν_2 *are right invariant Baire measures then there exist constants* c_1 *and* c_2 *such that* $c_1\nu_1 = c_2\nu_2$. *(Note that Baire measures are regular. For us a measure is nonnegative.)*

We now develop a theory for invariant measures induced on $\mathfrak{G}/\mathbf{H}$. For more generality see the last chapter of Halmos (1950). The projection map π (cf. (3.2.1)) induces a measure $\mu\pi^{-1}$ on the Baire sets of $\mathfrak{G}/\mathbf{H}$ from a measure μ on $\mathfrak{G}$ by means of the definition

$$\mu(\pi^{-1}(\mathbf{W})) = (\mu\pi^{-1})(\mathbf{W}). \tag{3.3.4}$$

For (3.3.4) to be meaningful when μ is a Baire measure we need to show that $\pi^{-1}(\mathbf{W})$ is a Baire subset. At this point we suppose the group is σ-compact and $\mathbf{H}$ is compact so that both $\mathfrak{G}$ and $\mathfrak{G}/\mathbf{H}$ are Baire sets. If f: $\mathfrak{G}/\mathbf{H} \to \mathbb{R}$ is continuous then the composition $f \circ \pi: \mathfrak{G} \to \mathbb{R}$ is continuous and for Borel subsets of the real numbers $\mathbf{A}$, $(f \circ \pi)^{-1}(\mathbf{A}) = \pi^{-1}(f^{-1}(\mathbf{A}))$ is thus a Baire set. Thus if $\mathfrak{C}$ is the set of all subsets $\mathbf{W}$ of $\mathfrak{G}/\mathbf{H}$ such that $\pi^{-1}(\mathbf{W})$ is a Baire subset of $\mathfrak{G}$ then $\mathfrak{C}$ contains all open and compact Baire sets and is a monotone class. Thus $\mathfrak{C}$ contains all the Baire subsets of $\mathfrak{G}/\mathbf{H}$. We summarize this in

Lemma 3.3.3. *Suppose* $\mathfrak{G}$ *is a* σ*-compact locally compact Hausdorff space, and* $\mathbf{H}$ *is a compact subgroup. The map* π^{-1} *maps the Baire subsets of* $\mathfrak{G}/\mathbf{H}$ *into the Baire subsets of* $\mathfrak{G}$. *The induced measure* $\mu\pi^{-1}$ *defined by* (3.3.4) *is a well defined Baire measure.*

$\mathfrak{G}$ acts as a transformation group on $\mathfrak{G}/\mathbf{H}$ by means of the action

$$\bar{g}(x \cdot \mathbf{H}) = (gx) \cdot \mathbf{H}. \tag{3.3.5}$$

The action $\bar{g}$ is well defined since $(gx) \cdot \mathbf{H} = (gy) \cdot \mathbf{H}$ if and only if $(gx)^{-1}(gy) \in \mathbf{H}$ if and only if $x^{-1}y \in \mathbf{H}$ if and only if $x \cdot \mathbf{H} = y \cdot \mathbf{H}$. Therefore

$$\bar{g} \circ \pi = \pi \circ g. \tag{3.3.6}$$

It is clear that the composition of mappings satisfies $\overline{h \circ g} = \bar{h} \circ \bar{g}$, and that $\bar{e}$ is the identity of the induced group. Therefore the mapping $\mathfrak{G} \to \bar{\mathfrak{G}}$ is a group homomorphism. In particular $\pi^{-1}\bar{g}^{-1} = g^{-1}\pi^{-1}$ and $\bar{g}^{-1} = \overline{(g^{-1})}$. Last, from (3.3.6) it follows that

$$\mu(\pi^{-1}\bar{g}(\mathbf{W})) = \mu(g\pi^{-1}(\mathbf{W})) = \mu(\pi^{-1}(\mathbf{W})). \tag{3.3.7}$$

Lemma 3.3.4. *$\mathfrak{G}$ acts as a transformation group on $\mathfrak{G}/\mathbf{H}$ by means of the actions defined in* (3.3.5). *The group $\overline{\mathfrak{G}}$ of actions on $\mathfrak{G}/\mathbf{H}$ is a homomorphic image of $\mathfrak{G}$ under the mapping $g \to \bar{g}$ which satisfies* (3.3.6). *If* $\mathbf{H}$ *is a compact subgroup the measure $\mu\pi^{-1}$ induced by the left invariant Haar measure μ for $\mathfrak{G}$ is invariant under the actions of $\overline{\mathfrak{G}}$ on* $\mathfrak{G}/\mathbf{H}$. (*See also Lemma* 3.3.6.)

Remark 3.3.5. $\bar{g}$ acts as the identity element if and only if $(gx)\cdot\mathbf{H} = x\cdot\mathbf{H}$ for all $x \in \mathfrak{G}$. This holds if and only if $x^{-1}gx \in \mathbf{H}$ for all $x \in \mathfrak{G}$. It is clear that $\bigcap_{x\in\mathfrak{G}} x\mathbf{H}x^{-1}$ is a normal subgroup of $\mathfrak{G}$ and in most of the examples consider the only proper normal subgroups are $\{e\}$, and in some cases, $\mathbf{D}(n)$. If the only proper normal subgroup is $\{e\}$ then, of course, $\mathfrak{G}$ and $\overline{\mathfrak{G}}$ are isomorphic.

Lemma 3.3.6. *If μ is a regular Baire measure on $\mathfrak{G}$ and* $\mathbf{H}$ *is a compact subgroup of $\mathfrak{G}$ then the induced measure $\bar{\mu}$ is a regular Baire measure on* $\mathfrak{G}/\mathbf{H}$. *We assume $\mathfrak{G}$ is σ-compact.*

PROOF. We have already shown in Lemma 3.3.3 that $\bar{\mu}$ is defined on the Baire subsets of $\mathfrak{G}/\mathbf{H}$. Lemma 3.3.3 implies and it is shown in the proof of Lemma 3.3.7 that if $\mathbf{W}$ is a compact Baire subset of $\mathfrak{G}/\mathbf{H}$ then $\bar{\mu}(\mathbf{W}) < \infty$. By Halmos (1950), Section 52, Theorem G, the Baire measure $\bar{\mu}$ is inner and outer regular, hence is regular. □

Lemma 3.3.7. *If* $\mathbf{H}$ *is a compact subgroup of the σ-compact group $\mathfrak{G}$ and μ is a Baire measure on $\mathfrak{G}$ then the induced measure $\bar{\mu}$ is a Baire measure on* $\mathfrak{G}/\mathbf{H}$. *If μ is a Haar measure then $\bar{\mu}$ gives positive mass to open sets.*

PROOF. Much of the proof is already contained in Lemma 3.3.3 and Lemma 3.3.6. Let $\mathbf{W} \subset \mathfrak{G}/\mathbf{H}$ and $\mathbf{W}$ be compact. By Lemma 3.2.2 there exists a compact set $\mathbf{W}_1$ with $\pi(\mathbf{W}_1) = \mathbf{W}$. Since $\mathbf{W}_1$ is compact so is $\mathbf{W}_1\cdot\mathbf{H}$. Then there exists a compact Baire set $\mathbf{W}_2$ such that $\mathbf{W}_1\cdot\mathbf{H} \subset \mathbf{W}_2 \subset \mathfrak{G}$. Then $\bar{\mu}(\mathbf{W}) = \mu(\pi^{-1}(\pi(\mathbf{W}_1))) \le \mu(\mathbf{W}_2) < \infty$. If μ gives positive mass to open sets and $\mathbf{W} \subset \mathfrak{G}/\mathbf{H}$ is open then $\bar{\mu}(\mathbf{W}) = \mu(\pi^{-1}(\mathbf{W})) > 0$. □

Theorem 3.3.8 (Uniqueness). *Suppose $\mathfrak{G}$ has a countable base for the open sets. If* $\mathbf{H}$ *is a compact subgroup of $\mathfrak{G}$ then invariant Baire measures on* $\mathfrak{G}/\mathbf{H}$ *are uniquely determined up to multiplicative constants.*

PROOF. Note at the onset that $\mathfrak{G}/\mathbf{H}$ is locally compact space. For if $\bar{g}$ has a neighborhood $\mathbf{W}$, then $\pi^{-1}(\mathbf{W})$ is an open set and contains a point g_1 with

$\pi(g_1) = \bar{g}$. Choose a neighborhood $\mathbf{V}$ of g_1 such that $\mathbf{V}^c$ is compact, and $V^c \subset \pi^{-1}(\mathbf{W})$. Then $\bar{g} \in \pi(\mathbf{V}) \subset \pi(\mathbf{V}^c) \subset \mathbf{W}$ and $\pi(\mathbf{V})$ is an open set, by Lemma 3.2.1. Thus $\pi(\mathbf{V})^c$ is compact since in a Hausdorff space a closed subset of a compact set is compact.

Thus both $\mathfrak{G}$ and $\mathfrak{G}/\mathbf{H}$ are metrizable. The proof following depends heavily on the fact that in a locally compact metric space with a countable base for the open sets, the Baire sets and the Borel sets are the same.

Let $\bar{\mu}$ be a regular invariant nonnegative measure on the Borel subsets of $\mathfrak{G}/\mathbf{H}$ (invariant for the group $\overline{\mathfrak{G}}$), so that $\bar{\mu}$ is finite valued on compact subsets. Let $\mu_{\mathbf{H}}$ be a nonzero Haar measure on the group $\mathbf{H}$. Since $\mathbf{H}$ is a compact metric space, the Baire and Borel subsets of $\mathbf{H}$ are the same. Since $\mathbf{H}$ is compact, $\mu_{\mathbf{H}}(\mathbf{H}) < \infty$. Let $\mathbf{U}$ be a Borel subset of $\mathfrak{G}$ and $1_{\mathbf{U}}$ be the indicator function of $\mathbf{U}$. We consider the integral $\int 1_{\mathbf{U}}(xh)\mu_{\mathbf{H}}(dh)$. By Fubini's theorem, since the map $(x, h) \to xh$ is jointly continuous, the integrand is jointly measurable and the integral is a measurable function of x. Clearly by invariance of the measure $\mu_{\mathbf{H}}$, if $yh_0 = x$ and $h_0 \in \mathbf{H}$ then

$$\int_{\mathbf{H}} 1_{\mathbf{U}}(xh)\mu_{\mathbf{H}}(dh) = \int_{\mathbf{H}} 1_{\mathbf{U}}(yh)\mu_{\mathbf{H}}(dh). \tag{3.3.8}$$

Thus (3.3.8) defines a function f of variables $\bar{x} \in \mathfrak{G}/\mathbf{H}$ and $\mathbf{U}$ given by

$$f(\bar{x}, \mathbf{U}) = \int_{\mathbf{H}} 1_{\mathbf{U}}(xh)\mu_{\mathbf{H}}(dh). \tag{3.3.9}$$

We need to know that if $\mathbf{U}$ is a Borel subset of $\mathfrak{G}$ then $f(\ ,\mathbf{U})$ is a $\mathfrak{G}/\mathbf{H}$ Borel measurable function. We now prove this. If g is a continuous real valued function on $\mathfrak{G}$ with compact support $\mathbf{C}$ then

$$\int_{\mathbf{H}} g(xh)\mu_{\mathbf{H}}(dh) = f(\bar{x}, g) \tag{3.3.10}$$

vanishes outside the compact set $\mathbf{C} \cdot \mathbf{H}$. Therefore using the fact that $\mathfrak{G}$ is a metric space and using the bounded convergence theorem it follows that (3.3.10) is a continuous function of x. Thus, if $\mathbf{A}$ is a closed set of real numbers, $0 \notin A$, then the inverse images

$$\left\{x \,\middle|\, \int g(xh)\mu_{\mathbf{H}}(dh) \in \mathbf{A}\right\} = \mathbf{B}, \quad \text{and,} \quad \pi(\mathbf{B}), \tag{3.3.11}$$

are compact, hence measurable sets. It is easy to see that the inverse image under $f(\ ,g)$ of $\mathbf{A}$ is the second set of (3.3.11). Thus (3.3.10) is a Borel measurable function on $\mathfrak{G}/\mathbf{H}$. By successive applications of the monotone convergence theorem it now follows that if $\mathbf{U}$ is the intersection of a compact and an open set then (3.3.9) is a Borel measurable function on $\mathfrak{G}/\mathbf{H}$. The sets which are intersections of a compact and an open set, or which are finite unions of these, form a set ring. And the set of $\mathbf{U}$ such that (3.3.9) is measurable is clearly a monotone class. Hence (3.3.9) is measurable for the σ-closure, i.e., for all Borel subsets $\mathbf{U}$.

As a function of the variable $\mathbf{U}$, (3.3.9) is nonnegative, countably additive, and finite on compact sets. Define a set function v by

$$v(\mathbf{U}) = \int f(\bar{x}, \mathbf{U})\bar{\mu}(d\bar{x}). \tag{3.3.12}$$

Then

$$\begin{aligned} v(g\mathbf{U}) &= \int f(\bar{x}, g\mathbf{U})\bar{\mu}(d\bar{x}) = \int f(\bar{g}^{-1}\bar{x}, \mathbf{U})\bar{\mu}(d\bar{x}) \\ &= \int f(\bar{x}, \mathbf{U})\bar{\mu}(d\bar{x}) = v(\mathbf{U}). \end{aligned} \tag{3.3.13}$$

From (3.3.9), if $\mathbf{U}$ is compact then $f(\ ,\mathbf{U})$ has compact support so that $v(\mathbf{U}) < \infty$. Thus v is a left invariant measure, finite on compact subsets, and hence is zero or is a left invariant Haar measure on the group $\mathfrak{G}$. As a reference measure let μ be a nonzero left invariant Haar measure on $\mathfrak{G}$ so that there exists a constant c such that $cv = \mu$, provided v is not zero.

For Borel subsets of the form $\mathbf{U} = \mathbf{V}\cdot\mathbf{H}$ a direct computation in (3.3.9) shows that

$$f(\bar{x}, \mathbf{V}\cdot\mathbf{H}) = \int 1_{\mathbf{V}\cdot\mathbf{H}}(xh)\mu_{\mathbf{H}}(dh) = \mu_{\mathbf{H}}(\mathbf{H})1_{\pi(\mathbf{V}\cdot\mathbf{H})}(\bar{x}). \tag{3.3.14}$$

By choice of $\mu_{\mathbf{H}}$ the number $\mu_{\mathbf{H}}(\mathbf{H})$ is nonzero and finite. If $\bar{\mu}_1$ and $\bar{\mu}_2$ are two regular left invariant nonzero measures on the Borel subsets of $\mathfrak{G}/\mathbf{H}$ then the preceding discussion together with integration of (3.3.14) shows that there exist constants c_1 and c_2 such that

$$\begin{aligned} c_1\bar{\mu}_1(\pi(\mathbf{V}\cdot\mathbf{H}))\mu_{\mathbf{H}}(\mathbf{H}) &= c_1 v_1(\mathbf{V}\cdot\mathbf{H}) = c_2 v_2(\mathbf{V}\cdot\mathbf{H}) \\ &= c_2\bar{\mu}_2(\pi(\mathbf{V}\cdot\mathbf{H})\mathbf{W}\mu_{\mathbf{H}}(\mathbf{H}). \end{aligned} \tag{3.3.15}$$

The arbitrary compact subset of $\mathfrak{G}/\mathbf{H}$ has the form $\pi(\mathbf{V}) = \pi(\mathbf{V}\cdot\mathbf{H})$, as shown in Lemma 3.2.2. By regularity of the measures, $c_1\bar{\mu}_1 = c_2\bar{\mu}_2$. □

Example 3.3.9. Matrices $X \in \mathbf{GL}(n)$ (real entries) have factorizations $X = AS$ with $A \in \mathbf{O}(n)$ and $S \in \mathbf{S}(n)$. One such factorization is given by $A = X(X^tX)^{-1/2}$ and $S = (X^tX)^{1/2}$. Clearly any factorization must satisfy $S^2 = X^tX$ which is nonsingular. (Problem.) If T is another positive definite matrix and $S^2 = T^2$ it follows that $S = T$. It follows that for the compact subgroup $\mathbf{O}(n)$ each right coset contains a unique positive definite matrix. Thus $\mathbf{S}(n) \simeq \mathbf{GL}(n)/\mathbf{O}(n)$ and $\mathbf{S}(n)$ may be interpreted as a homogeneous or quotient space. A similar interpretation holds for $\mathbf{T}(n)$, and this example is discussed below.

The cosets of this example are right cosets, so $\mathbf{GL}(n)$ is to be interpreted as acting on the right. Thus $X = AS$ and if $B \in \mathbf{GL}(n)$ then XB has the symmetrix matrix component $(B^tX^tXB)^{1/2} = (B^tS^2B)^{1/2}$. This is different from the more usual group action $S \to B^tSB$. This second action arises as follows. The function $f(S) = S^2$ defined on $\mathbf{S}(n)$ is one to one. Let g_B be the function $g_B(S) = (B^t S^2 B)^{1/2}$. Then $(f g_B f^{-1})(S) = B^t S B$. The transformations $f g_B f^{-1}$ form a group isomorphic to the group of transformations g_B. Haar measures μ on $\mathbf{GL}(n)$ induce invariant measures $\bar{\mu}$ on $\mathbf{S}(n)$ and invariant measures $\bar{\mu}(f^{-1}(\cdot)) = \bar{v}$ under the group actions $f g_B f^{-1}$. Show that $g_B = g_C$ if and only if $B = \pm C$ so that $\mathbf{GL}(n)/\mathbf{A}$ is isomorphic to the group with elements g_B and is isomorphic to the group with elements $f g_B f^{-1}$, $\mathbf{A} = \{I, -I\}$.

After the theory of differential forms is developed in a later chapter we will see that the invariant measures $\bar{v} = \bar{\mu}(f^{-1}(\cdot))$ on $S(\mathbf{n})$ are given by differential forms

$$c \bigwedge_{j \le i} d s_{ij} / (\det S)^{(n+1)/2}, \quad \text{with } S = (S_{ij}), \tag{3.3.16}$$

where c is a constant. By Theorem 3.3.8 the invariant measures $\bar{\mu}$ are determined up to a multiplicative constant. Since to every measure $\bar{v}$ there is a uniquely determined $\bar{\mu}(f(\cdot))$, the measures invariant under the actions fgf^{-1} are uniquely determined up to a multiplicative constant. Therefore every invariant measure $\bar{v}$ is representable by a differential form (3.3.16).

If $X \in \mathbf{GL}(n)$ we write X_T to be the matrix in $\mathbf{T}(n)$ such that $X_T \in \mathbf{O}(n)X$. This amounts to using a Gram–Schmidt orthogonalization process on X. If $X_T \in \mathbf{T}(n)$ and $A \in \mathbf{O}(n)$ we let $f(X_T, A) \in \mathbf{T}(n)$ represent the coset $\mathbf{O}(n) X_T A$. Then

$$\mathbf{O}(n) X Y = \mathbf{O}(n) X_T A Y_T = \mathbf{O}(n) f(X_T, A) Y_T \tag{3.3.17}$$

and

$$f(X_T, A) Y_T \in \mathbf{T}(n)$$

represents the coset $\mathbf{O}(n)XY$. By invariance the induced measure satisfies

$$\bar{\mu}(\{T \mid T \in \mathbf{U}\}) = \bar{\mu}(\{T \mid f(T, A) \in \mathbf{U}\}). \tag{3.3.18}$$

This property may be verified directly from the differential form for the invariant measure $\bar{\mu}$.

In the next section we need the following lemma.

Lemma 3.3.10. *Suppose* $\mathbf{H}$ *is a compact subgroup of* $\mathfrak{G}$, μ *is a nonzero left invariant Haar measure for* $\mathfrak{G}$, *and* $\bar{\mu}$ *is the induced measure on* $\mathfrak{G}/\mathbf{H}$. *Let* $\bar{f}: \mathfrak{G}/\mathbf{H} \to \mathbb{R}$ *be measurable and* $f: \mathfrak{G} \to \mathbb{R}$ *be defined by* $\bar{f}(\pi(x)) = f(x)$. *Then*

$$\int f(x)\mu(dx) = \int \bar{f}(\bar{x})\bar{\mu}(d\bar{x}) = \int \bar{f}(\pi(x))\mu(dx). \tag{3.3.19}$$

PROOF. If $\bar{\mathbf{A}} \subset \mathfrak{G}/\mathbf{H}$ is a Baire set and $\bar{\mu}(\bar{\mathbf{A}}) < \infty$ then

$$\int 1_{\bar{\mathbf{A}}}(\bar{x})\bar{\mu}(d\bar{x}) = \bar{\mu}(\bar{\mathbf{A}}) = \mu(\pi^{-1}(\bar{\mathbf{A}}) = \int 1_{\bar{\mathbf{A}}}(\pi(x))\mu(dx). \tag{3.3.20}$$

Hence, for any simple functions f and $\bar{f}$, it follows that (3.3.19) holds. By monotone convergence it then follows at once that (3.3.19) holds for any integrable $\bar{f}$ and corresponding f. □

Remark 3.3.11. If $\mathfrak{M}$ is a manifold on which $\mathfrak{G}$ acts and if $\mathfrak{M}$ admits an invariant nonzero (positive) invariant measure μ, then the induced measure $\bar{\mu}$ on the orbit space $\mathfrak{M}/\mathfrak{G}$ satisfies (3.3.19).

3.4. Factorization of Measures

The results needed in this book are for locally compact spaces with countable base for the open sets. These spaces are metrizable and every compact subset of a metric space is both a Baire set and a Borel set. In this section we suppose $\mathfrak{H}_1$ and $\mathfrak{H}_2$ are locally compact metric spaces with countable base and $\mathfrak{H} = \mathfrak{H}_1 \times \mathfrak{H}_2$, the Cartesian product in the product topology. Let the group $\mathfrak{G}$ act as transformations on $\mathfrak{H}_1$ and define $\overline{\mathfrak{G}}$ on $\mathfrak{H}$ by

$$\text{if} \quad g \in \mathfrak{G}, \quad \text{then} \quad \bar{g}(h_1, h_2) = (gh_1, h_2). \tag{3.4.1}$$

Then $\mathfrak{G}$ and $\overline{\mathfrak{G}}$ are "canonically" isomorphic.

In the sequel we will say left invariant Baire measures on $\mathfrak{H}_1$ satisfy the *uniqueness property* if such measures form a one-dimensional space.

Theorem 3.4.1. *Suppose $\mathfrak{H}_1$ and $\mathfrak{H}_2$ are locally compact Hausdorff spaces with countable bases and $\mathfrak{H} = \mathfrak{H}_1 \times \mathfrak{H}_2$. Let $\mathfrak{G}$ and $\overline{\mathfrak{G}}$ be as above. Let μ be a regular nonzero Borel measure for $\mathfrak{H}$ and assume μ is left invariant under the actions of $\overline{\mathfrak{G}}$. Suppose left invariant regular Borel measures for $\mathfrak{H}_1$ have the uniqueness property. Then there exist regular Borel measures ν_1 and ν_2 defined on the Borel subsets of $\mathfrak{H}_1$ and $\mathfrak{H}_2$ respectively such that ν_1 is left invariant under the actions of $\mathfrak{G}$ and such that if $\mathbf{A} \subset \mathfrak{H}_1$ and $\mathbf{B} \subset \mathfrak{H}_2$ are Borel subsets then*

$$\mu(\mathbf{A} \times \mathbf{B}) = \nu_1(\mathbf{A})\nu_2(\mathbf{B}). \tag{3.4.2}$$

Proof. Since μ is not zero there must exist a compact set $\mathbf{B} \subset \mathfrak{H}_2$ and an open set $\mathbf{A}_0$ with compact closure in $\mathfrak{H}_1$ for which $0 < \mu(\mathbf{A}_0 \times \mathbf{B}) < \infty$. For by Fubini's theorem, if μ vanishes for all such compact sets then μ vanishes identically. Define $\mu_{\mathbf{B}}$ by $\mu_{\mathbf{B}}(\mathbf{A}) = \mu(\mathbf{A} \times \mathbf{B})$. Then $\mu_{\mathbf{B}}$ is a nonnegative countably additive measure on the Borel subsets of $\mathfrak{H}_1$ such that $\mu_{\mathbf{B}}(\mathbf{A}) < \infty$ for every compact subset $\mathbf{A}$ of $\mathfrak{H}$. Since $\mathfrak{H}_1$ has a countable base, the Borel and Baire sets coincide, so it follows that $\mu_{\mathbf{B}}$ is a regular Borel measure. Further, if $g \in \mathfrak{G}$ then

$$\mu_{\mathbf{B}}(g\mathbf{A}) = \mu((g\mathbf{A}) \times \mathbf{B}) = \mu(\bar{g}(\mathbf{A} \times \mathbf{B})) = \mu(\mathbf{A} \times \mathbf{B}), \tag{3.4.3}$$

so that $\mu_{\mathbf{B}}$ is a regular left invariant Borel measure not identically zero. Pick a nonzero regular left invariant measure ν_1 for $\mathfrak{H}_1$ as a reference measure.

Then we may use the uniqueness hypothesis to define a function f on the Borel subsets of $\mathfrak{H}_2$ by the definition

$$f(\mathbf{B})\nu_1(\mathbf{A}) = \mu_{\mathbf{B}}(\mathbf{A}) = \mu(\mathbf{A} \times \mathbf{B}). \tag{3.4.4}$$

The equation (3.4.4) clearly defines a nonnegative countably additive function f on the Borel subsets of $\mathfrak{H}_2$ and $f(\mathbf{B}) < \infty$ for compact sets $\mathbf{B}$. Therefore $\nu_2 = f$ is a regular measure on the Borel subsets of $\mathfrak{H}_2$. □

EXAMPLE 3.4.2. We continue Example 3.3.9. Justification of the use of differential forms will be made later.

$X \in \mathbf{GL}(n)$ has a decomposition $X = UT$ with $U \in \mathbf{O}(n)$ and $T \in \mathbf{T}(n)$, by use of a Gram–Schmidt orthogonalization process. Consequently a nonzero Haar measure μ on $\mathbf{GL}(n)$ (using both right and left invariance) factors into $\mu = \nu_1\nu_2$ with ν_1 a left invariant Haar measure on $\mathbf{O}(n)$ and ν_2 a right invariant Haar measure on $\mathbf{T}(n)$. As is well known (and discussed in Section 3.5) $\mathbf{O}(n)$ is a unimodular group and the right and left invariant Haar measures are the same. On the other hand, the group $\mathbf{T}(n)$ is not unimodular and the measure ν_2 is not left invariant. See Remark 3.5.10.

In terms of differential forms this becomes the following.

$$\begin{aligned} X = UT \quad &\text{and} \quad dX = (dU)T + U(dT); \\ U^t\, dX\, T^{-1} &= U^t + (dT)T^{-1}. \end{aligned} \tag{3.4.5}$$

By computation of a wedge product for the individual matrix entries

$$\bigwedge_{i=1}^{n} \bigwedge_{j=1}^{n} (U^t\, dX\, T^{-1})_{ij} = \varepsilon_1 \bigwedge_{i<j} (u_i^t\, du_j) \bigwedge_{j \le i} (dTT^{-1})_{ij}, \tag{3.4.6}$$

where u_i is the i-th column of U and ε_1 is a real number in absolute value 1. The left side of (3.4.6) is

$$(\det U)^n (\det T)^{-n} \bigwedge_{i=1}^{n} \bigwedge_{j=1}^{n} dx_{ij}, \quad X = (x_{ij}), \tag{3.4.7}$$

and

$$\det X = (\det T)(\det U)^{-1}. \tag{3.4.8}$$

Therefore

$$\left(\bigwedge_{i=1}^{n} \bigwedge_{j=1}^{n} (dx_{ij}) \right) \Big/ (\det X)^n = \varepsilon_1 \bigwedge_{i<j} (u_i^t\, du_j) \bigwedge_{j \le i} (dT\, T^{-1})_{ij}. \tag{3.4.9}$$

The differential forms in (3.4.9) are differential forms for invariant measures on $\mathbf{GL}(n)$, $\mathbf{O}(n)$, and $\mathbf{T}(n)$ respectively. We do not compute the correct sign as it is not needed in the integrations. Subsequent integrations will use unsigned measures so the absolute value of the differential forms will be used.

EXAMPLE 3.4.3. $\mathfrak{H}$ will be the set of $n \times k$ matrices with real entries such that the $k \times k$ minor consisting of the first k rows, called X_1, is nonsingular.

We set $\mathbf{H}_1 = \mathbf{GL}(k)$ and $\mathbf{H}_2$ the set of $(n-k) \times k$ matrices, so that the decomposition becomes

$$X = YG, \quad G \in \mathbf{GL}(k) \quad \text{and} \quad Y^t = (I_k, Z^t), \tag{3.4.10}$$

with I_k the $k \times k$ identity matrix and Z a $(n-k) \times k$ matrix. The group $\mathbf{GL}(k)$ acts on the right and the regular right invariant measure μ factors to $\mu = \nu_1 \nu_2$ where ν_2 is a Haar measure on $\mathbf{GL}(k)$.

In terms of differential forms we have

$$X = YG, \qquad dX = (dY)G + Y(dG), \tag{3.4.11}$$

and

$$(\det G)^{-n} \bigwedge_{i=1}^{n} \bigwedge_{j=1}^{k} dx_{ij} = \bigwedge_{i=1}^{n} \bigwedge_{j=1}^{k} (dY + Y\,dG\,G^{-1})_{ij}$$

$$= \varepsilon_2 \bigwedge_{i=1}^{k} \bigwedge_{j=1}^{k} (dG\,G^{-1})_{ij} \bigwedge_{i=k+1}^{n} \bigwedge_{j=1}^{k} dz_{ij}.$$

In order to obtain an invariant measure on the left side of (3.4.11) the required normalization is

$$(\det G)^{-n} (\det (I_k + Z^t Z))^{-n/2} = (\det X^t X)^{-n/2}, \tag{3.4.12}$$

so that

$$(\det X^t X)^{-n/2} \bigwedge_{i=1}^{n} \bigwedge_{j=1}^{n} dx_{ij} = \varepsilon_2 \frac{\left[\bigwedge_{i=k+1}^{n} \bigwedge_{j=1}^{k} dz_{ij} \bigwedge_{i=1}^{k} \bigwedge_{j=1}^{k} (dG\,G^{-1})_{ij}\right]}{\det (I_k + Z^t Z)^{n/2}}. \tag{3.4.13}$$

Example 3.4.4. We continue the discussion started in Example 3.3.9. Write $X = US$, $S \in \mathbf{S}(n)$ and $U \in \mathbf{O}(n)$, so that $S = (X^t X)^{1/2}$. $\mathbf{GL}(n)/\mathbf{O}(n)$ is now considered to be a space of cosets represented by S. Since this decomposition is unique we may think of $\mathbf{GL}(n) = \mathbf{O}(n) \times \mathbf{S}(n)$. Since Haar measure μ on $\mathbf{GL}(n)$ is left invariant, there is induced on $\mathbf{S}(n) \simeq \mathbf{GL}(n)/\mathbf{O}(n)$ an invariant measure ν_2, see Lemma 3.3.4. The induced measure on $\mathbf{O}(n) \cdot \mathbf{S}(n)$ factors to $\nu_1 \nu_2$, see Theorem 3.4.1, and ν_1 is left invariant, hence we may suppose ν_1 is Haar measure of unit mass on $\mathbf{O}(n)$. If $G \in \mathbf{GL}(n)$ then

$$XG = U(SG) = U(SG)(G^t S^2 G)^{-1/2} (G^t S^2 G)^{1/2}.$$

Thus a (Cartesian) product $\mathbf{A} \cdot S$ maps to $(\mathbf{A} \cdot V) \cdot S'$ where $V = (SG)(G^t S^2 G)^{-1/2}$ is orthogonal. Since Haar measure on $\mathbf{O}(n)$ is unimodular, see Lemma 3.5.3, and since $\mu(\mathbf{A} \cdot \mathbf{B} \cdot G) = \mu(\mathbf{A} \cdot \mathbf{B})$, it follows that $\nu_2(\{S \mid S \in \mathbf{B}'\}) = \nu_2(\{S \mid S \in \mathbf{B}\})$, where $\mathbf{A} = \mathbf{O}(n)$ and $\mathbf{B}'$ is $\mathbf{B}' = \{S' \mid \text{exists } S \in \mathbf{B} \text{ with } S' = (G^t S^2 G)^{1/2}\}$. That is, ν_2 is an invariant measure under the group action. In terms of differential forms, $U^t\,dX\,S^{-1} = U^t\,dU + dS\,S^{-1}$. The diagonal of $U^t\,dU$ is zero. For terms $i > j$, $(U^t\,dU + dS\,S^{-1})_{ij} \wedge (U^t\,dU + dS\,S^{-1})_{ji} = (U^t\,dU)_{ij} \wedge ((S^{-1}\,dS)_{ij} + (dS\,S^{-1})_{ij}) + (S^{-1}\,dS)_{ji} \wedge (S^{-1}\,dS)_{ij}$, where $(U^t\,dU)_{ij} \wedge (U^t\,dU)_{ji} = 0$ is used for orthogonal matrices.

Forming the wedge product over $i \geq j$ and setting terms of more than maximal degree equal zero, gives the expansion

$$\bigwedge_{i=1}^{n}\bigwedge_{j=1}^{n}(dX)_{ij}/(\det X^tX)^{n/2} = \varepsilon \bigwedge_{i>j}(U^t\,dU)_{ij}\bigwedge_{i>j}((dS\,S^{-1})_{ji} + (dS\,S^{-1})_{ij})\bigwedge_{i=1}^{n}(dS\,S^{-1})_{ii}, \tag{3.4.14}$$

the sign $\varepsilon = \pm 1$ determined by the rearrangements. Part of (3.4.14) can be further simplified using Lemma 6.5.5. Consequently the group action $S \to (G^tS^2G)^{1/2}$ has invariant measure with the differential form indicated in (3.4.14) while the group action $S \to G^tSG$ has invariant measure given by the differential form (3.3.16). We return to this subject in Chapter 5 where methods of Karlin (1960) are discussed.

A special case of the factorization of measures is given in Basu's lemma. See for example Lehmann (1959).

Theorem 3.4.5. *Assume that the family $\{P_\omega, \omega \in \Omega\}$ is dominated by a totally σ-finite measure μ, and that the statistic* $\mathbf{S}$ *is a boundedly complete sufficient statistic for ω. If the statistic* $\mathbf{T}$ *has a distribution not depending on ω then* $\mathbf{S}$ *and* $\mathbf{T}$ *are (stochastically) independent.*

PROOF. Since $P(\mathbf{T} \in \mathbf{B}) = \mathbf{E}_\omega(1_\mathbf{B}(\mathbf{T})) = \mathbf{E}_\omega(\mathbf{E}(1_\mathbf{B}(\mathbf{T})|\mathbf{S}))$, by assumption of being boundedly complete, if $\omega \in \Omega$ then $\mathbf{E}(1_\mathbf{B}(\mathbf{T})|\mathbf{S}) = P(\mathbf{T} \in \mathbf{B})$ almost surely P_ω. Therefore $P_\omega(\mathbf{S} \in \mathbf{A}$ and $\mathbf{T} \in \mathbf{B}) = \mathbf{E}_\omega(1_\mathbf{A}(\mathbf{S})1_\mathbf{B}(\mathbf{T})) = P(\mathbf{T} \in \mathbf{B})\mathbf{E}_\omega(1_\mathbf{A}(\mathbf{S})) = P_\omega(\mathbf{S} \in \mathbf{A})P(\mathbf{T} \in \mathbf{B})$. □

Remark 3.4.6. There is a converse result, which if correctly stated, requires an additional assumption, such as, if ω_1 and ω_2 are parameters there exists a measurable set $\mathbf{A}$ such that $P_{\omega_1}(\mathbf{A}) > 0$ and $P_{\omega_2}(\mathbf{A}) > 0$.

EXAMPLE 3.4.7. Let $\mathbf{X}$ and $\mathbf{Y}$ be real valued random variables with a joint density function $a^2f(a^2(x^2 + y^2))\,dx\,dy$. Then $\mathbf{X}^2 + \mathbf{Y}^2$ is sufficient, complete, and $P(\mathbf{X}/\mathbf{Y} \in \mathbf{A})$ does not depend on a. Hence by Theorem 3.4.5, $\mathbf{X}^2 + \mathbf{Y}^2$ and $\mathbf{X}/\mathbf{Y}$ are independent random variables.

3.5. Modular Functions

If μ is a left invariant and regular measure for a locally compact group $\mathfrak{G}$ and $g \in \mathfrak{G}$ then the measure ν defined by

$$\nu(\mathbf{A}) = \mu(A\mathbf{g}) \tag{3.5.1}$$

is a nonzero left invariant regular measure defined on the Borel subsets of $\mathfrak{G}$.

By the uniqueness theorem, Theorem 3.3.2, there exists a nonzero constant $m(g)$ such that if $\mathbf{A}$ is a Borel subset of $\mathfrak{G}$ then

$$\nu(\mathbf{A}) = m(g)\mu(\mathbf{A}) = \mu(\mathbf{A}g). \tag{3.5.2}$$

The function m is called the *modular function*. This section is used to state and prove a few needed properties of modular functions. A Haar measure and the associated group are said to be *unimodular* if the modular function is identically one. Otherwise we say the measure and/or the group are not unimodular.

Lemma 3.5.1. *Let f be a continuous function with compact support $\mathbf{C}$, $f\colon \mathfrak{G} \to \mathbb{R}$, where $\mathfrak{G}$ is a locally compact group. Given $\varepsilon > 0$ there exists a compact neighborhood $\mathbf{U}$ of e, the identity of $\mathfrak{G}$, such that*

$$\sup_{x \in \mathfrak{G}} \sup_{g \subset \mathbf{U}} |f(gx) - f(x)| < \varepsilon. \tag{3.5.3}$$

Proof. Suppose the contrary is true. To every compact neighborhood $\mathbf{U}$ of e, $\{(g, x) \mid |f(gx) - f(x)| \geq \varepsilon \text{ and } g \in \bar{\mathbf{U}}\}$ is a compact set contained in the compact set $\bar{\mathbf{U}} \times (\bar{\mathbf{U}}^{-1} \cdot \mathbf{C})$. These sets clearly have the finite intersection property. Hence there is a point (g, x) in the intersection of all these sets and this point must have $g = e$. Thus $f(x) - f(x) \neq 0$. This contradiction shows that the required compact neighborhood of e must exist. □

Remark 3.5.2. This lemma may be used to generalize the proofs of several earlier results in which the existence of a countable base was assumed.

Lemma 3.5.3. *The modular function m is a continuous group homomorphism of $\mathfrak{G}$ to $(0, \infty)$. If $f\colon \mathfrak{G} \to \mathbb{R}$ is a μ-integrable function then*

$$\int f(hg^{-1})\mu(dh) = m(g) \int f(h)\mu(dh). \tag{3.5.4}$$

If $\mathfrak{G}$ is compact then $m(g_1) = m(g_2) = 1$ for all $g_1, g_2 \in \mathfrak{G}$.

Proof. Choose a Baire set $\mathbf{A}$ such that $0 < \mu(\mathbf{A}) < \infty$. Then $m(g_1g_2)\mu(\mathbf{A}) = \mu(\mathbf{A}(g_1g_2)) = \mu((\mathbf{A}g_1)g_2) = m(g_1)m(g_2)\mu(\mathbf{A})$. This implies $m(g_1)m(g_2) = m(g_1g_2)$.

We now prove continuity of the function m. Let f be a continuous function with compact support, $f\colon \mathfrak{G} \to \mathbb{R}$. By Lemma 3.5.1 it follows that since μ restricted to compact sets is a totally finite measure,

$$\left(\lim_{g \to e} m(g)\right) \int f(h)\mu(dh) = \lim_{g \to e} \int f(hg^{-1})\mu(dh) = \int f(h)\mu(dh). \tag{3.5.5}$$

We have used here (3.5.4) which is proven in the next paragraph. As is well known, continuity of a group homomorphism at e implies continuity at all points.

To establish (3.5.4) we use a standard approximation argument. If $1_{\mathbf{A}}$ is the indicator function of the Borel subset $\mathbf{A}$ then

$$m(g)\mu(\mathbf{A}) = \int 1_{\mathbf{A}g}(h)\mu(dh) = \int 1_{\mathbf{A}}(hg^{-1})\mu(dh). \tag{3.5.6}$$

By linearity it follows that (3.5.4) holds for all simple functions and by monotone convergence it follows that (3.5.4) holds for all μ-integrable functions f.

If $\mathfrak{G}$ is a compact group then the function $f(g) = 1$ for all $g \in \mathfrak{G}$ is integrable and $0 < \mu(\mathfrak{G}) < \infty$. By (3.5.4) it follows that $m(g) = 1$ for all $g \in \mathfrak{G}$. □

Lemma 3.5.4. *If the group $\mathfrak{G}$ is Abelian then the modular function is identically equal to one.*

PROOF. $m(g)\mu(\mathbf{A}) = \mu(\mathbf{A}g) = \mu(\mathbf{A})$. □

EXAMPLE 3.5.5. The invariant measures on $\mathbf{GL}(n)$ given by the differential forms $\bigwedge_{i=1}^{n} \bigwedge_{j=1}^{n} (G^{-1}\,dG)_{ij}$ and $\bigwedge_{i=1}^{n} \bigwedge_{j=1}^{n} ((dG)G^{-1})_{ij}$ are the same since both differential forms compute to be $\bigwedge_{i=1}^{n} \bigwedge_{j=1}^{n} (dG)_{ij}/(\det G)^n$. Thus the Haar measure on $\mathbf{GL}(n)$ are *unimodular*. By Lemma 3.5.3, the Haar measures on $\mathbf{O}(n)$ are unimodular. If $\mathbf{D}(n)$ is the group of $n \times n$ diagonal matrices with positive diagonal entries, then $\mathbf{D}(n)$ is Abelian and by Lemma 3.5.4 the Haar measures are unimodular. The differential forms are, if $L = \operatorname{diag}(\lambda_1, \ldots, \lambda_n)$, then $\bigwedge_{i=1}^{n} (L^{-1}\,dL)_{ii} = \bigwedge_{i=1}^{n} (d\lambda_i/\lambda_i) = \bigwedge_{i=1}^{n} ((dL)L^{-1})_{ii}$, verifying unimodularity in this case.

The group $\mathbf{T}(n)$ is not unimodular. We discuss this now. Also, see Sections 7.2 and 7.3.

We make use of the uniqueness theorem for Haar measure and make an argument similar to that for Theorem 2.0.1.

Theorem 3.5.6. *Let μ be a Lebesgue measure on the $\frac{1}{2}n(n+1)$-dimensional space of the variables $t_{11}, t_{21}, t_{22}, \ldots, t_{i1}, \ldots, t_{ii}, \ldots, t_{n1}, \ldots, t_{nn}$. If $\mathbf{C}$ is a Borel subset of lower triangular matrices and $A = (a_{ij})$ is a lower triangular matrix then define $\nu(\mathbf{C}) = \mu(\mathbf{C}A)$. There exists a functional m such that $\mu(\mathbf{C}A) = m(A)\mu(\mathbf{C})$. m is given by*

$$m(A) = |a_{11}^n a_{22}^{n-1} \cdots a_{nn}|. \tag{3.5.7}$$

PROOF. The measure ν is an invariant measure for the additive group, $\nu(\mathbf{C} + B) = \mu((\mathbf{C} + B)A) = \mu(\mathbf{C}A + BA) = \nu(\mathbf{C})$. Therefore there exists a constant $m(A)$ such that $m(A)\mu(\mathbf{C}) = \mu(\mathbf{C}A)$ for all $A \in \mathbf{T}(n)$ and Borel subsets $\mathbf{C}$. Then $m(I_n) = 1$, $m \neq 0$, and m is multiplicative. By Lemma 6.6.3, $m(A) = 1$ if the diagonal of A is $1, \ldots, 1$. If A is a diagonal matrix then the matrix TA has ij-element $a_i t_{ij}$, so that as a linear transformation of $\mathbb{R}^{n(n+1)/2}$

the transformation has a diagonal matrix with determinant $a_{11}^n a_{22}^{n-1} \cdots a_{nn}$. We use Theorem 2.0.1 to obtain for A diagonal that $m(A) = (3.5.7)$. □

Remark 3.5.7. A similar argument shows that transformation from the left defines the functional m' such that

$$\begin{aligned} m'(A)m'(B) &= m'(AB); \\ m'(A)\mu(\mathbf{C}) &= \mu(A\mathbf{C}); \\ m'(A) &= |a_{11}a_{22}^2 \cdots a_{nn}^n|. \end{aligned} \tag{3.5.8}$$

Theorem 3.5.8. *The left and right invariant Haar measures for* $\mathbf{T}(n)$ *are absolutely continuous with respect to Lebesgue measure* μ *and have the density functions*

$$\int 1_{\mathbf{C}}(T)(m(T))^{-1}\mu(dT) \tag{3.5.9}$$

which is the right invariant measure, and

$$\int 1_{\mathbf{C}}(T)(m'(T))^{-1}\mu(dT) \tag{3.5.10}$$

which is the left invariant measure.

Partial Proof. We give a partial proof in which right invariance is verified for (3.5.9).

$$\begin{aligned} \int 1_{\mathbf{C}A}(T)(m(T))^{-1}\mu(dT) &= \int 1_{\mathbf{C}}(TA^{-1})(m(A))^{-1}(m(TA^{-1}))^{-1}\mu(dT) \\ &= \int 1_{\mathbf{C}}(T)\mu(dT). \end{aligned} \tag{3.5.11}$$

This argument uses the multiplicative property, (3.5.8), and the fact that $m(I_n) = 1$ to establish $m(A^{-1}) = m(A)^{-1}$. The change of variable in the integral (3.5.11) uses the condition $m(A)\mu(\mathbf{C}) = \mu(\mathbf{C}A)$ of (3.5.8). A similar argument can be used to establish left invariance of (3.5.10). □

Remark 3.5.9. For the right invariant measure ν given in (3.5.9), the modular function may be computed as follows.

$$\begin{aligned} \nu(A\mathbf{C}) &= \int 1_{\mathbf{C}}(A^{-1}T)(m(T))^{-1}\mu(dT) \\ &= (m(A))^{-1} \int 1_{\mathbf{C}}(A^{-1}T)(m(A^{-1}T))^{-1}\mu(dT) \\ &= (m'(A)/m(A))\nu(\mathbf{C}). \end{aligned} \tag{3.5.12}$$

Remark 3.5.10. In the study of complex valued normal random variables $\mathbf{z} = \mathbf{x} + \mathbf{i}\mathbf{y}$ such that $\mathbf{x}$ and $\mathbf{y}$ are independent normal (0, 1) random variables the analogue of $\mathbf{T}(n)$ is $\mathbf{CT}(n)$, the set of lower triangular matrices $\mathbf{T}$ with complex number entries, and real positive number entries on the diagonal The number of real variables is then

$$n^2 = 2 \times \tfrac{1}{2}n(n-1) + n. \tag{3.5.13}$$

By analogy to the derivation of (3.5.7) and (3.5.8), n^2-dimensional Lebesgue measure determines a multiplicative functional m, which for transformation from the left, is

$$m(A) = |a_{11}a_{22}^3 \cdots a_{nn}^{2n-1}|. \tag{3.5.14}$$

It follows that the left invariant measures on $\mathbf{CT}(n)$ relative to n^2-dimensional Lebesgue measure μ have density functions proportional to

$$(x_{11}x_{22}^3 \cdots x_{nn}^{2n-1})^{-1} \prod_{i \geq j} dx_{ij} \prod_{i > j} dy_{ij}. \tag{3.5.15}$$

3.6. Differential Forms of Invariant Measures on Matrix Groups

Some of the matrix groups of interest to us are named in Example 3.5.5. In all these examples the group operation is matrix multiplication. Given G, $H \in \mathfrak{G}$, then $(GH)_{ij} = \Sigma_{k=1}^n G_{ik}H_{kj}$ so that under the topology of coordinatewise convergence the map $(G, H) \to GH$ is clearly jointly continuous. The entries of G^{-1} are rational functions of the entries of G so the map $G \to G^{-1}$ is continuous in the above topology. Since the coordinatewise convergence topology is induced by embedding $\mathfrak{G}$ in an Euclidean space the topology is a locally compact topology. The topology is clearly a metric topology.

In the case of each group we specify a set $\mathbf{I}$ which determines a maximal set of local coordinates. Then if $G = (g_{ij})$ the differential form $\bigwedge_{(i,j) \in \mathbf{I}} dg_{ij}$ has maximal degree and $\bigwedge_{(i,j) \in \mathbf{I}} (G^{-1}\, dG)_{ij}$ gives the differential form of a left invariant (regular) measure, i.e., a Haar measure. See Example 3.5.5.

Because the set of alternating forms of maximal degree is one dimensional, see Theorem 6.2.13, it follows that $\bigwedge_{(i,j) \in \mathbf{I}} (H\, dG)_{ij} = m(H) \bigwedge_{(i,j)} (dG)_{ij}$ since both forms are of maximal degree. Successive substitutions shows $m(H_1)m(H_2) = m(H_1H_2)$ and that $m(I) = 1$. Further m must be a polynomial in the entries of H, hence is a continuous function. Last, the substitution αH establishes $m(\alpha H) = \alpha^{\#\mathbf{I}} m(H)$ where $\#\, \mathbf{I}$ is the number of elements in $\mathbf{I}$. These four properties together with the group structure usually determine the function m. Special cases are discussed in Section 6.6. For the orthogonal group continuity of m implies $m(H) = \pm 1$. For the group of lower triangular matrices, left multiplication, discussed here, requires $m(H) = h_{11}h_{22}^2 \cdots h_{nn}^n$ for $H \in \mathbf{T}(n)$ with diagonal elements $h_{11}, \ldots, h_{nn}$. For further discussion see Lemma 6.6.3, Theorem 6.6.4, Theorem 3.5.6 and following.

3.7. Cross-Sections

Suppose $\mathfrak{G}$ acts as a transformation group on a manifold $\mathfrak{H}$. Then $\mathfrak{H}$ may be factored via $\mathfrak{G}$ into the space $\mathfrak{H}/\mathfrak{G}$ of orbits. The problem of cross-sections is to choose from each orbit $\mathfrak{G}h$ a single element in some "nice" way. We let $\mathfrak{H}_1$ be the subset of chosen elements. $\mathfrak{H}_1$ is to be topologized in such a way that $\mathfrak{G} \times \mathfrak{H}_1$ is measure isomorphic, or homeomorphic, or diffeomorphic to $\mathfrak{H}$, depending on the problem.

The construction of cross-sections and their use in factoring measures in the sense of Section 3.4 has been discussed by Wijsman (1966, 1984), Koehn (1970), and Bondar (1976). See also Chapter 10 of this book. We do not pursue the abstract treatment of cross-sections since in the examples considered in this book there are obvious choices of cross-sections $\mathfrak{H}_1$. The methods of Karlin (1960) discussed in Chapter 5, although not universally applicable, will be seen to avoid the abstract consideration of cross-sections by introducing into integrals of functions of matrices obvious invariants. Also, use of differential forms, suggested already in this chapter and discussed in Chapters 6 and 7, often will give answers to problems without resort to a formal theory of cross-sections.

EXAMPLE 3.7.1. The action of $\mathbf{GL}(k)$ on the set $\mathbf{H}_0$ of $n \times k$ matrices given by multiplication on the right was discussed in Example 3.4.3. We let $\mathbf{N}$ be the subset of $\mathbf{H}_0$ consisting of those $n \times k$ matrices having a linearly dependent set of first k rows. Then $\mathbf{N}$ is clearly an invariant subset under the action of $\mathfrak{G}$, hence so is $\mathbf{H} = \mathbf{H}_0 - \mathbf{N}$. If $X \in \mathbf{H}$ we may use the factorization

$$X^t = G^t(I_k, Z^t), \tag{3.7.1}$$

with the notation the same as in Example 3.4.3. The matrix $(I_k, Z^t)^t$ is then the choice of a matrix in the orbit of X. We call this set $\mathbf{H}_1$ and give it the usual Euclidean topology. Then $\mathbf{H}_1$ is a Borel subset of $\mathbf{H}$ and has a locally compact topology. Note that $\mathbf{H}_1$ has nk-dimensional Lebesgue measure zero.

As is illustrated in Example 3.7.1, in general an invariant null set will be constructed having zero measure relative to a given measure which is left (or right) invariant. Thus μ is also an invariant measure on $\mathbf{H} = \mathbf{H}_0 - \mathbf{N}$. The residual set $\mathbf{H}$ will factor in an obvious way to

$$\mathbf{H} = \mathbf{H}_0 - \mathbf{N} \simeq \mathbf{H}_1 \times \mathfrak{G}. \tag{3.7.2}$$

By the results of Section 3.4 the measure μ then factors to $\mu = \nu_1\nu_2$ with ν_2 a right invariant measure for $\mathfrak{G}$, and in cases where it is meaningful, ν_1 a left invariant measure on $\mathbf{H}_1$. These factors are regular measures.

EXAMPLE 3.7.2. The process of factoring, described above, should be compared with inducing a measure on the quotient space $\mathfrak{G}/\mathbf{H}$ in the sense discussed in Sections 3.2 and 3.3. Consider $\mathfrak{G} = \mathbb{R} \times \mathbb{R}$ and $\mathbf{H} = \{0\} \times \mathbb{R}$, $\mu =$ Lebesgue measure on $\mathfrak{G}$. The induced measure is infinite on all open subsets of $\mathfrak{G}/\mathbf{H}$ while the factored measure is one-dimensional Lebesgue measure.

3.8. Solvability, Amenability

This section has been included because of its relevance to use of the Hunt–Stein theory as presented in Lehmann (1959). As a notation, in this section, $\mathbf{A} \oplus \mathbf{B}$ means the symmetric difference of sets defined by

$$\mathbf{A} \oplus \mathbf{B} = (\mathbf{A} - \mathbf{B}) \cup (\mathbf{B} - \mathbf{A}). \tag{3.8.1}$$

The hypotheses of the Hunt–Stein theory require the existence of a sequence $\{\mathbf{U}_n, n \geq 1\}$ of Borel subsets of the group $\mathfrak{G}$ satisfying if $n \geq 1$ then $\mathbf{U}_n \subset \mathbf{U}_{n+1}$, $0 < \mu(\mathbf{U}_n) < \infty$, and $\lim_{n\to\infty} \mathbf{U}_n = \mathfrak{G}$, where μ is a *right invariant* Haar measure on the group $\mathfrak{G}$. In addition the following condition must hold.

$$\text{If} \quad g \in \mathfrak{G} \quad \text{then} \quad \lim_{n\to\infty} \mu(\mathbf{U}_n g \oplus \mathbf{U}_n)/\mu(\mathbf{U}_n) = 0. \tag{3.8.2}$$

EXAMPLE 3.8.1. $\mathfrak{G} = \mathbb{R}$, $\mathbf{U}_n = [-n, n]$, $\mu =$ Lebesgue measure. Then if $g \geq 0$, $(\mathbf{U}_n + g) \oplus \mathbf{U}_n = [-n, -n + g) \cup (n, n + g]$ and this set has Lebesgue measure $2g$. Since $\lim_{n\to\infty} 2g/2n = 0$, condition (3.8.2) holds.

If the group $\mathfrak{G}$ acts on $\mathfrak{X}$, the point set of a measure space, such that the map $(g, x) \to g(x)$ is jointly measurable, then it is shown in Lehmann, op. cit., that for a bounded measurable function ψ

$$\lim_{n\to\infty} \int_{\mathbf{U}_n} \psi(gx)\mu(dg)/\mu(\mathbf{U}_n) = \Psi(x) \tag{3.8.3}$$

exists as a weak limit in $\mathbf{L}_\infty$ of $\mathfrak{X}$ and Ψ is a $\mathfrak{H}$ invariant function, called an invariant mean. See Bonder and Milnes (1981).

For some groups $\mathfrak{G}$ this process may be carried out in stages, as follows. Suppose $\mathfrak{G}$ has subgroups $\mathfrak{G}_1$ and $\mathfrak{G}_2$ such that $\mathfrak{G}_1 \cap \mathfrak{G}_2 = \{e\}$ and $\mathfrak{G}_1$ is a normal subgroup. If ψ is a $\mathfrak{G}_1$ invariant function and $g \in \mathfrak{G}_2$ then ψ_1 defined by $\psi_1(x) = \psi(gx)$ is again a $\mathfrak{G}_1$ invariant function. For $\psi_1(g_1 x) = \psi(gg_1 x) = \psi(g_1' gx) = \psi_1(x)$. In this argument we use the assumption that $\mathfrak{G}_1$ is a normal subgroup. It then follows that if μ is a Borel measure for $\mathfrak{G}_2$ such that $\int_{\mathfrak{G}_2} \psi(gx)\mu(dg)$ is meaningful, one expects this to also be a $\mathfrak{G}_1$ invariant function. Thus if (3.8.2) holds for $\mathfrak{G}_2$ also, a limit (3.8.3) may be taken to generate a $\mathfrak{G}_2$ invariant function that is also $\mathfrak{G}_1$, hence because of normality, $\mathfrak{G}$ invariant. The argument for this last step requires further specification of the measures for $\mathfrak{X}$. See Lehmann, op. cit.

EXAMPLE 3.8.2. Let the group $\mathfrak{G} = \mathbf{T}(n)$ and let $\mathbf{T}_1(n)$ be the subgroup of those matrices with diagonal elements equal 1. Then $\mathbf{T}_1(n)$ is a normal subgroup of $\mathbf{T}(n)$ and

$$\mathbf{T}(n) = \mathbf{T}_1(n) \cdot \mathbf{D}(n). \tag{3.8.4}$$

A decomposition of $\mathbf{T}_1(n)$ is given by the chain

$$\mathbf{T}_1(n) = \mathfrak{G} = \mathfrak{G}_n \supset \mathfrak{G}_{n-1} \supset \cdots \supset \mathfrak{G}_1 \tag{3.8.5}$$

together with subgroups $\mathfrak{H}_i$, $2 \le i \le n$ such that $\mathfrak{H}_i$ is isomorphic to the additive group of $\mathbb{R}^{i-1}$. Then

$$\mathfrak{G}_i = \mathfrak{H}_i \cdots \mathfrak{H}_i, \qquad 1 \le i \le n, \tag{3.8.6}$$

and $\mathfrak{H}_i$ is a normal subgroup of $\mathfrak{G}_i$, $2 \le i \le n$. This is easily seen from the identity

$$\begin{vmatrix} I & 0 \\ a^t & 0 \end{vmatrix} \begin{vmatrix} T & 0 \\ 0 & 1 \end{vmatrix} \begin{vmatrix} I & 0 \\ b^t & 1 \end{vmatrix} \begin{vmatrix} T^{-1} & 0 \\ 0 & 1 \end{vmatrix} \begin{vmatrix} I & 0 \\ -a^t & 1 \end{vmatrix} = \begin{vmatrix} I & 0 \\ b^t T^{-1} & 1 \end{vmatrix}. \tag{3.8.7}$$

EXAMPLE 3.8.3. Let $\mathfrak{G}$ be the set of 2×2 matrices $\begin{vmatrix} a & 0 \\ b & 1 \end{vmatrix}$ with the subgroups, $\mathfrak{G}_1$ the matrices $\begin{vmatrix} 1 & 0 \\ b & 1 \end{vmatrix}$, and $\mathfrak{G}_2$ the matrices $\begin{vmatrix} a & 0 \\ 0 & 1 \end{vmatrix}$. Then $\mathfrak{G}_1$ is a normal subgroup and (3.8.2) may be verified for $\mathfrak{G}_1$ as in Example 3.8.1. For (3.8.2) in the case of $\mathfrak{G}_2$ we let $\mathbf{U}_n^{(2)}$ be those matrices with $1/n \le a \le n$ and use the Haar measure for the multiplicative group on $(0, \infty)$. One may compute that

$$\begin{vmatrix} a & 0 \\ 0 & 1 \end{vmatrix} \begin{vmatrix} 1 & 0 \\ b & 1 \end{vmatrix} \begin{vmatrix} a^{-1} & 0 \\ 0 & 1 \end{vmatrix} = \begin{vmatrix} 1 & 0 \\ ba^{-1} & 1 \end{vmatrix}. \tag{3.8.8}$$

Then if $g_2 = \begin{vmatrix} a & 0 \\ 0 & 1 \end{vmatrix}$ and $a \ne 1$, then

$$\lim_{n\to\infty} \mu_1(g_2^{-1}\mathbf{U}_n^{(1)} \cap \mathbf{U}_n^{(1)})/\mu_1(\mathbf{U}_n^{(1)}) \ne 1. \tag{3.8.9}$$

Here $\mathbf{U}_n^{(1)} = [-n, n]$.

EXAMPLE 3.8.4. Another decomposition of $\mathbf{T}(n)$ results if the multiplicative part is taken out in n steps rather than simultaneously. In this decomposition the alternation of normal subgroups is more complex than in Example 3.8.2. We have

$$\mathbf{T}(n) = \mathfrak{G}_{2n-1} \supset \mathfrak{G}_{2n-2} \supset \cdots \supset \mathfrak{G}_1 \tag{3.8.10}$$

together with subgroups $\mathfrak{H}_i$, $1 \le i \le 2n-1$, such that if $i = 2j$ then $\mathfrak{H}_i$ is isomorphic to $\mathbb{R}^j$ while if $i = 2j-1$ then $\mathfrak{H}_i$ is isomorphic to the multiplicative group on $(0, \infty)$. Then

$$\mathfrak{G}_i = \mathfrak{H}_1 \cdot \cdots \cdot \mathfrak{H}_i, \qquad 1 \le i \le 2n - 1. \tag{3.8.11}$$

In this decomposition, at the odd-numbered steps, when a multiplicative subgroup is factored, the normal subgroup is $\mathfrak{G}_{2j}$, while at the even numbered steps when an additive subgroup is factored, the normal subgroup is the subgroup $\mathfrak{H}_{2j-1}$.

Remark 3.8.5. In statistical applications the group is usually thought of as acting from the left on points x of $\mathfrak{X}$. In the subject of *amenability* the group is usually considered as acting on itself. Conditions on groups which imply the existence, or which are equivalent to, the existence of an invariant mean, have been studied by many mathematicians. The survey, Bondar and Milnes (1981), lists many of the important conditions that have been studied, their equivalences, and contains a literature survey of mathematicians and their suggested conditions. Kiefer (1957) constructs a sequence of subsets giving an invariant mean for $\mathbf{T}(n)$. An example due to Stein and contained in Lehmann (1959) shows that $\mathbf{GL}(n)$ cannot have any invariant means. If the group is thought of as acting on functions, then invariant functions are fixed points of the group action. Theorems such as the Kakutani fixed point theorem, see Dunford and Schwartz (1958), which construct fixed points as limits of averages, can then be used to construct invariant means. This is an old idea, which was suggested to the author by Huber in (1965). Further development may be found in Bondar and Milnes (1981), Brown (1980), and Brown (1982).

CHAPTER 4

Wishart's Paper

4.0. Introduction

Some of the earlier writers, notably J. Wishart and R. A. Fisher, made extensive use of geometric reasoning in order to compute the volume elements that arise from changes of variable. We have selected the paper by Wishart (1928) to illustrate this type of argument. The Wishart density function so obtained plays a central role in the discussions which follow in later chapters, where different methods of obtaining density functions of maximal invariants are discussed. See especially Chapters 5, 10 and 11. Section 4.1 gives Wishart's argument. Section 4.2 explains an idea of James (1955a) by means of which the noncentral Wishart density may be obtained. The expression for the density function involves an integral over $\mathbf{O}(n)$ which cannot be evaluated in closed form by any known method. Section 4.3 illustrates the difficulty that one encounters when trying to obtain a *power series* expansion of the integral. This difficulty led to the development of zonal polynomials and the series for hypergeometric functions, mostly the work of A. T. James (see Chapters 12 and 13). The problems, Section 4.4, leave certain necessary details to the reader, with hints. Material relevant to some of the problems will be found in other chapters.

With minor changes of notation Wishart writes as follows. $\mathbf{X}$ is a random $n \times h$ matrix having joint density function $f(X^tX)$ relative to Lebesgue measure $dx_{11} \cdots dx_{nh}$ on nh-dimensional space. One wishes to find the density function of $\mathbf{X}^t\mathbf{X}$ expressed in terms of Lebesgue measure on the $\frac{1}{2}h(h+1)$-dimensional space of the variables $(\xi_{ij}) = X^tX$. Since f is already a function of the new variables the change of variable followed by integration out of the extra variables introduces a multiplying factor (i.e., volume element) and the desired density function has the form

$$g(X^tX)f(X^tX)\prod_{j\le i} d\xi_{ij}. \tag{4.0.1}$$

The function g is independent of the function f so that by computing g in one case we solve the problem of computing (4.0.1) in all similar problems. This fact was noted and used by James (1955a), which we discuss in Sections 4.2 and 4.3.

4.1. Wishart's Argument

Wishart thought of the problem as follows. $X_1, \ldots, X_h$ are the columns of X, are vectors in $\mathbb{R}^n$. We suppose $X_1, \ldots, X_h$ are linearly independent and leave the proof of this as Problem 4.4.4 at the end of this chapter. Let $\mathbf{S}_h$ be the linear span of $X_1, \ldots, X_h$ and let $Y_1, \ldots, Y_h$ be an orthonormal basis of $\mathbf{S}_h$ obtained from $X_1, \ldots, X_h$ by Gram–Schmidt orthogonalization, so that

$$X_i = a_{i1}Y_1 + \cdots + a_{ii}Y_i, \qquad 1 \le i \le h, \tag{4.1.1}$$

and setting $a_i^t = (a_{i1}, \ldots, a_{ii}, 0, \ldots, 0)$ we compute

$$X_i^tX_j = a_i^ta_j, \qquad 1 \le i, j \le h. \tag{4.1.2}$$

As is well known the h-dimensional parallelopiped with sides $X_1, \ldots, X_h$ has volume V_h given by

$$(\det X^tX)^{1/2} = (\det(a_i^ta_j))^{1/2} = V_h. \tag{4.1.3}$$

More generally if $1 \le i \le h$ we will write V_i for the volume of the i-dimensional parallelopiped determined by $X_1, \ldots, X_i$. Then the length of the perpendicular projection of X_h on the span of $X_1, \ldots, X_{h-1}$ is V_h/V_{h-1}.

In h-dimensional space introduce new coordinates for X_h defined by considering $X_1, \ldots, X_{h-1}$ as fixed and setting

$$\xi_{hi} = a_h^ta_i = \sum_{j=1}^{h} a_{hj}a_{ij}. \tag{4.1.4}$$

The Jacobian of this transformation is easily computed to be

$$\frac{\partial(\xi_{h1}, \ldots, \xi_{hh})}{\partial(a_{h1}, \ldots, a_{hh})} = \det\begin{vmatrix} a_{11} & 0 & \cdots & 0 \\ a_{21} & a_{22} & \cdots & 0 \\ \cdots & \cdots & \cdots & \cdots \\ 2a_{h1} & 2a_{h2} & \cdots & 2a_{hh} \end{vmatrix} = 2V_h. \tag{4.1.5}$$

We now discuss the problem of integrating out the extra variables. At the first step we consider $X_1, \ldots, X_{h-1}$ as fixed and introduce new variables $\xi_h^t = (\xi_{h1}, \ldots, \xi_{hh})$. In addition let us write

$$\xi_i^t = (\xi_{i1}, \ldots, \xi_{ih}), \qquad 1 \le i \le h. \tag{4.1.6}$$

If we fix $\xi_1, \ldots, \xi_h$ then we fix the lengths of $X_1, \ldots, X_h$ and the angles between them. In particular if X_h' and X_h are two possible choices of the h-th vector then

$$X_i^t(X_h' - X_h) = 0, \qquad 1 \le i \le h - i. \tag{4.1.7}$$

Therefore $X_h' - X_h$ is orthogonal to the span of $X_1, \ldots, X_{h-1}$. Further, any such X_h' when adjointed to $X_1, \ldots, X_{h-1}$ gives rise to a parallelopiped of volume V_h since this volume is given by (4.1.3). If $\mathbf{S}_{h-1}$ is the linear span of $X_1, \ldots, X_{h-1}$ and $\mathbf{S}_{h-1}^\perp$ is its orthogonal complement in n-dimensional space, if $\mathbf{S}_{h-1\,\lambda}^\perp$ are those vectors in $\mathbf{S}_{h-1}^\perp$ of length λ, where we take $\lambda = V_h/V_{h-1}$, then for fixed $X_1, \ldots, X_{h-1}$, the set of X_h' having coordinates $\xi_{h1}, \ldots, \xi_{hh}$ are just the sum of $a_{h1}Y_1 + \cdots + a_{h\,h-1}Y_{h-1}$ depending on $X_1, \ldots, X_{h-1}$ and the vectors in $\mathbf{S}_{h-1,\lambda}^\perp$. Thus the required volume element is the surface of this sphere. Note that $a_{hh} = V_h/V_{h-1}$.

The required surface area is given by

$$[2(V_h/V_{h-1})^{n-h}\pi^{(n-h+1)/2}]/\Gamma(\tfrac{1}{2}(n-h+1)). \tag{4.1.8}$$

Derivation of this formula is given as Problem 4.4.3. Therefore the multiplier introduced by the change of variables (4.1.4) followed by integration is

$$\frac{2\pi^{(n-h+1)/2}}{\Gamma(\tfrac{1}{2}(n-h+1))}\left|\frac{V_h}{V_{h-1}}\right|^{n-h}\frac{1}{2V_h} = \frac{\pi^{(n-h+1)/2}V_h^{n-h-1}}{\Gamma(\tfrac{1}{2}(n-h+1))V_{h-1}^{n-h}}. \tag{4.1.9}$$

Having made the change of variable for X_h we proceed inductively for $X_{h-1}, \ldots, X_1$ and obtain as an end result

$$\frac{\pi^{n/2}\cdots\pi^{(n-h+1)/2}V_h^{n-(h+1)}V_{h-1}^{n-h}V_1^{n-2}}{\Gamma(\tfrac{1}{2}n)\cdots\Gamma(\tfrac{1}{2}(n-h+1))V_{h-1}^{n-h}V_{h-2}^{n-(h-1)}V_0^{n-1}}\prod_{j\le i} d\xi_{ij}. \tag{4.1.10}$$

Theorem 4.1.1 (Wishart). *Let the random $n \times h$ matrix $\mathbf{X}$ have a density function $f(X^tX)$ relative to Lebesgue measure. Then the joint density function of the $h(h+1)/2$ random variables $\mathbf{S} = \mathbf{X}^t\mathbf{X}$ is*

$$\frac{\pi^{nh/2-h(h-1)/4}(\det S)^{(n-h-1)/2}f(S)}{\Gamma(\tfrac{1}{2}n)\cdots\Gamma(\tfrac{1}{2}(n-h+1))}\prod_{j\le i} ds_{ij}. \tag{4.1.11}$$

Remark 4.1.2. The reader should refer to Section 5.2 to obtain a different perspective of this result.

4.2. The Noncentral Wishart Density Function

Herz (1955) and James (1955a) at roughly the same time (they were fellow students at Princeton University) obtained expressions for the noncentral Wishart density function thereby generalizing previous partial results of Anderson (1946). In this section we review the idea of James.

The noncentral problem is the problem of finding the density function of $\mathbf{X}^t\mathbf{X}$ when $\mathbf{X}$ is a random $n \times h$ matrix with independently and identically distributed rows each normal (M, Σ), $M \neq 0$. In the sequel we let M be $n \times h$ so that $\mathbf{E}\mathbf{X} = M$. If $M \neq 0$ the joint density function of $\mathbf{X}$ is not a function of $\mathbf{X}^t\mathbf{X}$ and the discussion of Section 4.1 fails to apply. James, op. cit., has the following idea by means of which the noncentral problem may be reduced to a problem to which the results of Section 4.1 *do* apply.

Let $H_1, \ldots, H_k$ be $n \times n$ orthogonal matrices and $(\alpha_1, \ldots, \alpha_k)$ be a probability vector. Then

$$\sum_{i=1}^{h} \alpha_i P(H_i\mathbf{X} \in \mathbf{A}) \tag{4.2.1}$$

is a probability measure on the Borel subsets of $\mathbb{R}^{nh}$. The measure (4.2.1) induces a measure on $\mathbf{S}(n)$, the space of $n \times n$ positive definite matrices, by the rule, if $\mathbf{B} \subset \mathbf{S}(n)$ is a Borel subset then

$$\mathbf{A} = \{X | X^tX \in \mathbf{B}\}. \tag{4.2.2}$$

Then if $\mathbf{Y}$ has the probability measure (4.2.1) the random variable $\mathbf{Y}^t\mathbf{Y}$ has the probability measure

$$\begin{aligned} P(\mathbf{Y}^t\mathbf{Y} \in \mathbf{B}) &= P(\mathbf{Y} \in \mathbf{A}) = \sum_{i=1}^{k} \alpha_i P(H_i\mathbf{X} \in \mathbf{A}) \\ &= \sum_{i=1}^{k} \alpha_i P((H_i\mathbf{X})^t(H_i\mathbf{X}) \in \mathbf{B}) = (\alpha_1 + \cdots + \alpha_k)P(\mathbf{X}^t\mathbf{X} \in \mathbf{B}) \qquad (4.2.3) \\ &= P(\mathbf{X}^t\mathbf{X} \in \mathbf{B}). \end{aligned}$$

Since (4.2.3) holds for all finite convex mixtures, we take a sequence of probability vectors and orthogonal matrices tending in the limit to Haar measure of unit mass on $\mathbf{O}(n)$ and obtain

$$P(\mathbf{X}^t\mathbf{X} \in \mathbf{B}) = \int_{\mathbf{O}(n)} P(H\mathbf{X} \in \mathbf{A})\, dH, \tag{4.2.4}$$

where dH represents the Haar measure of unit mass.

This sketch is not, of course, a full proof. It is easy to check dH-integrability of $P(H\mathbf{X} \in \mathbf{A})$ and that the integral in (4.2.4) is a probability measure. Then the steps of (4.2.3) can be repeated under the integral to obtain (4.2.4). We now consider the technical points which arise. The measurability of $P(H\mathbf{X} \in \mathbf{A})$ as a function of H follows from the joint measurability of the map $(H, X) \to HX$. The integral (4.2.4) is a double integral with respect to a product measure so Fubini's theorem applies. Measurability of $P(H\mathbf{X} \in \mathbf{A}) = E\,1_{\mathbf{A}}(H\mathbf{X})$ then follows. As a second consideration, in the sequel we are interested in cases where $\mathbf{X}$ has a density function, say g. Suppose $\mathbf{Y}$ has density function

$$\int_{\mathbf{O}(n)} g(HX)\, dH. \tag{4.2.5}$$

Measurability as a function of X follows from the above remarks. It then follows that $\mathbf{Y}^t\mathbf{Y}$ and $\mathbf{X}^t\mathbf{X}$ have the same density function which is obtained from (4.2.5) by introducing appropriate normalization as discussed in Section 4.1. Some further details are given in Problems 4.4.18 and 4.4.19.

In this section we need to know that (4.2.5) is a function of $(X^tX)^{1/2}$. We now prove this in the case $h \le n$ of taking n observations. We ignore the set of Lebesgue measure zero on which the columns of $\mathbf{X}$ are linearly dependent, see Problem 4.4.4. Choose X_1 a $n \times (n-h)$ matrix such that the $n \times n$ matrix $Z = (X, X_1)$ is nonsingular. Further suppose that $X^tX_1 = 0$. Then $H = Z(Z^tZ)^{-1/2}$ is an orthogonal matrix and

$$H^tZ = (H^tX, H^tX_1) = \begin{vmatrix} (X^tX)^{1/2} & 0 \\ 0 & (X_1^tX_1)^{1/2} \end{vmatrix}. \tag{4.2.6}$$

We let $X_1 \to 0$ in such a way that thc corresponding H's converge to some H_0 ($\mathbf{O}(n)$ is a compact metric space) to obtain

$$H_0X = \begin{vmatrix} (X^tX)^{1/2} \\ 0 \end{vmatrix}. \tag{4.2.7}$$

By invariance of Haar measure ($\mathbf{O}(n)$ is unimodular), (4.2.5) becomes

$$\int_{\mathbf{O}(n)} g(HX)\,dH = \int_{\mathbf{O}(n)} g(HH_0X)\,dH = \int_{\mathbf{O}(n)} g\left(H \begin{vmatrix} (X^tX)^{1/2} \\ 0 \end{vmatrix}\right) dH. \tag{4.2.8}$$

Let $g(X) = f((X-M)^t(X-M))$. The change of variable $Z = HX$ has Jacobian ± 1 so $\mathbf{Z}$ has density function

$$f(Z^tZ - Z^tHM - M^tH^tZ + M^tM), \tag{4.2.9}$$

and integration with respect to H by Haar measure of unit mass on $\mathbf{O}(n)$ gives

$$\int_{\mathbf{O}(n)} f(Z^tZ - Z^tHM - H^tM^tZ + M^tM)\,dH. \tag{4.2.10}$$

By our earlier discussion if a random variable $\mathbf{Z}$ has density function (4.2.10) then $\mathbf{X}^t\mathbf{X}$ and $\mathbf{Z}^t\mathbf{Z}$ have the same probability law. By the discussion of Section 4.1, the density function of the probability law is simply the function (4.2.10) multiplied by a normalization that represents integration out of the remaining variables. It follows that $\mathbf{X}^t\mathbf{X}$ has density function

$$\frac{\pi^{nh/2 - h(h-1)/4}}{\Gamma(\frac{1}{2}n)\cdots\Gamma(\frac{1}{2}(n-h+1))}(\det X^tX)^{(n-h-1)/2} \\ \times \int_{\mathbf{O}(n)} f(X^tX - X^tHM - M^tH^tX + M^tM)\,dH. \tag{4.2.11}$$

Specifically for the multivariate normal density function with $\Sigma = I_n$, and $\mathbf{E}\mathbf{X} = M$, a $n \times h$ matrix,

$$f((X-M)^t(X-M)) = (2\pi)^{-nh/2}\exp(-\tfrac{1}{2}\operatorname{tr} M^tM) \times \exp(-\tfrac{1}{2}\operatorname{tr} X^tX)\exp(\operatorname{tr} X^tM). \tag{4.2.12}$$

The noncentral Wishart density is, then,

$$\frac{\pi^{-h(h-1)/4}(\det X^tX)^{(n-h-1)/2}}{2^{nh/2}\Gamma(\frac{1}{2}n)\cdots\Gamma(\frac{1}{2}(n-h+1))} \times \exp(-\tfrac{1}{2}\operatorname{tr}(X^tX+M^tM))\int_{\mathbf{O}(n)}\exp(\operatorname{tr} X^tHM)\,dH. \tag{4.2.13}$$

The integral appearing in (4.2.13) can be phrased as

$$\int_{\mathbf{O}(n)}\exp(\operatorname{tr}(XM^tMX^t)^{1/2}H)\,dH \tag{4.2.14}$$

and this evaluation of (4.2.13) is equivalent to computing the Laplace transform of Haar measure. The answer in simple closed form is unknown. Transforms of functions of a matrix argument was part of the subject treated by Herz (1955). (4.2.14) was explicitly evaluated by Anderson (1946) when XM^t is a rank two matrix. James (1955b) has given in power series an evaluation of (4.2.14) in the case XM^t is a rank three matrix. See Section 4.3 for a summary of this work. Subsequent papers by James (1964) and Constantine (1963) on zonal polynomials give a different type of series evaluation of (4.2.14) as an infinite weighted sum of zonal polynomials.

The integral appearing in (4.2.13) can also be transformed to

$$\int_{\mathbf{O}(n)}\exp\left(\operatorname{tr}\left|(X^tX)^{1/2},0\right|H\left|\begin{matrix}(M^tM)^{1/2}\\0\end{matrix}\right|\right)dH. \tag{4.2.15}$$

In the current literature it is more usual to set $X^tX = S = (s_{ij})$ since integration is by Lebesgue measure on the variables s_{ij}, $i \geq j$. The form (4.2.15) makes the dependence on S explicit.

4.3. James on Series, Rank 3

Our source is James (1955b). We set

$$\Psi(Z) = \int_{\mathbf{O}(n)}\exp(\operatorname{tr} H^tZ)\,dH, \tag{4.3.1}$$

$Z \in \mathbf{GL}(n)$. Since Haar measure on $\mathbf{O}(n)$ is unimodular, see Lemma 3.5.3, and since $\operatorname{tr} AB = \operatorname{tr} BA$, from (4.3.1) we obtain

$$\int_{\mathbf{O}(n)}\exp(\operatorname{tr} H^tZ)\,dH = \int_{\mathbf{O}(n)}\exp(\operatorname{tr} H^tH_1ZH_2)\,dH, \tag{4.3.2}$$

and

$$\Psi(Z) = \Psi(H_1 Z H_2) \quad \text{for all } H_1 \text{ and } H_2 \text{ in } \mathbf{O}(n).$$

We take Z nonsingular and set $H_1 = H_2(Z^tZ)^{-1/2}Z^t$ and choose H_2 so that $H_2^t(Z^tZ)^{1/2}H_2 = D$ is a diagonal matrix. From (4.3.1) and (4.3.2) we obtain

$$\Psi(Z) = \Psi(D), \tag{4.3.3}$$

that is, the value of $\Psi(Z)$ depends only on the eigenvalues of $(Z^tZ)^{1/2}$. Since

$$\Psi(D) = \Psi(P^tDP), \qquad P \in \mathbf{O}(n) \text{ a permutation matrix,} \tag{4.3.4}$$

it follows that $\Psi(Z)$ is a symmetric function of the eigenvalues of $(Z^tZ)^{1/2}$. Since $\exp(\operatorname{tr} H^tD)$ has an infinite series expansion, it follows that $\Psi(D)$ is an entire function of the n eigenvalues of D.

The method of James (1955b) was to derive differential equations for Ψ by taking special types of matrices Z. For example, if

$$\Delta = (\partial^2/\partial z_{11}^2) + \cdots + (\partial^2/\partial z_{n1}^2), \qquad \text{then} \quad \Psi = \Delta\Psi.$$

From these differential equations recursion relations were developed. We don't pursue this further but give James's result.

Theorem 4.3.1 (James (1955b)). *Let $\lambda_1, \ldots, \lambda_n$ be the eigenvalues of $(Z^tZ)^{1/2}$ and define*

$$\xi(\lambda, \ldots, \lambda_n) = \int_{\mathbf{O}(n)} \exp(\operatorname{tr} H^tZ)\, dH. \tag{4.3.5}$$

Then

$$\xi(\lambda_1, \lambda_2, \lambda_3, 0, \ldots, 0) = \Sigma_{j_1 j_2 j_3} \frac{\Gamma(\tfrac{1}{2}n)\Gamma(\tfrac{1}{2}(n-1))\Gamma(\tfrac{1}{2}(n-2))(n+j_1+2j_2+4j_3-3)!}{2^{2j_1+4j_2-6j_3} j_1!j_2!j_3!\Gamma(\tfrac{1}{2}h+j_1+2j_2+3j_3-3)} \times \frac{\lambda_1^{j_1}\lambda_2^{j_2}\lambda_3^{j_3}}{\Gamma(j_3+\tfrac{1}{2}(n-2))\Gamma(j_2+2j_3+\tfrac{1}{2}(n-1))(n+j_1+2j_2+3j_3-3)!} \tag{4.3.6}$$

James noted and is quoted by Herz, op. cit., as saying the computation got out of hand if $\lambda_4 \neq 0$ was allowed. Please check the original source before using (4.3.6).

4.4. Related Problems

The calculations as presented assume $\mathbf{X}_1, \ldots, \mathbf{X}_n$ are linearly independent except on a set of Lebesgue measure zero. We develop several such results here as problems. First, write $V_n(r)$ as the volume of the set $\{(x_1, \ldots, x_n) | x_1^2 + \cdots + x_n^2 \le r^2\}$ and $A_{n-1}(r)$ as the surface area of the shell $\{(x_1, \ldots, x_n) | x_1^2 + \cdots + x_n^2 = r^2\}$. Required in the first step, (4.1.8), was the quantity

$A_{n-h-1}(V_h/V_{h-1})$. Problems 4.4.1 to 4.4.3 present a calculation of this area and volume. Problems 4.4.4 to 4.4.7 are about sets of measure zero. Problems 4.4.8 to 4.4.10 continue the change of variable calculation. Problems 4.4.11 to 4.4.14 are about noncentral chi-square random variables. Problems 4.4.15 to 4.4.18 are about integrals of polynomials in the entries of orthogonal matrices with respect to Haar measure. Problems 4.4.18 and 4.4.19 are about the density functions of integrated probability measures.

PROBLEM 4.4.1. Show that if $\mathbf{A}$ is the set of $x_1 \geq 0, \ldots, x_{n-1} \geq 0$, together with $x_1^2 + \cdots + x_{n-1}^2 \leq r^2$, then

$$\begin{aligned} V_n(r) &= 2^n \int_{\mathbf{A}} (r^2 - (x_1^2 + \cdots + x_{n-1}^2))^{1/2}\, dx_1 \cdots dx_{n-1} \\ &= 2^n r^n \prod_{i=1}^{n-1} \int_0^1 (1 - y^2)^{i/2}\, dy. \end{aligned} \tag{4.4.1}$$

HINT. Make successive changes of variable

$$x_j = y(r^2 - (x_1^2 + \cdots + x_{j-1}^2))^{1/2}, \qquad 2 \leq j \leq n. \tag{4.4.2}$$ □

PROBLEM 4.4.2. Show that

$$\begin{aligned} \int_0^1 (1 - y^2)^{i/2}\, dy &= \tfrac{1}{2} \int_0^1 t^{i/2}(1 - t)^{-1/2}\, dt \\ &= \tfrac{1}{2}\Gamma(\tfrac{1}{2}i + 1)\Gamma(\tfrac{1}{2})/\Gamma(\tfrac{1}{2}i + (\tfrac{3}{2})). \end{aligned} \tag{4.4.3}$$

HINT. Make the change of variable $t = 1 - y^2$. The result is a beta integral. □

PROBLEM 4.4.3. Show that

$$V_n(r) = r^n \pi^{n/2}/\Gamma(\tfrac{1}{2}n + 1). \tag{4.4.4}$$

Show that the area function satisfies

$$A_{n-1}(r) = (d/dr)\, V_n(r) = 2r^{n-1}\pi^{n/2}/\Gamma(\tfrac{1}{2}n). \tag{4.4.5}$$

A side issue to Wishart's calculation is the question, why is $A_{n-1}(r)$ the right quantity to use? We do not deal with this question.

The calculation as presented assumes $\mathbf{X}_1, \ldots, \mathbf{X}_n$ are linearly independent except on a set of Lebesgue measure zero. We develop several such results here as problems.

PROBLEM 4.4.4. Let $\mu_1, \ldots, \mu_n$ be regular σ-finite Borel measures on $\mathbb{R}^n$ each of which gives zero mass to proper hyperplanes of $\mathbb{R}^n$.

$$0 = (\mu_1 \times \cdots \times \mu_n)\{(x_1, \ldots, x_n) | x_1, \ldots, x_n \in \mathbb{R}^n, \det(x_1, \ldots, x_n) = 0\} \tag{4.4.6}$$

Hint. Fix $x_1, \ldots, x_{n-1}$ and expand the determinant on the last column. The equation (4.4.6) then defines a hyperplane in $\mathbb{R}^n$ which has μ_n measure zero. Use Fubini's theorem. □

Problem 4.4.5. Continuation of Problem 4.4.4. Let $h \leq n$. If $x_1, \ldots, x_h$ are dependent with positive $\mu_1 \times \cdots \times \mu_h$ measure then $\det(x_1, \ldots, x_n) = 0$ positive $\mu_1 \times \cdots \times \mu_n$ measure.

Problem 4.4.6. Continuation of Problem 4.4.4. The event that $\Sigma_{i=1}^{q} x_i x_i^t$ have a double eigenvalue λ such that $\lambda \neq 0$ has $\mu_1 \times \cdots \times \mu_q$ measure zero.

Hint. By induction on q. The case $q = 1$ is the empty event. If λ is a (random) double eigenvalue then $\Sigma_{i=1}^{q-1} x_i x_i^t - \lambda I_n$ is a singular $n \times n$ matrix. Choose $U \in \mathbf{O}(n)$ depending measurably on $x_1, \ldots, x_{q-1}$ so that

$$U\left(\sum_{i=1}^{q-1} x_i x_i^t\right) U^t = D \tag{4.4.7}$$

is a $n \times n$ diagonal matrix. Then in the (random) (i, i)-position on the diagonal of D occurs λ. One of the following two cases must hold. Show that each has zero measure. □

Case 1. $\Sigma_{i=1}^{q-1} x_i x_i^t$ has λ as a double eigenvalue.

Case 2. If Case 1 fails to hold then for some eigenvector a for λ, $a^t U x_q \neq 0$. From (4.4.7) it then follows that the i-th entry of Ux_q is zero, so that the event in question is a subset of $\bigcup_{i=1}^{n} \{x \mid (Ux)_i = 0\}$. By hypotheses on μ_n, this set has zero measure. Use Fubini's theorem.

Problem 4.4.7. Let $p \geq 1$, $q \geq 0$, and $p + q \geq n$. Suppose $x_1, \ldots, x_p \in \mathbb{R}^n$ and $y_1, \ldots, y_q \in \mathbb{R}^n$ and $\mu_1, \ldots, \mu_{p+q}$ are σ-finite regular Borel measures on the Borel subsets of $\mathbb{R}^n$, each giving zero measure to proper hyperplanes of $\mathbb{R}^n$. The equation

$$0 = \det\left(\lambda_0 \sum_{i=1}^{p} x_i x_i^t + \sum_{i-1}^{q} y_i y_i^t\right) \tag{4.4.8}$$

is said to have a double root (at least) λ_0 if there exists two linearly independent vectors a_1 and a_2 such that

$$0 = \left(\lambda_0 \sum_{i=1}^{p} x_i x_i^t + \sum_{i=1}^{q} y_i y_i^t\right) a_i, \qquad i = 1, 2. \tag{4.4.9}$$

Show the event that (4.4.8) have at least a double root $\lambda_0 \neq 0$ has $\mu_1 \times \cdots \times \mu_{p+q}$ measure zero.

Hint. Any double root λ_0 of (4.4.8) must be at least a single root of

$$\lambda_0 \sum_{i=1}^{p} x_i x_i^t + \sum_{i=1}^{q-1} y_i y_i^t. \tag{4.4.10}$$

Choose $G \in \mathbf{GL}(n)$ such that

$$G\left(\sum_{i=1}^{p} x_i x_i^t\right)G^t = D, \qquad \text{a diagonal matrix,} \tag{4.4.11}$$

and

$$G\left(\sum_{i=1}^{q-1} y_i y_i^t\right)G^t = I_n - D.$$

See Problem 8.5.13. Let $z_q = Gy_q$. The case $\lambda_0 = 1$ is Problem 4.4.4. We suppose $\lambda_0 \neq 1$. For a root λ_0 of (4.4.10) suppose the (i, i)-element of D is $1/(1 - \lambda_0)$. □

Case 1. Let $a \in \mathbb{R}^n$ be a solution of (4.4.9) such that $z_q^t a \neq 0$. Since the (i, i)-position of $\lambda_0 D + I_n - D$ is zero, if $z_q^t = (z_{q1}, \ldots, z_{qn})$ it follows that $z_{qi} = 0$. This says that the i-th row of G into y_q is zero.

Case 2. If a is any solution of (4.4.9) then $a^t z_q = 0$. Then there exists nonzero vectors a such that $a^t z_q = 0$ and $(\lambda_0 D + (I_n - D))a = 0$. That is,

$$\lambda_0 \sum_{i=1}^{p} x_i x_i^t + \sum_{i=1}^{q-1} y_i y_i^t \tag{4.4.12}$$

has a double root not zero. Make an induction on q with the first step for the induction being $p + q = n$. See the next case.

Case 3. In the case the $p + q = n$ we may choose $G \in \mathbf{GL}(n)$ such that

$$Gx_i = e_i \quad \text{and} \quad Gy_i = e_{p+j}, \qquad 1 \leq i \leq p \quad \text{and} \quad 1 \leq j \leq q. \tag{4.4.13}$$

Then,

$$G\left(\lambda_0 \sum_{i=1}^{p} x_i x_i^t + \sum_{i=1}^{q} y_i y_i^t\right)G^t = \begin{vmatrix} \lambda_0 & & & \\ & \lambda_0 & & \\ \cdots & \cdots & \cdots & \cdots \\ & & 1 & \\ & & & 1 \end{vmatrix}. \tag{4.4.14}$$

If $\lambda_0 \neq 0$ then this matrix is nonsingular. This establishes the first step of the induction.

PROBLEM 4.4.8. Let the real valued random variables $\mathbf{X}_1, \ldots, \mathbf{X}_n$ have joint density function relative to Lebesgue measure of the form

$$f(x_1^2 + \cdots + x_n^2), \qquad f \text{ a continuous function.} \tag{4.4.15}$$

Let $\mathbf{Y} = (\mathbf{X}_1^2 + \cdots + \mathbf{X}_n^2)^{1/2}$. Show that $\mathbf{Y}$ has a density function of the form

$$2y^{n-1}\pi^{n/2}f(y^2)/\Gamma(\tfrac{1}{2}n). \tag{4.4.16}$$

Use this to derive the density function of a χ_n^2 random variable.

HINT. Show that $P(y < \mathbf{Y} \le y + s) \approx f(y^2)(V_n(y+s) - V_n(y))$. Divide by s and take a limit as $s \to 0$. Use (4.4.5). □

PROBLEM 4.4.9. Substitute in (4.1.11) the joint normal density function of Chapter 2, means zero, covariance matrix Σ. That is, write the Wishart density function.

PROBLEM 4.4.10. See Problem 4.4.8. Suppose $\mathbf{X}_1, \ldots, \mathbf{X}_n$ have a joint density function $f(x_1^2 + \cdots + x_n^2)$. Let $\mathbf{Y} = \mathbf{X}_1^2 + \cdots + \mathbf{X}_n^2$. All random variables here are real valued. Then

$$\lim_{s \to 0} s^{-1} P(y < \mathbf{Y} \le y + s) = \frac{y^{(n/2)-1} \pi^{n/2} f(y)}{\Gamma(\frac{1}{2}n)}. \tag{4.4.17}$$

PROBLEM 4.4.11. Let $\mathbf{X}$ have density function $(2\pi)^{-n/2} \exp(-\frac{1}{2}\Sigma_{i=1}^n (x_i - a_i)^2)$. Let $a = \frac{1}{2}\Sigma_{i=1}^n a_i^2$. Let $\mathbf{Y} = \mathbf{X}^t\mathbf{X}$ and apply the ideas of this chapter. Show that the density function of $\mathbf{Y}$ is

$$\frac{y^{(n/2)-1} e^{-y/2} e^{-a}}{2^{n/2}\Gamma(\frac{1}{2}n)} \int_{\mathbf{O}(n)} \exp(b^t H x)\, dH, \tag{4.4.18}$$

where $b^t = (a_1, \ldots, a_n)$.

PROBLEM 4.4.12. Compare (4.4.18) with the noncentral chi-square density (2.3.10). Infer that

$$\int_{\mathbf{O}(n)} \exp(b^t H x)\, dH = \sum_{j=0}^{\infty} \frac{\Gamma(\frac{1}{2}n) r^j a^j}{\Gamma(\frac{1}{2}(n+2j)) 2^j j!}, \tag{4.4.19}$$

with $r = x^t x$, and $a = \frac{1}{2} b^t b = \Sigma_{i=1}^n \frac{1}{2} a_i^2$.

PROBLEM 4.4.13. Use the fact that there exists H_1 and $H_2 \in \mathbf{O}(n)$ such that $b^t H_1 = (|b|, 0, \ldots, 0)$ and $x^t H_2 = (|x|, 0, \ldots, 0)$. Show that

$$\int_{\mathbf{O}(n)} \exp(b^t H x)\, dH = \int_{\mathbf{O}(n)} \exp(|x|\,|b| H_{11})\, dH, \tag{4.4.20}$$

where H_{11} is the (1, 1)-element of H. Expand in a power series and match coefficients. In this way obtain,

$$\text{if } \ j \ge 0, \qquad \int_{\mathbf{O}(n)} H_{11}^{2j}\, dH = \frac{\Gamma(\frac{1}{2}n)\Gamma(2j+1)}{\Gamma(\frac{1}{2}(n+2j))\Gamma(j+1)2^{2j}}. \tag{4.4.21}$$

Problems 4.4.14 to 4.4.17 are an evaluation of integrals $\int_{\mathbf{O}(n)} f(H)\, dH$ where f is a polynomial. The identities following are based on the following observation.

$$\text{If } \ P, Q \in \mathbf{O}(n) \ \text{ then } \quad \int_{\mathbf{O}(n)} f(H)\, dH = \int_{\mathbf{O}(n)} f(PHQ)\, dH. \tag{4.4.22}$$

Observe that if $P \in \mathbf{O}(n)$ is a diagonal matrix with all diagonal entries $= 1$ except for the (1, 1)-entry which is -1, then for the function $f(H) = H_{11}^m$

$$\int_{\mathbf{O}(n)} H_{11}^m \, dH = (-1)^m \int_{\mathbf{O}(n)} H_{11}^m \, dH. \tag{4.4.23}$$

The identity (4.4.23) implies that if m is odd then $\int_{\mathbf{O}(n)} H_{11}^m \, dH = 0$.

More generally, if P is diagonal and Q induces the permutation σ of $1, \ldots, n$ on the rows (multiplication from the left by Q) then

$$\begin{aligned} \int_{\mathbf{O}(n)} H_{11}^{m_1} \cdots H_{nn}^{m_n} \, dH &= \int_{\mathbf{O}(n)} H_{\sigma(1)1}^{m_1} \cdots H_{\sigma(n)n}^{m_n} \, dH \\ &= \int_{\mathbf{O}(n)} H_{\sigma(1)\sigma(1)}^{m_1} \cdots H_{\sigma(n)\sigma(n)}^{m_n} \, dH, \end{aligned} \tag{4.4.24}$$

and similarly for the sign changes induced by P.

PROBLEM 4.4.14. Let $m = m_1 + \cdots + m_n$. Then

$$\int_{\mathbf{O}(n)} H_{11}^{2m_1} \cdots H_{1n}^{2m_n} \, dH = \frac{m!(2m_1)!}{(2m)!m_1!} \cdots \frac{(2m_n)!}{m_n!} \int_{\mathbf{O}(n)} H_{11}^{2m} \, dH. \tag{4.4.25}$$

HINT. Let $P \in \mathbf{O}(n)$ have a first row $u_1, \ldots, u_n$ and let the function f be $f(H) = H_{11}^m$. Then $(u_1^2 + \cdots + u_n^2)^m \int_{\mathbf{O}(n)} H_{11}^{2m} \, dH = \int_{\mathbf{O}(n)} (u_1 H_{11} + \cdots + u_n H_{1n})^{2m} \, dH$. Expand both sides using the multinomial theorem and match coefficients. Note that both sides are homogeneous of degree $2m$ and hence the identity holds for *all* real numbers $u_1, \ldots, u_n$. □

PROBLEM 4.4.15. Let $q = q_1 + \cdots + q_n$. Then

$$\int_{\mathbf{O}(n)} H_{11}^{2m} H_{22}^{2q_2} \cdots H_{2n}^{2q_n} \, dH = \frac{q!}{(2q)!} \frac{(2q_2)!}{q_2!} \cdots \frac{(2q_n)!}{q_n!} \int_{\mathbf{O}(n)} H_{11}^{2m} H_{22}^{2q} \, dH. \tag{4.4.26}$$

PROBLEM 4.4.16.

$$\int_{\mathbf{O}(n)} H_{11}^{2m-2} \, dH = (1 + (n-1)/(2m-1)) \int_{\mathbf{O}(n)} H_{11}^{2m} \, dH. \tag{4.4.27}$$

HINT. Integrate $H_{11}^{2m-2}(H_{11}^2 + \cdots + H_{n1}^2) = H_{11}^{2m-2}$. Use (4.4.25) to obtain an identity. □

PROBLEM 4.4.17. Show that (4.4.21) and (4.4.27) agree. Problem 4.4.18 and Problem 4.4.19 develop details of the theory for density functions referred to in Section 4.2.

PROBLEM 4.4.18. If the $n \times h$ random variable $\mathbf{X}$ has density function g_1 relative to Lebesgue measure on $\mathbb{R}^{nh}$ then the random variable $\mathbf{Y} = H\mathbf{X}$,

$H \in \mathbf{GL}(n)$, has density function

$$g_2(X) = g_1(H^{-1}X)/|\det H|^h, \tag{4.4.28}$$

where $|\det H|$ means the absolute value of the determinant.

HINT. $\mathrm{P}(\mathbf{Y} \in \mathbf{A}) = \int 1_{\mathbf{A}}(HX) g_1(H^{-1}HX)\, dX$. Make the change of $Y = HX$ and compute the Jacobian of the transformation. □

PROBLEM 4.4.19. If the $n \times h$ random matrix $\mathbf{X}$ has density function g then the probability measure

$$\int_{\mathbf{O}(n)} P(H\mathbf{X} \in \mathbf{A})\, dH \tag{4.4.29}$$

has density function

$$\int_{\mathbf{O}(n)} g(H^t X)\, dH. \tag{4.4.30}$$

It is assumed that the density functions are integrated by Lebesgue measure.

HINT. Write (4.4.30) as a double integral using (4.4.29). Apply Fubini's theorem. □

CHAPTER 5

The Fubini-Type Theorems of Karlin

5.0. Introduction

In 1960 Karlin lectured on multivariate analysis and notes in six sets were written. The material of this chapter is a reworking of the 1960 notes in order to bring notations into conformity with others used in this book, to rephrase proofs, and to eliminate some material. The emphasis in the following sections is on method rather than result.

The basic idea, factorization of invariant measures, coupled with the uniqueness theorem, Theorem 3.3.8, leads to expression of integrals as iterated integrals, the innermost integral taking out unwanted variables. In this way, except for constants of proportionality, the noncentral t-density is obtained (Section 5.1), the central Wishart density is obtained (Section 5.2), the generalized t-density is obtained (Section 5.4), the density of the conditional covariance matrix is obtained (Section 5.5), and for the sample covariance matrix $\begin{vmatrix} \mathbf{S}_{11} & \mathbf{S}_{12} \\ \mathbf{S}_{21} & \mathbf{S}_{22} \end{vmatrix}$ the density of $\mathbf{W}\mathbf{S}_{12}\mathbf{S}_{22}^{-1}\mathbf{S}_{21}\mathbf{W}^t$ is obtained (Section 5.6), where $\mathbf{W}$ is a "square root" of $\mathbf{S}_{11}$ as explained in Remark 5.2.7. In Section 5.3, for the sake of completeness, differential forms are used to factor $\bigwedge_{i\leq j}(dS)_{ij}/|S|^{(k+1)/2}$ into a product of Haar measure on $\mathbf{O}(k)$ and a noninvariant measure on $\mathbf{D}(k)$, the diagonal giving the eigenvalues of $\mathbf{S}$. This factorization does not seem to be obtainable by Karlin's methods. The methods of this chapter should be compared to those of Chapter 6 on differential forms as applied in Chapter 9, and to those of Chapter 10 on the use of cross-sections. Generally, if the methods of Chapter 5 can be applied, then results can be obtained with less machinery, i.e., in a more elementary fashion.

The constants of proportionality are computed by integration of joint normal density functions. The use of the joint normal density function to determine constants will occur several places in this book.

Following is a sketch of the concept. A group $\mathfrak{G}$ acts on a space $\mathfrak{X}$. A subgroup $\mathbf{H}$ of $\mathfrak{G}$ is given, $\mathbf{H} = \{g | gx_0 = x_0\}$. In this chapter it is assumed that the group $\mathfrak{G}$ acts transitively on $\mathfrak{X}$, and for each x, $\{g | gx = x \text{ and } g \in \mathfrak{G}\}$ is a compact subgroup of $\mathfrak{G}$. Relative to the preferred x_0, the subgroup $\mathbf{H}$ may be used to establish $\mathfrak{X} \simeq \mathfrak{G}/\mathbf{H}$. By Theorem 3.3.8 (uniqueness) the invariant measures on $\mathfrak{X}$ are a one-dimensional space.

In order to justify this construction we suppose the map $g \to gx$ is for each x a continuous map. Then the map $f: \mathfrak{G}/\mathbf{H} \to \mathfrak{X}$ defined by $f(\pi(g)) = gx_0$ is one-to-one on $\mathfrak{G}/\mathbf{H}$. If $\mathbf{C}_1 \subset \mathfrak{G}/\mathbf{H}$ and $\mathbf{C} \subset \mathfrak{G}$ is compact with $\pi(\mathbf{C}) = \mathbf{C}_1$, see Lemma 3.2.2, then $f(\mathbf{C}_1) = f(\pi(\mathbf{C}))$ is a compact subset of $\mathfrak{X}$. Consequently the σ-algebra of Borel subsets of $\mathfrak{G}/\mathbf{H}$ is mapped into the σ-algebra of Borel subsets of $\mathfrak{X}$. A regular invariant measure ν on the Borel subsets of $\mathfrak{X}$ then defines a regular measure $\nu(f(\mathbf{B}))$ on the Borel subsets $\mathbf{B}$ of $\mathfrak{G}/\mathbf{H}$. Because $hf(\pi(g)) = h(gx_0) = f(\pi(hg)) = f(\bar{h}\pi(g))$, it follows that $\nu(f(\bar{h}\mathbf{B})) = \nu(hf(\mathbf{B})) = \nu(f(\mathbf{B}))$. Consequently $\nu(f(\ \ ))$ is a regular invariant measure on $\mathfrak{G}/\mathbf{H}$ and by Theorem 3.3.8 this measure is uniquely determined up to multiplicative constant. Since f is one-to-one, ν is determined up to multiplicative constant.

The idea is to construct an invariant linear functional L of functions $f: \mathfrak{X} \to \mathbb{R}$ by iterated integration. Such a construction has already been used to prove Theorem 3.3.8. Once invariance is established we show in each case the existence of a continuous function f such that $f(x) > 0$ for all x and such that $L(f) < \infty$. This implies $L(1_{\mathbf{A}}) < \infty$ for all compact sets $\mathbf{A}$ and hence that $L(1_{\mathbf{A}})$ defines a regular invariant set function. By the uniqueness theorem L is determined up to proportionality. A second and prototype example is given in Section 5.1. The group is a subgroup of the two-dimensional affine group, used to construct the density function of a ratio. A multivariate analogue of this construction occurs in Section 5.4.

Remark on Notations. In this chapter, in order to abbreviate, $|A|$ will be used for the absolute value of the determinant of A rather than $|\det A|$, the notation used in other parts of the book.

The computations of this chapter are dependent on knowing the density function of various Haar measures relative to Lebesgue measure. One case has already been worked out and we use this as a model for the following discussion. See Theorem 3.5.9 about the left and right invariant Haar measures on $\mathbf{T}(n)$. We apply the method used in Section 3.5 to the following problems.

$$X \text{ is } n \times k, \qquad X \to XG \quad \text{with} \quad G \in \mathbf{GL}(k); \tag{5.0.1}$$

$$S \in \mathbf{S}(k), \qquad S \to GSG^t \quad \text{with} \quad G \in \mathbf{GL}(k). \tag{5.0.2}$$

In the case of (5.0.1) Lebesgue measure μ on $\mathbb{R}^{nk}$ is additively invariant so that $\mu(\mathbf{C}G)$ defines a set function that is also additively invariant, i.e.,

$\mu((\mathbf{C} + X)G) = \mu(\mathbf{C}G)$. Hence the uniqueness theorem for Haar measure implies the existence of a function $m: \mathbf{GL}(k) \to (0, \infty)$ such that $m(G_1 G_2) = m(G_1)m(G_2)$, $m(I_k) = 1$, and, in view of Lemma 3.5.3, m is continuous. Replacement of G by αG, $\alpha \in \mathbb{R}$, gives $m(\alpha G) = |\alpha|^{nk} m(G)$. By Lemma 6.6.2, $m(G) = |G|^n$. Then, as in the proof of Theorem 3.5.9 it follows that the measure

$$\int 1_{\mathbf{C}}(X)/m(X^t X)^{1/2} \mu(dX) \tag{5.0.3}$$

is an invariant measure.

In the case of (5.0.2) Lebesgue measure μ on $\mathbb{R}^{k(k+1)/2}$ is additively invariant so we obtain

$$\mu(G\mathbf{C}G^t) = m(G)\mu(\mathbf{C}). \tag{5.0.4}$$

Here the homogeneity is $m(\alpha G) = |\alpha|^{k(k+1)} m(G)$ so that by Lemma 6.6.2 $m(G) = |G|^{(k+1)}$. Then the invariant measure is

$$\int 1_{\mathbf{C}}(S)/m(S)^{1/2} \prod_{i \geq j} (dS)_{ij}. \tag{5.0.5}$$

Theorem 5.0.1. *An invariant measure for $n \times k$ matrices X under right multiplication by $G \in \mathbf{GL}(k)$ is*

$$\int 1_{\mathbf{C}}(X)|X^t X|^{-n/2} \prod_{i=1}^{n} \prod_{j=1}^{k} (dX)_{ij}. \tag{5.0.6}$$

An invariant measure on $\mathbf{S}(k)$ for the transformations $S \to GSG^t$ is

$$\int 1_{\mathbf{C}}(S)|S|^{-(k+1)/2} \prod_{i \geq j} (dS)_{ij}. \tag{5.0.7}$$

5.1. The Noncentral t-Density

We consider the following two-dimensional example. $\mathfrak{X} = \{(x, y) | x \in \mathbb{R}$ and $y \in (0, \infty)\}$. The group acting on $\mathfrak{X}$ is taken to be part of the affine group,

$$(x, y) \to (x, y) \begin{vmatrix} a & 0 \\ ab & a \end{vmatrix} \tag{5.1.1}$$

which has as a normal subgroup the diagonal matrices aI_2. The Jacobian of the substitution $(x, y) \to (ax + aby, ay)$ is a^2 so $dx\,dy/y^2$ defines an invariant measure on $\mathfrak{X}$. Given a function $f: \mathfrak{X} \to \mathbb{R}$, consider the linear functional

$$Lf = \int_{-\infty}^{\infty} du \int_{0}^{\infty} g^{-1} f(gu, g)\, dg. \tag{5.1.2}$$

We show by a direct calculation that Lf is an invariant functional by integrating $g^{-1}f(aug + ab, ag)$. Here the value is

$$\begin{aligned}\int_0^\infty g^{-1}\,dg \int_{-\infty}^\infty f(ag(u + (b/g)), ag)\,du \\ = \int_{-\infty}^\infty du \int_0^\infty g^{-1}f(agu, ag)\,dg = Lf.\end{aligned} \tag{5.1.3}$$

Consequently there exists a universal constant, i.e., independent of f, such that

$$Lf = \int_{-\infty}^\infty \int_0^\infty f(x, y)y^{-2}\,dx\,dy = c\int_{-\infty}^\infty du \int_0^\infty g^{-1}f(gu, g)\,dg. \tag{5.1.4}$$

It should be remembered here and in the sequel that x, y, u, g are variables of integration and are not otherwise related. The expression (5.1.4) is a function of f, not x, y.

We now verify that $c = 1$. Take for f the function $f(x, y) = y^{n+1}e^{-y}x^2e^{-x^2/2}$ so that the left side of (5.1.4) is $\int_{-\infty}^\infty x^2e^{-x^2/2}\,dx \int_{-\infty}^\infty y^{n-1}e^{-y}\,dy = (2\pi)^{1/2}\Gamma(n)$. The right side of (5.1.4) is $\int_{-\infty}^\infty du \int_0^\infty g^{-1}(gu)^2e^{-(gu)^2/2}g^{n+1}e^{-g}\,dg = (2\pi)^{1/2}\int_0^\infty g^{n-1}e^{-g}\,dg = (2\pi)^{1/2}\Gamma(n)$. Thus $c = 1$.

The existence of f everywhere positive having finite integral guarantees the linear functional Lf defines a regular invariant measure.

Theorem 5.1.1.

$$Lf = \int_{-\infty}^\infty \left(\int_0^\infty f(x, y)y^{-2}\,dy\right)dx = \int_{-\infty}^\infty du \int_0^\infty g^{-1}f(gu, g)\,dg$$

defines an invariant linear functional of functions f under the transitive group action $(x, y) \to (ax + aby, ay)$.

We now verify that if $Lf = 1$ then $\int_0^\infty g^{-1}f(gu, g)\,dg$ is the density function of $\mathbf{x}/\mathbf{y}$. In fact, $P(\mathbf{x}/\mathbf{y} \in \mathbf{A}) = \int\int 1_{y\mathbf{A}}(x)f(x, y)y^{-2}\,dx\,dy = \int du \int g^{-1}1_{g\mathbf{A}}(gu)f(gu, g)\,dg = \int_{-\infty}^\infty 1_{\mathbf{A}}(u)\,du \int_0^\infty g^{-1}f(gu, g)\,dg$, as was to be shown.

The purpose in this section has been to give an elementary example of the type of calculation that occurs in the following sections. Although we do not derive it, using the chi-square density and Theorem 5.1.1, one may write immediately in integral form an expression for the noncentral t-density.

5.2. The Wishart Density Function

The subject of this section is Karlin's first Fubini-type theorem and its application to the derivation of the Wishart density function. The space $\mathfrak{X}$ will be the set of $n \times k$ matrices of rank $k \leq n$. Notationally it is more convenient to write results in terms of k-frames. We let $\mathfrak{A}$ be the set of $n \times k$ matrices U such that $U^tU = I_k$. Thus if $S \in \mathbf{S}(k)$ the product US is defined.

Lemma 5.2.1. *Let μ be the left invariant measure of unit mass on $\mathfrak{A}$ (see Lemma 3.3.4, Lemma 3.3.6, Lemma 3.3.7 and Theorem 3.3.8). If $V \in \mathbf{O}(k)$ then the measure ν defined by $\nu(\mathbf{A}) = \mu(\mathbf{A} \cdot V)$ is μ, that is, $\mu = \nu$.*

PROOF. ν is clearly a left invariant measure since $(UV)^t(UV) = I_k$. Therefore $\mathbf{A} \cdot V$ is a subset of $\mathfrak{A}$. By Theorem 3.3.8, $\mu(\mathbf{A} \cdot V) = m(V)\mu(\mathbf{A})$ and the modular function m is a group homomorphism to $[0, \infty)$. By the arguments of Section 3.5, m is continuous, hence $m(V) = 1$ for all V. □

We use the result of Lemma 5.2.1 to construct an invariant linear functional $L(f)$ defined on functions $f: \mathfrak{X} \to \mathbb{R}$ where $\mathfrak{X}$ is the set of $n \times k$ matrices of rank k. For notational purposes we will write dU for the left invariant measure of unit mass on $\mathfrak{A}$, and dS for Lebesgue measure on the $\frac{1}{2}k(k+1)$ variables of $S \in \mathbf{S}(k)$. The invariant measure on $\mathbf{S}(k)$ that we use is given by the differential form $\bigwedge_{i \geq j} (dS)_{ij}/|S|^{(k+1)/2}$ (see Theorem 5.0.1). If $S \in \mathbf{S}(k)$ then the map $S \to S^{1/2}$ is measurable. See the construction of this map for operators on Hilbert space given in Section 8.3, where $S^{1/2}$ is obtained as a limit of polynomials in S. An alternative is the construction given in Section 14.1 on inequalities.

Theorem 5.2.2 (First Fubini-Type Theorem of Karlin (1960)).

$$\begin{aligned} L(f) &= c_{nk} \int_{\mathbf{S}(k)} dS/|S|^{(k+1)/2} \int_{\mathfrak{A}} f(US^{1/2})\, dU \\ &= \int |X^tX|^{-n/2} f(X)\, dX \end{aligned} \tag{5.2.1}$$

defines a nonzero invariant linear functional of functions $f: \mathfrak{X} \to \mathbb{R}$. If f is a probability density function relative to

$$\bigwedge_{i=1}^{n} \bigwedge_{j=1}^{k} (dX)_{ij}/|X^tX|^{n/2} \tag{5.2.2}$$

then $c_{nk} \int_{\mathfrak{A}} f(US^{1/2})\, dU$ is the density function of $\mathbf{X}^t\mathbf{X}$ relative $\bigwedge_{i \geq j} (dS)_{ij}/|S|^{(k+1)/2}$. The constant c_{nk} is

$$c_{nk} = \pi^{nk/2 - k(k-1)/4} \Big/ \prod_{j=0}^{k-1} \Gamma\left(\tfrac{1}{2}(n-j)\right). \tag{5.2.3}$$

PROOF. If $G \in \mathbf{GL}(k)$ we write (fg) for the function defined by $(fg)(X) = f(XG)$. Then define $L(f)$ by

$$L(f) = \int_{\mathbf{S}(k)} dS/|S|^{(k+1)/2} \int_{\mathfrak{A}} f(US^{1/2})\, dU. \tag{5.2.4}$$

If $V \in \mathbf{O}(n)$ and $(vf)(X) = f(VX)$ defines vf then the invariance $L(vf) = L(f)$ is clear. We have from (5.2.4) that $L(fg)$ will involve $US^{1/2}G = U(S^{1/2}G)(G^tSG)^{-1/2}(G^tSG)^{1/2} = UV(G^tSG)^{1/2}$, where $V \in \mathbf{O}(k)$. By Lemma 5.2.1, $\int_{\mathfrak{A}} f(US^{1/2}G)\,dU = \int_{\mathfrak{A}} f(U(G^tSG)^{1/2})\,dU$. Then by invariance of the measure on $\mathbf{S}(k)$ it now follows that $L(fg) = L(f)$.

The group $\mathbf{O}(n) \times \mathbf{GL}(k)$ acts on $\mathfrak{X}$ by the action $X \to UXG$ and this is a transitive action. We model the group by picking a convenient $X_0 = \begin{vmatrix} I_k \\ 0 \end{vmatrix}$. Then $U_1X_0G_1 = U_2X_0G_2$ implies $G_1^tG_1 = G_2^tG_2$ and hence that there exists $V \in \mathbf{O}(k)$ with $VG_1 = G_2$. Thus $U_1\begin{vmatrix} I_k \\ 0 \end{vmatrix} = U_2\begin{vmatrix} V \\ 0 \end{vmatrix}$. Thus the subgroup that fixes X_0 is isomorphic to $\mathbf{O}(k) \times \mathbf{O}(n-k)$. By Theorem 3.3.8 the invariant measures on $\mathfrak{X}$ are a one-dimensional space.

The invariant linear functional L defines an invariant measure on the Borel subsets of $\mathfrak{X}$ by the definition $\mu(\mathbf{A}) = L(1_{\mathbf{A}})$. We will show below that this measure is finite for compact sets $\mathbf{A}$ and hence is a regular measure. Since the measure with differential form $\bigwedge_{i=1}^{n}\bigwedge_{j=1}^{k}(dX)_{ij}/|X^tX|^{n/2}$ is a regular invariant measure for the group, there exists a constant $c_{nk} > 0$ such that (5.2.1) holds.

To complete the proof we verify that $L(f)$ determines a regular measure and that the constant and probability statements are correct. First, we exhibit an everywhere positive function having finite integral. The choice

$$f(X) = |X^tX|^{n/2}\exp(-\tfrac{1}{2}\operatorname{tr} X^tX) \tag{5.2.5}$$

results in a finite integral on both sides of (5.2.1) and the identity

$$c_{nk}\int_{\mathbf{S}(k)} |S|^{(n-k-1)/2}\exp(-\tfrac{1}{2}\operatorname{tr} S)\,dS = (2\pi)^{nk/2}. \tag{5.2.6}$$

Thus $L(1_{\mathbf{A}}f)$ is finite for all compact sets $\mathbf{A}$ and since $L(f) > 0$ invariance implies every open set has positive measure.

Thus the measure defined by $L(1_{\mathbf{A}})$ is a regular invariant Baire measure on $\mathfrak{X}$ and the uniqueness result applies.

To verify the probability statement, in the spirit of Section 4.2, let $\{X \mid X^tX \in \mathbf{B}\} = \{X \mid X \in \mathbf{A}\}$. Then $P(\mathbf{X}^t\mathbf{X} \in \mathbf{B}) = \int 1_{\mathbf{A}}(X)|X^tX|^{-n/2}f(X)\,dX = c_{nk}\int |S|^{-(k+1)/2}\,dS \int 1_{\mathbf{A}}(US^{1/2})f(US^{1/2})\,dU$. By construction $S^{1/2} \in \mathbf{A}$ if and only if $S \in \mathbf{B}$ so that

$$P(\mathbf{X}^t\mathbf{X} \in \mathbf{B}) = \int_{\mathbf{B}} |S|^{-(k+1)/2}\,dS \int_{\mathfrak{A}} f(US^{1/2})\,dU. \tag{5.2.7}$$

The constant c_{nk}, being independent of f, was computed in Section 4.1. To illustrate alternative methods we reevaluate the constant using the Bartlett decomposition. See Problem 6.7.6, and Section 11.0. By Problem 6.7.6 the substitution $S = TT^t$, T lower triangular, has Jacobian $2^k t_{11}^k t_{22}^{k-1} \cdots t_{kk}$. Thus, setting $t_i = t_{ii}$,

$$\begin{aligned}(2\pi)^{nk/2} &= c_{nk}2^k \int (t_1 \cdots t_k)^{n-k-1} t_k \cdots t_1^k \exp(-\tfrac{1}{2}\Sigma t_{ij}^2) \prod_{j\le i} dt_{ij} \\ &= c_{nk}2^k(2\pi)^{k(k-1)/4} \prod_{j=1}^{k} \int_0^\infty t^{(n-j)} \exp(-\tfrac{1}{2}t^2)\, dt.\end{aligned} \tag{5.2.8}$$

We use here the identity

$$\int_0^\infty t^k \exp(-\tfrac{1}{2}t^2)\, dt = 2^{(k-1)/2}\Gamma(\tfrac{1}{2}(k+1)). \tag{5.2.9}$$

which may be obtained by the change of variable $s = t^2$. Thus substitution into (5.2.8) gives the value (5.2.3). □

Remark 5.2.3. There is an alternative way of constructing an invariant linear functional L, namely,

$$L(f) = \int_{\mathbf{S}(k)} |S|^{-(k+1)/2}\, dS \int_{\mathbf{O}(n)} f\left(U \begin{vmatrix} S^{1/2} \\ 0 \end{vmatrix}\right) \tag{5.2.10}$$

where the $n \times k$ matrix $\begin{vmatrix} S^{1/2} \\ 0 \end{vmatrix}$ has $S \in \mathbf{S}(k)$. In this form integration is over $\mathbf{O}(n)$ rather than over the set of k-frames. The checking of invariance proceeds as before.

Lemma 5.2.4. *If the $n \times k$ random matrix $\mathbf{X}$ has density function $g(X) = g(UX)$ for $U \in \mathbf{O}(n)$ relative to $|X^tX|^{-n/2}\, dX$ then $(\mathbf{X}^t\mathbf{X})^{1/2}$ and $\mathbf{X}(\mathbf{X}^t\mathbf{X})^{-1/2}$ are independent random matrices and $\mathbf{X}(\mathbf{X}^t\mathbf{X})^{-1/2}$ has the uniform distribution on the manifold of k-frames $\mathfrak{A}$.*

PROOF.

$$\begin{aligned}&P((\mathbf{X}^t\mathbf{X})^{1/2} \in \mathbf{B} \text{ and } \mathbf{X}(\mathbf{X}^t\mathbf{X})^{-1/2} \in \mathbf{A}) \\ &\quad = \int 1_{\mathbf{A}}(X(X^tX)^{-1/2}) 1_{\mathbf{B}}((X^tX^{1/2}))|X^tX|^{-n/2} g(X)\, dX \\ &\quad = \int_{\mathbf{O}(n)} dU \int 1_{\mathbf{A}}(UX(X^tX)^{-1/2}) 1_{\mathbf{B}}((X^tX)^{1/2})|X^tX|^{-n/2} g(X)\, dX \\ &\quad = \left[\int_{\mathbf{O}(n)} 1_{\mathbf{A}}\left(U\begin{vmatrix} I_k \\ 0 \end{vmatrix}\right) dU\right] P((\mathbf{X}^t\mathbf{X})^{1/2} \in \mathbf{B}).\end{aligned}$$

We use here the facts that $g(UX) = g(X)$ and that the absolute Jacobian of the substitution $X \to UX$ is one. Thus dU as Haar measure of unit mass may be used. □

Remark 5.2.5. If instead we consider $\mathbf{O}(n) \times \mathbf{T}(k)$ acting on $\mathfrak{X}$ by the action $X \to UXT^t$, this is again a transitive group. Further $U_1X_0T_1^t = U_2X_0T_2^t$ for

$X_0 = \begin{vmatrix} I_k \\ 0 \end{vmatrix}$ requires $T_1 T_1^t = T_2 T_2^t$ so that $T_1 = T_2$ and thus $U_1 \begin{vmatrix} I_k \\ 0 \end{vmatrix} = U_2 \begin{vmatrix} I_k \\ 0 \end{vmatrix}$. Thus the subgroup which fixes X_0 is isomorphic to $\mathbf{O}(n-k)$. If v is a left invariant Haar measure for $\mathbf{T}(k)$, then $\int v(dT) \int f\left(U \begin{vmatrix} T^t \\ 0 \end{vmatrix}\right) dU$ defines an invariant linear functional of integrable functions f so that one has the following theorem.

Theorem 5.2.6. *There exists a universal constant $c > 0$ whose value is determined by the choice of the Haar measure v, such that*

$$\int_{\mathfrak{X}} f(X)|X^t X|^{-n/2}\, dX = c \int_{\mathbf{T}(k)} v(dT) \int_{\mathfrak{U}} f\left(U \begin{vmatrix} T^t \\ 0 \end{vmatrix}\right) dU.$$

In Section 5.4 and Section 5.7 we will need the idea of a "square root" of a Wishart matrix. This is not the usual square root.

Remark 5.2.7. Let the $k \times k$ matrix $\mathbf{W}$ have density function $|W^t W|^{-k/2} g(W^t W)$ relative to Lebesgue measure. By the first Fubini-type theorem, Theorem 5.2.2, if follows that

$$\begin{aligned} P(\mathbf{W}^t\mathbf{W} \in \mathbf{A}) &= \int 1_{\mathbf{A}}(W^t W)|W^t W|^{-k/2} g(W^t W)\, dW \\ &= c_{kk} \int 1_{\mathbf{A}}(S)|S|^{-(k+1)/2} g(S)\, dS. \end{aligned} \tag{5.2.11}$$

Then relative to the invariant measure $|S|^{-(k+1)/2}\, dS$ the random variable $\mathbf{W}^t\mathbf{W}$ has density function $c_{kk} g(S)$ and we call $\mathbf{W}$ a square root of $\mathbf{W}^t\mathbf{W}$. For the choice

$$g(W^t W) = (2\pi)^{-nk/2}(c_{nk}/c_{kk})|W^t W|^{(n-k)/2} \exp\left(-\tfrac{1}{2} \operatorname{tr} W^t W\right) \tag{5.2.12}$$

we will call $\mathbf{W}$ a square root of the Wishart (n, k) matrix $\mathbf{W}^t\mathbf{W}$. In making an interpretation of (5.2.12) we need the following lemma.

Lemma 5.2.8. *Let $\mathbf{S} \in \mathbf{S}(k)$ have density function $g(S)$ relative to $|S|^{-(k+1)/2}\, dS$. If $\mathbf{U} \in \mathbf{O}(k)$ has a uniform distribution then $\mathbf{X} = \mathbf{U}\mathbf{S}^{1/2}$ has density function relative to k^2-dimensional Lebesgue measure given by*

$$c_{kk}^{-1} g(X^t X)|X|^{-k}, \tag{5.2.13}$$

where $|X|$ is the absolute value of the determinant of X.

Proof. By Theorem 5.2.2, integration of (5.2.13) shows this function to be a density function. Let $\mathbf{A} \subset \mathbf{O}(k)$ and $\mathbf{B} \subset \mathbf{S}(k)$. Let $\mathbf{Y}$ have the density function (5.2.13). Then

$$
\begin{aligned}
P(\mathbf{Y}\in\mathbf{A}\cdot\mathbf{B}) &= P(\mathbf{Y}(\mathbf{Y}^t\mathbf{Y})^{-1/2}\in\mathbf{A} \text{ and } (\mathbf{Y}^t\mathbf{Y})^{1/2}\in\mathbf{B})\\
&= c_{kk}^{-1}\int 1_{\mathbf{A}}(Y(Y^tY)^{-1/2})1_{\mathbf{B}}((Y^tY)^{1/2})|Y|^{-k}g(Y^tY)\,dY\\
&= \int_{\mathbf{O}(k)} dU\int 1_{\mathbf{A}}(U)1_{\mathbf{B}}(S^{1/2})|S|^{-(k+1)/2}g(S)\,dS \qquad (5.2.14)\\
&= P(\mathbf{U}\in\mathbf{A})P(\mathbf{S}^{1/2}\in\mathbf{B}) = P(\mathbf{X}\in\mathbf{A}\cdot\mathbf{B}).
\end{aligned}
$$

This argument should be compared with the proof of Lemma 5.2.4. Since the collection of finite disjoint unions of rectangles $\mathbf{A}\cdot\mathbf{B}$ is a set ring dense in the σ-algebra of measurable sets, cf. Halmos (1950) on Fubini's theorem, it follows by approximation that $\mathbf{X}$ and $\mathbf{Y}$ have the same probabilities, hence $\mathbf{X}$ has (5.2.13) as density function. □

Remark 5.2.9. Consequently if $\mathbf{W}$ is Wishart (n, k), with positive semidefinite square root $\mathbf{W}^{1/2}$, then the $k \times k$ matrix $\mathbf{X} = \mathbf{U}\mathbf{W}^{1/2}$ will have the density function (5.2.12) and $P((\mathbf{X}^t\mathbf{X})\in\mathbf{A}) = P(\mathbf{W}\in\mathbf{A})$.

It should be noted that the density function of $\mathbf{S}^{1/2}$ when $\mathbf{S}$ is Wishart will involve a factor $\exp(-\frac{1}{2}\operatorname{tr} S^2)$. None of the methods of this chapter will obtain $\exp(-\frac{1}{2}\operatorname{tr} S^2)$ as a factor.

Remark 5.2.10. The Wishart density function is obtained by making in $(2\pi)^{-nk/2}|X^tX|^{n/2}\exp(-\frac{1}{2}\operatorname{tr} X^tX)$ the substitution $X = S^{1/2}$, then multiplying by $c_{nk}|S|^{-(k+1)/2}$. The result is

$$
\left[2^{nk/2}\pi^{k(k-1)/4}\prod_{j=0}^{k-1}\Gamma(\tfrac{1}{2}(n-j))\right]^{-1}|S|^{(n-k-1)/2}\exp(-\tfrac{1}{2}\operatorname{tr} S). \qquad (5.2.15)
$$

Remark 5.2.11. The square root $\mathbf{Y}$ of a Wishart (n, k) matrix with density function (5.2.11) may be decomposed into $\mathbf{Y} = \mathbf{U}\mathbf{T}^t$ with $\mathbf{U}\in\mathbf{O}(k)$ and $\mathbf{T}\in\mathbf{T}(k)$. Then as explained in Remark 5.2.5, the variables U integrate out and the density function of T relative to the left invariant measure on $\mathbf{T}(k)$ given in Theorem 3.5.8 is seen to be

$$
d_{kk}(2\pi)^{-nk/2}(c_{nk}/c_{kk})|T^tT|^{n/2}\exp(-\tfrac{1}{2}\operatorname{tr} T^tT). \qquad (5.2.16)
$$

The constant d_{nk} can be found from the integral

$$
\begin{aligned}
(2\pi)^{nk/2} &= \int \exp(-\tfrac{1}{2}X^tX)\,dX\\
&= d_{nk}\int dU\int (t_{11}t_{22}^2\cdots t_{kk}^k)^{-1}|T^tT|^{n/2}e^{-(1/2)\operatorname{tr} T^tT}\prod_{i\geq j} dt_{ij} \qquad (5.2.17)
\end{aligned}
$$

Thus

$$
2^k c_{nk} = d_{nk}. \qquad (5.2.18)
$$

Remark 5.2.12. For matrices $\mathbf{Z}$ of complex valued random variables the decomposition $\mathbf{Z} = \mathbf{U}\mathbf{T}^t$ results in $\mathbf{T} \in \mathbf{CT}(k)$ defined in Remark 3.5.10 and in $\mathbf{U}$ with random columns from a unitary matrix. The density function corresponding to (5.2.16) in the complex case relative to the left invariant measure (3.5.15) is then

$$((2\pi)^{k(k-1)} \prod_{j=0}^{k-1} \Gamma(n-j))^{-1} (t_{11} \cdots t_{kk})^{n+1} e^{-(1/2)\,\mathrm{tr}\, T^tT}. \tag{5.2.19}$$

Note that in both the real and in the complex case, when expressed as a density function relative to left invariant Haar measure, the density function is a function of the eigenvalues of T alone.

Theorem 5.2.13. *Assume* $\mathbf{U}$ *is distributed as Haar measure of unit mass on* $\mathbf{O}(k)$, *and that* $\mathbf{T} \in \mathbf{T}(k)$ *has a density function* $f(\lambda(T^tT))$ *relative to left invariant Haar measure* ν *on* $\mathbf{T}(k)$. *Suppose* $\mathbf{U}$ *and* $\mathbf{T}$ *are independently distributed. Then, the random variables* $\mathbf{X} = \mathbf{U}\mathbf{T}^t$ *and* $\mathbf{Y} = \mathbf{T}\mathbf{U}$ *have the same density function relative to Lebesgue measure on* $\mathbf{GL}(k)$ *which is*

$$d_{kk}^{-1} |X^tX|^{-k/2} f(\lambda(X^tX)). \tag{5.2.20}$$

PROOF. If $\mathbf{A} \subset \mathbf{O}(k)$ and $\mathbf{B} \subset \mathbf{T}(k)$ are measurable then

$$\begin{aligned} P(\mathbf{X} \in \mathbf{A}\mathbf{B}^t) &= P(\mathbf{U} \in \mathbf{A}) P(\mathbf{T} \in \mathbf{B}) \\ &= \int 1_{\mathbf{A}}(U)\, dU \int 1_{\mathbf{B}}(T) f(\lambda(T^tT)) \nu(dT) \\ &= \iint 1_{\mathbf{A}\mathbf{B}^t}(UT^t) f(\lambda(T^tT)) \nu(dT) \\ &= d_{kk}^{-1} \int 1_{\mathbf{A}\mathbf{B}^t}(X) f(\lambda(X^tX)) |X^tX|^{-k/2}\, dX. \end{aligned} \tag{5.2.21}$$

Since the set ring of disjoint unions of sets $\mathbf{A}\mathbf{B}^t$ is dense in the Borel subsets of $\mathbf{GL}(k)$, as follows from Fubini's theorem, (5.2.20) then follows. To finish the proof we calculate the density function of $\mathbf{Y}$. Here

$$\begin{aligned} P(\mathbf{Y} \in \mathbf{B}\mathbf{A}) &= \iint 1_{\mathbf{B}\mathbf{A}}(TU) f(\lambda(TT^t)) \nu(dT) \\ &= d_{kk}^{-1} \iint 1_{\mathbf{B}\mathbf{A}}(Y) f(\lambda(YY^t)) |YY^t|^{-k/2}\, dY. \end{aligned} \tag{5.2.22}$$

Since $f(\lambda(YY^t)) = f(\lambda(Y^tY))$ and $|YY^t| = |Y^tY|$ the result follows. □

Remark 5.2.14. The analogue for $\mathbf{U}$ unitary and $\mathbf{T} \in \mathbf{CT}(k)$ holds. Namely, $\mathbf{U}\mathbf{T}^*$ and $\mathbf{T}\mathbf{U}$ have the same density function.

5.3. The Eigenvalues of the Covariance Matrix

By Problem 4.4.6 the set of $S \in \mathbf{S}(k)$ with two or more equal eigenvalues is a set of zero Lebesgue measure. The eigenvalue functions $\lambda_1(S) \geq \lambda_2(S) \geq \cdots \geq \lambda_k(S)$ are continuous functions since $\lambda_1(S) + \cdots + \lambda_j(S)$ is a measurable and locally bounded convex function of S, and $\mathbf{S}(k)$ is an open convex subset of the set of $k \times k$ symmetric matrices. This holds for $1 \leq j \leq k$. Convexity follows from the matrix inequalities of Section 14.1, and continuity follows from arguments of Hardy, Littlewood, and Polya (1952).

Write $\mathbf{S}_<(k)$ for the subset of $\mathbf{S}(k)$ with pairwise distinct eigenvalues and $\mathbf{D}_<(k)$ for the space of diagonal matrices with distinct diagonal entries in decreasing order. The factorization $S = UDU^t$ with $D \in \mathbf{D}_<(k)$ is unique except for sign changes of the columns of $U \in \mathbf{O}(k)$ since $UDU^t = VDV^t$ implies U^tV is a diagonal orthogonal matrix. Thus there are 2^k possible choices of U solving $S = UDU^t$. By Theorem 3.4.1 the measure $\mu(\mathbf{A}) = \int_{\mathbf{A}} |S|^{-(k+1)/2}\, dS$ factors to

$$\mu = \nu_1 \nu_2 \tag{5.3.1}$$

where ν_1 is a left invariant measure on $\mathbf{O}(k)$ factored by sign changes, and ν_2 is a measure on $\mathbf{D}_<(k)$.

Although μ is an invariant measure there seems to be no obvious way to construct an invariant linear functional in the manner discussed in Sections 5.1 and 5.2. We argue as follows. Let $\mathbf{D}$ be a measurable subset of $\mathbf{D}_<(k)$. Take $D = \operatorname{diag}(d_{11}, \ldots, d_{nn})$ and

$$\mathbf{A} = \{S \mid \text{exists } D \in \mathbf{D} \text{ and } U \in \mathbf{O}(k) \text{ with } S = UDU^t\}. \tag{5.3.2}$$

By the factorization $\mu(\mathbf{A}) = 2^k \nu_1(\mathbf{O}(n))\nu_2(\mathbf{D})$. Elimination of the factor 2^k requires restriction of U to have some property such as all positive elements in the first row. We turn to the use of differential forms as developed in Chapter 6 and Section 9.3. Use $dS = (dU)DU^t + U(dD)U^t + UD(dU^t)$. Since $0 = d(UU^t)$ the matrix $U\,dU^t$ is skew-symmetric so that

$$\begin{aligned} \bigwedge_{i \leq j} (dS)_{ij} &= \varepsilon \bigwedge_{i=1}^{k} (dD)_{ii} \bigwedge_{i<j} (U^t(dU)D - DU^t(dU))_{ij} \\ &= \varepsilon \prod_{i<j} (d_{ii} - d_{jj}) \bigwedge_{i=1}^{k} (dD)_{ii} \bigwedge_{i<j} (U^t dU)_{ij}, \end{aligned} \tag{5.3.3}$$

where ε is ± 1. On the set $\mathbf{A}$ defined in (5.3.2) the factor $\prod_{i<j}(d_{ii} - d_{jj}) > 0$ so that with $\mathbf{A}$ as defined in (5.3.2),

$$2^k \int_{\mathbf{A}} \bigwedge_{i \leq j} (dS)_{ij} = \int_{\mathbf{D}} \prod_{i<j} (d_{ii} - d_{jj}) \bigwedge_{i=1}^{k} (dD)_{ii} \int_{\mathbf{O}(k)} \bigwedge_{i<j} (U^t dU)_{ij}. \tag{5.3.4}$$

The last integral in (5.3.4) is evaluated in (7.8.6) and has the value $2^k \pi^{k(k+1)/4} / \prod_{i=1}^{k} \Gamma(\tfrac{1}{2} i)$.

Thus, expressed as an iterated integral we have the following theorem.

Theorem 5.3.1.

$$\int_{\mathbf{A}} f(S)\,dS = \left(\pi^{k(k+1)/4} \Big/ \prod_{i=1}^{k} \Gamma(\tfrac{1}{2}i)\right) \times \int \prod_{i<j} (d_{ii} - d_{jj}) \bigwedge_{i=1}^{k} (dD)_{ii} \int_{\mathbf{O}(k)} f(UDU^t)\,dU \tag{5.3.5}$$

where the last integral is with respect to Haar measure of unit mass over $\mathbf{O}(k)$ *and* $\mathbf{A}$ *is defined in* (5.3.2).

For a related calculation see Section 9.3.

Theorem 5.3.1 is the only result used in this chapter that is not obtained from the uniqueness theorem for invariant measures. The method of cross-sections discussed in Chapter 10 will also suffice to show there exists a measure ν_2 but the method of cross-sections does not lead to a determination of ν_2 relative to Lebesgue measure.

5.4. The Generalized T

In this section we obtain the second Fubini-type theorem of Karlin and apply it. The model that underlies the construction of this section is as follows. $\mathbf{X}$ is a $h \times k$ random matrix independent of the $k \times k$ positive definite random matrix $\mathbf{S}$. In the transformations of normal random variables we think of the observations being transformed by

$$\mathbf{X}, \mathbf{S} \to (\mathbf{X} + A)B, B^t\mathbf{S}B \tag{5.4.1}$$

where A is $h \times k$ and $B \in \mathbf{GL}(k)$. For the calculation of the distribution of the generalized T-statistic this group action is unnecessarily large and we consider instead $\mathbf{Y} \in \mathbf{GL}(k)$ and the group action

$$\mathbf{X}, \mathbf{Y} \to (\mathbf{X} + A)B, \mathbf{Y}B \tag{5.4.2}$$

where A is $h \times k$ and $B \in \mathbf{GL}(k)$. In the application $\mathbf{Y}$ will become a square root of a Wishart random matrix as described in (5.2.11).

Lemma 5.4.1. *The group action described in* (5.4.2) *is transitive and one-to-one.*

PROOF. If $((X + A_1)B_1, YB_1) = ((X + A_2)B_2, YB_2)$, since Y is nonsingular, $B_1 = B_2$. Thus $A_1 = A_2$. To show the group acts transitively, solve the equations $(X + A)B = X'$ and $YB = Y'$ by $B = Y^{-1}Y'$ and $A = X'(Y^{-1}Y')^{-1} - X$. □

Recall in the sequel that $|Y|$ means the absolute value of the determinant of Y.

Lemma 5.4.2. *The measure* $|Y|^{-(h+k)}\,dX\,dY$ *is invariant under the group action* (5.4.2).

PROOF. The assertion follows directly from the computation of the Jacobian of a transformed variable. See the comments in Section 5.0. □

In reading Theorem 5.4.3 remember that X, Y, U, C are variables of integration and that it is not asserted that $XY^{-1} = U$.

Theorem 5.4.3. $L(f) = \int dU \int f(UC, C)|C|^{-k}\,dC$, *where* dU *stands for Lebesgue measure on* $\mathbb{R}^{hk}$ *and* dC *for Lebesgue measure on* $\mathbb{R}^{k^2}$, *defines an invariant linear functional. Therefore by uniqueness, there exists a constant* $d_{hk} > 0$ *with*

$$\iint f(X, Y)|Y|^{-(h+k)}\,dX\,dY = d_{hk}\int dU \int f(UC, C)|C|^{-k}\,dC. \quad (5.4.3)$$

The constant $d_{hk} = 1$ *for all choices of* h, k.

PROOF. The reader may find it helpful to refer back to the one-dimensional example stated in Theorem 5.1.1. In the present proof the function $(gf)(X, Y) = f((X + A)B, YB)$ so that

$$\begin{aligned} L(gf) &= \int dU \int (gf)(UC, C)|C|^{-k}\,dC \\ &= \int |C|^{-k}\,dC \int f((U + AC^{-1})CB, CB)\,dU \\ &= \int dU \int f(U(CB), CB)|C|^{-k}\,dC \\ &= L(f). \end{aligned} \quad (5.4.4)$$

By the uniqueness theorem, there is a universal constant d_{hk} such that

$$\int f(X, Y)|Y|^{-(h+k)}\,dX\,dY = d_{hk}\int dU \int f(UC, C)|C|^{-k}\,dC. \quad (5.4.5)$$

A proof that $d_{hk} = 1$ is given in the sequel. □

We make an application of Theorem 5.4.3 to the density function $h(X)g(Y^tY)$ where g is defined in (5.2.11), so that **Y** is the square root of a Wishart (n, k) random matrix. **X** is $h \times k$ and we give **X** the joint normal density function $h(X) = (2\pi)^{-hk/2}\exp(-\frac{1}{2}\operatorname{tr} X^tX)$. Then by Theorem 5.4.3,

$$
\begin{aligned}
&\int h(X)g(Y^tY)|Y^tY|^{(h+k)/2}|Y|^{-(h+k)}\,dX\,dY \\
&\quad = d_{hk}\int dU \int h(UC)g(C^tC)|C^tC|^{(h+k)/2}|C|^{-k}\,dC \\
&\quad = d_{hk}(c_{nk}/c_{kk})\iint (2\pi)^{-(n+h)k/2}|C|^{(n+h)} \\
&\qquad \times \exp(-\tfrac{1}{2}\operatorname{tr}(C^tU^tUC + C^tC))|C|^{-k}\,dC\,dU \\
&\quad = d_{hk}(c_{nk}/c_{kk})(2\pi)^{-(n+h)k/2}\int |I + U^tU|^{-(n+h)/2}\,dU \\
&\qquad \times \int |C^tC|^{(n+h-k)/2}\exp(-\tfrac{1}{2}\operatorname{tr} C^tC)\,dC.
\end{aligned}
\tag{5.4.6}
$$

By Theorem 5.2.2, the first Fubini-type theorem, see (5.2.7),

$$
\begin{aligned}
&\int |C^tC|^{(n+h)/2}\exp(-\tfrac{1}{2}\operatorname{tr} C^tC)|C|^{-k}\,dC \\
&\quad = c_{kk}\int |S|^{(n+h)/2}\exp(-\tfrac{1}{2}\operatorname{tr} S)|S|^{-(k+1)/2}\,dS \\
&\quad = (2\pi)^{(n+h)k/2}(c_{kk}/c_{n+hk}).
\end{aligned}
$$

Therefore (5.4.6) is equal to

$$
(d_{hk}c_{nk}/c_{n+hk})\int |I + U^tU|^{-(n+h)/2}\,dU, \tag{5.4.7}
$$

the integral being with respect to Lebesgue measure on the hk variables of U. If in the integration (5.4.6)–(5.4.7) we replace $h(X)g(Y^tY)$ by $1_A(XY^{-1})h(X)g(Y^tY)$ then we obtain

$$
P(\mathbf{XY}^{-1} \in \mathbf{A}) = \int_{\mathbf{A}} (d_{hk}c_{nk}/c_{n+hk})|I + U^tU|^{-(n+h)/2}\,dU, \tag{5.4.8}
$$

which proves the following theorem.

Theorem 5.4.4. *If* $\mathbf{X}$ *has a joint normal density,* $\mathbf{Y}$ *has density* (5.2.11), *and* $\mathbf{X}$ *and* $\mathbf{Y}$ *are independent, then* $\mathbf{XY}^{-1}$ *has density function relative to Lebesgue measure given in* (5.4.8).

To obtain the density function of $\mathbf{X}(\mathbf{Y}^t\mathbf{Y})^{-1}\mathbf{X}^t$, the generalized T-statistic, use the first Fubini-type theorem. We treat first the case that U is $h \times k$ with $h \le k$ so that the form is

$$g(U)|U^tU|^{-k/2}\,dU = c_{kh}\int |S|^{-(h+1)/2}\,dS\int g(VS^{1/2})\,dV. \tag{5.4.9}$$

We take

$$g(U) = (d_{hk}c_{nk}/c_{n+hk})|U^tU|^{k/2}|I+U^tU|^{-(n+h)/2} \tag{5.4.10}$$

and obtain for the density function

$$d_{hk}(c_{nk}c_{kh}/c_{n+hk})|S|^{(n-h-1)/2}|I+S|^{-(n+h)/2} \tag{5.4.11}$$

relative to Lebesgue measure on $\mathbf{S}(h)$.

The universal constant d_{hk} may be determined as follows. In the derivation of the density (5.4.11) $\mathbf{X}$ is a $h \times k$ matrix of hk mutually independent normal $(0, 1)$ random variables and $\mathbf{Y}^t\mathbf{Y}$ is Wishart (n, k). Thus with $\mathbf{X}$ and $\mathbf{Y}$ independent of each other, almost surely $\lim_{n\to\infty}\mathbf{X}(\mathbf{Y}^t\mathbf{Y}/n)^{-1}\mathbf{X}^t = \mathbf{X}\mathbf{X}^t$ has a Wishart (k, h) density function. In the change of variable $S = T/n$, where T and S are $h \times h$ matrices, because the measure $|S|^{-(h+1)/2}\,dS$ is invariant, the transformed density is

$$d_{hk}(c_{nk}c_{kh}/c_{n+hk})n^{-hk/2}|S|^{(k-h-1)/2}|I+S/n|^{-(n+h)/2}. \tag{5.4.12}$$

For $S \in \mathbf{S}(h)$ let U^tSU, $U \in \mathbf{O}(h)$, be the diagonal matrix with entries $\lambda_1(S) \geq \lambda_2(S) \geq \cdots \geq \lambda_h(S)$. Then $|I+S/n|^{-(n+h)/2} = \prod_{i=1}^h (1+\lambda_i(S)/n)^{-(n+h)/2}$ which converges boundedly to $\exp(-\frac{1}{2}\operatorname{tr} S)$. The ratios $c_{nk}c_{kh}/c_{n+hk}$ by writing the factorials and dividing are easily seen to be

$$\pi^{-h(h-1)/4}\left[\prod_{j=0}^{k-1}\{\Gamma(\tfrac{1}{2}(n+h-j))/\Gamma(\tfrac{1}{2}(n-j))\}\right]\Big/\prod_{j=0}^{h-1}\Gamma(\tfrac{1}{2}(k-j)). \tag{5.4.13}$$

Application of Stirlings formula,

$$\Gamma(x) \approx (2\pi)^{1/2}x^{x-1/2}e^{-x}, \tag{5.4.14}$$

gives the result that

$$\lim_{n\to\infty} n^{-hk/2}\prod_{j=0}^{k-1}\Gamma(\tfrac{1}{2}(n+h-j))/\Gamma(\tfrac{1}{2}(n-j)) = 2^{-hk/2}. \tag{5.4.15}$$

It follows by the dominated convergence theorem that for compact sets $\mathbf{A}$ with nonvoid interior,

$$\begin{aligned} &P(\mathbf{X}^t\mathbf{X} \in \mathbf{A}) \\ &\quad = \lim_{n\to\infty} P(\mathbf{X}^t(\mathbf{Y}^t\mathbf{Y}/n)^{-1}\mathbf{X}^t \in \mathbf{A}) \\ &\quad = d_{hk}\left[2^{hk/2}\pi^{h(h-1)/4}\prod_{j=0}^{h-1}\Gamma(\tfrac{1}{2}(k-j))\right]^{-1}\int_{\mathbf{A}} |S|^{(k-h-1)/2}\exp(-\tfrac{1}{2}\operatorname{tr} S)\,dS. \end{aligned} \tag{5.4.16}$$

Hence $d_{hk} = 1$. In the case that U is $h \times k$ and $h > k$ then the coefficient of (5.4.11) becomes $d_{hk}(c_{nk}c_{hk}/c_{n+hk})$ and as before c_{nk}/c_{n+hk} converges to

$(2\pi)^{hk/2}$. Since we now have a Wishart (h, k) matrix $d_{hk} = 1$ follows again. We may then rephrase (5.4.5) as a theorem.

Theorem 5.4.5. *Let X be $h \times k$ and Y be $k \times k$. Then*

$$\int f(X, Y)|Y|^{-(h+k)}\,dX\,dY = \int dU \int f(UC, C)|C|^{-k}\,dC \qquad (5.4.17)$$

where dX, dY, dU, dC mean integration by Lebesgue measure on the respective spaces, and $|\;|$ means absolute value of the determinant.

And, as a result of the calculations which precede Theorem 5.4.5, we obtain for normal random matrices

Theorem 5.4.6. *The density function of $\mathbf{X}(\mathbf{Y}^t\mathbf{Y})^{-1}\mathbf{X}^t = \mathbf{S}$ relative to the measure dS is one of the following:*
If $h < k$ then $\mathbf{S}$ is $h \times h$ and the density function is

$$(c_{nk}c_{kh}/c_{n+hk})|S|^{(k-h-1)/2}|I + S|^{-(n+h)/2}. \qquad (5.4.18)$$

If $h \geq k$ then U in (5.4.17) *is $k \times k$ and $\mathbf{S}$ is $k \times k$ and the density function is*

$$(c_{nk}c_{hk}/c_{n+hk})|S|^{(h-k-1)/2}|I + S|^{-(n+h)/2}. \qquad (5.4.19)$$

5.5. Remarks on Noncentral Problems

The Fubini-type theorems give invariant linear functionals that may be evaluated for all integrable functions. In Chapter 5, to illustrate method using examples of normal random variables, we have taken $\Sigma = I_k$ and means zero. Substitution of nonstandard density functions allows derivation of the density functions for noncentral problems. Theorem 5.1.1 leads at once to an expression for the noncentral t-density, while Theorem 5.4.4 and Theorem 5.4.6 may be used to obtain the density function of the generalized noncentral T-statistic. This was done by Karlin (1960) who used

$$\begin{aligned} f(X, Y) = (2\pi)^{-hk/2}|\Sigma|^{-h/2}|Y^tY|^{(h+k)/2}g(Y^tY) \\ \times \exp\left(-\tfrac{1}{2}\operatorname{tr}\Sigma^{-1}(X-\mu)^t(X-\mu)\right), \end{aligned} \qquad (5.5.1)$$

where $\mathbf{X}$ is $h \times k$, with independent rows each multivariate normal (μ, Σ), and g, given by (5.2.11), corresponds to a Wishart (n, k) random variable. Invariance shows that the density of $\mathbf{XY}^{-1}$ depends on the noncentrality parameter $\eta = \mu\Sigma^{-1/2}$ but not on μ or Σ directly. One obtains from Theorem 5.4.4 the following theorem.

Theorem 5.5.1. *If $\mathbf{X}$ and $\mathbf{Y}$ are as in* (5.5.1) *then the noncentral density function of $\mathbf{U} = \mathbf{XY}^{-1/2}$ is*

$$(c_{nk}/c_{kk})(2\pi)^{-hk/2}\int |C|^{n+h-k}\exp(-\tfrac{1}{2}\mathrm{tr}((UC-\eta)^t(UC-\eta)+C^tC)\,dC \tag{5.5.2}$$

where $\eta = \mu\Sigma^{-1/2}$ *is the* $h \times k$ *matrix of noncentrality parameters.*

We do not pursue the investigation of possible changes of variable to modify the form of (5.5.2) but note that the invariance of the measure $|C|^{-k}\,dC$ can be used to integrate out $(U^tU+I)^{1/2}$ and obtain the form

$$(c_{nk}/c_{kk})(2\pi)^{-hk/2}\exp(-\tfrac{1}{2}\mathrm{tr}\,\eta^t\eta)|U^tU+I|^{-(n+h)/2}$$
$$\times\int |C|^{n+h-k}\exp(\mathrm{tr}\,C^t(U^tU+I)^{-1/2}U^t\eta)\exp(-\tfrac{1}{2}\mathrm{tr}\,C^tC)\,dC. \tag{5.5.3}$$

Using orthogonal invariance of the Haar measure used to integrate (5.5.3) the integral becomes

$$\int |C|^{n+h-k}\exp(-\tfrac{1}{2}\mathrm{tr}\,C^tC)$$
$$\times\exp(C^t((U^tU+I)^{-1/2}U^tU(U^tU+I)^{-1/2})^{1/2}(\eta^t\eta)^{1/2})\,dC. \tag{5.5.4}$$

Thus the noncentral density of $\mathbf{XY}^{-1}$ depends only on the variables

$$U^tU \quad \text{and} \quad (\eta^t\eta)^{1/2} = (\Sigma^{-1/2}\mu^t\mu\Sigma^{-1/2})^{1/2}, \tag{5.5.5}$$

both of which are $k \times k$ matrices.

The noncentral $T_0^2 = \mathrm{tr}\,\mathbf{X}(\mathbf{Y}^t\mathbf{Y})^{-1}\mathbf{X}^t$ has been studied by Constantine (1966) who defined Laguerre polynomials of a matrix argument and used them to give a series expansion of the density function of T_0^2.

The density function of $\mathbf{X}(\mathbf{Y}^t\mathbf{Y})^{-1}\mathbf{X}^t$ has been given by James (1964), page 485, in terms of a hypergeometric function of type ${}_1F_1$. This may be expressed as a series of zonal polynomials, thus answering a question raised by Karlin (1960).

5.6. The Conditional Covariance Matrix

We consider here $k \times k$ symmetric matrices S and write $S = \begin{vmatrix} S_{11} & S_{12} \\ S_{21} & S_{22} \end{vmatrix}$ where S_{11} is $p \times p$ and S_{22} is $q \times q$. If $A \in \mathbf{GL}(p)$, $C \in \mathbf{GL}(q)$, and B is $q \times p$, then the matrix group $\mathfrak{G}$ of matrices $\begin{vmatrix} A & 0 \\ B & C \end{vmatrix} = T$ acts on S by TST^t in a transitive manner. The set of T leaving the identity matrix fixed is the set of T with $B = 0$, $A^tA = I$, and $C^tC = I$, hence is isomorphic to $\mathbf{O}(p) \times \mathbf{O}(q)$. By Theorem 3.3.8 this group induces a one-dimensional family of invariant measures, which, since $|S|^{-(k+1)/2}\,dS$ is invariant under the stated transformations, must give a measure in the family. Consequently the following lemma holds.

Lemma 5.6.1. *The measures on* $\mathbf{S}(k)$ *invariant under the transformations* $S \to TST^t$ *are multiples of the invariant measure determined by* $|S|^{-(k+1)/2}\, dS$, *hence are invariant under the full linear group.*

Theorem 5.6.2.

$$\int |S|^{-(k+1)/2} f(S)\, dS = \alpha \int |R|^{-(p-q+1)/2}\, dR \int |T|^{-(p-q+1)/2}\, dT \times \int f\left(\begin{vmatrix} I & 0 \\ C^t & I \end{vmatrix} \begin{vmatrix} R & 0 \\ 0 & T \end{vmatrix} \begin{vmatrix} I & C \\ 0 & I \end{vmatrix}\right) dC \tag{5.6.1}$$

where dC *is pq-dimensional Lebesgue measure on matrices* C *and* $|S|^{-(k+1)/2}\, dS$, $|R|^{-(p+1)/2}\, dR$, *and* $|T|^{-(q+1)/2}\, dT$ *refer to the Haar measures on the respective sets of symmetric matrices.* $\alpha = 1$.

Proof. We establish that the right side of (5.6.1) defines an invariant linear functional $L(f)$. Then we use the uniqueness argument. Observe that $\begin{vmatrix} A & 0 \\ B & D \end{vmatrix} \begin{vmatrix} I & 0 \\ C^t & I \end{vmatrix} = \begin{vmatrix} A & 0 \\ B + DC^t & D \end{vmatrix}$ and that $B + DC^t = D(D^{-1}B + C^t)$ so that if f is transformed and Lebesgue measure on C is used then the integral becomes

$$\alpha |D|^{-p} |A|^{q} \int |R|^{-(p-q+1)/2}\, dR \int |T|^{-(p+q+1)/2}\, dT \times \int f\left(\begin{vmatrix} I & 0 \\ C^t & I \end{vmatrix} \begin{vmatrix} ARA^t & 0 \\ 0 & DTD^t \end{vmatrix} \begin{vmatrix} I & C \\ 0 & I \end{vmatrix}\right) dC$$
$$= \alpha L(f).$$

Hence invariance is established. □

The following are maximal invariants under subgroups of the transformation group being considered.

$$S_{11}, S_{12}S_{22}^{-1}S_{11} \quad \text{as a pair.} \tag{5.6.2}$$

$$S_{11}, S_{22} - S_{21}S_{11}^{-1}S_{12} \quad \text{as a pair.} \tag{5.6.3}$$

$$S_{22} - S_{21}S_{11}^{-1}S_{12}. \tag{5.6.4}$$

The roots of the equation $0 = |S_{12}S_{22}^{-1}S_{21} - \lambda S_{11}|$. (5.6.5)

The nonzero roots of (5.6.5) are the squares of the canonical correlations. See Section 8.2 and in particular (8.2.39).

PROOF THAT $\alpha = 1$. We integrate a Wishart (n, k) density function to determine the constant α. Here we use the function

$$f(S) = |S|^{-(n-k-1)/2} \exp(-\tfrac{1}{2} \operatorname{tr} S).$$

Theorem 5.6.2 requires calculation of

$$\operatorname{tr} \begin{vmatrix} I & 0 \\ C^t & I \end{vmatrix} \begin{vmatrix} R & 0 \\ 0 & T \end{vmatrix} \begin{vmatrix} I & C \\ 0 & I \end{vmatrix} = \operatorname{tr}(R + T + C^tRC),$$

while $\det \begin{vmatrix} R & 0 \\ 0 & T \end{vmatrix}^{n/2} = |R|^{n/2}|T|^{n/2}$. Thus the inner integral is

$$\begin{aligned} &|R|^{n/2}|T|^{n/2} \int \exp(-\tfrac{1}{2} \operatorname{tr}(R + T + C^tRC))\, dC \\ &\quad = (2\pi)^{pq/2}|R|^{(n-q)/2}|T|^{n/2} \exp(-\tfrac{1}{2} \operatorname{tr}(R + T)). \end{aligned} \tag{5.6.6}$$

Then the full integral on the right side of (5.6.1) is

$$\begin{aligned} &\alpha(2\pi)^{pq/2} \int |R|^{(n-p-1)/2}|T|^{(n-q-p-1)/2} \exp(-\tfrac{1}{2} \operatorname{tr}(R + T))\, dR\, dT \\ &\quad = \alpha 2^{nk/2} \pi^{k(k-1)/4} \prod_{j=0}^{k-1} \Gamma(\tfrac{1}{2}(n-j)). \end{aligned} \tag{5.6.7}$$

The integral on the left side of (5.6.1) is by (5.2.6)

$$\int |S|^{(n-k-1)/2} \exp(-\tfrac{1}{2} \operatorname{tr} S)\, dS = 2^{nk/2} \pi^{k(k-1)/4} \prod_{j=0}^{k-1} \Gamma(\tfrac{1}{2}(n-j)).$$

Therefore $\alpha = 1$. □

Corollary 5.6.3. *When means are zero and the covariance matrix is $\Sigma = I$ then $\mathbf{S}_{11}$ and $\mathbf{S}_{22} - \mathbf{S}_{21}\mathbf{S}_{11}^{-1}\mathbf{S}_{12}$ are independent random variables and $\mathbf{S}_{22} - \mathbf{S}_{21}\mathbf{S}_{11}^{-1}\mathbf{S}_{12}$ is Wishart $(n - p, q)$.*

Remark 5.6.4. Another proof of Corollary 5.6.3 is obtained in Chapter 11, Problems 11.11.8 to 11.11.12. These problems were abstracted from Wijsman (1958).

The calculations have been done with $\Sigma = I_k$. For the general covariance matrix the inner integral of (5.6.1) now becomes

$$\int \exp\left(-\tfrac{1}{2} \operatorname{tr} \Sigma^{-1} \begin{vmatrix} I & 0 \\ C^t & I \end{vmatrix} \begin{vmatrix} R & 0 \\ 0 & T \end{vmatrix} \begin{vmatrix} I & C \\ 0 & I \end{vmatrix}\right) dC. \tag{5.6.8}$$

The invariant under integration over C changes $\Sigma^{-1} = \begin{vmatrix} A_{11} & A_{12} \\ A_{21} & A_{22} \end{vmatrix}$ into

$$\begin{vmatrix} A_{11} - A_{12}A_{22}^{-1}A_{21} & 0 \\ 0 & A_{22} \end{vmatrix} = \begin{vmatrix} \Sigma_{11}^{-1} & 0 \\ 0 & (\Sigma_{22} - \Sigma_{21}\Sigma_{11}^{-1}\Sigma_{12})^{-1} \end{vmatrix}.$$

See Problems 8.5.14 and 8.5.15. Therefore the variables R and T will separate and $\mathbf{S}_{11}$, $\mathbf{S}_{22} - \mathbf{S}_{21}\mathbf{S}_{11}^{-1}\mathbf{S}_{12}$ are again Wishart but with covariances Σ_{11} and $\Sigma_{22} - \Sigma_{21}\Sigma_{11}^{-1}\Sigma_{12}$ in the respective cases.

5.7. The Invariant $\mathbf{S}_{22}^{-1/2}\mathbf{S}_{21}\mathbf{S}_{11}^{-1}\mathbf{S}_{12}\mathbf{S}_{22}^{-1/2}$

The eigenvalues of this matrix are squares of the canonical correlations. See Section 8.2. The density function is obtained in this section by use of the Fubini-type theorems of Sections 5.2 and 5.4 together with a Jacobian calculation. If one replaces $\mathbf{S}_{11}^{-1/2}$ by $\mathbf{W}^{-1}$ where $\mathbf{W}$ is a "square root" of $\mathbf{S}_{11}$ as described in Remark 5.2.6, then the density function of $\mathbf{V} = \mathbf{W}^{-1}\mathbf{S}_{12}\mathbf{S}_{22}^{-1/2}$ is obtainable and we compute it in this section. The desired matrix is $\mathbf{V}^t\mathbf{V}$ and its density function can then be obtained by use of Theorem 5.2.2. The result, see (5.7.13), is a multivariate beta density function. An alternative computation results if one also replaces $\mathbf{S}_{22}$ by a square root $\mathbf{Z}$ so that one works with $\mathbf{W}^{-1}\mathbf{S}_{12}\mathbf{Z}^{-1}$, returning, so to speak, to first principles. Instead we follow the sixth set of notes, Karlin (1960), and use the second Fubini-type theorem, Theorem 5.4.3, to get at $\mathbf{W}^{-1}\mathbf{S}_{12}\mathbf{S}_{22}^{-1/2}$ directly.

We suppose $S = \begin{vmatrix} S_{11} & S_{12} \\ S_{21} & S_{22} \end{vmatrix}$, with $S_{12}^t = S_{21}$, and that S_{11} is $p \times p$, S_{22} is $q \times q$, so that S_{12} is $p \times q$, and assume $p \geq q$ with $p + q = k$. In the following we make the identifications:

$$W \text{ is a square root of } S_{11}; \tag{5.7.1}$$

$$S_{12} = X \quad \text{and} \quad S_{22} = R. \tag{5.7.2}$$

Let $\mathbf{A}$ be a set of $p \times q$ matrices such that $X \in \mathbf{A}$ if and only if $UX \in \mathbf{A}$, $U \in \mathbf{O}(p)$. Then by the first Fubini-type theorem, with W replaced by $S^{1/2}U$,

$$\begin{aligned} I(\mathbf{A}) &= \iint dR\,dX \int 1_{\mathbf{A}}(W^{-1}XR^{-1/2})|WW^t|^{-(p+q)/2} h(WW^t, X, R)\,dW \\ &= c_{pp} \iint dR\,dX \int 1_{\mathbf{A}}(S^{-1/2}XR^{-1/2})|S|^{-(p+q+1)/2} h(S, X, R)\,dS. \end{aligned} \tag{5.7.3}$$

In (5.7.3) the integrals with respect to dR and dS are over $\mathbf{S}(p)$ and $\mathbf{S}(q)$, while dW and dX refer to Lebesgue measure on the pq variables of W or X. In the computation of examples

$$|S|^{-(p+q+1)/2} h(S, X, R) \tag{5.7.4}$$

will be a joint density function, say, of covariances.

To proceed with formal calculations, apply the second Fubini-type theorem, Section 5.4, to the first line of (5.7.3). Replace W and X by C and CU respectively, U a $p \times q$ matrix and C a $p \times p$ matrix, to obtain

$$I(\mathbf{A}) = \iint dR\,dU \int 1_{\mathbf{A}}(UR^{-1/2})|C|^{-p}h(CC^t, CU, R)\,dC. \tag{5.7.5}$$

In (5.7.5) make the substitution of $V = UR^{-1/2}$ with Jacobian $|R|^{p/2}dV = dU$ to obtain

$$I(\mathbf{A}) = \int 1_{\mathbf{A}}(V)\,dV \iint |R|^{p/2}|C|^{-p}h(CC^t, CVR^{1/2}, R)\,dC\,dR. \tag{5.7.6}$$

The calculation leading to (5.7.6) from the first line of (5.7.3) holds for all sets **A**, whereas the equivalence of the two lines of (5.7.3) assumes **A** is orthogonally invariant.

Theorem 5.7.1. *If* $|WW^t|^{-(p+q)/2}h(WW^t, X, R)$ *is the joint density function of* $\mathbf{W}$, $\mathbf{X}$, $\mathbf{R}$ *then* $\mathbf{V} = \mathbf{W}^{-1}\mathbf{X}\mathbf{R}^{-1/2}$ *has density function relative to Lebesgue measure* dV *given by*

$$\iint |R|^{p/2}|C|^{-p}h(CC^t, CVR^{1/2}, R)\,dC\,dR. \tag{5.7.7}$$

Then, the density function of $\mathbf{V}^t\mathbf{V} = \mathbf{R}^{-1/2}\mathbf{X}^t(\mathbf{W}\mathbf{W}^t)^{-1}\mathbf{X}\mathbf{R}^{-1/2}$ *may be obtained from application of the first Fubini-type theorem. The result is*

$$c_{pq}\iiint |R|^{p/2}|C|^{-p}|T|^{(p-q-1)/2}h(CC^t, CUT^{1/2}R^{1/2}, R)\,dC\,dR\,dU \tag{5.7.8}$$

where $U \in \mathfrak{A}$ *is a k-frame, and* V *in (5.7.6) is replaced by* $UT^{1/2}$.

In applications (5.7.4) will be a Wishart density function with parameters n, k and covariance matrix $\Sigma = \begin{vmatrix} \Sigma_{11} & \Sigma_{12} \\ \Sigma_{21} & \Sigma_{22} \end{vmatrix}$ having inverse $A = \begin{vmatrix} A_{11} & A_{12} \\ A_{21} & A_{22} \end{vmatrix}$. Thus

$$|S|^{-(p+q+1)/2}h(S, X, R) = c|Z|^{(n-k-1)/2}|\Sigma|^{-n/2}\exp(-\tfrac{1}{2}\operatorname{tr} AZ) \tag{5.7.9}$$

where

$$p + q = k, \qquad c^{-1} = 2^{nk/2}\pi^{k(k-1)/4}\prod_{j=0}^{k-1}\Gamma(\tfrac{1}{2}(n-j)),$$
$$Z = \begin{vmatrix} S & X \\ X^t & R \end{vmatrix} \quad \text{and} \quad |Z| = |S|\,|R - X^tS^{-1}X|. \tag{5.7.10}$$

Thus the integrand of (5.7.8) becomes

$$c_{pp}^{-1}c_{pq}c|\Sigma|^{-n/2}|C|^{n-p}|R|^{(n-q-1)/2}|T|^{(p-q-1)/2} \times |I - T|^{(n-k-1)/2}\exp(-\tfrac{1}{2}\operatorname{tr} AZ). \tag{5.7.11}$$

In the central case that $\Sigma_{12} = 0$ and $A_{12} = 0$, with $S = CC^t$,

$$\exp(-\tfrac{1}{2}\operatorname{tr} AZ) = \exp(-\tfrac{1}{2}\operatorname{tr}(\Sigma_{11}^{-1}CC^t + \Sigma_{22}^{-1}R)). \tag{5.7.12}$$

Then in (5.7.8) U integrates out as do C and R, with factors

$$c_{pp}^{-1}\int |C|^{n-p}\exp(-\tfrac{1}{2}\operatorname{tr} CC^t)\,dC = 2^{np/2}\pi^{p(p-1)/4}\prod_{j=0}^{p-1}\Gamma(\tfrac{1}{2}(n-j));$$
$$\int |R|^{(n-q-1)/2}\exp(-\tfrac{1}{2}\operatorname{tr} R)\,dR = 2^{nq/2}\pi^{q(q-1)/4}\prod_{j=0}^{q-1}\Gamma(\tfrac{1}{2}(n-j)). \tag{5.7.13}$$

We summarize in a theorem.

Theorem 5.7.2. *The joint density function of* $\mathbf{T} = \mathbf{R}^{-1/2}\mathbf{X}^t\mathbf{S}^{-1}\mathbf{X}\mathbf{R}^{-1/2}$ *in the Wishart* (n, k) *case, where* $\mathbf{S}$ *is* $p \times p$, $\mathbf{R}$ *is* $q \times q$, *and* $\mathbf{X}$ *is* $p \times q$, *is*

$$\pi^{-q(q-1)/4}\prod_{j=0}^{q-1}\frac{\Gamma(\frac{1}{2}(n-j))}{\Gamma(\frac{1}{2}(p-j))\Gamma(\frac{1}{2}(n-p-j))} \times |T|^{(p-q-1)/2}|I-T|^{(n-k-1)/2}. \tag{5.7.14}$$

Remark 5.7.3. The density function (5.7.13) is known as the multivariate beta density function. The version (5.7.13) agrees with the form (10.3.8) obtained in Section 10.3 with reference to a different problem.

5.8. Some Problems

PROBLEM 5.8.1. If $\mathbf{X}$ is a $k \times 1$ joint normal $(0, \Sigma)$ random vector then $\mathbf{E}\exp(-\frac{1}{2}\operatorname{tr}\mathbf{X}^tS\mathbf{X}) = (\det(I - 2\Sigma S))^{-1/2}$.

PROBLEM 5.8.2. Why does the first Fubini-type theorem, Theorem 5.2.2, apply here?

PROBLEM 5.8.3. If $\mathbf{X}$ is an $n \times k$ matrix of independently and identically distributed normal $(0, \Sigma)$ row vectors then the $k \times k$ random matrix $\mathbf{X}^t\mathbf{X}$ has moment generating function $(\det(I - 2\Sigma S))^{-n/2}$, valid if $n \geq 1$.

PROBLEM 5.8.4. In the notation of differential forms, $0 = d(AA^{-1}) = (dA)A^{-1} + Ad(A^{-1})$. Therefore $d(A^{-1}) = A^{-1}\,dA\,A^{-1}$.

PROBLEM 5.8.5. The matrix with ij-element $\partial \ln\det A/\partial a_{ij}$ is the matrix A^{-1}.

PROBLEM 5.8.6. If $\mathbf{X}$ is as in Problem 5.8.1 then $\mathbf{E}\,\mathbf{X}^t\mathbf{X} = \Sigma$, and if $\mathbf{X}$ is as in Problem 5.8.3 then $\mathbf{E}\,\mathbf{X}^t\mathbf{X} = n\Sigma$.

PROBLEM 5.8.7. Use (13.3.21) and express the generating function of Problem 5.8.3 as an infinite series of zonal polynomials.

Problem 5.8.8. Refer to (5.6.2)–(5.6.5) and determine the subgroups having these as maximal invariants.

Problem 5.8.9. Transform pairs $(S, T) \in \mathbf{S}(k) \times \mathbf{S}(k)$ by $(S, T) \to (ASA^t, ATA^t)$, $A \in \mathbf{GL}(k)$. Show that a maximal invariant is the set of roots of $|S - \lambda T| = 0$.

Remark 5.8.10. Related is the result that there exists $A \in \mathbf{GL}(k)$ such that $ASA^t = I_k$ and ATA^t is a diagonal matrix. See Problem 8.5.13.

CHAPTER 6

Manifolds and Exterior Differential Forms

6.0. Introduction

Our discussion of differential forms is spread over the next four chapters. The fundamentals of the algebraic theory and theory of integration of differential forms form the substance of Chapter 6. This material is a necessary expansion of the introductory material in James (1954). Chapter 7 is concerned with explicit derivation of differential forms for invariant measures on various manifolds. Although somewhat premature, in case the manifold is a locally compact metric matrix group, the basic idea has already been discussed in Chapter 3, Section 3.6. We discuss several manifolds, the Stiefel and Grassman manifolds, which are not groups, in Chapter 7. These examples require some long detailed calculations and this material has been segregated into a separate chapter. The fact that differential forms are so useful in the derivation of density functions is due to the fact that the manifold of $n \times h$ matrices naturally decomposes into a topological product of manifolds on which groups of transformations act. The theory of Sections 3.3 and 3.4 suggests that the invariant measures should factor. Since Lebesgue measure is absolutely continuous relative to the invariant measures, Lebesgue measure also factors. Chapter 8 discusses the well-known methods of decomposing a $n \times h$ matrix. Chapter 9 factors Lebesgue measure over several of these products of manifolds thereby obtaining density functions of several multivariate statistics. Think of the use of differential forms as a method which yields normalized density functions. Chapter 5 discusses an alternative approach due to Karlin (1960) which produces a normalized density function of many, but not all, of the classical multivariate statistics. An exception, discussed in Section 5.3, provides another example of the use of differential forms. Also, although Chapter 5 tries to avoid use of Jacobians, univariate changes and the Bartlett decomposition are used. In certain parts of inference probability ratios only are needed, not normalized density functions. In these cases the full force of Chapters 5 to 9 is not needed and a more direct approach due to Stein (1956c) may be used to compute the density function of a maximal invariant and the desired probability ratio. See Chapter 10.

Chapter 6 is not totally self-contained and missing details may be found in Dieudonne (1972) and Helgason (1962). Notably, the manifolds considered here are analytic manifolds, but we do not prove this. We do need the fact that the manifolds considered are C_2, but again we do not prove the existence of local coordinates with the necessary differentiability on the overlaps of the charts.

6.1. Basic Structural Definitions and Assumptions

We start with a topological space $\mathfrak{M}$ which is a locally compact Hausdorff space. To each point $x \in \mathfrak{M}$ is assigned an open set $\mathbf{U}_x \subset \mathfrak{M}$ such that $x \in \mathbf{U}_x$. Associated with the pair x, $\mathbf{U}_x$ is a function $\varphi(x, \ ): \mathbf{U}_x \to \mathbb{R}^n$ which is a homeomorphism of $\mathbf{U}_x$ onto $\mathbb{R}^n$. We assume that the integer n is the same for all $x \in \mathfrak{M}$, and *n is called the dimension of the manifold* $\mathfrak{M}$. The mapping $\varphi(x, \ )$ is called a coordinate chart at x.

The definition of a manifold requires the following consistency relations. If $x \in \mathbf{U}_{x_1} \cap \mathbf{U}_{x_2}$ then the function

$$\varphi(x_1, \varphi^{-1}(x_2, \ )) \tag{6.1.1}$$

is a homeomorphism of its domain into its range $\mathbb{R}^n$. The differentiability of a manifold is described by specification of the differentiability of the functions (6.1.1). An analytic manifold is one in which the functions (6.1.1) are analytic functions of n real variables. The manifold is said to be C_k if the functions (6.1.1) are k-fold continuously differentiable.

The *local coordinates* of a point $y \in \mathbf{U}_x$ are the numbers

$$\varphi(x, y)^t = (\varphi_1(x, y), \ldots, \varphi_n(x, y)). \tag{6.1.2}$$

A function $f: \mathbf{U}_x \to \mathbb{R}$ is said to be a C_k function *locally at* x if the function

$$f(\varphi^{-1}(x, \ )): \mathbb{R}^n \to \mathbb{R} \tag{6.1.3}$$

is k-fold continuously differentiable. In the sequel we use the *local coefficient rings* $\mathfrak{C}_{x,k}$ of all functions $f: \mathbf{U}_x \to \mathbb{R}$ which are C_k functions. The local coefficient rings are the coefficient rings used in the construction of multilinear forms. It should be noted that these rings have multiplicative units, which is important for the construction of canonical bases of modules. Unless otherwise stated we assume that our manifolds are C_2 and that the coefficient rings are $\mathfrak{C}_{x,2}$, $x \in \mathfrak{M}$.

We list several basic examples of manifolds. $\mathbb{R}^n$ is an n-dimensional analytic manifold using, if $x \in \mathbb{R}^n$ then $\mathbf{U}_x = \mathbb{R}^n$ and $\varphi(x, y) = y$, $y \in \mathbb{R}^n$.

The *sphere* $\{(x_1, \ldots, x_n) | x_1^2 + \cdots + x_n^2 = 1\}$ is an $(n-1)$-dimensional analytic manifold. To each point x let $\mathbf{U}_x$ be all points of the sphere making an angle of absolute value $< \pi/2$ with x. Project $\mathbf{U}_x$ onto $\mathbb{R}^{n-1}$ by taking the $(n-1)$-dimensional hyperplane tangent at x. If $y \in \mathbf{U}_x$ then $\varphi(x, y)$ is defined to be the intersection of the hyperplane with the line through the points 0 and y.

$\mathbf{O}(n)$, *the orthogonal group*. Here the manifold is the set of $n \times n$ orthogonal matrices with real entries, in the topology of coordinatewise convergence. If we interpret $H \in \mathbf{O}(n)$ as a point of $\mathbb{R}^{n^2}$ then the map $H \to \det H$ is a continuous function. The nonsingular matrices thus are an open subset of $\mathbb{R}^{n^2}$, and $\mathbf{O}(n)$ is a closed bounded subset of $\mathbb{R}^{n^2}$, hence a compact topological space.

The condition $H^t H = I_n$ gives $\frac{1}{2}n(n+1)$ algebraic relations so one expects the dimension of the manifold to be $\frac{1}{2}(n-1)n$. We do not give an explicit construction of local coordinates for this example.

6.2. Multilinear Forms, Algebraic Theory

In developing the algebraic theory we let $\mathfrak{C}$ be a coefficient ring with Abelian multiplication and multiplicative unit. We choose a free basis $e_1, \ldots, e_n$ for a n-dimensional free module

$$\mathbf{E} = \left\{x \,|\, x = \sum_{i=1}^{n} c_i e_i, c_1, \ldots, c_n \subset \mathfrak{C}\right\}. \tag{6.2.1}$$

We will write $\mathbf{M}(\mathbf{E}^q, \mathfrak{C})$ for the space of multilinear q-forms on $\mathbf{E}^q = \mathbf{E} x \cdots x \mathbf{E}$. Of special importance is the fact that since $\mathbf{E}$ is a free module, $\mathbf{M}(\mathbf{E}, \mathfrak{C}) = \mathbf{M}(\mathbf{E}^1, \mathfrak{C})$ has a canonical basis $u_1, \ldots, u_n$ defined by the conditions of linearity together with

$$\begin{aligned} u_i(e_j) = \delta_{ij} &= 1 \quad \text{if} \quad i = j \\ &= 0 \quad \text{if} \quad i \neq j. \end{aligned} \tag{6.2.2}$$

Permutations act on $\mathbf{M}(\mathbf{E}^q, \mathfrak{C})$ as follows. Let σ be a permutation of $1, \ldots, q$. If $f \in \mathbf{M}(\mathbf{E}^q, \mathfrak{C})$ and $x_1, \ldots, x_q \in \mathbf{E}$ then

$$(\sigma f)(x_1, \ldots, x_q) = f(x_{\sigma(1)}, \ldots, x_{\sigma(q)}). \tag{6.2.3}$$

Subject to this definition σ acts as a linear transformation of $\mathbf{M}(\mathbf{E}^q, \mathfrak{C})$. The product of permutations σ and τ as linear transformations is $\sigma\tau$, that is

$$((\sigma\tau)f)(x_1, \ldots, x_q) = (\sigma(\tau(f)))(x_1, \ldots, x_q). \tag{6.2.4}$$

The permutation operators are used to define the *alternating operators*.

To make the definition we need the sign function $\varepsilon(\sigma)$ of a permutation. We define

$$\begin{aligned}\varepsilon(\sigma) &= 1 \quad \text{if } \sigma \text{ is the product of an even number of transpositions;}\\ &= -1 \quad \text{if } \sigma \text{ is the product of an odd number of transpositions.}\end{aligned} \tag{6.2.5}$$

We assume that the reader knows that the sign function is well defined and note that it follows from (6.2.5) that if σ and τ are permutations of $1, \ldots, q$ then

$$\varepsilon(\sigma\tau) = \varepsilon(\sigma)\varepsilon(\tau). \tag{6.2.6}$$

A linear operator A (we use the same letter for each q), $A: \mathbf{M}(\mathbf{E}^q, \mathfrak{C}) \to \mathbf{M}(\mathbf{E}^q, \mathfrak{C})$ may be defined by, if $f \in \mathbf{M}(\mathbf{E}^q, \mathfrak{C})$ then

$$(Af) = (q!)^{-1}\Sigma_\sigma\, \varepsilon(\sigma)(\sigma f), \tag{6.2.7}$$

and make the definition

$$\mathbf{A}^q = \{f \,|\, f \in \mathbf{M}(\mathbf{E}^q, \mathfrak{C}), Af = f\} = \mathbf{A}^q(\mathbf{E}, \mathfrak{C}). \tag{6.2.8}$$

In words, $\mathbf{A}^q$ is the set of alternating q-forms.

Lemma 6.2.1. *If $f \in \mathbf{M}(\mathbf{E}^q, \mathfrak{C})$ then $A(Af) = Af$.*

Proof.
$$\begin{aligned}A(Af) &= (q!)^{-1}\,\Sigma_\sigma\, \varepsilon(\sigma)\sigma((q!)^{-1}\,\Sigma_\tau\, \varepsilon(\tau)(\tau f))\\ &= (q!)^{-2}\,\Sigma_\sigma\, \Sigma_\tau\, \varepsilon(\sigma\tau)(\sigma\tau) f\\ &= (q!)^{-1}\,\Sigma_\sigma\, Af = Af.\end{aligned}$$ □

Lemma 6.2.2. *$\sigma A = \varepsilon(\sigma)A$.*

Proof.
$$\begin{aligned}(\sigma A)f &= \sigma(\Sigma_\tau\,(q!)^{-1}\varepsilon(\tau)(\tau f)) = \Sigma_t\,(q!)^{-1}\varepsilon(\tau)(\sigma\tau)(f)\\ &= \varepsilon(\sigma)\,\Sigma_\tau\,(q!)^{-1}\varepsilon(\sigma)\varepsilon(\tau)(\sigma\tau)(f) = \varepsilon(\sigma)Af.\end{aligned}$$ □

Definition 6.2.3. If $f \in \mathbf{M}(\mathbf{E}^q, \mathfrak{C})$ and $g \in \mathbf{M}(\mathbf{E}^r, \mathfrak{C})$ then the function fg is defined by

$$(fg)(x_1, \ldots, x_{q+r}) = f(x_1, \ldots, x_q)g(x_{q+1}, \ldots, x_{q+r}). \tag{6.2.9}$$

Definition 6.2.4. If $f \in \mathbf{M}(\mathbf{E}^q, \mathfrak{C})$ and $g \in \mathbf{M}(\mathbf{E}^r, \mathfrak{C})$ then $f \wedge g$ is defined by

$$f \wedge g = A(fg). \tag{6.2.10}$$

It is customary to use the same symbols $\wedge$ and A for all choices of $q \geq 1$ and $r \geq 1$. With this understanding it is meaningful to say that $\wedge$ is an associative operation.

Lemma 6.2.5. *Let* $f \in \mathbf{M}(\mathbf{E}^q, \mathfrak{C})$, $g \in \mathbf{M}(\mathbf{E}^r, \mathfrak{C})$ *and* $h \in \mathbf{M}(\mathbf{E}^s, \mathfrak{C})$. *Then*

$$(f \wedge g) \wedge h = f \wedge (g \wedge h) = A(fgh). \tag{6.2.11}$$

PROOF. $(f \wedge g)h = (A(fg))h = (\Sigma_\tau ((q+r)!)^{-1} \varepsilon(\tau)\tau(fg))h$, where τ runs over all permutations of $1, \ldots, q+r$. Let τ' be the permutation of $1, \ldots, q+r+s$ leaving $q+r+1, \ldots, q+r+s$ fixed and acting as τ on $1, \ldots, q+r$. Then

$$\begin{aligned}\Sigma_\sigma \varepsilon(\sigma)\sigma((f \wedge g)h) &= \Sigma_\sigma \Sigma_\tau \varepsilon(\sigma)\varepsilon(\tau)((q+r)!)^{-1}\sigma(\tau(fg)h) \\ &= \Sigma_\sigma \Sigma_\tau \varepsilon(\sigma\tau')((q+r)!)^{-1}(\sigma\tau')(fgh) \\ &= (q+r+s)!A(fgh).\end{aligned}$$

A similar argument will show that $f \wedge (g \wedge h) = A(fgh)$. □

Lemma 6.2.6. *The operation* $\wedge$ *on* $A^q(\mathbf{E}, \mathfrak{C}) \times A^r(\mathbf{E}, \mathfrak{C}) \to A^{q+r}(\mathbf{E}, \mathfrak{C})$ is a bilinear operation.

Lemma 6.2.7. *If* $f \in \mathbf{M}(\mathbf{E}^q, \mathfrak{C})$ *and* $g \in \mathbf{M}(\mathbf{E}^r, \mathfrak{C})$ *then*

$$f \wedge g = (-1)^{qr}(g \wedge f). \tag{6.2.12}$$

PROOF. Let τ be the permutation

$$\tau = \begin{vmatrix} 1 & 2 & & r & r+1 & \cdots & q+r \\ q+1 & q+2 & \cdots & q+r & 1 & \cdots & q \end{vmatrix}. \tag{6.2.13}$$

The proof depends on knowing that $\varepsilon(\tau) = (-1)^{qr}$. Note that if σ is the cycle $(1, 2, \ldots, q+r)$ then $\tau = \sigma^q$. Then since $\varepsilon(\sigma) = (-1)^{q+r-1}$, it follows that $\varepsilon(\tau) = (-1)^{(q+r-1)q} = (-1)^{qr}$. Then

$$\begin{aligned}(q+r)!f \wedge g &= \Sigma_\sigma \varepsilon(\sigma) f(x_{\sigma(1)}, \ldots, x_{\sigma(q)}) g(x_{\sigma(q+1)}, \ldots, x_{\sigma(q+r)}) \\ &= \Sigma_\sigma \varepsilon(\sigma) g(x_{\sigma(q+1)}, \ldots, x_{\sigma(q+r)}) f(x_{\sigma(1)}, \ldots, x_{\sigma(q)}) \\ &= \Sigma_\sigma \varepsilon(\sigma) g(x_{\sigma\tau(1)}, \ldots, x_{\sigma\tau(q+r)}) f(x_{\sigma\tau(r+1)}, \ldots, x_{\sigma\tau(r+q)}) \\ &= \varepsilon(\tau) \Sigma_\sigma \varepsilon(\sigma\tau)((\sigma\tau)(gf)) = (-1)^{qr}(q+r)!g \wedge f.\end{aligned} \tag{6.2.14}$$

□

Lemma 6.2.8. *If* $f \in \mathbf{M}(\mathbf{E}^q, \mathfrak{C})$ *and* q *is an odd integer then* $f \wedge f = 0$.

PROOF. By (6.2.12) $f \wedge f = (-1)^{q^2} f \wedge f$. □

Lemma 6.2.9. *If* $m \geq 1$ *is an integer, if* $f_1, \ldots, f_m \in \mathbf{M}(\mathbf{E}, \mathfrak{C})$ *and* $x_1, \ldots, x_m \in \mathbf{E}$, *then*

$$(f_1 \wedge f_2 \wedge \ldots \wedge f_m)(x_1, \ldots, x_m) = (m!)^{-1} \det(f_i(x_j)). \tag{6.2.15}$$

Proof. We make an induction on m. (6.2.15) is clearly true if $m = 1$. Let a permutation σ_i be defined by

$$\sigma_i = \begin{vmatrix} 1 & 2 & \cdots & i & i+1 & \cdots & m \\ i & 1 & \cdots & i-1 & i+1 & \cdots & m \end{vmatrix}. \tag{6.2.16}$$

Then $\varepsilon(\sigma_i) = (-1)^{i+1}$. We compute

$$\begin{aligned}(m!)&(f_1 \wedge \cdots \wedge f_m)(x_1, \ldots, x_m) \\ &= \Sigma_\tau \varepsilon(\tau) f_1(x_{\tau(1)})(f_2 \wedge \cdots \wedge f_m)(x_{\tau(2)}, \ldots, x_{\tau(m)}) \qquad (6.2.17) \\ &= \Sigma_i f_1(x_i) \sum_{\tau(1)=i} \varepsilon(\tau)(\tau\sigma_i^{-1})(f_2 \wedge \cdots \wedge f_m)(x_1, \ldots, x_{i-1}, x_{i+1}, \ldots, x_m).\end{aligned}$$

As τ runs over permutations of $1, \ldots, m$ such that $\tau(1) = i$, we have $\tau\sigma_i^{-1}$ runs over all permutations of $1, \ldots, i-1, i+1, \ldots, m$. Therefore

$$\begin{aligned}\sum_{\tau(1)=i} &\varepsilon(\tau)(\tau\sigma_i^{-1})(f_2 \wedge \cdots \wedge f_m)(x_1, \ldots, x_{i-1}, x_{i+1}, \ldots, x_m) \\ &= \varepsilon(\sigma_i)((m-1)!)(A(f_2 \wedge \cdots \wedge f_m))(x_1, \ldots, x_{i-1}, x_{i+1}, \ldots, x_m) \\ &= (-1)^{i+1}((m-1)!)(f_2 \wedge \cdots \wedge f_m)(x_1, \ldots, x_{i-1}, x_{i+1}, \ldots, x_m).\end{aligned} \tag{6.2.18}$$

Apply the inductive hypothesis to the last line of (6.2.18), substitute the result in (6.2.17), and the Lemma follows. □

Lemma 6.2.10. *Let $f_1, \ldots, f_m \in \mathbf{M}(\mathbf{E}, \mathfrak{C})$ and $x_{ij} \in \mathfrak{C}$, $1 \le i, j \le m$. Then*

$$\begin{aligned}(x_{11}f_1 + \cdots + x_{1m}f_m) \wedge (x_{21}f_1 + \cdots x_{2m}f_m) \wedge \cdots \\ \wedge (x_{m1}f_1 + \cdots + x_{mm}f_m) \\ = (\det(x_{ij}))f_1 \wedge \cdots \wedge f_m.\end{aligned} \tag{6.2.19}$$

Proof. Expand (6.2.19) to a sum of m^m terms. Using Lemma 6.2.8 and (6.2.12), the only nonzero terms are the terms in which m distinct functions f_i occur. A term

$$x_{1\sigma(1)} \cdots x_{m\sigma(m)} f_{\sigma(1)} \wedge \cdots \wedge f_{\sigma(m)} = \varepsilon(\sigma) x_{1\sigma(1)} \cdots x_{m\sigma(m)} f_1 \wedge \cdots \wedge f_m.$$

Summing over these $m!$ terms gives the result (6.2.19). □

We close Section 6.2 by computing the dimensions of the alternating algebras $A^q(\mathbf{E}, \mathfrak{C})$. We use the canonical basis of $\mathbf{M}(\mathbf{E}, \mathfrak{C})$ defined in (6.2.2) to construct bases of the various spaces. A linear functional $f \in \mathbf{M}(\mathbf{E}, \mathfrak{C})$ can be expressed as

$$f = \sum_{i=1}^{n} c_i u_i, \; c_1, \ldots, c_n \in \mathfrak{C}. \tag{6.2.20}$$

Given $f_1, \ldots, f_q$ with $f_j = \Sigma_{i=1} c_{ij} u_i$ then

$$f_1 \cdots f_q = \Sigma_{i_1} \Sigma_{i_2} \cdots \Sigma_{i_q} c_{i_1 1} \dots c_{i_q q} u_{i_1} \dots u_{i_q}. \tag{6.2.21}$$

Thus the set of terms $u_{i_1} \cdots u_{i_q}$ span $\mathbf{M}(\mathbf{E}^q, \mathfrak{C})$ where repetition of indices is allowed.

To prove linear independence suppose

$$\Sigma_{i_1} \cdots \Sigma_{i_q} c_{i_1 \cdots i_q} u_{i_1} \cdots u_{i_q} = 0. \tag{6.2.22}$$

Evaluate this multilinear form at $e_{j_1}, \dots, e_{j_q}$. All these terms vanish except for one term so the value is $c_{j_1 \cdots j_q} = 0$. The multiplicative unit of $\mathfrak{C}$ is used here.

We summarize the discussion in a Lemma.

Lemma 6.2.11. $\mathbf{M}(\mathbf{E}^q, \mathfrak{C})$ *has dimension* n^q *where* n *is the dimension of* $\mathbf{E}$. *A canonical basis for this space is the set of* n^q *terms of the form* $u_{i_1} \cdots u_{i_q}$.

In order to obtain a basis of $A^q(\mathbf{E}, \mathfrak{C})$ we use the following lemma. It is easily verified and a formal proof is omitted. See Lemmas 6.2.1 and 6.2.2.

Lemma 6.2.12. *The following identities hold.*

$$\begin{aligned} A(\sigma f) &= \varepsilon(\sigma) A f. \\ A(f(Ag)) &= f \wedge g. \end{aligned} \tag{6.2.23}$$

We now compute the dimension of an alternating algebra. The general element of $\mathbf{M}(\mathbf{E}^q, \mathfrak{C})$ can be expressed as

$$f = \Sigma_{i_1} \cdots \Sigma_{i_q} f(e_{i_1}, \dots, e_{i_q}) u_{i_1} \cdots u_{i_q}. \tag{6.2.24}$$

Using the linearity of A and (6.2.23) it follows that

$$Af = \Sigma_{i_1} \cdots \Sigma_{i_q} f(e_{i_1}, \dots, e_{i_q}) u_{i_1} \wedge \cdots \wedge u_{i_q}. \tag{6.2.25}$$

Then the terms $u_{i_1} \wedge \cdots \wedge u_{i_q}$ span $A^q(\mathbf{E}, \mathfrak{C})$. Reordering the factors in such a wedge product at most changes the sign so we may suppose $i_1 < i_2 < \cdots < i_q$. Conversely, if

$$0 = \sum_{i_1 < \cdots < i_q} \sum c_{i_1 \cdots i_q} u_{i_1} \wedge \cdots \wedge u_{i_q}, \tag{6.2.26}$$

then evaluated at $e_{j_1}, \dots, e_{j_q}$ we find

$$\begin{aligned} &\text{if } j_1, \dots, j_q \text{ is not a permutation of } i_1, \dots, i_q \\ &\text{then } u_{i_1} \wedge \cdots \wedge u_{i_q}(e_{j_1}, \dots, e_{j_q}) = 0. \text{ If } \sigma \text{ is a} \\ &\text{permutation of } 1, \dots, q \text{ such that } i_{\sigma(1)} = j_1, \dots, i_{\sigma(q)} \\ &= j_q, \text{ then } u_{i_1} \wedge \cdots \wedge u_{i_q}(e_{j_1}, \dots, e_{j_q}) = \varepsilon(\sigma). \end{aligned} \tag{6.2.27}$$

The relations (6.2.27) imply that all coefficients in (6.2.26) vanish. We summarize this in a theorem.

Theorem 6.2.13. *Let* $\mathbf{E}$ *have dimension* n. *The space* $A^q(\mathbf{E}, \mathfrak{C})$ *has dimension* $\binom{n}{q}$, $1 \leq q \leq n$. *If* $q > n$ *then the space* $A^q(\mathbf{E}, \mathfrak{C}) = \{0\}$. *If* $1 \leq q \leq n$ *a canonical basis for* $A^q(\mathbf{E}, \mathfrak{C})$ *is the set of multilinear forms* $u_{i_1} \wedge \cdots \wedge u_{i_q}$, $i_1 < \cdots < i_q$.

6.3. Differential Forms and the Operator *d*

The abstract algebraic theory is now applied when $\mathfrak{C}$ is a local coefficient ring $\mathfrak{C}_{x,k}$ as described following (6.3.1). In the sequel we suppose to each $x \in \mathfrak{M}$ there is assigned the canonical basis elements of $\mathbf{E}_x$, and of $M(\mathbf{E}_x, \mathfrak{C}_{x,k})$ the canonical basis is

$$u_1^x, \ldots, u_n^x \quad \text{satisfying} \quad u_i^x(e_j^x) = \delta_{ij}. \tag{6.3.1}$$

From Theorem 6.2.13 the terms $u_{i_1}^x \wedge \cdots \wedge u_{i_q}^x$ are a basis of $A^q(\mathbf{E}_x, \mathfrak{C}_{x,k})$ locally at x. We assume $k \geq 2$ but do not otherwise specify k.

We define an operator $d: A^q(\mathbf{E}, \mathfrak{C}) \to A^{q+1}(\mathbf{E}, \mathfrak{C})$ simultaneously for all $q \geq 0$ and $\mathfrak{C} = \mathfrak{C}_{x,k}$, $k \geq 1$.

Definition 6.3.1. If $f \in \mathfrak{C}_{x,k}$ then

$$df = \sum_{i=1}^{n} \frac{\partial f}{\partial p_i}(\varphi^{-1}(x, p_1, \ldots, p_n))|_{\varphi(x,\ )} u_i^x. \tag{6.3.2}$$

If $q \geq 1$ and $f u_{i_1}^x \wedge \cdots \wedge u_{i_q}^x \in A^q(\mathbf{E}_x, \mathfrak{C}_{x,k})$ then

$$d(f u_{i_1}^x \wedge \cdots \wedge u_{i_q}^x) = (df) \wedge u_{i_1}^x \wedge \cdots \wedge u_{i_q}^x. \tag{6.3.3}$$

Extend the definition of d by linearity to all of $A^q(\mathbf{E}_x, \mathfrak{C}_{x,k})$.

Note that by the basis Theorem 6.2.13 the extension of the definition of d by linearity is meaningful. In terms of Definition 6.3.1 we define a relation of equivalence, denoted "$\equiv$", on the intersections of $U_x \cap U_y$. If $p = \varphi(x, \varphi^{-1}(y, q))$ we define

Definition 6.3.2.

$$\Delta_{ij}(x, y, z) = \left.\frac{\partial p_i}{\partial q_j}\right|_{\varphi(y,z)}. \tag{6.3.4}$$

The function Δ_{ij} is defined for $z \in U_x \cap U_y$. p as a function of $q \in \mathbb{R}^n$ is $\mathbb{R}^n$ valued so $p^t = (p_1, \ldots, p_n)$. (6.3.4) is to be read as the partial derivative evaluated at $\varphi(y, z) \in \mathbb{R}^n$.

Definition 6.3.3. An element of $A^q(\mathbf{E}_x, \mathfrak{C}_{x,k})$, $q > 0$, will be called a differential form defined at x.

Definition 6.3.4. The equivalence relation "$\equiv$" is defined by the following rules. Functions are self equivalent. Next,

$$u_i^x \equiv \sum_{j=1}^{n} \Delta_{ij}(x, y, \;)u_j^y. \tag{6.3.5}$$

Last, a differential form defined at x is equivalent to a differential form defined at y if upon substitution of $\Sigma_{j=1}^n \Delta_{ij}(x, y, \;)u_j^y$ for u_i^x, $1 \le i \le n$, the differential form at x is transformed into the differential form at y on the set $U_x \cap U_y$. The coefficients being functions defined on $U_x \cap U_y$ do not require transformation in this definition.

Theorem 6.3.5. "$\equiv$" *is an equivalence relation.*

PROOF.

Reflexive. $\varphi(x, \varphi^{-1}(x, \;))$ is the identity map of $\mathbb{R}^n$ onto $\mathbb{R}^n$. Consequently the functions (6.3.4) and (6.3.5) are

$$\Delta_{ii}(x, y, z) = 1 \quad \text{for all} \quad z \in U_x;$$

$$\Delta_{ij}(x, y, z) = 0, \qquad i \neq j, \qquad \text{for all } z \in U_x; \tag{6.3.6}$$

and

$$u_i^x \equiv u_i^x.$$

Substitution into a differential form now clearly shows "$\equiv$" to be a reflexive relation.

Symmetric. To shorten the notations let

$$\psi(\;) = \varphi(x, \varphi^{-1}(y, \;)). \tag{6.3.7}$$

Then clearly

$$\psi^{-1}(\;) = \varphi(y, \varphi^{-1}(x, \;)). \tag{6.3.8}$$

If we write ψ_{ij} for the partial derivative of the i-th component of ψ with respect to the j-th variable of ψ, and similarly for ψ^{-1}, then the chain rule clearly requires

$$\delta_{ij} = \sum_{k=1}^{n} \psi_{ik}(\psi^{-1})_{kj}, \tag{6.3.9}$$

identically on the part of $\mathbb{R}^n$ on which the functions are defined. Making a double substitution, as required by Definition 6.3.4, shows "$\equiv$" to be a symmetric relation.

Transitive. To continue the notation of (6.3.7) we write

$$\psi = \varphi(x, \varphi^{-1}(y, \;)), \psi' = \varphi(y, \varphi^{-1}(z, \;)), \tag{6.3.10}$$

and

$$\psi'' = \varphi(x, \varphi^{-1}(z, \;)).$$

Here the chain rule clearly requires

$$\psi''_{ij} = \sum_{k=1}^{n} \psi_{ik}\psi'_{kj}. \tag{6.3.11}$$

A double substitution as required by Definition 6.3.4 shows "$\equiv$" to be transitive. □

The definition of "$\equiv$" was made for a single value of q, but in the sequel we assume the same symbol of equivalence applies to all the alternating algebras $A^1, \ldots, A^n$ simultaneously, and for all coefficient rings $\mathfrak{C}_{x,k}$, $k \geq 1$.

We now develop properties of the operator d. It should be noted that from Definitions 6.3.1 and 6.3.4,

$$u^x \equiv d\varphi(x, \varphi^{-1}(y, \ )), \tag{6.3.12}$$

where d is computed locally at y and both sides of (6.3.12) are n-vectors.

Theorem 6.3.6. *If $f \in A^q(\mathbf{E}_x, \mathfrak{C}_{x,2})$ and $g \in A^r(\mathbf{E}_x, \mathfrak{C}_{x,2})$ then the operator d is a linear operator. The following relations hold.*

$$d(f \wedge g) = (df) \wedge g + (-1)^q f \wedge (dg). \tag{6.3.13}$$

$$d(df) = d^2 f = 0. \tag{6.3.14}$$

PROOF. It is sufficient to verify (6.3.13) and (6.3.14) on the basis elements. Let $f = f_1 u^x_{i_1} \wedge \cdots \wedge u^x_{i_q}$ and $g = g_1 u^x_{j_1} \wedge \cdots \wedge u^x_{j_r}$. As a bilinear form $f \wedge g = f_1 g_1 u^x_{i_1} \wedge \cdots \wedge u^x_{j_1} \wedge \cdots \wedge u^x_{j_r}$, so that

$$\begin{aligned}
d(f \wedge g) &= d(f_1 g_1) \wedge u^x_{i_1} \wedge \cdots \wedge u^x_{j_r} \\
&= ((df_1)g_1 + f_1(dg_1)) \wedge u^x_{i_1} \wedge \cdots \wedge u^x_{j_r} \\
&= ((df_1) \wedge u^x_{i_1} \wedge \cdots \wedge u^x_{i_q}) \wedge ((g_1)u^x_{j_1} \wedge \cdots \wedge u^x_{j_r}) \\
&\quad + (-1)^q((f_1)u^x_{i_1} \wedge \cdots \wedge u^x_{i_q}) \wedge ((dg_1)u^x_{j_1} \wedge \cdots \wedge u^x_{j_r}) \\
&= (df) \wedge g + (-1)^q f \wedge (dg).
\end{aligned} \tag{6.3.15}$$

In (6.3.15) we use the fact that $d(f_1 g_1) = d(f_1)g_1 + f_1(dg_1) \in A^1$, as follows at once from (6.3.2).

To show $d^2 = 0$, we compute d^2 on a basis element.

$$\begin{aligned}
d^2(f_1 u^x_{i_1} \wedge \cdots \wedge u^x_{i_q}) &= d(df_1 \wedge u^x_{i_1} \wedge \cdots \wedge u^x_{i_q}) \\
&= \left(\sum_{j=1}^{n} \left(d\left(\frac{\partial f}{\partial x_j} u^x_j\right)\right) \wedge u^x_{i_1} \wedge \cdots \wedge u^x_{i_q}\right) \\
&= \left(\sum_{j=1}^{n} \sum_{k=1}^{n} \frac{\partial^2 f}{\partial x_k \partial x_j} u^x_k \wedge u^x_j\right) \wedge u^x_{i_1} \wedge \cdots \wedge u^x_{i_q} = 0.
\end{aligned} \tag{6.3.16}$$

This follows since the terms $u^x_k \wedge u^x_k = 0$ while if $k \neq j$ then $u^x_k \wedge u^x_j =$

$-u_j^x \wedge u_k^x$. Since f is assumed to have continuous second partial derivatives,

$$\frac{\partial^2 f}{\partial x_k \partial x_j} = \frac{\partial^2 f}{\partial x_j \partial x_k}. \qquad \square$$

We state an obvious lemma.

Lemma 6.3.7. *If differential forms ω_1 and ω_2 are defined locally at x, ω_3 and ω_4 are defined locally at y, and $\omega_1 \equiv \omega_3, \omega_2 \equiv \omega_4$, then $\omega_1 \wedge \omega_2 \equiv \omega_3 \wedge \omega_4$.*

Theorem 6.3.8. *Let ω_1 and ω_2 be differential forms such that ω_1 is defined locally at x, and ω_2 is defined locally at y, and $\omega_1 \equiv \omega_2$ on $\mathbf{U}_x \cap \mathbf{U}_y$. Then $d\omega_1 \equiv d\omega_2$.*

Proof. We first verify the theorem for zero forms. Thus ω represents a function defined on $\mathbf{U}_x \cap \mathbf{U}_y$. Locally at x,

$$d\omega = \sum_{i=1}^{n} \frac{\partial \omega}{\partial p_i}(\varphi^{-1}(x, \ ))\big|_{\varphi(x, \ )} u_i^x, \tag{6.3.17}$$

and locally at y,

$$d\omega = \sum_{i=1}^{n} \frac{\partial \omega}{\partial p_i}(\varphi^{-1}(y, \ ))\big|_{\varphi(y, \ )} u_i^y. \tag{6.3.18}$$

Use of the substitutions (6.3.6) and the chain rule shows that the differential forms exhibited in (6.3.17) and (6.3.18) are equivalent 1-forms.

In order to simplify subsequent computations we make the following observation. Suppose p_i is the i-th coordinate of $\varphi(x, \varphi^{-1}(y, \ ))$. We suppose this is a function of $q \in \mathbb{R}^n$, and $q^t = (q_1, \ldots, q_n)$. Then $\left.\frac{\partial p_i}{\partial q_j}\right|_{\varphi(y, \ )} = \Delta_{ij}(x, y, \ )$ is defined on $\mathbf{U}_x \cap \mathbf{U}_y$. Locally at y we compute

$$d\varphi(x, \varphi^{-1}(y, \varphi(y, \ ))) = \sum_{j=1}^{n} \Delta_{ij}(x, y, \ ) u_j^y \tag{6.3.19}$$

and

$$0 = d^2\varphi(x, \varphi^{-1}(y, \varphi(y, \ ))) = \sum_{j=1}^{n} (d\Delta_{ij}(x, y, \ )) \wedge u_j^y.$$

Given a basis element $f u_{i_1}^x \wedge \cdots \wedge u_{i_q}^x$ equivalent to a differential form ω locally at y, we have

$$\omega = f\left(\sum_{j=1}^{n} \Delta_{i_1 j}(x, y, \ ) u_j^y\right) \wedge \cdots \wedge \left(\sum_{j=1}^{n} \Delta_{i_q j}(x, y, \ ) u_j^y\right). \tag{6.3.20}$$

We now use (6.3.13) and (6.3.19) and Lemma 6.3.7. Then, computed locally at y,

$$\begin{aligned} d\omega &= (df) \wedge \left(\sum_{j=1}^{n} \Delta_{i_1 j}(x, y, \ ) u_j^y \right) \wedge \cdots \wedge \left(\sum_{j=1}^{n} \Delta_{i_q j}(x, y, \ ) u_j^y \right) \\ &\equiv (df) \wedge u_{i_1}^x \wedge \cdots \wedge u_{i_q}^x = d(f u_{i_1}^x \wedge \cdots \wedge u_{i_q}^x). \end{aligned} \tag{6.3.21}$$

The last part of (6.3.21) is computed locally at x, finishing the proof. □

Corollary 6.3.9. *If $f\colon \mathfrak{M} \to \mathbb{R}$ is a globally defined 0-form which is continuously differentiable then df is a locally defined 1-form, local at x for all $x \in \mathfrak{M}$, representing equivalent forms on the overlaps of charts.*

6.4. Theory of Integration

Definition 6.4.1. *If $\omega_1 = f u_1^x \wedge \cdots \wedge u_n^x$ is defined locally at x and if $\mathbf{A} \subset \mathbf{U}_x$* is a Borel subset of $\mathbf{U}_x$ then

$$\int_{\mathbf{A}} \omega_1 \underset{\text{def}}{=} \int_{\varphi(x, \mathbf{A})} f(\varphi^{-1}(x, (p_1, \ldots, p_n))) dp_1 \cdots dp_n. \tag{6.4.1}$$

If ω is a general n-form, since by Theorem 6.2.13 the dimension of $\mathbf{A}^n(\mathbf{E}, \mathfrak{C})$ is one, ω has the form of ω_1.

Lemma 6.4.2. *Let ω_1 and ω_2 be differential n-forms such that ω_1 is defined locally at x, and ω_2 is defined locally at y, and $\omega_1 \equiv \omega_2$. If the determinants* $\det(\Delta_{ij}(x, y, z))$ *are everywhere positive on $\mathbf{U}_x \cap \mathbf{U}_y$ and if $\mathbf{C}$ is a Borel subset of $\mathbf{U}_x \cap \mathbf{U}_y$ then*

$$\int_{\mathbf{C}} \omega_1 = \int_{\mathbf{C}} \omega_2. \tag{6.4.2}$$

PROOF. We let $\omega_1 = f u_1^x \wedge \cdots \wedge u_n^x$. By (6.3.5), by the condition of equivalence, and by the algebraic relations (6.2.19),

$$\begin{aligned} \omega_2 &= f\left(\sum_{j=1}^{n} \Delta_{ij}(x, y, z) u_j^y \right) \wedge \cdots \wedge \left(\sum_{j=1}^{n} \Delta_{nj}(x, y, z) u_j^y \right) \\ &= f(\det(\Delta_{ij}(x, y, z))) u_1^y \wedge \cdots \wedge u_n^y. \end{aligned} \tag{6.4.3}$$

Therefore using the notation of (6.3.4) that $p = \varphi(x, \varphi^{-1}(y, q))$, we obtain

$$\int_{\mathbf{C}} \omega_2 = \int_{\varphi(y, \mathbf{C})} f(\varphi^{-1}(y, q)) \det\left(\frac{\partial p_i}{\partial q_j} \right) dq_1 \cdots dq_n. \tag{6.4.4}$$

Since $\varphi^{-1}(x, p) = \varphi^{-1}(x, \varphi(x, \varphi^{-1}(y, q))) = \varphi^{-1}(y, q)$ we obtain from (6.4.4) that

$$\int_{\mathbf{C}} \omega_2 = \int_{\varphi(x, \mathbf{C})} f(\varphi^{-1}(x, p))\, dp_1 \cdots dp_n = \int_{\mathbf{C}} \omega_1. \tag{6.4.5}$$

□

Definition 6.4.3. A globally defined differential n-form $\omega = \{w_x, x \in \mathfrak{M}\}$ is a set of locally defined differential n-forms satisfying

$$\text{if} \quad x, y \in \mathfrak{M} \quad \text{then} \quad \omega_x \equiv \omega_y. \tag{6.4.6}$$

Definition 6.4.4. If ω is a globally defined n-form then $|\omega|$ is defined locally by $|\omega| = |\omega_x| = |f| u_1^x \wedge \cdots \wedge u_n^x$.

Theorem 6.4.5. *If ω is a globally defined n-form on a $\mathbf{C}_2$ manifold $\mathfrak{M}$ having countable base for the open sets then $\int_{\mathbf{C}} \omega$ defines a countably additive signed measure which is a regular Borel measure.*

PROOF. The proof that $\int_{\mathbf{C}} \omega$ defines a measure is an obvious application of Lemma 6.4.2. This application uses the countable cover of $\mathfrak{M}$ to obtain a σ-finite measure on the manifold. The integral over subsets $\mathbf{C}$ will be finite for compact subsets $\mathbf{C}$. It follows from Halmos (1950) that the measure is a regular measure (we assume the manifold is a locally compact Hausdorff space, so that given a countable base for the open sets the Baire and Borel subsets are the same. □

The integrations in subsequent chapters will be purely formal calculations. It is hoped that the machinery of Chapter 6 will provide sufficient theory to justify the applications. In the integrations made a frequent situation is the following. Globally defined functions $a_1, \ldots, a_n$ are given and the globally defined differential form is

$$\omega = f(a_1, \ldots, a_n)\, da_1 \wedge \cdots \wedge da_n, \tag{6.4.7}$$

where on a n-dimensional manifold the differential form (6.4.7) is of maximal degree and hence is integrable locally when $da_1, \ldots, da_n$ are computed locally. If we have a Borel subset $\mathbf{C} \subset \mathbf{U}_x$ let $\mathbf{C}'$ be the corresponding set in the range of $(a_1, \ldots, a_n)$. Then if we set $b_i(x, p_1, \ldots, p_n) = a_i(\varphi^{-1}(x, p_1, \ldots, p_n))$, $1 \le i \le n$, it follows that

$$\begin{aligned} \int_{\mathbf{C}} \omega &= \int_{\varphi(x,\mathbf{C})} f(b_1, \ldots, b_n) \det \left| \frac{\partial b_i}{\partial p_j} \right| dp_1 \cdots dp_n \\ &= \int_{\mathbf{C}'} f(a_1, \ldots, a_n)\, da_1 \cdots da_n. \end{aligned} \tag{6.4.8}$$

In (6.4.8) the variables $a_1, \ldots, a_n$ are now formal variables of integration for the n-fold integral over a subset of $\mathbb{R}^n$.

In some examples the map $x \leftrightarrow (a_1(x), \ldots, a_n(x))$ is not one-to-one, this condition failing on a null set $\mathbf{N}$ which is seen to satisfy, if $x \in \mathfrak{M}$ then

$$\int_{\mathbf{U}_x \cap \mathbf{N}} u_1 \wedge \cdots \wedge u_n = 0. \tag{6.4.9}$$

In such problems the null set $\mathbf{N}$ is usually ignored.

6.5. Transformation of Manifolds

We suppose $\mathfrak{M}_1$ and $\mathfrak{M}_2$ are n-dimensional manifolds and that $f: \mathfrak{M}_1 \to \mathfrak{M}_2$ is a homeomorphism. The basic assumption here is that if $g: \mathfrak{M}_2 \to \mathbb{R}$ is a C_2-function then the composition $g \circ f: \mathfrak{M}_1 \to \mathbb{R}$ is also a C_2-function.

As just noted f induces a map F of $\mathfrak{M}_2$ 0-forms to $\mathfrak{M}_1$ 0-forms. We extend this map to all differential forms subject to the requirement $dF = Fd$, where in each case the operator d is to be computed in the appropriate local coordinates.

If $y = f(x)$ then for z near x we have

$$\varphi_2(y, f(z)): \mathfrak{M}_1 \to \mathbb{R}^n \tag{6.5.1}$$

gives local coordinates of $f(z)$ on $\mathfrak{M}_2$ near y. We want

$$\varphi_2(y, f(\varphi_1^{-1}(x, \ ))) \tag{6.5.2}$$

to be twice continuously differentiable where defined. If

$$(p_1, \ldots, p_n) = \varphi_1(x, z)^t \tag{6.5.3}$$

and

$$(q_1, \ldots, q_n) = \varphi_2(y, f(z))^t = \varphi_2(f(x), f(z))^t$$

then $p_1, \ldots, p_n$ are functions defined on $\mathfrak{M}_1$ as are $q_1, \ldots, q_n$. Locally near x we may compute $dq_1, \ldots, dq_n$.

Definition 6.5.1. Define F on 0-forms by

$$(Fg)(x) = g(f(x)). \tag{6.5.4}$$

Extend F to basis elements of r-forms by the definition

$$F(gv_{i_1}^y \wedge \cdots \wedge v_{i_r}^y) = (Fg)\, dq_{i_1} \wedge \cdots \wedge dq_{i_r} \tag{6.5.5}$$

where $dq_1, \ldots, dq_n$ are computed locally at x. Extend F to be a linear transformation of differential forms.

Theorem 6.5.2.

$$dF = Fd. \tag{6.5.6}$$

Proof. By linearity of d and F it is sufficient to consider basis terms, $gv_{i_1}^y \wedge \cdots \wedge v_{i_r}^y$. Using (6.3.13) and (6.3.14)

$$d(F(gv_{i_1}^y \wedge \cdots \wedge v_{i_r}^y)) = d(Fg) \wedge dq_{i_1} \wedge \cdots \wedge dq_{i_r}. \tag{6.5.7}$$

Also

$$\begin{aligned} d(Fg) = d(g(f(\ ))) &= \sum_{i=1}^n \sum_{j=1}^n \frac{\partial g}{\partial q_i} \frac{\partial q_i}{\partial p_j} u_j^x \\ &= \sum_{i=1}^n \frac{\partial g}{\partial q_i} dq_i. \end{aligned} \tag{6.5.8}$$

Also

$$F(dg \wedge v_{i_1}^y \wedge \cdots \wedge v_{i_r}^y) = F\left(\left(\sum_{i=1}^n \frac{\partial g}{\partial q_i} v_i^y\right) \wedge v_{i_1}^y \wedge \cdots \wedge v_{i_r}^y\right)$$
$$= \left(\sum_{i=1}^n \frac{\partial g}{\partial q_i} dq_i\right) \wedge dq_{i_1} \wedge \cdots \wedge dq_{i_r}. \tag{6.5.9}$$

Thus (6.5.6) holds. □

The transformation F of differential forms then extends to a transformation of measures obtained by integration of n-forms.

Definition 6.5.3. If ω is a globally defined n-form on $\mathfrak{M}$, and μ is defined by $\mu(\mathbf{C}) = \int_{\mathbf{C}} \omega$, then $F^{-1}\mu$ is defined by

$$(F^{-1}\mu)(\mathbf{C}) = \mu(f^{-1}(\mathbf{C})). \tag{6.5.10}$$

Theorem 6.5.4. *Let f be a transformation of manifolds as described above, $f: \mathfrak{M}_1 \to \mathfrak{M}_2$. Suppose the Jacobians of the transformations* (6.5.2) *are positive. If ω is an n-form on $\mathfrak{M}_2$ then*

$$\int_{\mathbf{C}} F(\omega) = \int_{f(\mathbf{C})} \omega. \tag{6.5.11}$$

Proof. Locally for a basis element $g v_1^y \wedge \cdots \wedge v_n^y$ the transformation maps this n-form into $(Fg) \det \left|\frac{\partial q_i}{\partial p_j}\right| u_1^x \wedge \cdots \wedge u_n^x$. Let $g' = g(\varphi^{-1}(y,\ ))$. Then

$$\int_{\mathbf{C}} (Fg) \det \left|\frac{\partial q_i}{\partial p_j}\right| dp_1 \cdots dp_n$$
$$= \int_{\mathbf{C}} g'(q_1(p_1, \ldots, p_n), \ldots, q_n(p_1, \ldots, p_n)) \det \left|\frac{\partial q_i}{\partial p_j}\right| dp_1 \cdots dp_n$$
$$= \int_{f(\mathbf{C})} g'(q_1, \ldots, q_n)\, dq_1 \cdots dq_n = \int_{f(\mathbf{C})} \omega. \tag{6.5.12}$$

By linearity the result follows for all local n-forms. Since $\mathfrak{M}_1$ and $\mathfrak{M}_2$ are assumed to be locally compact separable Hausdorff spaces, one may choose a countable cover of sets $\mathbf{U}_x$. The theorem then follows by a countable additivity argument. □

Definition 6.5.5. If $f: \mathfrak{M} \to \mathfrak{M}$ is a transformation of manifolds with induced mapping F of differential forms, then a differential form $\omega = \{\omega_x, x \in \mathfrak{M}\}$ which is globally defined is said to be invariant if,

$$\text{if} \quad x \in \mathfrak{M} \quad \text{then} \quad F(\omega_x) = \omega_{f^{-1}(x)}. \tag{6.5.13}$$

Theorem 6.5.6. *If $f: \mathfrak{M} \to \mathfrak{M}$ is a transformation of manifolds and the globally defined differential form ω is invariant then the measure μ defined by*

$\mu(\mathbf{C}) = \int_{\mathbf{C}} \omega$ *satisfies*

$$\mu = F\mu. \tag{6.5.14}$$

PROOF. Locally, if $\mathbf{C} \subset \mathbf{U}_x$ and $y = f(x)$ and $f(\mathbf{C}) \subset \mathbf{U}_y$, then by (6.5.11)

$$\begin{aligned} \mu(f(\mathbf{C})) &= \int_{f(\mathbf{C})} \omega_y = \int_{\mathbf{C}} F(\omega_y) = \int_{\mathbf{C}} F(\omega_{f(x)}) \\ &= \int_{\mathbf{C}} \omega_x = \mu(\mathbf{C}). \end{aligned} \tag{6.5.15}$$

Since f is a homeomorphism the sets

$$\mathbf{U}_x \cap f^{-1}(\mathbf{U}_{f(x)}) \tag{6.5.16}$$

each contain the index point x, hence are nonempty open sets and $\mathfrak{M}$ thus has a countable subcover $\{\mathbf{U}_{x_i} \cap f^{-1}(\mathbf{U}_{f(x_i)}), i \geq 1\}$. Thus we may construct a measurable partition $\mathbf{B}_i$, $i \geq 1$, of $\mathfrak{M}$ such that if $i \geq 1$ then

$$\mathbf{B}_i \subset \mathbf{U}_{x_i}. \tag{6.5.17}$$

Given a Borel subset $\mathbf{C}$ then (6.5.15) applies to $\mathbf{C} \cap \mathbf{B}_i$, $i \geq 1$, so that

$$\mu(f(\mathbf{C})) = \sum_{i=1}^{\infty} \mu(f(\mathbf{C} \cap \mathbf{B}_i)) = \sum_{i=1}^{\infty} \mu(\mathbf{C} \cap \mathbf{B}_i) = \mu(\mathbf{C}). \tag{6.5.18}$$ □

6.6. Lemmas on Multiplicative Functionals

The following results are directed to the determination of formulas for multiplicative functions m in several special cases. These results are related to determination of modular functions. See Section 3.5.

Lemma 6.6.1. *Let m be a function of $n \times n$ matrices with real entries such that $m(I_n) \neq 0$ and $r > 0$ is a real number such that*

for each $A \in \mathbf{GL}(n)$, $m(A)$ is a homogeneous polynomial of degree r in the entries of A; (6.6.1)

if $A, B \in \mathbf{GL}(n)$ then $m(AB) = m(A)m(B)$. (6.6.2)

Then r/n is an integer and

$$m(A) = (\det A)^{(r/n)}. \tag{6.6.3}$$

PROOF. If I_n is the $n \times n$ identity matrix then $m(I_n) = m(I_n)^2 = m(I_n)^3$ so $m(I_n) = 1$ follows. Then $1 = m(I_n) = m(AA^{-1}) = m(A)m(A^{-1})$ so that $m(A) = 1/m(A^{-1})$. Since m is continuous, $m(\mathbf{O}(n))$ is a compact set of real numbers that is a subgroup. Hence $m(\mathbf{O}(n)) = \{-1, 1\}$ or $m(\mathbf{O}(n)) = \{1\}$.

Next, take A_1 to be the diagonal matrix with d in the (1, 1)-position and elsewhere on the diagonal entries = 1. Let A_i be obtained from A_1 by permutation of the (1, 1)-entry into the (i, i)-position so that $A_i = QA_1Q^t$ for some permutation matrix Q. It follows that $m(A_1) = m(A_i)$, since

$$m(A_i) = m(Q)m(A_1)m(Q^t) \quad = m(A_1). \tag{6.6.4}$$

Then the diagonal matrix dI_n factors into $dI_n = A_1 \cdots A_n$ and

$$d^r = m(dI_n) = m(A_1 \cdots A_n) = (m(A_1))^n. \tag{6.6.5}$$

Since $m(A_1)$ is a polynomial in the variable d, it follows that r/n is an integer and

$$m(A_1) = d^{r/n}. \tag{6.6.6}$$

The identities (6.6.5) and (6.6.6) clearly imply that if A is a diagonal matrix then (6.6.3) holds for A.

If A is a symmetric matrix then there exists $U \in \mathbf{O}(n)$ such that UAU^t is a diagonal matrix, and by the result just obtained, if A is also nonsingular then $m(A) = (\det A)^{r/n}$. See Section 8.1. For the arbitrary $X \in \mathbf{GL}(n)$, we may write $X = AS$ with $A = X(X^tX)^{-1/2} \in \mathbf{O}(n)$ and $S = (X^tX)^{1/2} \in \mathbf{S}(n)$. Then the above implies

$$m(X) = m(A)m(S) = (\pm m(A))(\det X)^{r/n} \tag{6.6.7}$$

since $\det(X^tX)^{1/2} = \pm\det X$ depending on the sign choice of the square root.

Since $A \in \mathbf{O}(n)$ implies $m(A) = \pm 1$, it follows that if $x \in \mathbf{GL}(n)$ then $m(X) = \pm(\det X)^{r/n}$. Note that $\{X \mid \det X > 0\}$ is an open subset of $\mathbb{R}^{n^2}$ so one may choose an open neighborhood of I_n, call it $\mathbf{U}$, such that if $X \in \mathbf{U}$ then $|m(X) - (\det X)^{r/n}| < \frac{1}{2}$ because $|m(X) - 1| < \frac{1}{4}$ and $|(\det X)^{r/n} - 1| < \frac{1}{4}$. We use here the continuity of the functions. This clearly implies that if $X \in \mathbf{U}$ then $m(X) = (\det X)^{r/n}$. Since these polynomials agree on an open set they are everywhere equal. □

Lemma 6.6.2. *Let m be a function of $n \times n$ matrices with real entries such that $m(I_n) \neq 0$ and such that*

$$\textit{m is a positive homogeneous function of degree } r; \tag{6.6.8}$$

$$m(\mathbf{O}(n)) \textit{ is a bounded set of real numbers}; \tag{6.6.9}$$

$$\textit{if } \; A, B \in \mathbf{GL}(n) \; \textit{ then } \; m(AB) = m(A)m(B). \tag{6.6.10}$$

Then

$$m(A) = |\det A|^{r/n}. \tag{6.6.11}$$

Proof. As before $m(I_n) = m(I_n)^2$ so that since m is nonnegative, $m(I_n) = 1$. $m(\mathbf{O}(n))$ is a bounded group of real numbers, so since m is nonnegative, $m(\mathbf{O}(n)) = \{1\}$. Then, if A is a diagonal matrix, as in the preceding proof, $m(A) = |\det A|^{r/n}$. Also, if $X \in \mathbf{GL}(n)$ then $X = X(X^tX)^{-1/2}(X^tX)^{1/2}$ and

$X(X^tX)^{-1/2} \in \mathbf{O}(n)$ so that $m(X) = m((X^tX)^{1/2})$. Choose $U \in \mathbf{O}(n)$ so that $U(X^tX)^{1/2}U^t = \operatorname{diag}(\lambda_1, \ldots, \lambda_n)$. Then $m(X) = m(\operatorname{diag}(\lambda_1, \ldots, \lambda_n)) = (\lambda_1 \cdots \lambda_n)^{r/n} = |\det(X^tX)^{1/2}|^{r/n} = |\det X|^{r/n}$. See Theorem 8.1.1 about diagonalization of symmetric matrices. The existence of the square root also follows. □

The following discussion is aimed at computation of Jacobians for lower triangular substitutions.

Lemma 6.6.3. *Let* $m: \mathbf{T}(n) \to \mathbb{R}$ *satisfy*

$$m(I_n) = 1 \quad \textit{and if} \quad T \in \mathbf{T}(n) \quad \textit{then} \quad m(T) \neq 0; \tag{6.6.12}$$

$$\textit{if} \quad S, T \in \mathbf{T}(n) \quad \textit{then} \quad m(S)m(T) = m(ST). \tag{6.6.13}$$

Then $m(T) = 1$ *for all* $T \in \mathbf{T}(n)$ *such that the diagonal of* T *is* $1, \ldots, 1$.

Proof. By induction on n. The case $n = 1$ is clear. Consider the matrix product

$$\begin{vmatrix} I_{n-1} & 0 \\ 0 & \alpha \end{vmatrix} \begin{vmatrix} I_{n-1} & 0 \\ b^t & 1 \end{vmatrix} \begin{vmatrix} I_{n-1} & 0 \\ 0 & \alpha^{-1} \end{vmatrix} = \begin{vmatrix} I_{n-1} & 0 \\ \alpha b^t & 1 \end{vmatrix}. \tag{6.6.14}$$

Therefore using (6.6.12) and (6.6.13) it follows that

$$m\left(\begin{vmatrix} I_{n-1} & 0 \\ b^t & 1 \end{vmatrix}\right) = m\left(\begin{vmatrix} I_{n-1} & 0 \\ \alpha b^t & 1 \end{vmatrix}\right), \tag{6.6.15}$$

valid for all real $\alpha \neq 0$. Since

$$\begin{vmatrix} I_{n-1} & 0 \\ b^t & 1 \end{vmatrix}^2 = \begin{vmatrix} I_{n-1} & 0 \\ 2b^t & 1 \end{vmatrix}, \tag{6.6.16}$$

it follows from (6.6.13) and (6.6.15) that

$$m\left(\begin{vmatrix} I_{n-1} & 0 \\ b^t & 1 \end{vmatrix}\right) = 1 \quad \text{for all } b \in \mathbb{R}^{n-1}. \tag{6.6.17}$$

If $A \in \mathbf{T}(n-1)$ then we may define $m'(A) = m\left(\begin{vmatrix} A & 0 \\ 0 & 1 \end{vmatrix}\right)$. Then m' satisfies (6.6.12) and (6.6.13). By inductive hypothesis, if A has diagonal $1, \ldots, 1$, then $m'(A) = 1$. From the identity

$$\begin{vmatrix} A & 0 \\ 0 & 1 \end{vmatrix} \begin{vmatrix} I_{n-1} & 0 \\ b^t & 1 \end{vmatrix} = \begin{vmatrix} A & 0 \\ b^t & 1 \end{vmatrix}, \tag{6.6.18}$$

the conclusion of the lemma follows from (6.6.13). □

For the lower triangular matrices T, transformation by $A \in \mathbf{T}(n)$ leads to the differential form of maximal degree

$$\bigwedge_{i\geq j}(d(TA))_{ij} = m(A)\bigwedge_{i\geq j}(dT)_{ij}, \tag{6.6.19}$$

where $m \neq 0$, $m(I_n) = 1$, and $m(AB) = m(A)m(B)$. By Lemma 6.6.3 if the diagonal of A is $1, \ldots, 1$ then $m(A) = 1$. In addition, if $A = \operatorname{diag}(a_1, \ldots, a_n)$ then direct evaluation of (6.6.19) shows

$$m(\operatorname{diag}(a_1, \ldots, a_n)) = a_1^n a_2^{n-1} \cdots a_n. \tag{6.6.20}$$

Hence,

Theorem 6.6.4. *If the functional* $m: \mathbf{T}(n) \to \mathbb{R}$ *satisfies* (6.6.12), (6.6.13) *and* (6.6.20) *then*

$$m(A) = a_{11}^n a_{22}^{n-1} \cdots a_{nn}, \tag{6.6.21}$$

where the diagonal of A is $a_{11}, \ldots, a_{nn}$.

In the computation of $\bigwedge_{i\geq j}(dS)_{ij} = \bigwedge_{i\geq j}(d(TT^t))_{ij}$ it is necessary to look at the functional m defined by, if $A \in \mathbf{T}(k)$ then $\bigwedge_{i\geq j}(d(TA^t))_{ij} = m(A)\bigwedge_{i\geq j}(dT)_{ij}$. This functional is multiplicative, polynomial, homogeneous of degree $\frac{1}{2}k(k+1)$, and $m(I_k) = 1$. If $A = \operatorname{diag}(a_1, \ldots, a_n)$ then a direct evaluation shows $m(A) = a_1^k a_2^{k-1} \cdots a_k$ so that by Theorem 6.6.4, if $A \in \mathbf{T}(k)$ then $m(A) = a_{11}^k a_{22}^{k-1} \cdots a_{kk}$.

The change of variable $S = TT^t$ leads to the differential form

$$\bigwedge_{i\geq j}(dS)_{ij} = \bigwedge_{i\geq j}(dT\,T^t + T\,dT^t)_{ij}. \tag{6.6.22}$$

If we consider the functional

$$\bigwedge_{i\geq j}(dT\,A^t + A\,dT^t)_{ij} = k(A)\bigwedge_{i>j}(dT)_{ij}, \tag{6.6.23}$$

where $k(A)$ exists because the differential form is of maximal degree, then, the substitution $T \to TB^t$ shows

$$k(AB) = k(A)m(B), \tag{6.6.24}$$

where m is given in (6.6.20). Thus $k(B) = k(I_n)m(B)$ and we evaluate $k(I_n)$. Here $\bigwedge_{i\geq j}(dT + dT^t)_{ij} = 2^n \bigwedge_{i\geq j}(dT)_{ij}$ since $(dT)_{ij} = 0$ if $i < j$. Therefore,

$$\bigwedge_{i\geq j}(dS)_{ij} = 2^n t_{11}^n t_{22}^{n-1} \cdots t_{nn} \bigwedge_{i\geq j}(dT)_{ij}. \tag{6.6.25}$$

This result is derived in a different way in Problem 6.7.6. See also Section 7.3.

The following results are concerned with quadratic expressions in the entries of a symmetric matrix S.

Lemma 6.6.5. *Let S and A be $n \times n$ symmetric matrices and define m by*

$$m(A)\bigwedge_{i\geq j}(dS)_{ij} = \bigwedge_{i\geq j}(A\,dS + dS\,A)_{ij}. \tag{6.6.26}$$

Then

$$m(A) = \prod_{i \geq j} (\lambda_i(A) + \lambda_j(A)), \tag{6.6.27}$$

where $\lambda(A)$ *is the eigenvalue function.*

PROOF. By Problem 6.7.7, if $U \in \mathbf{O}(n)$ then

$$\begin{aligned}
(\det U)^{n+1} m(A) \bigwedge_{i \geq j} (dS)_{ij} &= (\det U)^{n+1} \bigwedge_{i \geq j} (A\, dS + dS\, A)_{ij} \\
&= \bigwedge_{i \geq j} [U(A\, dS + dS\, A)U^t]_{ij} \\
&= \bigwedge_{i \geq j} [(UAU^t)d(USU^t) + d(USU^t)(UAU^t)]_{ij} \qquad (6.6.28) \\
&= m(UAU^t) \bigwedge_{i \geq j} (d(USU^t))_{ij} \\
&= (\det U)^{n+1} m(UAU^t) \bigwedge_{i \geq j} (dS)_{ij}.
\end{aligned}$$

Thus the value of m depends only on the eigenvalues of A. If $A = \operatorname{diag}(\lambda_1, \ldots, \lambda_n)$ then the ij-element of $A\, dS$ and $dS\, A$ are $\lambda_i (dS)_{ij}$ and $\lambda_j (dS)_{ij}$. Thus in case of a diagonal matrix A

$$\bigwedge_{i \geq j} (A\, dS + dS\, A)_{ij} = \prod_{i \geq j} (\lambda_i(A) + \lambda_j(A)) \bigwedge_{i \geq j} (dS)_{ij}. \tag{6.6.29}$$

Hence (6.6.27) holds. □

Corollary 6.6.6.

$$\bigwedge_{i \geq j} (dS^2)_{ij} = \prod_{i \geq j} (\lambda_i(S) + \lambda_j(S)) \bigwedge_{i \geq j} (dS)_{ij}. \tag{6.6.30}$$

Lemma 6.6.7. *Let* S *be* $n \times n$ *skew-symmetric and* A *be* $n \times n$ *symmetric and define* m *by*

$$m(A) \bigwedge_{i > j} (dS)_{ij} = \bigwedge_{i > j} (A\, dS - dS\, A)_{ij}. \tag{6.6.31}$$

Then

$$m(A) = \prod_{i > j} (\lambda_i(A) - \lambda_j(A)). \tag{6.6.32}$$

PROOF. The same argument as for Lemma 6.6.5 except that $+$ is changed to $-$. □

6.7. Problems

PROBLEM 6.7.1. If $(p_1, \ldots, p_n)$ are local coordinates of z near x and $(q_1, \ldots, q_n)$ are local coordinates of z near y, so that $z \in \mathbf{U}_x \cap \mathbf{U}_y$, then

$$fu^x_{i_1} \wedge \cdots \wedge u^x_{i_r} \quad \text{and} \quad f(dp_{i_1}) \wedge \cdots \wedge (dp_{i_r}) \tag{6.7.1}$$

are equivalent r-forms where the latter is computed at y.

PROBLEM 6.7.2. Let $u_i \wedge \cdots \wedge \underline{u_i} \wedge \cdots \wedge u_n = u_1 \wedge \cdots \wedge u_{i-1} \wedge u_{i+1} \wedge \cdots \wedge u_n$. The join (wedge product) of $n-1$ 1-forms

$$\left(\sum_{i=1}^n b_{1i}u_i\right) \wedge \cdots \wedge \left(\sum_{i=1}^n b_{(n-1)i}u_i\right) = \sum_{i=1}^n \tilde{a}_i u_1 \wedge \cdots \wedge \underline{u_i} \wedge \cdots \wedge u_n, \tag{6.7.2}$$

where the coefficients $\tilde{a}_i$ are given by the determinants, $1 \le i \le n$:

$$\tilde{a}_i = \det \begin{vmatrix} b_{11} & \cdots & b_{1\,i-1} & b_{1\,i+1} & \cdots & b_{1n} \\ \vdots & & \vdots & \vdots & & \vdots \\ b_{n-1\,1} & \cdots & b_{n-1\,i-1} & b_{n-1\,i+1} & \cdots & b_{n-1\,n} \end{vmatrix}. \tag{6.7.3}$$

PROBLEM 6.7.3. Continuation of Problem 6.7.2. Let the b_{ij} be functions of $a_1, \ldots, a_n$ such that the matrix

$$\begin{vmatrix} a_1 & a_2 & \cdots & a_n \\ b_{11} & b_{12} & & b_{1n} \\ \vdots & \vdots & & \vdots \\ b_{n-1\,1} & b_{n-1\,2} & \cdots & b_{n-1\,n} \end{vmatrix} \tag{6.7.4}$$

is an orthogonal matrix with determinant $= \varepsilon$. Compute $da_1, \ldots, da_n$ locally at p. Then

$$\begin{aligned} &\left(\sum_{i=1}^n b_{1i}\, da_i\right) \wedge \cdots \wedge \left(\sum_{i=1}^n b_{n-1\,i}\, da_i\right) \\ &\quad = \varepsilon \sum_{i=1}^n a_i(-1)^{i+1}\, da_1 \wedge \cdots \wedge \underline{da_i} \wedge \cdots \wedge da_n. \end{aligned} \tag{6.7.5}$$

PROBLEM 6.7.4. Continue Problem 6.7.3. Since $a_1^2 + \cdots + a_n^2 = 1$, we find

$$\begin{aligned} da_n &= \sum_{i=1}^{n-1} \frac{-a_i\, da_i}{(1 - a_1^2 - \cdots - a_{n-1}^2)^{1/2}} \\ &= \sum_{i=1}^{n-1} -a_i\, da_i/a_n. \end{aligned} \tag{6.7.6}$$

Substitution into (6.7.5) shows (6.7.5) to be equal to

$$\frac{(-1)^{n+1}\varepsilon\, da_1 \wedge \cdots \wedge da_{n-1}}{(1 - a_1^2 - \cdots - a_{n-1}^2)^{1/2}}. \tag{6.7.7}$$

PROBLEM 6.7.5. Let $\mathfrak{M} = \mathfrak{M}_1 = \mathfrak{M}_2 = \mathbb{R}^{nh}$ with local coordinates given globally by $n \times h$ matrices X. Let A be a $n \times h$ matrix and define $f(X) = AX$.

Let the canonical ordering of 1-forms u_{ij} be $\bigwedge_{j=1}^{h} \bigwedge_{i=1}^{n} u_{ij}$. Show a nh-form $\omega = g \bigwedge_{j=1}^{h} \bigwedge_{i=1}^{n} u_{ij}$ transforms to

$$\begin{aligned} F\omega &= (g \circ f) \bigwedge_{j=1}^{h} \bigwedge_{i=1}^{n} \left(\sum_{k=1}^{n} a_{ik} u_{kj} \right) \\ &= (\det A)^h (g \circ f) \bigwedge_{j=1}^{h} \bigwedge_{i=1}^{n} u_{ij}. \end{aligned} \tag{6.7.8}$$

PROBLEM 6.7.6.

$$\text{If} \quad T = \begin{vmatrix} t_{11} & 0 & \cdots & 0 \\ t_{21} & t_{22} & \cdots & 0 \\ \cdots & \cdots & \cdots & \cdots \\ t_{n1} & t_{n2} & \cdots & t_{nn} \end{vmatrix} \quad \text{and} \quad S = (s_{ij}) = TT^t,$$

then wanted is the Jacobian of the substitution $t_{ij} \to s_{ij}$, $1 \le j \le i$, $1 \le i \le n$. The value of the Jacobian is not a determinant and this problem is not especially tractable to exterior algebra. However

$$s_{nj} = t_{n1} t_{j1} + \cdots t_{nj} t_{jj}, \quad 1 \le j \le n, \tag{6.7.9}$$

so for these n variables the Jacobian of the substitution is

$$\begin{aligned} \frac{\partial(s_{n1}, \ldots, s_{nn})}{\partial(t_{n1}, \ldots, t_{nn})} &= \det \begin{vmatrix} t_{11} & 0 & \cdots & 0 \\ \cdots & \cdots & \cdots & \cdots \\ 2t_{n1} & 2t_{n2} & \cdots & 2t_{nn} \end{vmatrix} \\ &= 2t_{11} t_{22} \cdots t_{nn}. \end{aligned} \tag{6.7.10}$$

By induction show that the required Jacobian is

$$2^n t_{11}^n t_{22}^{n-1} \cdots t_{nn}. \tag{6.7.11}$$

PROBLEM 6.7.7. Let S be a $n \times n$ symmetric matrix and let $A \in \mathbf{GL}(n)$. Set $T = ASA^t$. If the entries of S are differential forms then so are the entries of T. Show

$$\bigwedge_{j \le i} t_{ij} = m(A) \bigwedge_{j \le i} s_{ij}, \tag{6.7.12}$$

where $m(A)$ is a homogeneous polynomial of degree $n(n + 1)$ in the variables (a_{ij}). Show that m must satisfy $m(AB) = m(A)m(B)$ for all A, $B \in \mathbf{GL}(n)$ so that by Lemma 6.6.1, $m(A) = (\det A)^{n+1}$.

PROBLEM 6.7.8. Let S be a $n \times n$ symmetric positive definite matrix and $T = S(I_n + S)^{-1}$. Show that T is a symmetric matrix. Compute

$$dT(I_n + S) = dS \tag{6.7.13}$$

and show that $I_n = (I_n - T)(I_n + S)$. Then

$$dS = (I_n - T)^{-1} \, dT(I_n - T)^{-1}. \tag{6.7.14}$$

By Problem 6.7.7 show

$$\bigwedge_{j\le i} ds_{ij} = (\det(I_n - T))^{-(n+1)} \bigwedge_{j\le i} dt_{ij}. \tag{6.7.15}$$

PROBLEM 6.7.9. Continue Problems 6.7.7 and 6.7.8. Let A, $T \in \mathbf{T}(n)$ be nonsingular lower triangular matrices and let $S = AT$. Show the Jacobian of the substitution $t_{ij} \to (AT)_{ij} = s_{ij}$ is

$$a_{11}a_{22}^2 \cdots a_{nn}^n. \tag{6.7.16}$$

Also, note that

$$\bigwedge_{j\le i} ds_{ij} = m(A) \bigwedge_{j\le i} dt_{ij} \tag{6.7.17}$$

where $m(A) = a_{11}a_{22}^2 \cdots a_{nn}^n$. Thus m is a homogeneous polynomial in the entries of A such that $m(A)m(B) = m(AB)$ but $m(A) \neq (\det A)^r$. Thus Lemma 6.6.1 fails if the function m is defined on a proper subgroup of $\mathbf{GL}(n)$.

PROBLEM 6.7.10. Let $f\colon \mathfrak{M}_1 \to \mathfrak{M}_2$ be a transformation of manifolds with induced mapping F of differential forms. Show

$$F(a \wedge b) = (Fa) \wedge (Fb). \tag{6.7.18}$$

HINT. Since F is a linear transformation it is sufficient to take $a = g_1 v_{i_1}^y \wedge \cdots \wedge v_{i_r}^y$ and $b = g_2 v_{j_1}^y \wedge \cdots \wedge v_{j_s}^y$. □

PROBLEM 6.7.11. Let $\mathfrak{M}_1 = \mathfrak{M}_2 = \mathbf{S}(n)$ and the transformation of manifolds f be $f(S) = S^{-1}$. Compute the following transformed differential form:

$$F\left(\bigwedge_{j\le i} (dS)_{ij}/(\det S)^{(n+1)/2}\right). \tag{6.7.19}$$

Compare your answer with Problem 7.10.6.

CHAPTER 7

Invariant Measures on Manifolds

7.0. Introduction

In this chapter we discuss the action of matrix groups on various manifolds. Mostly conclusions will not be stated as formal theorems except in the last sections of the chapter.

The different manifolds are described below as we treat them. It is our purpose to derive differential forms for regular invariant measures. As suggested in Section 3.4 it is the regular invariant measures which enter into the factorization of measures, and in many examples discussed in Chapter 9 integration out of extra variables is equivalent to integration of the differential form over the entire manifold. The various differential forms introduced in Chapter 3, namely (3.3.16), (3.4.9), (3.4.14), Example 3.5.5, and Section 3.6, are justifiable on the basis of results contained in Chapters 6 and 7.

7.1. $\mathbb{R}^{nh}$

We consider the set of $n \times h$ matrices X and let the group action be multiplication on the right by $h \times h$ matrices $A \in \mathbf{GL}(h)$. If $Y = XA$ then

$$\bigwedge_{j=1}^{h} \bigwedge_{i=1}^{n} dy_{ij} = m(A) \bigwedge_{j=1}^{h} \bigwedge_{i=1}^{n} dx_{ij} \tag{7.1.1}$$

where we write $Y = (y_{ij})$, $X = (x_{ij})$ and use the fact that the differential forms are of maximal degree so that the space of alternating forms of degree nh has dimension equal one, determining the constant $m(A)$. The function m then satisfies the hypotheses of Lemma 6.6.1. Therefore

$$m(A) = (\det A)^{nh/h} = (\det A)^n. \tag{7.1.2}$$

In terms of globally defined canonical basis elements u_{ij} for the 1-forms, the differential form

$$\omega = (\det X^t X)^{-n/2} \bigwedge_{j=1}^{h} \bigwedge_{i=1}^{n} u_{ij} \tag{7.1.3}$$

is an invariant form. The transformation $X \to XA = f(X)$ replaces $(u_{i1}, \ldots, u_{1h})$ by $(u_{i1}, \ldots, u_{ih})\, A$ and by (6.2.19) we obtain

$$\left(\sum_{k=1}^{h} u_{ik} a_{k1}\right) \wedge \cdots \wedge \left(\sum_{k=1}^{h} u_{ik} a_{kh}\right) = (\det A) u_{i1} \wedge \cdots \wedge u_{ih}. \tag{7.1.4}$$

From (7.1.3) and (7.1.4) and Section 6.5 the transformed differential form is

$$F\omega = (\det (XA)^t (XA))^{-n/2} (\det A)^n \bigwedge_{j=1}^{h} \bigwedge_{i=1}^{n} u_{ij} = \omega. \tag{7.1.5}$$

By Theorem 6.5.6 the differential form ω defines an invariant measure on the Borel subsets of $\mathbb{R}^{nh}$. The more usual way, of course, of writing the density function of this measure, as used in Chapter 5, is

$$\prod_{j=1}^{h} \prod_{i=1}^{n} dx_{ij} / (\det X^t X)^{n/2}. \tag{7.1.6}$$

The measures given by integration of differential forms are absolutely continuous relative to Lebesgue measure since the integral is defined locally by homeomorphisms with Euclidean space. Hence the measures are given by density functions integrated by Lebesgue measure. Derivation of the density (7.1.6) does not require use of differential forms. Existence of the multiplicative function in (7.1.2) was shown in Section 5.0 using the uniqueness of Haar measures for the additive group of $\mathbb{R}^{nh}$. See Theorem 5.0.1 and the discussion which precedes it, Theorem 2.0.1 and its proof, Example 3.4.2, and Example 3.4.3. In these examples, proofs and discussions, differential forms or the uniqueness theorems are used to derive density functions of invariant measures. As is explained in Section 3.6 a multiplicative functional necessarily arises due to the fact that the alternating forms of maximal degree are a one-dimensional space.

7.2. Lower Triangular Matrices, Left and Right Multiplication

Left multiplication and right multiplication are different problems, leading to different differential forms, due to the fact that the group $\mathbf{T}(h)$ is not unimodular. See Section 3.5.

We consider transformations $T \to AT$, with $A, T \in \mathbf{T}(h)$. From Problem 6.7.9 with $S = AT$ we find

$$\bigwedge_{j\le i} ds_{ij} = a_{11}a_{22}^2 \cdots a_{hh}^h \bigwedge_{j\le i} dt_{ij}. \tag{7.2.1}$$

$\mathbf{T}(h)$ is an open simply connected subset of $\mathbb{R}^{h(h+1)/2}$ so there exists a one-to-one C_∞ mapping which maps $\mathbf{T}(h)$ onto $\mathbb{R}^{h(h+1)/2}$ and this gives a global definition of coordinates, i.e., only one chart function is needed. Consequently in terms of globally defined 1-forms u_{ij} the differential form

$$\omega = \bigwedge_{j\le i} u_{ij}/(t_{11}t_{22}^2 \cdots t_{hh}^h) \tag{7.2.2}$$

is an invariant differential form. To show this, set $u_{ij} = dt_{ij}$ and obtain

$$F\omega = (a_{11} \cdots a_{hh}^h) \bigwedge_{j\le i} dt_{ij}/(a_{11}t_{11}) \cdots (a_{hh}t_{hh})^h = \omega. \tag{7.2.3}$$

An alternative way of obtaining these results is presented in Section 6.6, Lemma 6.6.3, and in connection with a discussion of modular functions, Section 3.5, Theorem 3.5.6 and following. In these discussions a multiplicative functional m is defined by

$$\bigwedge_{j\le i} (AT)_{ij} = m(A) \bigwedge_{j\le i} (dT)_{ij}. \tag{7.2.4}$$

The discussion preceding Remark 3.5.7 shows that this functional is

$$m(A) = a_{11}a_{22}^2 \cdots a_{hh}^h. \tag{7.2.5}$$

Multiplication on the right, $S = TA$, results in

$$\bigwedge_{j\le i} (dS)_{ij} = a_{11}^h a_{22}^{h-1} \cdots a_{hh} \bigwedge_{j\le i} (dT)_{ij}. \tag{7.2.6}$$

Therefore the differential form (see (7.2.2))

$$\omega = \bigwedge_{j\le i} (dT)_{ij}/(t_{11}^h t_{22}^{h-1} \cdots t_{hh}) \tag{7.2.7}$$

when integrated gives a right invariant measure. Use of the multiplicative functional

$$\bigwedge_{j\le i} (dTA)_{ij} = m(A) \bigwedge_{j\le i} (dT)_{ij} \tag{7.2.8}$$

is discussed in Theorem 3.5.6, the proof of which depends on Lemma 6.6.3 and Theorem 6.6.4.

The idea of using multiplicative functionals leads to an elementary derivation of the modular function for the right invariant measures on $\mathbf{T}(h)$. See Theorem 3.5.10.

7.3. $\mathbf{S}(h)$

If one views $S \in \mathbf{S}(h)$ as being in $\mathbb{R}^{h(h+1)/2}$ by taking the entries $(S)_{ij}$ with $j \le i$ then it follows from Theorem 8.1.2 and Theorem 8.1.3 that $\mathbf{S}(h)$ is an open subset of $\mathbb{R}^{h(h+1)/2}$. Further since $\Sigma_{j\le i}(T)_{ij}^2 = \operatorname{tr} TT^t = \operatorname{tr} S$ it follows that the map $T \to TT^t = S$ is a homeomorphism that is $\mathbf{C}_\infty$. Consequently **S**(h) is simply connected and a single chart function is sufficient.

Consideration of Lebesgue measure and the additive group on $\mathbb{R}^{h(h+1)/2}$ yields the density function for an invariant measure as being

$$(\det S)^{-(h+1)/2} \prod_{j \le i} (dS)_{ij}. \tag{7.3.1}$$

See Theorem 5.0.1 and the discussion which precedes it.

In the factorization of the $n \times h$ matrix $X = US^{1/2}$ it follows that $S = X^t X$ and the h-frame $U = X(X^t X)^{-1/2}$. This factorization assumes X has rank h and that $n \ge h$. The set $\mathbf{N} = \{X | \operatorname{rank} X < h\}$ has Lebesgue measure zero. See Problems 4.4.4 and 4.4.5. On $\mathbb{R}^{nh} - \mathbf{N}$ Lebesgue measure factors to Haar measure of unit mass on the homogeneous space of h-frames times a measure on the symmetric matrices which is not invariant. See Example 3.3.9 and Example 3.4.4 where this question is discussed, and Theorem 5.2.2 whose statement shows how to establish invariance.

As a subset of $\mathbb{R}^{h(h+1)/2}$ calculation of an integral $\int_{\mathbf{S}(h)} f(S) \prod_{j \le i} (dS)_{ij}$ would require expression of this integral as an iterated integral, which does not seem to be tractable. The calculation is invariably done by the change of variable $S = TT^t$ with $T \in \mathbf{T}(h)$. This substitution is examined as a direct Jacobian calculation in Problem 6.7.6. As a problem of manipulating differential forms see the discussion preceding Theorem 6.6.5, especially the lines (6.6.22) to (6.6.25). Use of (6.6.25) together with univariate integrations that reduce evaluations of integrals to gamma functions, yield most, if not all, the constants of normalization and constants of proportionality used in this book.

In the spirit of Chapter 5, the group **T**(h) acts transitively on **S**(h) by the action $S \to TST^t$. The subgroup that leaves I_h fixed is $\{I_h\}$. This implies that the measures on **S**(h) invariant for this group are a one-dimensional set, see Theorem 3.3.8, and since (7.3.1) gives an invariant measure, the measures which are invariant for **T**(h) are already invariant for **GL**(h) and are proportional to the measure that results from integration of (7.3.1). Since

$$\int_{\mathbf{S}(h)} (\det S)^{-(h+1)/2} f(S) \prod_{j \le i} (dS)_{ij} \tag{7.3.2}$$

and

$$\int_{\mathbf{T}(h)} t_{11} t_{22}^2 \cdots t_{hh}^h f(TT^t) \prod_{j \le i} (dT)_{ij} \tag{7.3.3}$$

both define invariant linear functionals of integrable f, it follows from the uniqueness theorems, as discussed in Section 5.0, that

$$c t_{11} t_{22}^2 \cdots t_{hh}^h \prod_{j \le i} (dT)_{ij} = (\det S)^{-(h+1)/2} \prod_{j \le i} (d)_{ij}. \tag{7.3.4}$$

That $c = 2^h$ is computable using Jacobians or differential forms, but does not seem to be obtainable from the uniqueness theorem for Haar measure. Hence it may be that one multivariate Jacobian calculation is unavoidable in the subject to calculation of multivariate density functions.

7.4. The Orthogonal Group $\mathbf{O}(n)$

We let α be a generic point of $\mathbf{O}(n)$ and $a_{ij}(\alpha)$ be the (i,j)-entry function. Since $\mathbf{O}(n)$ is compact local coordinates cannot be globally defined (the continuous image of a compact space is a compact subset of Euclidean space). If we map $\mathbf{O}(n)$ into $\mathbb{R}^{n(n-1)/2}$ using the functions $\{a_{ij}, 1 \le j < i \le n\}$ this map is one-to-one into and hence is a homeomorphism into $\mathbb{R}^{n(n-1)/2}$. Call this mapping f. Clearly there exists an interior point $f(\alpha_0) \subset f(\mathbf{O}(n))$ and we may choose an open rectangle $\mathbf{U} \subset f(\mathbf{O}(n))$ with $f(\alpha_0) \in \mathbf{U}$. Use a C_∞ map of $\mathbf{U}$ onto $\mathbb{R}^{n(n-1)/2}$ to obtain a chart function of α_0 over the neighborhood $\mathbf{U}_{\alpha_0} = f^{-1}(\mathbf{U})$. For the arbitrary point of $\mathbf{O}(n)$ use translations $\mathbf{U}_\alpha = \alpha\alpha_0^{-1}(\mathbf{U}_{\alpha_0})$ and make the obvious definition of a chart. This makes $\mathbf{O}(n)$ into a C_∞ manifold.

The question of local coordinates is not treated in more detail here. We suppose it has been shown that the functions a_{ij} are C_∞ functions of the local coordinates. The differential form of interest to us is

$$\omega = \bigwedge_{j \le i} a_i^t \, da_j \tag{7.4.1}$$

where the (column) vectors $a_1, \ldots, a_n$ are the columns of A so that

$$A = (a_1, \ldots, a_n) \in \mathbf{O}(n). \tag{7.4.2}$$

The functions $a_1, \ldots, a_n$ are globally defined so that ω is a globally defined n-form. See Section 6.4.

We begin our calculations by noting that

$$I_n = A^t A \quad \text{and} \quad 0 = dI_n = (dA)^t A + A^t(dA). \tag{7.4.3}$$

Therefore $A^t\,dA$ is a skew-symmetric matrix.

Let $H \in \mathbf{O}(n)$ and define a transformation $f: \mathbf{O}(n) \to \mathbf{O}(n)$ by $A(f(\alpha)) = HA(\alpha)$. Let F be the induced map of differential forms. By Theorem 6.5.2, $dF = Fd$ (when computed locally) so that

$$H\,dA = d(HA) = d(F(A)) = F(dA). \tag{7.4.4}$$

Therefore

$$(FA)^t(d(F(A))) = A^t H^t H\,dA = A^t\,dA. \tag{7.4.5}$$

By Definition 6.5.5 and Theorem 6.5.6 the differential form ω of (7.4.1) defines a left invariant measure on $\mathbf{O}(n)$ which by the construction of these measures locally must be of finite mass on $\mathbf{O}(n)$. The measure is therefore regular, is a Haar measure, and by Chapter 3, is uniquely determined by its total mass and must also be a right invariant measure since $\mathbf{O}(n)$ is unimodular, see Lemma 3.5.3.

There is a question whether the signed measure so determined can change sign. The positive and negative parts of an invariant measure are easily seen to be invariant, and must be regular measures, hence Haar measures.

Consequently the difference of the positive and negative parts is a Haar measure or is the negative of a Haar measure.

As an exercise we verify the right invariance. Changing our definition, define f by $A(f(\alpha)) = A(\alpha)H$ and let F be the induced map. Then $FA = AH$ and

$$d(FA) = F(dA) = (dA)H, \tag{7.4.6}$$

and

$$(FA)^t(dFA) = H^tA^t\,dA\,H. \tag{7.4.7}$$

Write $H = (h_{ij})$ so that the (i,j)-entry of (7.4.7) is

$$\sum_{k_1=1}^{n}\sum_{k_2=1}^{n} h_{k_1i}(a_{k_1}^t\,da_{k_2})h_{k_2j}. \tag{7.4.8}$$

Using the skew symmetry property $a_i^t\,da_j + a_j^t\,da_i = 0$, from (7.4.8) we obtain

$$\bigwedge_{j\le i}(H^tA^t\,dA\,H)_{ij} = m(H)\bigwedge_{j\le i} a_i^t\,da_j, \tag{7.4.9}$$

where the function m is a polynomial homogeneous of degree $n(n-1)$. (7.4.9) in fact holds for $H \in \mathbf{GL}(n)$ rather than just for $H \in \mathbf{O}(n)$. Using successive substitutions we see that $m(H_1H_2) = m(H_1)m(H_2)$ so that $m(H) = (\det H)^{n-1}$, by Lemma 6.6.1. If H is an orthogonal matrix then $m(H) = \pm 1$. Therefore if $n-1$ is odd and $\det H = -1$, right multiplication by H changes orientation and changes the sign of the differential form. The absolute value of the differential form is unchanged.

The original source of this material is James (1954), Theorems 4.2 and 4.3.

7.5. Grassman Manifolds $\mathbf{G}_{k,n-k}$

The Grassman manifold $\mathbf{G}_{k,n-k}$ is defined to be the set of all k-dimensional hyperplanes in $\mathbb{R}^n$ containing 0. No globally defined system of local coordinates suffice to describe the manifold. In this book we assume local coordinates can be defined but we never explicitly use them.

Given a k-plane $\mathbf{P}$ let $a_1, \ldots, a_k$ be an orthonormal basis of $\mathbf{P}$, $a_1, \ldots, a_k \in \mathbb{R}^n$. Below we will construct $b_1, \ldots, b_{n-k} \in \mathbb{R}^n$ which are analytic functions of $a_1, \ldots, a_k$ such that the $n \times n$ matrix $a_1, \ldots, a_k, b_1, \ldots, b_{n-k}$ is in $\mathbf{O}(n)$. We consider the differential form

$$\omega = \bigwedge_{j=1}^{n-k}\bigwedge_{i=1}^{k} b_j^t\,da_i. \tag{7.5.1}$$

We will show this differential form locally about $\mathbf{P}$ is independent of the choices of $a_1, \ldots, a_k, b_1, \ldots, b_{n-k}$ and that the differential form is globally defined. The measure

$$\int_{\mathbf{C}} |\omega| \tag{7.5.2}$$

is a left invariant measure under the action of $\mathbf{O}(n)$ on $\mathbf{G}_{k,n-k}$.

Let $\mathbf{P}_0$ be a k-plane and let the $n \times k$ matrix X give local coordinates of $\mathbf{P}$ near $\mathbf{P}_0$ with the (i,j)-entry x_{ij} of X analytic in $\mathbf{P}$. For ease of discussion we suppose $X_0^t = (X_{01}^t, X_{02}^t)$ represents $\mathbf{P}_0$ where X_{01} is a $k \times k$ nonsingular matrix. Similarly we write $X^t = (X_1^t, X_2^t)$ with X_1 a $k \times k$ nonsingular matrix. Then X_1^{-1} near $\mathbf{P}_0$ is an analytic function of the local coordinates.

From X we pass analytically to $\begin{vmatrix} I_k \\ X_2 X_1^{-1} \end{vmatrix}$ and choose

$$Y = \begin{vmatrix} Y_1 \\ I_{n-k} \end{vmatrix}, \qquad Y_1 = -(X_1^{-1})^t X_2^t, \tag{7.5.3}$$

with Y_1 a $k \times (n-k)$ matrix. This defines Y and Y_1 analytically in terms of X and

$$Y^t \begin{vmatrix} I_k \\ X_2 X_1^{-1} \end{vmatrix} = 0. \tag{7.5.4}$$

We take as bases of $\mathbf{P}$ and $\mathbf{P}^\perp$ the columns of $\begin{vmatrix} I_k \\ X_2 X_1^{-1} \end{vmatrix}$ and Y. Apply the Gram–Schmidt process to each matrix. The operations involved are rational or are the taking of square roots, and since lengths are bounded away from zero, the operations will be analytic. The result of the Gram–Schmidt process is an orthonormal set of vectors.

Therefore we may suppose

$$\begin{aligned} &A \text{ is } n \times k,\ B \text{ is } n \times (n-k), \text{ and } (A, B) \in \mathbf{O}(n);\\ &\text{the columns of } A \text{ are a basis of } \mathbf{P};\\ &\text{the columns of } A, B \text{ are analytic functions of the}\\ &\text{local coordinates at } \mathbf{P}_0. \end{aligned} \tag{7.5.5}$$

Suppose $\tilde{A}$, $\tilde{B}$ is a second such representation. We show the differential form of (7.5.1) is unchanged. There are $H_1 \in \mathbf{O}(k)$ and $H_2 \in \mathbf{O}(n-k)$, H_1 and H_2 uniquely determined, such that

$$\tilde{A} = AH_1 \quad \text{and} \quad \tilde{B} = BH_2. \tag{7.5.6}$$

That is

$$H_1 = A^t\tilde{A} \quad \text{and} \quad H_2 = B^t\tilde{B}. \tag{7.5.7}$$

It is easy to check that H_1 and H_2 so defined are orthogonal matrices that satisfy (7.5.6) since AA^t and $\tilde{A}\tilde{A}^t$, BB^t and $\tilde{B}\tilde{B}^t$ are orthogonal projections of respective ranges $\mathbf{P}$ and $\mathbf{P}^\perp$. It follows that the entries of H_1 and H_2 are analytic functions of the local coordinates at $\mathbf{P}_0$. Therefore, passing from A, B to $\tilde{A}, B$

$$d\tilde{A} = dA\,H_1 + A\,dH_1 \quad \text{and} \quad B^t\,d\tilde{A} = B^t\,dA\,H_1. \tag{7.5.8}$$

Then

$$\bigwedge_{i=1}^{k} b_j^t\,d\tilde{a}_i = \bigwedge_{i=1}^{k} (B^t\,dA)_j H_i = \left(\bigwedge_{i=1}^{k} b_j^t\,da_i\right)(\det H_1). \tag{7.5.9}$$

In this notation H_i is the i-th column of H_1 and $(B^t\,dA)_j$ is the j-th row of $B^t\,dA$. By (6.2.19) the last step of (7.5.9) follows. Then

$$\bigwedge_{j=1}^{n-k}\bigwedge_{i=1}^{k} b_j^t\,da_i = (\det H_1)^{n-k} \bigwedge_{j=1}^{n-k}\bigwedge_{i=1}^{k} b_j^t\,d\tilde{a}_i. \tag{7.5.10}$$

Passing from $\tilde{A}$, B to $\tilde{A}$, $\tilde{B}$, using (7.5.6),

$$\tilde{B}^t\,d\tilde{A} = H_2^t B^t\,d\tilde{A}. \tag{7.5.11}$$

The columns of BH_2 are linear combinations of the columns of B. Then

$$\begin{aligned}\bigwedge_{j=1}^{n-k} \tilde{b}_j^t\,d\tilde{a}_i &= \bigwedge_{j=1}^{n-k} (H_2^t B^t)_j\,d\tilde{a}_i \\ &= (\det H_2) \bigwedge_{j=1}^{n-k} b_j^t\,d\tilde{a}_i.\end{aligned} \tag{7.5.12}$$

Therefore

$$\begin{aligned}\bigwedge_{j=1}^{n-k}\bigwedge_{i=1}^{k} \tilde{b}_j^t\,d\tilde{a}_i &= (\det H_2)^k \bigwedge_{j=1}^{n-k}\bigwedge_{i=1}^{k} b_j^t\,d\tilde{a}_i \\ &= (\det H_1)^{n-k}(\det H_2)^k \bigwedge_{j=1}^{n-k}\bigwedge_{i=1}^{k} b_j^t\,da_i.\end{aligned} \tag{7.5.13}$$

If we suppose the orientation of $\mathbf{P}$ and $\mathbf{P}^\perp$ is preserved in all representations (A, B) then $\det H_1 = \det H_2 = 1$.

We consider ω on the overlap of charts locally at $\mathbf{P}_0$ and charts locally at $\mathbf{Q}_0$. If $A_{\mathbf{P}_0}$, $B_{\mathbf{P}_0}$ represent $\mathbf{P}$ near $\mathbf{P}_0$ and $A_{\mathbf{Q}_0}$, $B_{\mathbf{Q}_0}$ represent $\mathbf{P}$ near $\mathbf{Q}_0$, then as in (7.5.6),

$$A_{\mathbf{P}_0} = A_{\mathbf{Q}_0} H_3 \quad \text{and} \quad B_{\mathbf{P}_0} = B_{\mathbf{Q}_0} H_4. \tag{7.5.14}$$

The uniqueness of H_3 and H_4 makes it possible to consider these matrices as analytic functions of the local coordinates. Repetition of the above calculations shows ω to be independent of the use of $A_{\mathbf{P}_0}$, $B_{\mathbf{P}_0}$ or $A_{\mathbf{Q}_0}$, $B_{\mathbf{Q}_0}$.

We now consider the action of $\mathbf{O}(n)$ on $\mathbf{G}_{k,n-k}$ by multiplication on the left. Let $\mathbf{P}$ locally near $\mathbf{P}_0$ be represented by (A, B). Take $H \in \mathbf{O}(n)$ and let $\mathbf{Q}_0$ be the image of $\mathbf{P}_0$ under H. The point $\mathbf{P}_0$ itself is represented by (A_0, B_0), the value of (A, B) at $\mathbf{P} = \mathbf{P}_0$, so $\mathbf{Q}_0$ can be represented by (HA_0, HB_0). The plane $\mathbf{P}$ maps to a plane $\mathbf{Q}$ which has columns of HA as basis and $\mathbf{Q}^\perp$ has the columns of HB as basis. The sequence

$$HA_0 \to A_0 \to A \to HA \tag{7.5.15}$$

gives analytic maps whereby the entries of HA and HB can be described analytically in terms of the local coordinates at $\mathbf{Q}_0$.

Let $f(\mathbf{P})$ be the image of $\mathbf{P}$ under the action of H and let F be the induced mapping of differential forms. Then $\mathbf{Q} = f(\mathbf{P})$ and we write $\tilde{A}$ for the matrix representing $\mathbf{Q} = f(\mathbf{P})$, with entries $\tilde{a}_{ij}$. Then

$$F(\tilde{a}_{ij})(\mathbf{P}) = \tilde{a}_{ij}(f(\mathbf{P})) \tag{7.5.16}$$

is the (i,j)-entry of HA. Similarly

$$F(\tilde{b}_{ij})(\mathbf{P}) = \tilde{b}_{ij}(f(\mathbf{P})) \tag{7.5.17}$$

is the (i,j)-entry of HB. Thus

$$B^t dA = (HB)^t d(HA) = (F(\tilde{b}_{ij}))^t (dF(\tilde{a}_{ij})) = F(\tilde{B}^t d\tilde{A}). \tag{7.5.18}$$

We use (6.7.18) here together with (6.5.6) to evaluate the join of functions (7.5.18). That is, ω at $\mathbf{P}_0$ is the image of ω at $\mathbf{Q}_0$ under F. By Theorem 6.5.6 the measure $\int_{\mathbf{C}} |\omega|$ is a (left) invariant measure under the action of $\mathbf{O}(n)$ on $\mathbf{G}_{k,n-k}$.

This discussion shows $\mathbf{G}_{k,n-k}$ to be $\mathbf{O}(n)$ factored by the compact subgroup $\mathbf{O}(k) \times \mathbf{O}(n-k)$ and the action of $\mathbf{O}(n)$ on $\mathbf{G}_{k,n-k}$ is multiplication of cosets on the left. By Theorem 3.3.8 the regular invariant measures are uniquely determined up to a multiplicative constant.

7.6. Stiefel Manifolds $\mathbf{V}_{k,n}$

We suppose A is a $n \times k$ matrix such that $A^t A = I_k$. The column of A determines a k-frame and A uniquely represents a point α of $\mathbf{V}_{k,n}$. Then the (i,j)-entry of A, say a_{ij}, is a function of the points $\alpha \in \mathbf{V}_{k,n}$, and this function when evaluated is $a_{ij}(\alpha)$. The functions a_{ij} are globally defined but cannot be used as local coordinates since the ranges $a_{ij}(\mathbf{V}_{k,n})$ are compact sets of real numbers. As in the case of $\mathbf{O}(n)$ the local coordinates do not explicitly enter the calculations. So we do not consider the construction of local coordinates.

The group action considered is $\mathbf{O}(n)$ acting on the left. We show that if $A \in \mathbf{V}_{k,n}$, if $(A, B) \in \mathbf{O}(n)$, then

$$\omega = \bigwedge_{j=1}^{n-k} \bigwedge_{i=1}^{k} b_j^t \, da_i \bigwedge_{i<j} a_j^t \, da_i \tag{7.6.1}$$

is the differential form of an invariant measure, where B depends analytically on A. To do this, the entire discussion for Grassman manifolds can be carried over here to show that

(i) the value of ω is independent of the choice of B except for sign changes due to change of orientation;
(ii) on the overlaps $\mathbf{U}_{\alpha_1} \cap \mathbf{U}_{\alpha_2}$ the differential form ω is self-equivalent;

(iii) left invariance follows at once for $\bigwedge_{i<j} a_j^t da_i$ as explained in Section 3.6, while left invariance of $\bigwedge_{j=1}^{n-k} \bigwedge_{i=1}^{k} b_j^t da_i$ is the left invariance established for Grassman manifolds in Section 7.5.

7.7. Total Mass on the Stiefel Manifold, $k = 1$

The Stiefel manifold may be thought of as a quotient space $\mathbf{O}(n)/\mathbf{O}(n-k)$ since if $(A, B_1) \in \mathbf{O}(n)$ and if $(A, B_2) \in \mathbf{O}(n)$ then $\begin{vmatrix} I_k & A^t B_1 \\ B_2^t A & B_2^t B_1 \end{vmatrix} \in \mathbf{O}(n)$ which implies $A^t B_1 = 0$ and $B_2^t A = 0$. Thus $B_2^t B_1 \in \mathbf{O}(n-k)$. By the uniqueness Theorem 3.3.8, the family of regular invariant measures form a 1-dimensional space, i.e., invariant measures on the quotient space $\mathbf{O}(n)/\mathbf{O}(n-k)$.

In integration of $\omega = \bigwedge_{j=1}^{n-k} b_j^t da$, i.e., the case $k = 1$, we note that $b_1, \ldots, b_{n-1}$ are defined as functions of a which is a function of local coordinates near a given point. But the differential form ω is globally defined and is independent of the choice of $a, b_1, \ldots, b_{n-1}$. It follows from Chapter 6 that if we write $a^t = (a_1, \ldots, a_n)$ and $a_n = (1 - a_1^2 - \cdots - a_{n-1}^2)^{1/2}$ and

$$\omega = f(a_1, \ldots, a_{n-1})\, da_1 \wedge \cdots \wedge da_{n-1} \tag{7.7.1}$$

then

$$\omega = \int f(a_1, \ldots, a_{n-1})\, da_1 \cdots da_{n-1}. \tag{7.7.2}$$

See (6.4.8). The elimination of extra variables in ω follows from Problems 6.7.2, 6.7.3 and 6.7.4, which show that

$$\omega = (-1)^{n+1} \varepsilon (1 - a_1^2 - \cdots - a_{n-1}^2)^{-1/2}\, da_1 \wedge \cdots \wedge da_{n-1} \tag{7.7.3}$$

where $\varepsilon = \pm 1$. Consequently allowing for $\varepsilon = 1$, i.e., $a_n > 0$, and $\varepsilon = -1$, i.e., $a_n < 0$,

$$\int |\omega| = 2 \int \cdots \int_{x_1^2 + \cdots + x_{n-1}^2 < 1} \frac{dx_1 \cdots dx_{n-1}}{(1 - x_1^2 - \cdots - x_{n-1}^2)^{1/2}}. \tag{7.7.4}$$

Make the change of variable

$$x_{n-1} = y(1 - x_1^2 - \cdots - x_{n-2}^2)^{1/2} \tag{7.7.5}$$

to obtain

$$\begin{aligned} \int |\omega| &= 2 \int_{-1}^{1} (1 - y^2)^{-1/2}\, dy \int \cdots \int_{x_1^2 + \cdots + x_{n-2}^2 < 1} dx_1 \cdots dx_{n-2} \\ &= 2\pi \text{ volume } ((n-2)\text{-ball}) = 2\pi^{n/2}/\Gamma(\tfrac{1}{2}n). \end{aligned} \tag{7.7.6}$$

See Problem 4.4.3. As a notation used in subsequent calculations we define

$$A_n = \text{area of the unit shell in } \mathbb{R}^n \\ = 2\pi^{n/2}/\Gamma(\tfrac{1}{2}n) = \int |\omega|. \tag{7.7.7}$$

7.8. Mass on the Stiefel Manifold, General Case

Wanted is $\int |\omega|$, where ω is given in (7.6.1). Write

$$\omega = \varepsilon_1\left(\bigwedge_{j=1}^{n-k}\bigwedge_{i=1}^{k-1} b_j^t\,da_i \bigwedge_{j<i\le k-1} a_j^t\,da_i\right)\bigwedge_{j=1}^{n-k} b_j^t\,da_k \bigwedge_{i=1}^{k-1} a_i^t\,da_k. \tag{7.8.1}$$

$\varepsilon_i = \pm 1$ is a sign introduced by the transpositions made. Except for a possible change of sign, the last part of (7.8.1) is the differential form of the case $k = 1$ treated in Section 7.7, which depends only on a_k. The remaining variables $b_1, \ldots, b_{n-k}, a_1, \ldots, a_{k-1}$ are in $a_k^\perp$ of dimension $n-1$. We fix a_k and integrate out the remaining variables as follows, to obtain a recursion relation.

Let $H \in \mathbf{O}(n)$ have a_k^t as the n-th row and transform variables to

$$(Hb_1, \ldots, Hb_{n-k}, Ha_1, \ldots, Ha_{k-1}) = (\tilde{b}_1, \ldots, \tilde{b}_{n-k}, \tilde{a}_1, \ldots, \tilde{a}_{k-1}). \tag{7.8.2}$$

Each of these vectors is zero in the n-th position and therefore each of these vectors may be considered as being in $\mathbb{R}^{n-1}$. The differential form

$$\bigwedge_{j=1}^{n-k}\bigwedge_{i=1}^{k-1} \tilde{b}_j^t\,d\tilde{a}_i \bigwedge_{j<i\le k-1} \tilde{a}_j^t\,d\tilde{a}_i = F\left(\bigwedge_{j=1}^{n-k}\bigwedge_{i=1}^{k-1} b_j^t\,da_i \bigwedge_{j<i\le k-1} a_j^t\,da_i\right) \tag{7.8.3}$$

is invariant under the transformation of manifolds and is the differential form for the invariant measure on $\mathbf{V}_{k-1,n-1}$. Therefore

$$(\text{mass of } \mathbf{V}_{k-1,n-1})2\pi^{n/2}/\Gamma(\tfrac{1}{2}n) = \text{mass } \mathbf{V}_{k,n}. \tag{7.8.4}$$

Induction using (7.7.7) and (7.8.4) gives the following result.

Theorem 7.8.1. *If ω is given in* (7.6.1) *then*

$$\text{mass } \mathbf{V}_{k,n} = \int_{\mathbf{V}_{k,n}} |\omega| = A_n A_{n-1} \cdots A_{n-k+1}, \tag{7.8.5}$$

where $A_i = 2\pi^{i/2}/\Gamma(\frac{1}{2}i)$, $1 \le i \le n$. In particular, if $k = n$, and $\mathbf{V}_{k,n} = \mathbf{O}(n)$, then

$$\int_{\mathbf{O}(n)} |\omega| = A_n \cdots A_1. \tag{7.8.6}$$

In formula (7.8.6) *$A_1 = 2$ is the area of the unit shell in $\mathbb{R} = \mathbb{R}^1$.*

Compare this result with (5.2.3), not the same but similar. The source for the section, James (1954).

7.9. Total Mass on the Grassman Manifold $\mathbf{G}_{k,n-k}$

The group action considered is $\mathbf{O}(n)$ and in this section $\mathbf{G}_{k,n-k}$ may be thought of as the quotient space

$$\mathbf{O}(n)/\mathbf{O}(k) \times \mathbf{O}(n-k). \tag{7.9.1}$$

Then Haar measure on $\mathbf{O}(n)$ induces an invariant measure of finite total mass on $\mathbf{G}_{k,n-k}$ which is uniquely determined up to a multiplicative constant. See Theorem 3.3.8.

$\mathbf{G}_{k,n-k}$ may also be thought of in terms of $\mathbf{O}(k)$ acting on $\mathbf{V}_{k,n}$, the Stiefel manifold of k-frames by the action $A \to AU$, $U \in \mathbf{O}(k)$. The orbit of A under this action is the set of all k-frames which generate a fixed hyperplane $\mathbf{P}$, so that $\mathbf{G}_{k,n-k}$ is a homogeneous space $\mathbf{V}_{k,n}/\mathbf{O}(k)$.

Recall the discussion at (7.5.5). If $\mathbf{P}$ is a plane near $\mathbf{P}_0$ and $\mathbf{P}$ is represented by (A, B) in the local coordinates at $\mathbf{P}_0$, then the orbit in $\mathbf{V}_{k,n}$ is the set of AU, $U \in \mathbf{O}(k)$. Therefore the local coordinates in $\mathbf{G}_{k,n-k}$ together with local coordinates in $\mathbf{O}(k)$ give local coordinates in the Stiefel manifold $\mathbf{V}_{k,n}$. Therefore if $A \in \mathbf{G}_{k,n-k}$ and $H = AU \in \mathbf{V}_{k,n}$ then

$$dH = A(dU) + (dA)U, \tag{7.9.2}$$

and

$$B^t(dH) = (B^tA)(dU) + B^t(dA)U = B^t(dA)U.$$

Also

$$H^t(dH) = U^tA^tA(dU) + U^tA^t(dA)U = U^t(dU) + U^tA^t(dA)U. \tag{7.9.3}$$

From (7.9.2) and using (6.2.19) we obtain

$$\bigwedge_{j=1}^{n-k} \bigwedge_{i=1}^{k} b_j^t\, dh_i = (\det U)^{n-k} \bigwedge_{j=1}^{n-k} \bigwedge_{i=1}^{k} b_j^t\, da_i. \tag{7.9.4}$$

Since $H = AU$ we find for the i-th column h_i of H and u_i of U that

$$dh_i = (dA)u_i + A(du_i) \tag{7.9.5}$$

and

$$h_j^t\, dh_i = u_j^t A^t((dA)u_i + A(du_i)) = u_j^t\, du_i + u_j^t A^t\, dA\, u_i.$$

In forming a join of these expressions we use the observation that

$$\left(\bigwedge_{j=1}^{n-k} \bigwedge_{i=1}^{k} b_j^t\, da_i\right)(u_j^t A^t\, dA\, u_i) = 0, \tag{7.9.6}$$

since this is a $(1 + (n - k)k)$-form on a manifold of dimension $(n - k)k$. Consequently from (7.9.4) and (7.9.5) we obtain

$$(\det U)^{n-k} \bigwedge_{j=1}^{n-k} \bigwedge_{i=1}^{k} b_j^t da_i \bigwedge_{i<j} u_j^t du_i = \bigwedge_{j=1}^{n-k} \bigwedge_{i=1}^{k} b_j^t dh_i \bigwedge_{i<j} h_j^t dh_i, \tag{7.9.7}$$

where the right side of (7.9.7) is the differential form of an invariant measure on $\mathbf{V}_{k,n}$. Integration of the absolute value gives

Theorem 7.9.1.

$$A_n \cdots A_{n-k+1} = \left(\int_{\mathbf{G}_{k,n-k}} |\omega| \right) A_k \cdots A_1. \tag{7.9.8}$$

Source, James (1954).

7.10. Problems

Problem 7.10.1. For the group $\mathbf{T}(n)$ of nonsingular lower triangular matrices with positive diagonal elements, let $(T^{-1}\,dT)_{ij}$ be the (i,j)-element of $T^{-1}\,dT$. Then the differential form $\omega = \bigwedge_{j \le i} (T^{-1}\,dT)_{ij}$ is an invariant differential form under the transformations $T \to T_0 T$, $T_0 \in \mathbf{T}(n)$. Therefore $\int_{\mathbf{C}} |\omega|$ is a left invariant Haar measure.

Problem 7.10.2. Continue Problem 7.10.1. If the (i,j)-element of $((dT)T^{-1})$ is s_{ij} then the differential form $\omega = \bigwedge_{j \le i} ds_{ij}$ defines a right invariant Haar measure for $\mathbf{T}(n)$.

Problem 7.10.3. Continue Problem 7.10.2. Use (6.7.10) and (6.7.16) to compute the modular functions for $\mathbf{T}(n)$.

Problem 7.10.4. For the group $\mathbf{GL}(n)$ of nonsingular $n \times n$ matrices X, the differential form $\omega = \bigwedge_{i=1}^{n} \bigwedge_{j=1}^{n} ((XX^t)^{-1} X_j)^t\, dX_i$ is a left invariant differential form, where the j-th column of X is X_j.

Problem 7.10.5. Use the decomposition, if $X \in \mathbf{GL}(n)$ then $X = TA^t$ with $T \in \mathbf{T}(n)$, $A \in \mathbf{O}(n)$ and T and A uniquely determined. See Chapter 8. Then

$$dX = T(dA)^t + (dT)A^t, \tag{7.10.1}$$

and

$$A^t((XX^t)^{-1})^t(dX)A = (dA^t)A + T^{-1}(dT).$$

Use the fact that the (i,j)-element of $T^{-1}\,dT$ is zero if $j > i$. Therefore

$$\bigwedge_{i=1}^{n} \bigwedge_{j=1}^{n} (((XX^t)^{-1}X)^t\, dX)_{ij} = \varepsilon_1 \bigwedge_{i<j} ((dA^t)A)_{ij} \bigwedge_{j \le i} (T^{-1}(dT))_{ij}. \tag{7.10.2}$$

The sign ε_1 is determined by the number of transpositions needed to obtain the arrangement on the right side of (7.10.2). Note that since $I_n = A^t A$ it follows that $0 = A^t(dA) + (dA)^t A$. Then for a sign ε_2,

$$\bigwedge_{i=1}^{n} \bigwedge_{j=1}^{n} (((XX^t)^{-1}X)^t \, dX)_{ij} \varepsilon_2 \bigwedge_{i<j} (a_i^t \, da_j) \bigwedge_{j \le i} (T^{-1} \, dT)_{ij}. \tag{7.10.3}$$

Problem 7.10.6. Let μ be an invariant regular Borel measure for the Borel subsets of $\mathbf{S}(n)$. Define $v(\mathbf{A}) = \mu(\{S \mid S^{-1} \in \mathbf{A}\}) = \mu(\mathbf{A}^{-1})$. Show that v is an invariant measure defined on the Borel subsets of $\mathbf{S}(n)$. Therefore, by uniqueness there is a number δ such that $v = \delta\mu$. Show that $\delta = 1$. Recall Example 3.3.9 in which the transformations $S \to X^t S X$ are discussed. This is not the action of $\mathbf{GL}(n)$ on the coset space $\mathbf{GL}(n)/\mathbf{O}(n)$ but a uniqueness theorem holds. See the discussion in Example 3.3.9 and also Problem 6.7.11.

CHAPTER 8

Matrices, Operators, Null Sets

8.0. Introduction

Originally this chapter was to bridge the discussion of Chapters 6 and 7 on differential forms and the applications in Chapter 9 of differential forms to matrix problems made by James. As the book has grown, material on matrix decompositions, Section 8.1, and an inventory of problems, Section 8.5, has accumulated. This material has been found to be needed by most chapters of the book. Likewise the problems in Chapter 4 on sets of measure zero, although basic, did not cover the needed basics, and Section 8.4 has evolved. The complicated subject of canonical correlations has been segregated into Section 8.2. And under the disguise of being infinite dimensional matriccs, Scction 8.3 covers some operator theory and some results on Hilbert space valued Gaussian random variables, a subject related to contemporary literature on estimation.

Sections 8.1 and 8.3 are complex and were written with little connective material. The following comments are intended to give the reader a guide to the material found in these sections.

Section 8.1 starts with material on spectral decomposition of matrices, Theorem 8.1.1. This is followed by constructive proofs of the factorizations of $\mathbf{S}(n)$ and $\mathbf{GL}(n)$ by $\mathbf{O}(n)$ and $\mathbf{T}(n)$, Theorems 8.1.2 through 8.1.5. There then follows results on the *singular value decomposition*, Lemma 8.1.6 to Theorem 8.1.11. *Generalized inverses* are closely related to the singular value decomposition, Definition 8.1.12 and Theorem 8.1.13. The section closes with two algebraic versions of Cochran's theorem.

The crucial results from operator theory needed in this book are, the existence of a square root of a positive semidefinite operator, and, the existence of a complete orthonormal system of eigenvectors for compact operators. Section 8.3 starts with a discussion of these results. In finite dimensions the iteration to the square root given by Newton's method is sufficient to prove the existence of the square root and that the map $A \to A^{1/2}$ is a nondecreasing function in the ordering of semidefinite matrices. See the discussion in Chapter 14 of Loewner's theory. In infinite dimensions a different argument seems necessary and we have used the Riesz–Sz.–Nage argument for existence, and adapted the Newton method argument to prove monotonicity. The existence proof shows that the map $A \to A^{1/2}$ is measurable, a result widely used in this book. A construction of the eigenvectors of a compact operator is given. This argument is partially repeated later to show that the covariance operator of a Gaussian process with values in a Hilbert space is a bounded and trace class operator. The trace class operators form an ideal in the ring of bounded operators. Likewise the set of Hilbert Schmidt operators is an ideal. We only partially prove this. Brownian motion is discussed as an example of a Gaussian process. If A and B are bounded self-adjoint positive semidefinite compact operators then operators $A(A+B)^{-1}$ arise in the discussion of Bayes estimators of the drift of a Brownian motion. It is shown that operators $A(A+B)^{-1}$ may be unbounded whereas $A^{1/2}(A+B)^{-1/2}$ is a bounded operator. Mandelbaum (1983) shows that if $\mathbf{X}$ is a Hilbert space valued Gaussian random variable (process) the "right" class of linear estimators $A\mathbf{X}$ arise from possibly unbounded but *measurable* linear operators A. We do not develop this theory but do give a few results about linear Bayes estimators of the drift.

The author is indebted to A. Mandelbaum and J. Malley for many discussions that helped form Chapter 8.

8.1. Matrix Decompositions

This section consists of a sequence of isolated results, all of which are needed.

Theorem 8.1.1.

(1) *If U is a unitary matrix there exists a unitary matrix V such that VUV^* is a diagonal matrix.*
(2) *If T is a lower triangular matrix with no two diagonal elements the same then there exists a lower triangular matrix T_1 such that $T_1TT_1^{-1}$ is a diagonal matrix.*
(3) *If S is symmetric with real entries then there exists an orthogonal matrix U such that USU^t is a diagonal matrix.*
(4) *If S is Hermitian then there exists a unitary matrix U such that USU^* is a diagonal matrix.*

Proof. All matrices are to be thought of as being $n \times n$. The proof of each assertion is by induction on n. All four statements are obvious if $n = 1$.

Inductive step for (1). Over the complex numbers the polynomial $\det(U - \lambda I_n) = 0$ has a root, λ_0, and the rows of the singular matrix $U - \lambda_0 I_n$ have a nontrivial orthogonal complement x, i.e., $Ux^* = \lambda_0 x^*$, x^* the complex conjugate of x. Then $(1/\lambda_0)x^* = U^*x^*$ since $U^*U = I_n$. Write (x, y) for the complex inner product. Then if $(x, y) = 0$ it follows that $(x, Uy) = (U^*x, y) = (1/\lambda_0)(x, y) = 0$. Thus the orthogonal complement of x reduces U.

Inductive step for (2). Write $T = \begin{vmatrix} T_0 & 0 \\ a^t & t_{nn} \end{vmatrix}$. The equation $(x^t, y)T = t_{nn}(x^t, y)$ becomes $x^t T_0 + ya^t = t_{nn}x^t$, $t_{nn}y = t_{nn}y$. Since $T_0 - t_{nn}I$ is nonsingular, $x^t = -(T_0 - t_{nn}I)^{-1}ya^t$ is solvable and the remaining eigenvectors are determined by T_0.

Inductive step for (3). The unit sphere is compact so $\lambda_0 = \sup_{\|x\|=1}(Sx, x)$ is assumed at some x_0. If y is perpendicular to x_0 and $\|y\| = 1$ then $(1 - t)^{1/2}y + t^{1/2}x_0$ is a unit vector so that $(S((1 - t)^{1/2}y + t^{1/2}x_0), (1 - t)^{1/2}y + t^{1/2}x_0) \le (Sx_0, x_0)$. Expand the left side and compute the derivative with respect to t to obtain for the derivative $(Sx_0, x_0) - (Sy, y) + (t^{-1/2} - (1 - t)^{-1/2})(Sy, x_0)$, which must be negative for t near one. Since $(Sx^0, x_0) - (Sy, y) \ge 0$ it follows that $(Sy, x_0) \le 0$. This holds if y is replaced by $-y$, so $-(Sy, x_0) \le 0$. Thus $(Sy, x_0) = 0$ and S is reduced by the orthogonal complement of x_0. Then if $(x_0, y) = 0$ it follows that $(Sx_0, y) = (x_0, Sy) = 0$. Hence $Sx_0 = cx_0$ for some c, and thus $c = \lambda_0$.

Inductive step for (4). The argument is the same as for (3). In part (3) the assumption of symmetry is expressed as $(Sx, y) = (x, Sy)$. In part (4) the assumption of Hermitian would be expressed as $(Sx, y) = (x, Sy)$ for the complex inner product. □

A $n \times n$ matrix $S = S^t$ is said to be positive definite (real entries) if $x^t Sx > 0$ for all $n \times 1$ vectors $x \ne 0$. By taking x to have coordinates $i_1, \ldots, i_h$ nonzero, the other coordinates zero, it follows that the submatrix $S_0 = S(i_1, \ldots, i_h)$ is also positive definite. Since $S_0 = S_0^t$, by Theorem 8.1.1 there exists an $h \times h$ orthogonal matrix U such that US_0U^t is a diagonal matrix. Thus if $y \ne 0$ is $h \times 1$ with entries $y_1, \ldots, y_h$ we have $0 < y^t(US_0U^t)y = \Sigma_{i=1}^h y_i^2(US_0U^t)_{ii}$. Thus the diagonal entries of US_0U^t are all positive and the determinant $\det(US_0U^t) = \det(S_0) > 0$. Similar comments apply in the complex case to Hermitian matrices.

Theorem 8.1.2. *If the $n \times n$ Hermitian matrix S is positive definite the all principal submatrices $S(i_1, \ldots, i_h)$ are positive definite and* $\det(S(i_1, \ldots, i_h)) > 0$. *In particular the diagonal of S consists of positive numbers.*

Theorem 8.1.3. *Let S be a $n \times n$ Hermitian matrix with (i,j)-entry S_{ij}, $1 \le i, j \le n$. If $1 \le k \le n$ implies*

$$\det\begin{vmatrix} S_{11} & \cdots & S_{1k} \\ S_{k1} & \cdots & S_{kk} \end{vmatrix} = \det(S(1, \ldots, k)) > 0 \tag{8.1.1}$$

then there exists a uniquely determined lower triangular matrix $T \in \mathbf{T}(n)$ such that

$$TT^* = S. \tag{8.1.2}$$

If the entries of S are real then the entries of T are real. In all cases the diagonal of T consists of real numbers.

PROOF. By induction on the number of rows n. The result is obvious for 1×1 matrices. Let the result hold for $(m-1) \times (m-1)$ matrices and suppose

$$T_{m-1} \in \mathbf{T}(m-1), \quad \text{and} \quad (T_{m-1}T_{m-1}^*)_{ij} = s_{ij}, \quad \text{for } 1 \le i, j \le m-1, \tag{8.1.3}$$

where we write T^* for the conjugate transpose of T. Let

$$T = \begin{vmatrix} T_{m-1} & \cdots & 0 \\ t_{m1} & \cdots & t_{mm} \end{vmatrix} \tag{8.1.4}$$

so that $(TT^*)_{ij} = s_{ij}$. This gives equations

$$s_{mj} = t_{m1}\bar{t}_{j1} + \cdots + t_{mj}\bar{t}_{jj}. \tag{8.1.5}$$

Since t_{jj} is real and positive, $1 \le j \le m-1$, these equations have a solution. Write $x = |t_{m1}^2| + \cdots + |t_{mm}^2|$ so that

$$TT^* = \begin{vmatrix} s_{11} & \cdots & s_{1m} \\ s_{m1} & \cdots & x \end{vmatrix}. \tag{8.1.6}$$

By construction

$$\det TT^* = (t_{11} \cdots t_{mm})^2, \tag{8.1.7}$$

and

$$\det T_{m-1}T_{m-1}^* = (t_{11} \cdots t_{m-1\, m-1})^2 > 0.$$

Since (8.1.1) holds for $k = m$, it follows there is a number t_{mm} real and positive such that $\det TT^* = \det S(1, \ldots, m)$. Since $\det TT^*$ is linear in the variable x, there is a unique value of x such that the two sides of (8.1.6) have the same value. This is $x = s_{mm}$. Thus the equation $t_{11}^2 + \cdots + t_{mm}^2 = s_{mm}$ is solvable and $T = T_m$ has been constructed. That completes the inductive step. □

Theorem 8.1.4. *Let X and Y be $n \times k$ matrices of real entries and of full rank $k \le n$ such that $X^tX = Y^tY$. Then there exists an $n \times n$ orthogonal matrix U with $X = UY$.*

PROOF. X^tX has the same rank as X, see Lemma 11.10.7. Since X has rank k, X^tX is nonsingular. Then $X(X^tX)^{-1/2}$ and $Y(Y^tY)^{-1/2}$ are $n \times k$ matrices with orthonormal columns and there exists $U \in \mathbf{O}(n)$ with $X(X^tX)^{-1/2} = UY(Y^tY)^{-1/2}$. Since $(X^tX)^{1/2} = (Y^tY)^{1/2}$ it follows that $X = UY$. The existence of a positive definite square root $(X^tX)^{1/2}$ is shown in Theorem 8.3.2 and again in Section 14.1. □

Theorem 8.1.5. *Let X and Y be $n \times k$ matrices of real entries and rank $h \leq k < n$ such that $X^tX = Y^tY$. There exists a $n \times n$ orthogonal matrix U with $X = UY$.*

Proof. Permute the columns of X so that we may suppose without loss of generality write $X = (X_1, X_2)$ with X_1 of rank h, and $X_1A = X_2$ where A is $h \times (k - h)$. Then since $X_1^tX_1$ is nonsingular $A = (X_1^tX_1)^{-1}X_1^tX_2$ is uniquely determined by the inner products X^tX.

Make the same permutation of the columns of Y so that without loss of generality we may write $Y = (Y_1, Y_2)$. Then $Y_1^tY_1 = X_1^tX_1$ is of full rank h and $Y_1B = Y_2$ must hold since Y has rank h. Then $B = (Y_1^tY_1)^{-1}Y_1^tY_2 = (X_1^tX_1)^{-1}X_1^tX_2 = A$. By Theorem 8.1.4 there exists $U \in \mathbf{O}(n)$ with $X_1 = UY_1$. Then $X_2 = X_1A = UY_1A = UY_2$. Thus $X = U(Y_1, Y_2) = UY$, as was to be shown. □

The proof of Theorem 8.1.7 requires a lemma which we now state and prove.

Lemma 8.1.6. *If $A \in \mathbf{O}(n) \cap \mathbf{T}(n)$ then $A = I_n$.*

Proof. If $A \in \mathbf{T}(n)$ and $1 \leq i < j \leq n$ then $a_{ij} = 0$. Since $A \in \mathbf{O}(n)$ then all the entries in the last column are zero except a_{nn} so it follows $a_{nn}^2 = 1$. Since the elements of $\mathbf{T}(n)$ have *positive* diagonal elements, $a_{nn} = 1$. This implies that the last row of A is $(0, 0, \ldots, 1)$. Hence, similarly, $a_{n-1\,n-1} = 1$, etc. The obvious backward induction now shows that $A = I_n$. □

Theorem 8.1.7. *If $X \in \mathbf{GL}(n)$ then*

$$X = A_1T_1 = A_2T_2^t = T_3A_3 = T_4^tA_4, \qquad \text{with} \quad A_1, \ldots, A_4 \in \mathbf{O}(n) \quad \text{and} \quad T_1, \ldots, T_4 \in \mathbf{T}(n). \tag{8.1.8}$$

Each of the factorizations of X is uniquely determined.

Proof. Given that every $X \in \mathbf{GL}(n)$ is expressible as $X = AT$, $A \in \mathbf{O}(n)$ and $T^t \in \mathbf{T}(n)$, the other three decompositions follow from consideration of X^t, X^{-1}, and $(X^t)^{-1}$, each of which maps $\mathbf{GL}(n)$ onto $\mathbf{GL}(n)$.

The Gram–Schmidt orthogonalization process, see Problem 8.5.1, produces an upper triangular matrix T such that the columns of XT^{-1} are orthonormal, i.e., $XT^{-1} = A \in \mathbf{O}(n)$ and $X = AT$ with $T^t \in \mathbf{T}(n)$.

It remains to prove uniqueness. If $A_1T_1 = A_2T_2$ then $A_2^{-1}A_1 = T_2T_1^{-1}$.

Since $((T_2 T_1)^{-1})^t \in \mathbf{T}(n)$ and $(A_2^{-1} A_1) \in \mathbf{O}(n)$, by the lemma, Lemma 8.1.6, it follows that $T_1 = T_2$ and $A_1 = A_2$. □

Corollary 8.1.8. *If $X \in \mathbf{GL}(n)$ there exists $A \in \mathbf{O}(n)$, $D \in \mathbf{D}(n)$ (a diagonal matrix) and $T^t \in \mathbf{T}(n)$ having diagonal elements equal one, such that*

$$X = ADT. \tag{8.1.9}$$

The next proof requires a preliminary lemma.

Lemma 8.1.9. *Let $D \in \mathbf{D}(n)$ and suppose the diagonal elements of D are pairwise distinct. If A is $n \times n$ and $DA = AD$, then A is a diagonal matrix.*

Proof. If the (i, i)-entry of D is d_i and if $A = (a_{ij})$, then the (i, j)-entry of AD is $a_{ij} d_j$ and the (i, j)-entry of DA is $d_i a_{ij}$. Since $AD = DA$ it follows that $0 = a_{ij}(d_i - d_j)$. By hypothesis, if $i \neq j$ then $d_i \neq d_j$. Hence $a_{ij} = 0$. □

Theorem 8.1.10. *Let X be a $n \times k$ matrix of rank $k \leq n$. Assume $X^t X$ has k distinct eigenvalues. There exists matrices A, D, and G such that A is $n \times k$ and*

$$A^t A = I_k, \qquad D \in \mathbf{D}(k), \qquad G \in \mathbf{O}(k), \qquad \text{and} \quad X = ADG. \tag{8.1.10}$$

If the diagonal entries of D_2 and D_3 are positive and in decreasing order of magnitude and if $X = A_2 D_2 G_2 = A_3 D_3 G_3$ are two such factorizations of X satisfying (8.1.10) *then*

$$\begin{aligned} &D_2 = D_3, \qquad G_3 G_2^t = D_1 \quad \text{is a diagonal orthogonal matrix,} \\ &\text{and} \quad A_3 = A_2 D_1. \end{aligned} \tag{8.1.11}$$

Proof. The matrices X and $X^t X$ have the same rank since the matrix entries are real numbers. Thus $X^t X$ is positive definite and

$$A = X(X^t X)^{-1/2} \tag{8.1.12}$$

is a $n \times k$ matrix satisfying $A^t A = I_k$. Then, choose $G \in \mathbf{O}(k)$ such that $D = G(X^t X)^{1/2} G^t$ is a diagonal matrix with the entries in decreasing order of magnitude. Then the existence of a factorization follows from

$$X = X(X^t X)^{-1/2} G^t G (X^t X)^{1/2} G^t G = ADG. \tag{8.1.13}$$

Given two decompositions $X = A_2 D_2 G_2 = A_3 D_3 G_3$ satisfying (8.1.10) then

$$X^t X = G_2^t D_2^2 G_2 = G_3^t D_3^2 G_3. \tag{8.1.14}$$

By hypothesis the diagonal entries of the diagonal matrices D_2^2, D_3^2 are in order of decreasing magnitude so that $D_2^2 = D_3^2$ follows since the diagonal entries of D_2 and D_3 are the eigenvalues of $(X^t X)^{1/2}$ and are positive. Therefore $D_2 = D_3$. Set $G_3 G_2^t = D_1$. This orthogonal matrix clearly satisfies $DD_1 =$

$D_1 D$, where from the above, $D = D_2 = D_3$ follows. By Lemma 8.1.9, since the diagonal entries of D are pairwise distinct, it follows that D_1 is a diagonal orthogonal matrix. Further $G_3 = D_1 G_2$. Thus

$$X = A_2 D G_2 = A_2 D D_1 G_2, \tag{8.1.15}$$

and since $DD_1 = D_1 D$, cancellation of the factor DG_2 from both sides yields $A_2 = A_3 D_1$. Since $D_1^2 = I_k$ it is also true that $A_2 D_1 = A_3$. □

In the computation of probability density functions relative to Lebesgue measure, the event (rank X) $< k$ is a set of nk-dimensional Lebesgue measure zero. See Problem 4.4.5. Problem 4.4.6 shows the event that two nonzero eigenvalues of $X^t X$ be equal is an event of Lebesgue measure zero. Thus, except on a set of Lebesgue measure zero, Theorem 8.1.10 is applicable in these problems. Theorem 8.1.11 stated in the sequel is primarily of algebraic interest rather than having direct application to the subject of this book. The decomposition is often called the *singular value decomposition* of X, and the eigenvalues of $(X^t X)^{1/2}$ the *singular values*.

Theorem 8.1.11. *Let X be a $n \times k$ matrix (of real entries) of* rank r *and let $X^t X$ have nonzero eigenvalues $\lambda_1 \geq \lambda_2 \geq \cdots \geq \lambda_r$. Then*

> *$X = ADG$, with A a $n \times r$ matrix, $D \in \mathbf{D}(r)$, and G a $r \times k$ matrix, such that $A^t A = I_r$, $GG^t = I_r$, and $D =$ diag $(\sqrt{\lambda_1}, \ldots, \sqrt{\lambda_r})$. Given a second factorization of $X = A_1 D_1 G_1$ with the diagonal entries of D_1 is decreasing order, then $D = D_1$ and there exists $H \in \mathbf{O}(r)$ such that* (8.1.16)

$$HD = DH \quad \text{and} \quad A_1 H = A, \qquad HG = G_1. \tag{8.1.17}$$

Remark. Since the diagonal entries of D are not necessarily pairwise distinct, it no longer follows from (8.1.17) that H is a *diagonal* orthogonal matrix.

Proof. Let $Y_1, \ldots, Y_r$ be an orthonormal set such that $Y_1, \ldots, Y_r \in \mathbb{R}^k$ and

$$(X^t X) Y_i = \lambda_i Y_i, \qquad 1 \leq i \leq r. \tag{8.1.18}$$

Then $\Sigma_{i=1}^r Y_i Y_i^t$ is a projection matrix of rank r that maps $\mathbb{R}^k$ onto the range of $X^t X$. It is easy to see that the range of $X^t X$ is the same as the linear span of the rows of X and therefore if $a \in \mathbb{R}^n$ then

$$(a^t X) \sum_{i=1}^r Y_i Y_i^t = a^t X. \tag{8.1.19}$$

In particular

$$X = X \sum_{i=1}^r Y_i Y_i^t = X \left| \frac{Y_1}{\sqrt{\lambda_1}}, \ldots, \frac{Y_r}{\sqrt{\lambda_r}} \right| \begin{vmatrix} \sqrt{\lambda_1} & \cdots & 0 \\ \vdots & & \vdots \\ 0 & \cdots & \sqrt{\lambda_r} \end{vmatrix} \begin{vmatrix} Y_1^t \\ \vdots \\ Y_r^t \end{vmatrix}. \tag{8.1.20}$$

It is easily verified that the matrices

$$A = X((\lambda_1)^{-1/2} Y_1, \ldots, (\lambda_r)^{-1/2} Y_r) \quad \text{and} \quad G^t = (Y_1, \ldots, Y_r) \tag{8.1.21}$$

satisfy (8.1.16). We write $X = ADG$.

Given two representations $X = ADG = A_1 D_1 G_1$ then

$$X^t X = G_1^t D_1^2 G_1 = G^t D^2 G, \quad \text{and} \tag{8.1.22}$$

$$\lambda_i Y_i = (X^t X) Y_i = G_1^t D_1^2 (G_1 Y_i). \tag{8.1.23}$$

This implies $G_1 Y_i \neq 0$ and since $G_1 G_1^t = I_r$, we obtain

$$\lambda_i (G_1 Y_i) = D_1^2 (G_1 Y_i). \tag{8.1.24}$$

That is, $\lambda_1, \ldots, \lambda_r$ are eigenvalues of D_1^2, and since D_1^2 is a diagonal matrix with entries in order of decreasing magnitude,

$$D^2 = D_1^2 \quad \text{and} \quad D = D_1 \quad \text{follows.} \tag{8.1.25}$$

By construction the columns of G^t are nonzero eigenvectors for the nonzero eigenvalues of $X^t X$. By (8.1.22) it follows that the columns of G_1^t likewise are an orthonormal set of eigenvectors for the nonzero eigenvalues of $X^t X$. Therefore there is an orthogonal matrix $H \in \mathbf{O}(r)$ such that $HG = G_1$. To see this, let $\tilde{G}$ be a $k \times k$ orthogonal matrix whose first r rows are G, and $\tilde{G}_1$ be a $k \times k$ orthogonal matrix whose first r rows are G_1. Then let $\tilde{H} \in \mathbf{O}(k)$ satisfy $\tilde{H}\tilde{G} = \tilde{G}_1$. Since

$$\tilde{G}(X^t X) = \begin{vmatrix} D^2 & 0 \\ 0 & 0 \end{vmatrix} \tilde{G} \quad \text{and} \quad \tilde{G}_1(X^t X) = \begin{vmatrix} D^2 & 0 \\ 0 & 0 \end{vmatrix} \tilde{G}_1 \tag{8.1.26}$$

if we write

$$H = \begin{vmatrix} \tilde{H}_{11} & \tilde{H}_{12} \\ \tilde{H}_{21} & \tilde{H}_{22} \end{vmatrix}, \tag{8.1.27}$$

then

$$\begin{vmatrix} D^2 & 0 \\ 0 & 0 \end{vmatrix} = \tilde{G}_1 (X^t X) \tilde{G}_1^t = (\tilde{H}\tilde{G})(X^t X)(\tilde{H}\tilde{G})^t = \tilde{H} \begin{vmatrix} D^2 & 0 \\ 0 & 0 \end{vmatrix} \tilde{H}^t. \tag{8.1.28}$$

Since D is a nonsingular matrix, this implies $\tilde{H}_{21} = 0$, hence that $\tilde{H}_{22} = I_{k-r}$, and $\tilde{H}_{12} = 0$. Therefore $\tilde{H}_{11} G = G_1$ and to obtain the assertion above we now call $\tilde{H}_{11}$ the matrix $H \in \mathbf{O}(r)$.

From (8.1.22) it follows that since $GG^t = I_r$,

$$X = ADG = A_1 DHG \quad \text{and} \quad AD = A_1 DH. \tag{8.1.29}$$

Thus

$$D^2 = DA^t AD = H^t DA_1^t A_1 DH = H^t D^2 H. \tag{8.1.30}$$

Hence H commutes with D^2, and hence with D. Since D is nonsingular,

(8.1.29) and (8.1.30) imply

$$A = A_1 H. \tag{8.1.31}$$

□

Definition 8.1.12. If X is a $n \times k$ matrix the generalized inverse X^+ of X is the $k \times n$ matrix

$$X^+ = G^t D^{-1} A^t, \tag{8.1.32}$$

where X has the factorization $X = ADG$ satisfying (8.1.16).

Theorem 8.1.13 (Penrose (1955)). *The generalized inverse X^+ satisfies the following properties:*

$$X^+X \quad \text{is a } k \times k \text{ orthogonal projection (i.e., is symmetric);} \tag{8.1.33}$$

$$XX^+ \quad \text{is a } n \times n \text{ orthogonal projection;} \tag{8.1.34}$$

$$XX^+X = X \quad \text{and} \quad X^+XX^+ = X^+. \tag{8.1.35}$$

Furthermore, if another $k \times n$ matrix X^0 has properties (8.1.33) *to* (8.1.35) *then $X^0 = X^+$.*

Proof. It is easily verified that the matrices $X = ADG$ and $X^+ = G^t D^{-1} A^t$ satisfy (8.1.33) to (8.1.35) provided A, D, and G satisfy (8.1.16). Suppose X^0 satisfies the hypothesis. Then

$$X^0 X = X^0(XX^+X) = (X^0X)(X^+X). \tag{8.1.36}$$

Take transposes of (8.1.36) and use the symmetry of the matrices involved. Then

$$X^0X = (X^0X)^t = (X^+X)^t(X^0X)^t = X^+(XX^0X) = X^+X. \tag{8.1.37}$$

Similarly

$$XX^0 = XX^+. \tag{8.1.38}$$

Then

$$X^+ = X^+(XX^+) = X^+(X^0) = (X^+X)X^0 = (X^0X)X^0 = X^0. \tag{8.1.39}$$

□

The results given in Appendix VI of *The Analysis of Variance*. Scheffe (1959), refer to quadratic forms and independence of normally distributed random variables. In this presentation we abstract the coefficient matrix from the quadratic form and state some similar results about matrices.

A $n \times n$ matrix P with real entries will be called an *orthogonal projection* if $P^tP = P$ and $P^2 = P$; in the case of complex number entries, $P^* = \bar{P}^t = P$ and $P^2 = P$.

Theorem 8.1.14. *Let $P_1, \ldots, P_h$ be $n \times n$ orthogonal projection matrices such that $P_1 + \cdots + P_h = I_n$. Then the following are equivalent.*

(1) $\operatorname{rank} P_1 + \cdots + \operatorname{rank} P_h = n$;
(2) *there exists a $n \times n$ orthogonal (unitary in the complex case) matrix U such that U^tP_iU is diagonal, and $P_iP_j = P_jP_i = 0$, $1 \le i \ne j \le h$ (U^*P_iU in the complex case).*

PROOF. (2) *implies* (1) is almost immediate. If U^tP_iU is a diagonal matrix then the diagonal entries are 0 or 1. Since $P_iP_j = 0$ the locations of the non-zero diagonal entries are disjoint sets of integers. Since $P_1 + \cdots + P_h = I_n$, the union of these integer sets is $\{1, \ldots, n\}$, hence the sum of the ranks is n.

(1) *implies* (2). $\operatorname{rank}(P_1 + \cdots + P_{h-1}) \le \Sigma_{i=1}^{h-1} \operatorname{rank} P_i = n - \operatorname{rank} P_h$. Therefore $n \le \operatorname{rank}(I_n - P_h + P_h) \le \operatorname{rank}(I_n - P_h) + \operatorname{rank} P_h \le n$. Let $r_i = \operatorname{rank} P_i$, $1 \le i \le n$. Thus there is a $n \times n$ orthogonal matrix U such that U^tP_hU is diagonal. Suppose U so chosen that

$$U^tP_hU = \begin{vmatrix} 0 & 0 \\ 0 & I_{r_h} \end{vmatrix}.$$

Then

$$\sum_{i=1}^{h-1} (U^tP_iU) = \begin{vmatrix} I_{n-r_h} & 0 \\ 0 & 0 \end{vmatrix}.$$

Since the matrices P_i are positive semidefinite, so is $\Sigma_{i=1}^{h-1}(U^tP_iU)$. Using $x^t = (0, \ldots, 0, y_1, \ldots, y_{r_h})$ then $x^tU^tP_iUx \ge 0$ and $\Sigma_{i=1}^{h-1} x^tU^tP_iUx = 0$. This implies the p, q-entry of U^tP_iU is zero if $p > n \quad r_h$ or $q > n - r_h$.

Consequently a backward induction applies. There exists V orthogonal with $V^t(U^tP_iU)V$ a diagonal matrix, $1 \le i \le h - 1$, where

$$V = \begin{vmatrix} V_0 & 0 \\ 0 & I_{r_h} \end{vmatrix}$$

and $V_0V_0^t$ is the identity matrix. Consequently $V^t(U^tP_hU)V = U^tP_hU$. By consideration of ranks, as in the proof that (2) implies (1), it now follows that $P_iP_j = P_jP_i = 0$ if $1 \le i \le j \le h$. □

Theorem 8.1.15. *Let $A_1, \ldots, A_h$ be symmetric matrices with real number entries such that $A_1 + \cdots + A_h = I_n$. The following are equivalent.*

(1) $\operatorname{rank} A_1 + \cdots + \operatorname{rank} A_h = n$;
(2) $A_i = A_i^2$ *for all i, and Theorem* 8.1.14 *applies.*

PROOF. (2) *implies* (1) is immediate. We examine (1) implies (2). Since $\operatorname{rank}(I_n - A_h) = \operatorname{rank}(A_1 + \cdots + A_{h-1}) \le \operatorname{rank} A_1 + \cdots + \operatorname{rank} A_{h-1} =$

$n - \operatorname{rank} A_h$, it follows that $n = \operatorname{rank}(I_n - A_h + A_h) \le \operatorname{rank}(I_n - A_h) + \operatorname{rank} A_h \le n$. By Problem 8.5.17 it follows that $A_h = A_h^2$. Similarly $A_j = A_j^2$, $1 \le j \le h - 1$. By Theorem 8.1.14 the result now follows. □

8.2. Canonical Correlations

We begin with a statement of the problem in terms of the population parameters. Later in this section the problem is stated in terms of sample quantities. Assume

$$\mathbf{X} \text{ is } 1 \times p, \qquad \text{and} \quad \mathbf{Y} \text{ is } 1 \times q. \tag{8.2.1}$$

The assumption about the moments of $\mathbf{X}$ and $\mathbf{Y}$ is

$$\begin{aligned} &\mathbf{E}\mathbf{X} = \mathbf{E}\mathbf{Y} = 0; \\ &\mathbf{E}\mathbf{X}^t\mathbf{X} = \Sigma_{11}, \qquad \mathbf{E}\mathbf{Y}^t\mathbf{Y} = \Sigma_{22}, \qquad \text{and} \quad \mathbf{E}\mathbf{X}^t\mathbf{Y} = \Sigma_{12}. \end{aligned} \tag{8.2.2}$$

The problem is to choose $a \in \mathbb{R}^p$ and $b \in \mathbb{R}^q$ which maximize the correlation between $\mathbf{X}a$ and $\mathbf{Y}b$. Since the expectations of $\mathbf{X}$ and $\mathbf{Y}$ are zero, the quantity to be maximized is

$$(a^t\Sigma_{12}b)/(a^t\Sigma_{11}a)^{1/2}(b^t\Sigma_{22}b)^{1/2}. \tag{8.2.3}$$

The solution to this problem can be obtained from the Cauchy–Schwarz inequality, which we now do. Define

$$\alpha = \Sigma_{11}^{1/2}a, \qquad \text{and} \quad \beta = \Sigma_{22}^{1/2}b \tag{8.2.4}$$

and normalize a and b so that

$$\alpha^t\alpha = 1 \quad \text{and} \quad \beta^t\beta = 1. \tag{8.2.5}$$

Since the expression (8.2.3) for the correlation is homogeneous of degree zero in each of a and b, the normalization (8.2.5) can be assumed to hold. Let

$$A = \Sigma_{11}^{-1/2}\Sigma_{12}\Sigma_{22}^{-1/2}. \tag{8.2.6}$$

Then the transformed problem is to maximize, subject to (8.2.5), the quantity

$$\alpha^t A \beta. \tag{8.2.7}$$

By the Cauchy–Schwarz inequality,

$$\alpha^t A\beta \le (\alpha^t AA^t\alpha)^{1/2}(\beta^t\beta)^{1/2} = (\alpha^t AA^t\alpha)^{1/2} \tag{8.2.8}$$

with equality if and only if $A^t\alpha$ and β are proportional. The condition (8.2.5) then gives

$$\beta = A^t\alpha/(\alpha^t AA^t\alpha)^{1/2} \tag{8.2.9}$$

and

$$\alpha^t A\beta = (\alpha^t AA^t\alpha)^{1/2}.$$

Subject to $\alpha^t\alpha = 1$, to maximize

$$\alpha^t AA^t\alpha = \alpha^t\Sigma_{11}^{-1/2}\Sigma_{12}\Sigma_{22}^{-1}\Sigma_{12}^t\Sigma_{11}^{-1/2}\alpha \tag{8.2.10}$$

is to find the largest eigenvalue and corresponding eigenvector of the indicated matrix (see Theorem 8.1.1), which is

$$\Sigma_{11}^{-1/2}\Sigma_{12}\Sigma_{22}^{-1}\Sigma_{12}^t\Sigma_{11}^{-1/2}. \tag{8.2.11}$$

We let λ_1 be the largest eigenvalue of the matrix (8.2.11), α_1 be the corresponding eigenvector, and $\beta_1 = A^t\alpha_1/(\alpha_1^t AA^t\alpha_1)^{1/2}$ as in (8.2.9). The problem may now be repeated as follows. Choose $a_i \in \mathbb{R}^p$ and $b_i \in \mathbb{R}^q$ so that

$$\mathbf{E}(\mathbf{X}a_i)^t(\mathbf{X}a_j) = 0, \qquad \mathbf{E}(\mathbf{Y}b_i)^t(\mathbf{Y}b_j) = 0, \tag{8.2.12}$$

for

$$1 \le i < j \le \operatorname{rank} A,$$

and subject to (8.2.12) maximize the correlation between $\mathbf{X}a_i$ and $\mathbf{Y}b_i$, $1 \le i \le \operatorname{rank} A$. As in (8.2.4) and (8.2.6) we set

$$\alpha_i = \Sigma_{11}^{1/2}a_i, \quad \beta_i = \Sigma_{22}^{1/2}b_i, \quad \text{and} \quad A = \Sigma_{11}^{-1/2}\Sigma_{12}\Sigma_{22}^{-1/2}, \tag{8.2.13}$$

for

$$1 \le i \le \operatorname{rank} A,$$

and suppose the normalization is such that

$$\alpha_i^t\alpha_i = 1, \qquad \beta_i^t\beta_i - 1, \qquad 1 \le i \le \operatorname{rank} A. \tag{8.2.14}$$

Last, the condition that (8.2.12) holds says that

$$a_i^t\Sigma_{11}a_j = 0 \quad \text{and} \quad b_i^t\Sigma_{22}b_j = 0, \qquad 1 \le j < i, \tag{8.2.15}$$

or that

$$\alpha_i^t\alpha_j = 0 \quad \text{and} \quad \beta_i^t\beta_j = 0, \qquad 1 \le j < i. \tag{8.2.16}$$

Repetition of the first stage of the argument shows that, by the use of the Cauchy–Schwarz inequality, in order to maximize $\alpha_i^t A\beta_i$ one should take

$$\beta_i = A^t\alpha_i/(\alpha_i^t AA^t\alpha_i)^{1/2} \tag{8.2.17}$$

and

$$\alpha_i^t A\beta_i = (\alpha_i^t AA^t\alpha_i)^{1/2},$$

and seek α_i to maximize (8.2.17). We note that to maximize $\alpha_i^t AA^t\alpha_i$ subject to (8.2.16), where the α_j are eigenvectors of AA^t is equivalent to finding the largest eigenvalue having an eigenvector orthogonal to $\alpha_1, \ldots, \alpha_{i-1}$. Review the proof of Theorem 8.1.1. If rank $A = r$ and $\lambda_1 \ge \cdots \ge \lambda_r$ are the nonzero eigenvalues of AA^t then λ_i is the maximum value subject to the stated conditions and α_i is the corresponding eigenvector. The condition that

$$\beta_i^t\beta_j = 0 \text{ is the condition that } \alpha_i^t AA^t\alpha_j = 0 \tag{8.2.18}$$

which is automatically satisfied.

The numbers $\lambda_1^{1/2}, \ldots, \lambda_r^{1/2}$ are the correlation coefficients and have values between zero and one. They are the *canonical correlation coefficients* and if $\cos\theta_i = \lambda_i^{1/2}$, $1 \le i \le r$, then $\theta_1, \ldots, \theta_r$ are the *critical angles.*

We now give a definition of the corresponding *sample quantities.* The analogy is to make the correspondences

$$\mathbf{X}^t\mathbf{X} \text{ to } \Sigma_{11}, \qquad \mathbf{X}^t\mathbf{Y} \text{ to } \Sigma_{12}, \qquad \text{and } \mathbf{Y}^t\mathbf{Y} \text{ to } \Sigma_{22}. \tag{8.2.19}$$

Then

$$A = (\mathbf{X}^t\mathbf{X})^{-1/2}\mathbf{X}^t\mathbf{Y}(\mathbf{Y}^t\mathbf{Y})^{-1/2} \tag{8.2.20}$$

and

$$r = \operatorname{rank} A. \tag{8.2.21}$$

Outside an exceptional set of zero Lebesgue measure $\mathbf{X}$ and $\mathbf{Y}$ have full rank p and q respectively, see Section 8.4, and A is well defined. Clearly $r \le \min(p, q)$. We prove that $r = \min(p, q)$, except on a set of measure zero. See (8.2.30) and following.

We seek $a_1, \ldots, a_r$ and $b_1, \ldots, b_r$ satisfying

$$a_i^t(\mathbf{X}^t\mathbf{X})a_j = 0 \quad \text{and} \quad b_i^t(\mathbf{Y}^t\mathbf{Y})b_j = 0, \qquad 1 \le j < i. \tag{8.2.22}$$

Define

$$\alpha_i = (\mathbf{X}^t\mathbf{X})^{1/2}a_i \quad \text{and} \quad \beta_i = (\mathbf{Y}^t\mathbf{Y})^{1/2}b_i, \qquad 1 \le i \le r, \tag{8.2.23}$$

normalized so that

$$\alpha_i^t\alpha_i = \beta_i^t\beta_i = 1, \qquad 1 \le i \le r. \tag{8.2.24}$$

Subject to (8.2.22), (8.2.23) and (8.2.24) choose α_i and β_i to maximize

$$\alpha_i^t A\beta_i, \qquad 1 \le i \le r. \tag{8.2.25}$$

The analysis applied to the population quantities may now be applied. Choose

$$\beta_i = A^t\alpha_i/(\alpha_i^t AA^t\alpha_i)^{1/2} \tag{8.2.26}$$

and maximize

$$\alpha_i^t AA^t\alpha_i \tag{8.2.27}$$

by choosing α_i to be the eigenvector corresponding to the i-th largest eigenvalue λ_i. This guarantees orthogonality of $\alpha_1, \ldots, \alpha_r$, and $\beta_1, \ldots, \beta_r$.

Observe from (8.2.23) that

$$\begin{aligned}(\mathbf{X}a_i)^t(\mathbf{X}a_j) &= a_i^t(\mathbf{X}^t\mathbf{X})a_j = ((\mathbf{X}^t\mathbf{X})^{1/2}a_i)^t((\mathbf{X}^t\mathbf{X})^{1/2}a_j)\\ &= \alpha_i^t\alpha_j = 0 \text{ if } i \ne j \text{ and } = 1 \text{ if } i = j. \text{ Similarly,}\\ (\mathbf{Y}b_i)^t(\mathbf{Y}b_j) &= \beta_i^t\beta_j = 0 \text{ if } i \ne j \text{ and } = 1 \text{ if } i = j.\end{aligned} \tag{8.2.28}$$

Therefore

$$\mathbf{X}a_1, \ldots, \mathbf{X}a_r \quad \text{is an orthonormal set,} \tag{8.2.29}$$

and

$$\mathbf{Y}b_1, \ldots, \mathbf{Y}b_r \quad \text{is an orthonormal set.}$$

Last, we prove that $r = \min(p, q)$ except for a set of zero Lebesgue measure, Suppose $z^t A A^t z = 0$ and let

$$\tilde{z}^t = z^t(\mathbf{X}^t\mathbf{X})^{-1/2}\mathbf{X}^t. \tag{8.2.30}$$

Then

$$\tilde{z}^t\mathbf{Y}(\mathbf{Y}^t\mathbf{Y})^{-1}\mathbf{Y}^t\tilde{z} = 0. \tag{8.2.31}$$

The orthogonal projection $\mathbf{Y}(\mathbf{Y}^t\mathbf{Y})^{-1}\mathbf{Y}^t$ maps $\mathbb{R}^n$ onto the column space of $\mathbf{Y}$ so (8.2.31) together with (8.2.30) say that

$$\tilde{z} \in (\text{column space } \mathbf{Y})^\perp \cap (\text{column space } \mathbf{X}). \tag{8.2.32}$$

Let us assume $p \le q$. Then $p + (n - q) \le q + (n - q) = n$. Given $\mathbf{Y}$ fixed we may choose a basis of (column space $\mathbf{Y})^\perp$ and extend it to $n - p$ elements, say $W = W_1, \ldots, W_{n-p}$. Then $\tilde{z} \ne 0$ satisfying (8.2.32) implies

$$\det(\mathbf{X}, W) = 0, \tag{8.2.33}$$

which by Problem 4.4.5 has $\mathbb{R}^{np}$ dimensional Lebesgue measure zero. This holds for almost all $\mathbf{Y}$ so by Fubini's theorem the set of X, Y such that

$$(\text{column space } Y)^\perp \cap (\text{column space } X) \ne \{0\} \tag{8.2.34}$$

has $\mathbb{R}^{n(p+q)}$ Lebesgue measure zero.

We summarize the discussion of sample quantities in a theorem.

Theorem 8.2.1. *Let the random $n \times p$ matrix $\mathbf{X}$ and the random $n \times q$ matrix $\mathbf{Y}$ have a joint probability density function relative to Lebesgue measure and suppose $p \le q$. There exists vectors $a_1, \ldots, a_r \in \mathbb{R}^p$ and $b_1, \ldots, b_r \in \mathbb{R}^q$ such that $r = \min(p, q)$, and*

$$\begin{aligned} &\mathbf{X}a_1, \ldots, \mathbf{X}a_r \text{ is an orthonormal set;} \\ &\mathbf{Y}b_1, \ldots, \mathbf{Y}b_r \text{ is an orthonormal set;} \end{aligned} \tag{8.2.35}$$

and

$$\text{if } i \ne j \text{ then } a_i^t\mathbf{X}^t\mathbf{Y}b_j = 0.$$

The *canonical correlations* defined by

$$\lambda_i^2 = a_i^t\mathbf{X}^t\mathbf{Y}b_i, \qquad 1 \le i \le r, \tag{8.2.36}$$

are numbers between zero and one.

The number of integers i such that $a_i^t \mathbf{X}^t \mathbf{Y} b_i = 1$ is $\max(p+q-n, 0)$. (8.2.37)

The projection of $\mathbf{X}a_i$ on the column space of $\mathbf{Y}$ is $\mathbf{Y}(\mathbf{Y}^t\mathbf{Y})^{-1}\mathbf{Y}^t\mathbf{X}a_i$. The cosine of the angle between $\mathbf{X}a_i$ and its projection is $\lambda_i^{1/2} > 0$. Write $\cos\theta_i = \lambda_i^{1/2}$. $\theta_1, \ldots, \theta_r$ are the critical angles and $\lambda_1^{1/2}, \ldots, \lambda_r^{1/2}$ are the canonical correlations. (8.2.38)

The numbers $\lambda_1, \ldots, \lambda_r$ are the nonzero roots of the determinental equation

$$0 = |\lambda \mathbf{X}^t\mathbf{X} - (\mathbf{X}^t\mathbf{Y})(\mathbf{Y}^t\mathbf{Y})^{-1}(\mathbf{Y}^t\mathbf{X})|. \tag{8.2.39}$$

Related to this equation is the invariant discussed in Section 5.7.

A few additional properties are stated in Problems 8.5.4 and following, which describe a decomposition of X. The source of the material presented in this section is Anderson (1958).

8.3. Operators and Gaussian Processes

The discussion of this section is restricted to operators on a Hilbert space to itself and to Gaussian processes taking values in a Hilbert space. A more general theory is available in the literature which uses the dual of a Banach space and the resulting bilinear form. The first results of this section have to do with monotone operator sequences, strong convergence, and the existence of a (positive) square root. The positive square root is a uniquely determined operator. We do not prove this.

Lemma 8.3.1. *If $\{A_n, n \geq 1\}$ is a sequence of self-adjoint positive semidefinite operators such that for all x, $(A_n x, x)$ is an increasing sequence bounded by $\|x\|^2$, then $A_n \to A$ strongly and $A_n \leq A \leq I$ for all n.*

Proof. If B is self-adjoint and positive definite then $[x, y] = (Bx, y)$ defines a semi-innerproduct and $[x, y] \leq ([x, x][y, y])^{1/2} = ((Bx, x)(By, y))^{1/2}$. Then $\|(A_m - A_n)x\|^4 \leq ((A_m - A_n)x, x)((A_m - A_n)^2 x, (A_m - A_n)x)$. If $m > n$ so that $A_m - A_n \geq 0$ then $\|(A_m - A_n)x\|^4 \leq ((A_m x, x) - (A_n x, x))\,\|x\|^2$. Since $\lim_{n\to\infty}(A_n x, x)$ exists, it follows that $\lim_{m,n\to\infty}\|(A_m - A_n)x\| = 0$. Hence the sequence $A_n x$ is Cauchy and the limit $Ax = \lim_{n\to\infty} A_n x$ exists for all x. A clearly must be linear, and $A \geq A_n$ for all n, and since $A_n \leq I$ it follows that $A \leq I$. □

Theorem 8.3.2. *A bounded self-adjoint positive semidefinite operator A has a self-adjoint positive semidefinite square root $A^{1/2}$ which commutes will all polynomials of A and is a stong limit of polynomials of A.*

PROOF. Since A is bounded, $(Ax, x) \le \|A\|(x, x)$. Consequently $A/\|A\| \le I$ in the ordering of self-adjoint operators. We assume therefore that $A \le I$ and set $B = I - A$ and solve the equation $Y = \frac{1}{2}(B + Y^2)$ recursively by $Y_0 = 0$, $Y_1 = \frac{1}{2}B$, and $Y_{n+1} = \frac{1}{2}(B + Y_n^2)$. It follows immediately by induction that Y_n is a polynomial in B with positive coefficients. Further $Y_n \le I$ follows from $B \le I$ and $Y_{n-1} \le I$. Then $Y_{n+1} - Y_n = (Y_n - Y_{n-1})(Y_n + Y_{n-1})$ follows since the Y_n commute. $Y_1 - Y_0 = \frac{1}{2}B$ is a polynomial with positive coefficients, so by induction, if $Y_n - Y_{n-1}$ is such a polynomial then so is $Y_{n+1} - Y_n$. Note that $B \ge 0$ implies $(B^{2n}x, x) = (B^n x, B^n x) \ge 0$ and $(B^{2n+1}x, x) = (BB^n x, B^n x) \ge 0$. Therefore it follows that $I \ge Y_{n+1} \ge Y_n$ and $\lim_{n\to\infty} Y_n = Y$ exists in the strong sense, hence $\lim_{n\to\infty} Y_n^2 = Y^2$ exists in the strong sense, so that $Y = \frac{1}{2}(B + Y^2)$. Substitute $C = I - Y$ and obtain $C^2 = A$. By construction $C \ge 0$ since $I \ge Y$. [Uniqueness can be proven but is not done so here.] □

Lemma 8.3.3. *If A is a self-adjoint positive semidefinite operator then $\|A\|(Ax, x) \ge \|Ax\|^2$ for all x. This implies $\|A\| = \sup_{\|x\|=1}(Ax, x)$.*

PROOF. Define B by $B = \|A\|I - A$ so that B commutes with A. By the proof of Theorem 8.3.2, B commutes with $A^{1/2}$. Then $(Bx, Ax) = (A^{1/2}Bx, A^{1/2}x) = (BA^{1/2}x, A^{1/2}x) \ge 0$ since $B \ge 0$. Thus $0 \le \|A\|(x, Ax) - (Ax, Ax)$ or $\|Ax\|^2 \le (Ax, x)\|A\|$. Then $\|A\|^2 = \sup_{\|x\|=1}\|Ax\|^2 \le \sup_{\|x\|=1}(Ax, x)\|A\| \le (\sup_{\|x\|=1}\|Ax\|)\|A\|$ which proves the second statement. □

EXAMPLE 8.3.4. To show that the map $A \to A^2$ is not monotone take $A = \begin{vmatrix} 0 & 1 \\ 1 & 0 \end{vmatrix}$ and $B = \begin{vmatrix} \frac{1}{2} & \frac{1}{2} \\ \frac{1}{2} & \frac{1}{2} \end{vmatrix}$ so that $A^2 = I$ and $B^2 = B$. In this example $B - A \ge 0$ while $B^2 - A^2$ is a negative definite matrix.

Theorem 8.3.5. *The map $A \to A^{1/2}$ is a nondecreasing function of self-adjoint positive semidefinite operators.*

PROOF. Assume $A \ge 0$ is self-adjoint and that the existence of square roots is known. At first suppose $\varepsilon > 0$ and that $A \ge B \ge \varepsilon I$ so A^{-1} and B^{-1} exist. Take $A_0 = I$ and $B_0 = I$ and in parallel, $A_{n+1} = \frac{1}{2}(A_n + A_n^{-1}A)$ and $B_{n+1} = \frac{1}{2}(B_n + B_n^{-1}B)$. That is, we construct $A^{1/2}$ and $B^{1/2}$ using Newton's method. Then $B_1 = \frac{1}{2}(I + B) \le \frac{1}{2}(I + A)$. We show inductively that

$$B_n \le A_n \quad \text{and} \quad B_n B^{-1} \ge A_n A^{-1}. \tag{8.3.1}$$

In addition, note for self-adjoint C that if $(x, x) \ge (Cx, x)$ then with $x = C^{-1/2}y$, $(C^{-1}y, y) \ge (y, y)$, so $I \ge C$ implies $C^{-1} \ge I$. If (8.3.1) holds then $B_{n+1} = \frac{1}{2}(B_n + B_n^{-1}B) \le \frac{1}{2}(A_n + A_n^{-1}A) = A_{n+1}$. Also $A_n^{-1}A = AA_n^{-1}$ and $B_n^{-1}B = BB_n^{-1}$ so that $A_{n+1}A^{-1} = \frac{1}{2}(A_nA^{-1} + A_n^{-1}) \le \frac{1}{2}(B_nB^{-1} + B_n^{-1}) = B_{n+1}B^{-1}$. We now show that $\lim A_n$ and $\lim B_n$ exist as strong limits. In fact $A_{n+1}^2 - A = (A_n - A_n^{-1}A)^2/4 \ge 0$ for all n and $A_n - A_{n+1} = \frac{1}{2}(A_n^2 - A)A_n^{-1}$.

Since $A_n^{-1/2}$ exists, it follows that $A_n - A_{n+1} \geq 0$. Similarly $B_n \geq B_{n+1}$. Thus $A_1 \geq A_n \geq 0$ and this implies $\lim_{n\to\infty} A_n = A_0$ exists as a strong limit. Then $A_0 = \frac{1}{2}(A_0 + A_0^{-1}A)$ so that $A_0^2 = A$ and similarly $B_0^2 = B$, and $B_0 \leq A_0$.

For the general case of $A \geq 0$ and $B \geq 0$ take $A_\varepsilon = A + \varepsilon I$ and $B_\varepsilon = B + \varepsilon I$. Then $A_\varepsilon^{1/2}$ is monotone in ε by the above so $\lim_{\varepsilon \downarrow 0} A_\varepsilon$ exists as a strong limit and is $A^{1/2}$. Hence $A^{1/2} \geq B^{1/2}$ follows. □

Definition 8.3.6. A bounded self-adjoint operator A is said to be *compact* or completely continuous if bounded sequences are mapped by A into precompact sets.

Theorem 8.3.7. *If A is a self-adjoint positive semidefinite compact operator then there exist positive numbers $\lambda_1 \geq \lambda_2 \geq \cdots$ and orthonormal vectors e_1, $e_2, \ldots$ such that $Ae_i = \lambda_i e_i$. If x is orthogonal to all the e_i then $Ax = 0$. $\lambda_j > 0$ for all j and if the set of nonzero λ_j is infinite then $\lim_{j\to\infty} \lambda_j = 0$. $\lambda_j = \lambda_j(A)$ is called the j-th eigenvalue of A.*

PROOF. Choose a sequence of unit vectors x_n such that $\lim_{n\to\infty} (Ax_n, x_n) = \|A\|$. See Lemma 8.3.3. Then $\lim_{n\to\infty} \| \|A\| x_n - Ax_n\|^2 = \lim_{n\to\infty} (\|A\|^2 - 2\|A\|(x_n, Ax_n) + (Ax_n, Ax_n)) \leq 0$. By hypothesis of compactness we may assume $\lim_{n\to\infty} Ax_n$ exists. Then $\lim_{n\to\infty} \|A\| x_n$ exists and since $\|A\| \neq 0$, $x = \lim_{n\to\infty} x_n$ exists. Then $\|A\| x = Ax$. As in the proof of Theorem 8.1.1, if y is perpendicular to x then Ay is perpendicular to x. Proceed by induction with $\lambda_1(A) = \|A\|$ and $e_1 = x$. Obtain $\lambda_2(A) = \sup_{y \perp x} \|Ay\|/\|y\|$ with eigenvector e_2, etc. Either after some k steps y perpendicular to $e_1, \ldots, e_k$ implies $Ay = 0$ or else a countably infinite sequence $\lambda_k(A) > 0$ is determined. In the infinite case if $\lim_{k\to\infty} \lambda_k(A) \geq \delta > 0$ then $\lambda_k(A)^{-1} e_k$ is a bounded sequence of vectors and $\|A(\lambda_k(A)^{-1} e_k) - A(\lambda_j(A)^{-1} e_j)\| = 2$ for all j, k. That contradicts the compactness assumption. Hence $\lim_{k\to\infty} \lambda_k(A) = 0$. Let y be in the orthogonal complement of the $\{e_k, k \geq 1\}$. Then $\|Ay\| \leq \lambda_k(A)$ for all k so that $\|Ay\| = 0$ and $Ay = 0$. □

Lemma 8.3.8. *If A is self-adjoint, positive semidefinite, and if $\Sigma_{i=1}^\infty (Ae_i, e_i) < \infty$ for some orthonormal basis $\{e_i, i \geq 1\}$, then $A^{1/2}$ is a compact operator.*

PROOF. Given a bounded sequence $\{f_j, j \geq 1\}$ with $F = \sup \|f_j\|^2$, write $f_j = \Sigma_{i=1}^\infty \lambda_{ji} e_i$, so that $\Sigma_{i=1}^\infty \lambda_{ji}^2 \leq F$. Then $\|A^{1/2} f_j - A^{1/2} f_k\| \leq \Sigma_{i=1}^\infty |\lambda_{ij} - \lambda_{ik}| (Ae_i, e_i)^{1/2}$. A tail sum satisfies $(\Sigma_{i=h}^\infty |\lambda_{ij} - \lambda_{ik}|(Ae_i, e_i)^{1/2})^2 \leq 2F \Sigma_{i=h}^\infty (Ae_i, e_i)$. Choose h so that $(\varepsilon/2)^2 \geq 2F \Sigma_{i=h}^\infty (Ae_i, e_i)$. By a diagonalization argument choose A subsequence a_j of integers such that $\lim_{j\to\infty} \lambda_{ia_j} = y_i$ exists. Then for all $\varepsilon > 0$, $\limsup_{j,k\to\infty} \|A^{1/2} f_{a_j} - A^{1/2} f_{a_k}\| \leq \varepsilon/2$, which proves the result. □

Lemma 8.3.9. *Suppose A is self-adjoint, positive semidefinite, and $\Sigma_{i=1}^\infty (Ae_i, e_i) < \infty$ for some orthonormal basis $\{e_i, i \geq 1\}$. Let $\{f_i, i \geq 1\}$ be an orthonormal set such that $Af_i = \lambda_i(A) f_i$. Then $\Sigma_{i=1}^\infty \lambda_i(A) = \Sigma_{i=1}^\infty (Ae_i, e_i) = \Sigma_{i=1}^\infty (Ah_i, h_i)$ for every orthonormal basis $\{h_i, i > 1\}$.*

Proof. Write $e_i = \Sigma_{j=1}^{\infty} \gamma_{ij} f_j + e_i'$ where e_i' is orthogonal to the f_j. Then $Ae_i = \Sigma_{j=1}^{\infty} \gamma_{ij} A f_j + A e_i' = \Sigma_{j=1}^{\infty} \gamma_{ij} \lambda_j f_j$ and $(Ae_i, e_i) = \Sigma_{j=1}^{\infty} |\gamma_{ij}|^2 \lambda_j$. Since $\{e_i, i \geq 1\}$ is a complete orthonormal system the Fourier coefficient of f_j in this basis is $(e_i, f_j) = \gamma_{ij}$. Therefore $\Sigma_{j=1}^{\infty} |\gamma_{ij}|^2 = 1$. Thus $\infty > \Sigma_{i=1}^{\infty} (Ae_i, e_i) = \Sigma_{i=1}^{\infty} \Sigma_{j=1}^{\infty} |\gamma_{ij}|^2 \lambda_j = \Sigma_{j=1}^{\infty} \lambda_j$, the series being absolutely convergent.

Write $h_i = \Sigma_{j=1}^{\infty} \gamma_{ij}' f_j + h_i'$ so that as above $\Sigma_{i=1}^{\infty} (Ah_i, h_i) = \Sigma_{j=1}^{\infty} \Sigma_{i=1}^{\infty} |\gamma_{ij}'|^2 \lambda_j = \Sigma_{j=1}^{\infty} \lambda_j < \infty$. □

Definition 8.3.10. A self-adjoint positive semidefinite operator T is said to be of *trace class* if there exists a complete orthonormal system $\{f_i, i \geq 1\}$ such that $\Sigma_{i=1}^{\infty} (Tf_i, f_i) < \infty$. For the arbitrary bounded operator T, make the definition

$$\operatorname{trace} T = \operatorname{trace} (T^*T)^{1/2}. \tag{8.3.2}$$

T is said to be of *trace class* if $\operatorname{trace} (T^*T)^{1/2} < \infty$, which by Lemma 8.3.9 is a condition independent of the choice of basis. T is said to be *Hilbert–Schmidt* if T^*T is a trace class operator, or equivalently, $\Sigma_{i=1}^{\infty} (Tf_i, Tf_i) < \infty$ for some orthonormal basis.

Lemma 8.3.11. *In the set of Hilbert–Schmidt operators* $[S, T] = \Sigma_{i=1}^{\infty} (Se_i, Te_i)$ *defines an inner product that is independent of the choice of basis.*

Proof. $|(Se_i, Te_i)| \leq \|Se_i\| \|Te_i\|$ so $\Sigma_{i=1}^{\infty} |(Se_i, Te_i)| \leq (\Sigma_{i=1}^{\infty} \|Se_i\|^2 \Sigma_{i=1}^{\infty} \|Te_i\|^2)^{1/2} = (\operatorname{trace} S^*S \operatorname{trace} T^*T)^{1/2} < \infty$. Since $\Sigma_{i=1}^{\infty} \|Se_i\|^2$ and $\Sigma_{i=1}^{\infty} \|Te_i\|^2$ are independent of the choice of basis, and since the inner product can be defined by the norm $(4(x, y) = \|x + y\| - \|x - y\|)$, it follows $[S, T]$ is invariantly defined. $[S, T]$ is clearly nonnegative and bilinear. If $\Sigma_{i=1}^{\infty} ((S - T)e_i, (S - T)e_i) = 0$ then for all i, $\|(S - T)e_i\| = 0$ and $Se_i = Te_i$ for all i, so that $S = T$. □

Definition 8.3.12. The Hilbert–Schmidt norm is defined by

$$\|T\|_{HS}^2 = \operatorname{trace} T^*T. \tag{8.3.3}$$

Lemma 8.3.13. *The Hilbert–Schmidt operators are an ideal in the ring of bounded operators. The adjoint of a trace class operator is a trace class operator.*

Proof. We first examine the adjoint of T where T is a Hilbert–Schmidt operator. Then by definition T^*T is a trace class operator and by Lemma 8.3.8, $(T^*T)^{1/2}$ is a compact operator with eigenvalues $\lambda_i^{1/2}$ and eigenvectors x_i, $i \geq 1$. Then $T^*Tx_i = \lambda_i x_i$, which implies, the vectors Tx_i are mutually orthogonal eigenvectors of TT^*. Let λ be in the spectrum of TT^* and $\{y_n, n \geq 1\}$ be a sequence of unit vectors such that $\lim_{n\to\infty} \|TT^*y_n - \lambda y_n\| = 0$. Then $\lim_{n\to\infty} \|T^*T(T^*y_n) - \lambda(T^*y_n)\| = 0$ and since T^*T is a compact operator we may suppose (take a subsequence if necessary) that $\lim_{n\to\infty} T^*y_n$

exists. This implies $\lim_{n\to\infty} y_n = y$ exists so that $TT^*y = \lambda y$. In view of Lemma 8.3.3 it follows that if y is orthogonal to Tx_i for all i then $TT^*y = 0$. Hence TT^* is a compact operator and $(TT^*)^{1/2}$ has the same trace as $(T^*T)^{1/2}$. Clearly $\Sigma_{i=1}^{\infty}(STe_i, STe_i) = \Sigma_{i=1}^{\infty}\|STe_i\|^2 \le \|S\|^2\Sigma_{i=1}^{\infty}\|Te_i\|^2 < \infty$. From $(S^*T^*)^* = TS$ it then follows that ST and TS are both Hilbert–Schmidt whenever S is a bounded operator.

This argument applied to the adjoint of a trace class operator T shows T^* is of trace class with the same trace. □

In the sequel we will be mainly concerned with the $\mathbf{L}_2$-space of $[0, 1]$. A function $K(\cdot, \cdot)$ which satisfies $\int_0^1\int_0^1 |K(s,t)|^2\,ds\,dt < \infty$ defines a linear transformation of $\mathbf{L}_2$ by $(Tf)(s) = \int_0^1 K(s,t)f(t)\,dt$ and $\int_0^1 |Tf(s)|^2\,ds \le \|f\|_2^2\int_0^1\int_0^1 |K(s,t)|^2\,ds\,dt$. We have used absolute values to indicate that complex numbers are allowed. For the complex inner product $(f, g) = \int_0^1 f(t)\overline{g(t)}\,dt$, and for T as defined to be self adjoint requires $K(s,t) = \overline{K(t,s)}$. The basic theorem, stated here without proof, see Riesz–Sz.-Nagy (1955), states the following:

Theorem 8.3.14. *A self-adjoint $\mathbf{L}_2$ kernal K defines a self-adjoint operator T which has eigenvalues $\lambda_i(T)$ and orthonormal eigenfunctions e_i. If $g \in \mathbf{L}_2$ then almost surely $g = h + \Sigma_{i=1}^{\infty}(g, e_i)e_i$. In addition $Th = 0$ and almost surely $Tg = \Sigma_{i=1}^{\infty}\lambda_i(T)(g, e_i)e_i$.*

We consider Brownian motion $\mathbf{X}_t$ on the unit interval $[0, 1]$ and assume there is no drift and unit variance. Then if $s < t$, $\mathbf{E}\,\mathbf{X}_s\mathbf{X}_t = \mathbf{E}\,\mathbf{X}_s(\mathbf{X}_s + \mathbf{X}_{t-s}) = s$ because of the independence of increments. Then

$$s \wedge t = \min(s, t) = \mathbf{cov}(\mathbf{X}_s, \mathbf{X}_t). \tag{8.3.4}$$

If $f \in \mathbf{L}_2(0, 1)$ then by Fubini's theorem,

$$0 \le \mathbf{E}\int_0^1\int_0^1 f(s)\mathbf{X}_s f(t)\mathbf{X}_t\,ds\,dt = \int_0^1\int_0^1 s \wedge t f(s)f(t)\,ds\,dt. \tag{8.3.5}$$

The covariance kernal $s \wedge t$ thus defines a self-adjoint positive semidefinite operator on $L_2(0, 1)$. As noted above, see Theorem 8.3.14, the operator is compact. We now determine the eigenfunctions.

$$\lambda f(s) = \int_0^1 s \wedge t f(t)\,dt = \int_0^s t f(t)\,dt + s\int_s^1 f(t)\,dt \tag{8.3.6}$$

with $\lambda \ne 0$ requires f to be continuous, $f(0) = 0$, and hence that f be a $\mathbf{C}_\infty$ function. The first two derivatives are

$$\lambda f'(s) = \int_s^1 f(t)\,dt; \qquad \lambda f''(s) = -f(s). \tag{8.3.7}$$

This requires $f'(1) = 0$ and that an eigenfunction be $f(t) = \sin\lambda^{-1/2}t$. Then $f'(1) = 0$ requires $\cos\lambda^{-1/2} = 0$ so that $\lambda^{-1/2} = \frac{1}{2}(2n+1)\pi$ and $\lambda =$

$((n+\frac{1}{2})\pi)^{-2}$. To make the eigenfunctions orthonormal requires normalization by $\sqrt{2}$. From (8.3.7), if $\int_0^1 s \wedge t f(t)\,dt = 0$ then $0 = \int_s^1 f(t)\,dt$ holds for $0 \le s \le 1$ which implies $f = 0$ almost surely.

Theorem 8.3.15. *The convariance kernal $s \wedge t$ defines a self-adjoint positive definite operator A on $\mathbf{L}_2(0, 1)$ with eigenfunctions and eigenvalues*

$$\sqrt{2}\sin(n+\tfrac{1}{2})\pi s, \qquad ((n+\tfrac{1}{2})\pi)^{-2}. \tag{8.3.8}$$

The eigenfunctions are a complete orthonormal system for $\mathbf{L}_2(0, 1)$.

Definition 8.3.16. The subspace $\mathbf{H}_0$ consists of those absolutely continuous functions f such that $\int_0^1 (f'(t))^2\,dt < \infty$. For $f \in \mathbf{H}_0$

$$\begin{aligned} (f, g)_{\mathbf{H}_0} &= \int_0^1 f'(t)g'(t)\,dt, \qquad \|f\|_{\mathbf{H}_0}^2 = (f, f)_{\mathbf{H}_0}. \\ (Af)(s) &= \int_0^1 (s \wedge t)f(t)\,dt. \end{aligned} \tag{8.3.9}$$

Then integration by parts shows that for continuous functions f

$$(Af, g)_{\mathbf{H}_0} - \int_0^1 f(t)g(t)\,dt = \int_s^1 f(t)\,dt \int_0^s g'(t)\,dt \Big|_0^1 = 0. \tag{8.3.10}$$

Therefore the following holds.

Theorem 8.3.17. *If f and g are in $\mathbf{H}_0$ then*

$$(Af, f)_{\mathbf{H}_0} = \|A^{1/2}f\|_{\mathbf{H}_0}^2 = \|f\|_2^2; \tag{8.3.11}$$

and,

$$(Af, g)_{\mathbf{H}_0} = (f, Ag)_{\mathbf{H}_0} = (f, g) = \int_0^1 f(t)g(t)\,dt.$$

Thus $A^{1/2}: \mathbf{L}_2(0, 1) \to \mathbf{H}_0$ is an isometry.

Remark 8.3.18 and proof. Use (8.3.10) for continuous f and the fact that the continuous functions are dense in L_2. If $g(t) = \Sigma_{n=0}^{\infty} g_n \sqrt{2}\sin\lambda_n^{-1/2} t$ is an $\mathbf{L}_2$-function then $(A^{1/2}g)(t) = \Sigma_{n=0}^{\infty} \lambda_n^{1/2} g_n \sqrt{2}\sin\lambda_n^{-1/2} t$ and the derivative computed term by term is $\Sigma_{n=0}^{\infty} g_n \sqrt{2}\cos\lambda_n^{-1/2} t$ which is in $\mathbf{L}_2(0, 1)$. Thus $A^{1/2}$ will map a dense set of $\mathbf{L}_2$ functions to $\mathbf{H}_0$ and the isometry guarantees a unique extension to all $\mathbf{L}_2$. Also, $\|f\|_{H_0} = 0$ means $f'(t) = 0$ so that absolute continuity of f together with $f(0) = 0$ implies $f = 0$ identically. Hence (8.3.11) defines an inner product.

Note that if f is continuous then $\|Af\|_{\mathbf{H}_0}^2 = \int_0^1 ds(\int_s^1 f(t)\,dt)^2 = \int_0^1 (1-s)^2 (\int_s^1 f(t)/(1-s)\,dt)^2 \le \frac{1}{2}\int_0^1 (f(s))^2\,ds = \frac{1}{2}\|f\|_2^2$. Therefore A maps $\mathbf{L}_2$ to $\mathbf{H}_0$ boundedly. □

Theorem 8.3.19. *If* $\{\mathbf{X}_n, n \geq 0\}$ *are independently distributed normal* (0, 1) *random variables then, with* $\lambda_n = ((n + \frac{1}{2})\pi)^{-2}$,

$$\mathbf{X}_t = \sum_{n=0}^{\infty} \lambda_n^{1/2} \mathbf{X}_n \sqrt{2} \sin (n + \tfrac{1}{2})\pi t \tag{8.3.12}$$

is a representation of Brownian motion. Almost surely all sample paths are $\mathbf{L}_2$ *functions of* t.

Remark 8.3.20 and proof. $\int_0^1 \mathbf{E}(\Sigma_{n=0}^m \lambda_n^{1/2} \mathbf{X}_n \sqrt{2} \sin (n + \frac{1}{2})\pi t)^2 \, dt = \Sigma_{n=0}^m \lambda_n$. By Fatou's lemma, $\mathbf{E} \int_0^1 \mathbf{X}_t^2 \, dt \leq \liminf_{m \to \infty} \Sigma_{n=0}^m \lambda_n < \infty$. Hence almost surely sample paths are in $\mathbf{L}_2$ and for s and t fixed. $\mathbf{E}\mathbf{X}_s\mathbf{X}_t = \Sigma_{n=0}^{\infty} \lambda_n (\sqrt{2} \sin (n + \frac{1}{2})\pi s)(\sqrt{2} \sin (n + \frac{1}{2})\pi t) = s \wedge t$ so that the Gaussian process $\mathbf{X}_t$ has the covariance function of Brownian motion. □

If drift is introduced in the form $\mathbf{X}_n + \theta_n$ then $\mathbf{X}_t + \Sigma_{n=0}^{\infty} \theta_n \lambda_n^{1/2} \sin (n + \frac{1}{2})\pi t = \mathbf{X}_t + \theta_t$ is the resulting process. Shepp (1965) has shown that the translated random variables either have a distribution mutually singular with that of the $\mathbf{X}_n$, or else they have the same sets of measure zero. Singularity results if $\Sigma\theta_n^2 = \infty$, while mutual absolute continuity holds if $\Sigma\theta_n^2 < \infty$. From the definition of $\mathbf{H}_0$ (see definition 8.3.16), it follows that $\Sigma\theta_n^2 < \infty$ if and only if $\theta_t \in \mathbf{H}_0$, so

Theorem 8.3.21. *The measure of Brownian motion* $\mathbf{X}_t$ *and translated Brownian motion* $\mathbf{X}_t + \theta_t$ *are either mutually absolutely continuous or are mutually singular, depending on whether* $\theta_t \in \mathbf{H}_0$ (*i.e., whether* θ_t *has a square integrable derivative.*)

A function $\mathbf{X}$ defined on a probability space $(\Omega, \mathfrak{F}, \mu)$ with values in a Hilbert space $\mathbf{H}$ will be called a Gaussian process if the functions $\mathbf{X}(\cdot)$: $\Omega \to \mathbf{H}$ are Borel measurable and if given $h_1, \ldots, h_n \in \mathbf{H}$ the random variables $(\mathbf{X}, h_1), \ldots, (\mathbf{X}, h_n)$ have a joint normal distribution, this holding for all n and choices of $h_1, \ldots, h_n$. Then if $\{h_n, n \geq 1\}$ is a sequence of unit vectors and $h = \text{weak}_{n \to \infty} \lim h_n$, it follows that $\lim_{n \to \infty} (\mathbf{X}, h_n) = (\mathbf{X}, h)$ almost surely, hence in distribution. This implies $\sup_n \mathbf{E}(\mathbf{X}, h_n)^2 < \infty$. Thus $\sup_{\|h\|=1} \mathbf{E}(\mathbf{X}, h)^2 < \infty$. Then $\mathbf{E}(\mathbf{X}, e)(\mathbf{X}, f)$ defines a continuous, symmetric, bilinear form, and there exists a self-adjoint positive semi-definite operator B with $(Be, f) = \mathbf{E}(\mathbf{X}, e)(\mathbf{X}, f)$. B is called the covariance operator of the Gaussian process.

In the example of Brownian motion $(Ae, f) = \mathbf{E}(\mathbf{X}, e)(\mathbf{X}, f) = \iint (s \wedge t) e(s)f(t) \, ds \, dt$. A is a trace class operator. We show below that B, as defined above, must always be trace class. In general, a Gaussian process with the identity operator as covariance operator would require for an orthonormal basis $\{e_i, i \geq 1\}$ that the sequence $(\mathbf{X}, e_i)$ be a sequence of independently distributed normal (0, 1) random variables for which $\mathbf{X} = \Sigma_{i=1}^{\infty} (\mathbf{X}, e_i) e_i$.

Therefore $\|\mathbf{X}\|^2 = \Sigma_{i=1}^{\infty} (\mathbf{X}, e_i)^2$ is a chi-square with an infinite number of degrees of freedom and is therefore almost surely infinite. This shows that this particular example is not possible.

More generally if the covariance operator of $\mathbf{X}$ is B then choose a sequence $\{e_n, n \geq 1\}$ of unit vectors such that $\lim_{n\to\infty} (Be_n, e_n) = \sup_{\|f\|=1} \|Bf\|$. The random variables $(\mathbf{X}, e_n)$ are normal and if $e_n \to e$ weakly in the Hilbert space then since $\mathbf{X}$ takes values in $\mathbf{H}$, $(\mathbf{X}, e_n) \to (\mathbf{X}, e)$ almost surely. Since $\sup_n (Be_n, e_n) < \infty$ the fourth moments $\mathbf{E}(\mathbf{X}, e_n)^4 = 3(Be_n, e_n)$ are uniformly bounded. Thus $\lim_{n\to\infty} \mathbf{E}(\mathbf{X}, e_n)^2 = \mathbf{E}(\mathbf{X}, e)^2 = (Be, e)$. Therefore as argued for Lemma 8.3.3 and Theorem 8.3.7, there is an orthonormal set $\{e_n, n \geq 1\}$ with $Be_n = \lambda_n e_n$. Repeating the chi-square argument, $\|\mathbf{X}\|^2 \geq \Sigma_{i=1}^{\infty} (\mathbf{X}, e_i)^2$ is a sum of independent random variables convergent in distribution. If $\sigma_n^2 = \mathbf{E}(\mathbf{X}, e_n)^2$ then $(\mathbf{X}, e_n)^2$ has Fourier transform $(1 - 2it\sigma_n^2)^{-1}$ so the Fourier transform of the infinite sum is an infinite product which converges if and only if $\Sigma\sigma_n^2 < \infty$. That is,

$$\sum_{n=1}^{\infty} \lambda_n = \sum_{n=1}^{\infty} (Be_n, e_n) = \sum_{n=1}^{\infty} \mathbf{E}(\mathbf{X}, e_n)^2 = \sum_{n=1}^{\infty} \sigma_n^2 < \infty.$$

It follows that B must be a trace class operator. We summarize in

Theorem 8.3.22. *Let* $\mathbf{X}$ *be a Gaussian process with values in* $\mathbf{H}$ *almost surely. Let* $\mathbf{X}$ *have covariance operator B. Then B is a self-adjoint positive semidefinite trace class operator.*

We now begin consideration of $\mathbf{l}_2$ as the space $\mathbf{H}$.

Theorem 8.3.23. *If* $\{\mathbf{X}_n, n \geq 1\}$ *is a sequence of mean zero random variables such that* $\Sigma_{n=1}^{\infty} \mathbf{X}_n^2 < \infty$ *almost surely, and if for every n,* $\mathbf{X}_1, \ldots, \mathbf{X}_n$ *has a joint normal density, then* $\mathbf{X} = (\mathbf{X}_1, \mathbf{X}_2, \ldots)$ *is a Gaussian process with values in* $\mathbf{l}_2$. *The covariance operator B has the matrix form* $\mathbf{E}\mathbf{X}_i\mathbf{X}_j = b_{ij}$ *which is the ij-element of B. B is of trace class so* $\Sigma_{n=1}^{\infty} \mathbf{E}\mathbf{X}_n^2 = \operatorname{trace} B < \infty$.

Remark on the proof. For sequences $a_1, \ldots, a_h$ in $\mathbf{l}_2$ with only a finite number of nonzero coordinates it follows that $(\mathbf{X}, a_1), \ldots, (\mathbf{X}, a_h)$ has a joint normal distribution. A limiting argument as the number of nonzero coordinates goes to infinity shows that $(\mathbf{X}, a_1), \ldots, (\mathbf{X}, a_h)$ have a joint normal distribution for the arbitrary $a_1, \ldots, a_h$ in $\mathbf{l}_2$. Then, by Theorem 8.3.22 the result now follows.

Let $\{\mathbf{X}_n, n \geq 1\}$ be a sequence of independently distributed normal (0, 1) random variables and $\{\lambda_n, n \geq 1\}$ a real number sequence such that if $n \geq 1$ then $\lambda_n \geq \lambda_{n+1} > 0$ and $\Sigma_{n=1}^{\infty} \lambda_n < \infty$. Then $\mathbf{E}(\Sigma_{n=1}^{\infty} (\lambda_n^{1/2}\mathbf{X}_n)^2) = \Sigma_{n=1}^{\infty} \lambda_n < \infty$ so that almost all sample paths $\{\lambda_n^{1/2}\mathbf{X}_n, n \geq 1\}$ are in $\mathbf{l}_2$. Further by the orthogonality of the $\mathbf{X}_n$, $\mathbf{E}(\Sigma_{n=1}^{\infty} \lambda_n^{1/2}\mathbf{X}_n)^2 = \Sigma_{n=1}^{\infty} \lambda_n < \infty$ which implies $\Sigma_{n=1}^{\infty} \lambda_n^{1/2}\mathbf{X}_n$ converges with probability one. A parameter sequence $\{\theta_n, n \geq 1\}$, by Shepp (1965), gives rise to an equivalent probability P_θ as the distribution of $\{\lambda_n^{1/2}(\mathbf{X}_n + \theta_n), n \geq 1\}$ if and only if $\Sigma_{n=1}^\infty \theta_n^2 < \infty$. If P_θ is the probability measure of $\{\mathbf{X}_n + \theta_n, n \geq 1\}$ then we may write the density $dP_\theta/dP_0 = \exp(\Sigma_{i=1}^\infty \theta_i X_i - \Sigma_{i=1}^\infty \theta_n^2)$, this being defined with probability one. If $\tilde{\theta}_j = \{\theta_{ij}, i \geq 1\}$ are parameter sequences such that $\tilde{\theta} = \lim_{j\to\infty} \tilde{\theta}_j$ exists as an $\mathbf{l}_2$ limit then for nonnegative functions ϕ if follows from Fatou's lemma that $\liminf_{j\to\infty} \mathbf{E}_{\tilde{\theta}j}\phi = \liminf_{j\to\infty} \int \phi \exp(\Sigma_{i=1}^\infty \theta_{ij} X_i - \Sigma_{i=1}^\infty \theta_{ij}^2)\, dP_0 \geq \int \phi \exp(\tilde{\theta}\cdot X - \|\tilde{\theta}\|^2)\, dP_0$. This proves the following.

Lemma 8.3.24. *If ϕ is a nonnegative measurable function defined on the range of $\{\mathbf{X}_n, n \geq 1\}$ then $\mathbf{E}_\theta s$ is lower semicontinuous in parameter sequences θ such that $\|\theta\| < \infty$.*

Lemma 8.3.25. *If $\mathbf{H}_1$ is a subspace of $\mathbf{l}_2$ such that $\mathbf{H}_1$ is measurable and $P_0(\{\lambda_n^{1/2}\mathbf{X}_n, n \geq 1\} \in \mathbf{H}_1) > 0$ then $\mathbf{H}_1$ contains all parameters $\{\lambda_n^{1/2}\theta_n, n \geq 1\}$ such that $\Sigma_{n=1}^\infty \theta_n^2 < \infty$, that is, $\mathbf{H}_1 \supset \mathbf{H}_0$.*

Proof. We write $\theta = \{\theta_n, n \geq 1\}$. The equivalence of measures requires $P_{\theta/j}(\{\lambda_n^{1/2}\mathbf{X}_n, n \geq 1\} \in \mathbf{H}_1) > 0$, holding for $j \geq 1$. If the sequence $\{\lambda_n^{1/2}\theta_n, n \geq 1\}$ is not in H_0 then the hyperplanes $\{-\lambda_n^{1/2}\theta_n/j, n \geq 1\} + \mathbf{H}_1$ are pairwise disjoint so that $\Sigma_{j=1}^\infty P_{\theta/j}(\{\lambda_n^{1/2}\mathbf{X}_n, n \geq 1\} \in \mathbf{H}_1) < \infty$. By Lemma 8.3.24 on lower semicontinuity, $0 \leq P_0(\{\lambda_n^{1/2}\mathbf{X}_n, n \geq 1\} \in \mathbf{H}_1) \leq \liminf_{j\to\infty} P_{\theta/j}(\{\lambda_n^{1/2}\mathbf{X}_n, n \geq 1\} \in \mathbf{H}_1) = 0$. This contradiction shows that $\mathbf{H}_1$ must contain the parameter sequence. □

Theorem 8.3.26. *The family of measures P_θ is complete. That is, if $\mathbf{E}_\theta|\phi| < \infty$ and $\mathbf{E}_\theta \phi = 0$ for all $\theta \in \mathbf{H}_0$ then $\phi = 0$ almost surely P_θ, for all θ.*

Proof (due to Mandelbaum). Relative to parameter sequences θ with $\theta_i = 0$ for $i \geq n+1$, we have that

$$0 = \mathbf{E}_\theta \phi = \int \phi \exp \sum_{i=1}^n (\theta_i X_i - \theta_i^2)\, dP_0 = \mathbf{E}_\theta\, \mathbf{E}_0(\phi|\mathbf{X}_1, \ldots, \mathbf{X}_n).$$

The finite dimensional measures are complete so $\mathbf{E}_0(\phi|\mathbf{X}_1, \ldots, \mathbf{X}_n) = 0$ almost surely P_θ. This martingale converges almost surely to $\phi = \mathbf{E}_0(\phi|\mathbf{X}_1, \mathbf{X}_2, \ldots)$ so that $\phi = 0$ almost surely P_0. By equivalence of the measures the result now follows. □

Theorem 8.3.27. *If $\mathbf{H}_1$ is a subspace of $\mathbf{l}_2$ such that $\mathbf{H}_1$ is measurable then either $P_0(\{\lambda_n^{1/2}\mathbf{X}_n, n \geq 1\} \in \mathbf{H}_1) = 0$ or $= 1$. In the latter case $\mathbf{H}_1$ contains all parameter sequences $\theta = \{\lambda_n^{1/2}\theta_n, n \geq 1\}$ such that $\Sigma_{n=1}^\infty \theta_n^2 < \infty$, i.e., for Brownian motion with $\lambda_n = ((n + \frac{1}{2})\pi)^{-2}$, and $n \geq 0$, $\mathbf{H}_1 \supset \mathbf{H}_0$.*

Proof. The indicator function $1_{\mathbf{H}_1}$ is measurable. $\mathbf{E}_\theta 1_{\mathbf{H}_1}(\mathbf{X}) = \mathbf{E}_0 1_{\mathbf{H}_1}(\mathbf{X} + \theta) = \mathbf{E}_0 1_{-\theta+\mathbf{H}_1}(\mathbf{X})$. If this is positive then by Lemma 8.3.25, $-\theta + \mathbf{H}_1 = \mathbf{H}_1$

so that $\mathbf{E}_\theta(1_{\mathbf{H}_1}(\mathbf{X}) - \mathbf{E}_0 1_{\mathbf{H}_1}(\mathbf{X})) = 0$. By Theorem 8.3.26, $1_{\mathbf{H}_1}(\mathbf{X}) = \mathbf{E}_0 1_{\mathbf{H}_1}(\mathbf{X})$ almost surely P_θ, all θ. Thus $\mathbf{E}_0 1_{\mathbf{H}_1}(\mathbf{X}) = 1$ and by equivalence of measures the result follows for all parameter sequences. □

We now consider briefly Bayes estimators resulting from normally distributed parameters, i.e., a priori measures on the parameter space. In finite dimensions, if $\mathbf{X}$ and $\mathbf{Y}$ are jointly normal with zero means and covariances $\mathbf{EXX}^t = \Sigma_{xx}$, $\mathbf{EXY}^t = \Sigma_{xy}$, and $\mathbf{EYY}^t = \Sigma_{yy}$, then $\mathbf{E}(\mathbf{X} - \Sigma_{xy}\Sigma_{yy}^{-1}\mathbf{Y}) = 0$ so that $\mathbf{X} - \Sigma_{xy}\Sigma_{yy}^{-1}\mathbf{Y} = \mathbf{E}(\mathbf{X}|\mathbf{Y})$ has a joint normal distribution with covariance matrix $\Sigma_{xx} - \Sigma_{xy}\Sigma_{yy}^{-1}\Sigma_{yx}$.

For location parameters, if $\mathbf{X}$ and $\boldsymbol{\theta}$ are independent, jointly normal, covariances C, S, then the joint covariances of $\boldsymbol{\theta}$, $\mathbf{X} + \boldsymbol{\theta}$ are $\begin{vmatrix} S & S \\ S & S+C \end{vmatrix}$ so $\mathbf{E}(\boldsymbol{\theta}|\mathbf{X} + \boldsymbol{\theta})$ has covariance matrix $S - S(S + C)^{-1}S = S(S + C)^{-1}C$. To translate these calculations to infinite dimensions we will be concerned about the boundedness of $S(S + C)^{-1}$ as an operator. We show in Example 8.3.29 that this operator may be unbounded, whereas, $S^{1/2}(S + C)^{-1/2}$ is a bounded operator, as is shown next.

Lemma 8.3.28. *If A and B are bounded self-adjoint positive semidefinite compact operators on a Hilbert space* $\mathbf{H}$ *to* $\mathbf{H}$ *such that $(A + B)x = 0$ implies $x = 0$, then $A^{1/2}(A + B)^{-1/2}$ is a bounded operator of norm ≤ 1. The range of* $\mathbf{A}^{1/2}$ *is contained in the range of $(A + B)^{1/2}$.*

Proof. $(Ax, x) \geq 0$ and $(Bx, x) \geq 0$ so that since $A + B$ is self-adjoint, $(A + B)$ has a square root and $((A + B)^{1/2}x, (A + B)^{1/2}x) = ((A + B)x, x) \geq (Ax, x) = (A^{1/2}x, A^{1/2}x)$. That is, $\|(A + B)^{1/2}x\| \geq \|A^{1/2}x\|$ for all x. Let y be such that $x = (A + B)^{-1/2}y$ is defined. Since $A + B$ is compact and $(A + B)x = 0$ implies $x = 0$, there is a complete orthonormal system of eigenvectors so the set of y is dense in $\mathbf{H}$. We find $\|y\| = \|(A + B)^{1/2}(A + B)^{-1/2}y\| \geq \|A^{1/2}(A + B)^{-1/2}y\|$ on a dense subset. Therefore $A^{1/2}(A + B)^{-1/2}$ has a unique extension to $\mathbf{H}$ with norm ≤ 1. Then $\|A(A + B)^{-1/2}\| \leq \|A^{1/2}\|$ and $\|(A + B)^{-1/2}A^2(A + B)^{-1/2}\|^2 \leq \|A\| < \infty$. Let $C = A^{1/2}(A + B)^{-1/2}$. Then on a dense subset $C(A + B)^{1/2}y = A^{1/2}(A + B)^{-1/2}(A + B)^{1/2}y = A^{1/2}y$. Since the operators on both sides of this identity are bounded, $C(A + B)^{1/2} = A^{1/2}$ and $(A + B)^{1/2}C^* = A^{1/2}$ so that the range of $A^{1/2}$ is contained in the range of $(A + B)^{1/2}$. □

Example 8.3.29. To show that $A(A + B)^{-1}$ may be an unbounded operator we construct A and B as tri-diagonal matrices as follows. Let $P = \begin{vmatrix} 0 & g \\ 1/g & 0 \end{vmatrix}$ so that $P^2 = I$. To obtain the inverse of $\alpha P + \beta I$ let $I = (\alpha P + \beta I)(aP + bI)$ so that $a = -\alpha/(\beta^2 - \alpha^2)$ and $b = \beta/(\beta^2 - \alpha^2)$. Then $\alpha P(\alpha P + \beta I)^{-1} = (-\alpha^2 I + \alpha\beta P)/(\beta^2 - \alpha^2)$. Let $\Sigma_{i=1}^\infty \alpha_i^2 < \infty$ and $\Sigma_{i=1}^\infty \beta_i^2 < \infty$ and construct the matrices A and B by stringing out 2×2 matrices along the diagonal,

$$A = \begin{vmatrix} \alpha_1 P & & & & & \\ & \alpha_2 P & & & & \\ & & \alpha_3 P & & & \\ & & & \cdot & & \\ & & & & \cdot & \\ & & & & & \cdot \end{vmatrix} \quad \text{and} \quad B = \begin{vmatrix} \beta_1 I & & & & & \\ & \beta_2 I & & & & \\ & & \beta_3 I & & & \\ & & & \cdot & & \\ & & & & \cdot & \\ & & & & & \cdot \end{vmatrix}.$$

Then the matrices A and B are Hilbert–Schmidt and self-adjoint if $g = 1$. The matrix $A(A + B)^{-1}$ has 2×2 blocks $(-\alpha_i^2 I + \alpha_i \beta_i P)/(\beta_i^2 - \alpha_i^2)$ which become large as $\alpha_i - \beta_i \to 0$.

Bayes estimators that arise from Gaussian a priori measures are linear of the form $M\mathbf{X}$ where $\mathbf{X}$ is a Gaussian process with covariance operator C and $M = S(S + C)^{-1}$.

Lemma 8.3.30. *Suppose M is a bounded operator and $(I - M)^{-1}$ exists as a bounded operator. Then $S = (I - M)^{-1}MC$ solves $M = S(S + C)^{-1}$. The condition that S be self-adjoint is $MC = CM^*$, i.e., that MC is self-adjoint. S is positive as an operator if and only if MC is positive as an operator.*

Proof. We assume by the statement of the Lemma that $(S + C)^{-1}$ exists. From $S = (I - M)^{-1}MC$ follows $MC = (I - M)S$ and $M(S + C) = S$. And conversely, $S = S^*$ requires $MC(I - M^*) = (I - M)CM^*$ and $MC = CM^*$ follows. Note that $MC = S(S + C)^{-1}C$ is the covariance of $\boldsymbol{\theta} - M\mathbf{X}$, independent of $\mathbf{X}$, which is the conditional covariance of $\boldsymbol{\theta}$ given $\mathbf{X}$ so must be positive and of trace class if S is positive and of trace class. Conversely, if $\boldsymbol{\theta} - M\mathbf{X}$ is Gaussian with mean zero and covariance operator $MC = S(S + C)^{-1}C \geq 0$ and is independent of $M\mathbf{X} = S(S + C)^{-1}\mathbf{X}$ then $\boldsymbol{\theta}$ is Gaussian with covariance operator $S(S + C)^{-1}C + S(S + C)^{-1}S = S \geq 0$. □

Remark 8.3.31. If for numbers $0 \leq \alpha < 1$ the operators $(I - \alpha M)$ are invertible then we have solutions $S_\alpha = \alpha(I - \alpha M)^{-1}MC$, and if the condition $\alpha CM = \alpha M^*C$ holds for some $\alpha \neq 0$ then the self-adjoint condition is automatic for all α. Consequently the estimator $M\mathbf{X}$ is the pointwise limit of Bayes estimators $\alpha M\mathbf{X}$ such that for all θ, squared error loss, the limit of the risks is

$$\lim_{\alpha \uparrow 1} (\alpha^2 \operatorname{tr} MCM^* + \|(I - \alpha M)\theta\|^2) = \operatorname{tr} MCM^* + \|(I - M)\theta\|^2. \tag{8.3.13}$$

If the estimator $N\mathbf{X}$ with risk $\operatorname{tr} NCN^* + \|(I - N)\theta\|^2$ is as good as $M\mathbf{X}$ then by considering $\theta = 0$ and $\|\theta\| \to \infty$ it follows that

$$\operatorname{tr} NCN^* \le \operatorname{tr} MCM^*; \tag{8.3.14}$$

and,

$$\|(I - N)\theta\| \le \|(I - M)\theta\| \text{ for all } \theta.$$

Remark 8.3.32. Relative to the loss function $\|D^{1/2}(M\mathbf{X} - \theta)\|^2$ the risk is

$$\operatorname{tr} D^{1/2}MCM^*D^{1/2} + \|D^{1/2}(I - M)\theta\|^2. \tag{8.3.15}$$

Here $\operatorname{tr} D^{1/2}MCM^*D^{1/2} = \Sigma_{i=1}^{\infty} \|D^{1/2}MC^{1/2}e_i\|^2$ taken over an orthonormal basis. This is a convex function of M and in fact $\|D^{1/2}(M_1 + M_2)C^{1/2}e_i/2\|^2 = \frac{1}{2}(\|D^{1/2}M_1C^{1/2}e_i\|^2 + \|D^{1/2}M_2C^{1/2}e_i\|^2)$ for all i if and only if $D^{1/2}M_1C^{1/2}e_i = D^{1/2}M_2C^{1/2}e_i$ for all i which requires $M_1 = M_2$ provided $D^{1/2}$ is one-to-one and $C^{1/2}$ has dense range.

Note that since C is a trace class operator, by Lemma 8.3.13 $D^{1/2}MC^{1/2}$ is Hilbert–Schmidt so $D^{1/2}MCM^*D^{1/2}$ is always a trace class operator. In the case of estimators $\alpha M\mathbf{X}$, the risk is $\alpha^2 \operatorname{tr} D^{1/2}MCM^*D^{1/2} + \|D^{1/2}(I - \alpha M)\theta\|^2$ and as $\alpha \uparrow 1$ these numbers converge to the risk of $M\mathbf{X}$.

Theorem 8.3.33. *If $I - \alpha M$ is invertible for all numbers $0 \le \alpha < 1$, if $MC = CM^*$ and $MC \ge 0$, then within the class of linear estimators the estimator $M\mathbf{X}$ is admissible if it has finite risk, i.e., $\operatorname{tr} D^{1/2}MCM^*D^{1/2} < \infty$. In particular within the class of linear estimators $\mathbf{X}$ is admissible for the risk function (8.3.15) if and only if $\operatorname{tr} D^{1/2}CD^{1/2} < \infty$.*

Proof. Recall the inequalities established in (8.3.14). If $M\mathbf{X}$ is not admissible within the class of linear estimators then there exists M_1 and $\varepsilon > 0$ such that

$$\varepsilon + \operatorname{tr} D^{1/2}M_1CM_1^*D^{1/2} + \|D^{1/2}(I - M_1)\theta\|^2$$
$$\le \operatorname{tr} D^{1/2}MCM^*D^{1/2} + \|D^{1/2}(I - M)\theta\|^2.$$

However the difference of the Bayes risks of $M\mathbf{X}$ and $\alpha M\mathbf{X}$ is of order $1 - \alpha$ and hence converges to zero as $\alpha \uparrow 1$. This contradiction shows $\varepsilon > 0$ cannot exist. □

The next result, adapted from Rao (1976), gives a finite dimensional result that fails in infinite dimensions. It says in effect that if an estimator is admissible for some quadratic loss then it is admissible for all quadratic loss functions.

Lemma 8.3.34. *In finite (p-dimensions) let B and C be symmetric positive definite matrices. Let δ be admissible for the loss function $(\delta - \theta)^t B(\delta - \theta)$. Then δ is admissible for the loss function $(\delta - \theta)^t C(\delta - \theta)$.*

Proof. Let δ_0 be better than δ for the loss using C. Take $\gamma = \delta + F(\delta_0 - \delta)$ where $F = \alpha B^{-1}C$ and α is such that $\alpha C^{1/2}B^{-1}C^{1/2} \le I$ so that $\alpha CB^{-1}C \le C$.

Then by definition $BF = \alpha C$ and $F^t BF = \alpha^2 CB^{-1}C \le \alpha C$ so that

$$\begin{aligned}\mathbf{E}(\gamma-\theta)^t B(\gamma-\theta) &\le \mathbf{E}(\delta-\theta)^t B(\delta-\theta) + \alpha\mathbf{E}(\delta-\theta)^t C(\delta_0-\delta)\\ &\quad + \alpha\mathbf{E}(\delta_0-\delta)^t C(\delta-\theta) + \alpha\mathbf{E}(\delta_0-\delta)^t C(\delta_0-\delta)\\ &= \mathbf{E}(\delta-\theta)^t B(\delta-\theta) + \alpha\mathbf{E}(\delta_0-\theta)^t C(\delta_0-\delta)\\ &\quad + \alpha\mathbf{E}(\delta_0-\delta)^t C(\delta-\theta)\\ &= \mathbf{E}(\delta-\theta)^t B(\delta-\theta) + \alpha\mathbf{E}(\delta_0-\theta)^t C(\delta_0-\theta)\\ &\quad - \alpha\mathbf{E}(\delta-\theta)^t C(\delta-\theta)\\ &< \mathbf{E}(\delta-\theta)^t B(\theta-\delta). \qquad \square\end{aligned}$$

Bayes estimators that arise from Gaussian a priori measures are linear of the form $M\mathbf{X}$ where $\mathbf{X}$ has covariance operator C and $M = S(S+C)^{-1}$. As noted above in Example 8.3.29 the operator $S(S+C)^{-1}$ need not be a bounded operator. We look briefly at the situation when M is bounded, not necessarily Bayes, and consider estimation of $\mathbf{E}_\theta(\mathbf{X}, h) = (\theta, h)$ where the risk is computed with the special choice $D^{1/2} = C^{-1/2}$, see Remark 8.3.32 and (8.3.15). The object here is to try and improve the estimator by modification of M in the way Cohen (1965) improved on linear estimators of the multivariate mean vector. Relative to any orthonormal basis $\{e_n, n \ge 1\}$ one has that $\mathbf{X} = \Sigma_{n=1}^\infty (\mathbf{X}, e_n)e_n$ almost surely so that $\|\mathbf{X}\|^2 = \Sigma_{n=1}^\infty (\mathbf{X}, e_n)^2$. For the parameters θ we need $\|C^{-1/2}\theta\| < \infty$ and one obtains for the risk function

$$\operatorname{tr}(C^{-1/2}MC^{1/2})(C^{-1/2}MC^{1/2})^* + \|(C^{-1/2}MC^{1/2} - I)C^{-1/2}\theta\|^2 \tag{8.3.16}$$

which, to be finite, requires $N = C^{-1/2}MC^{1/2}$ to be a Hilbert–Schmidt operator. If the risk (8.3.16) is identically infinite then 0 is a better estimator. We show next that if the risk is finite for some θ and if N is not self-adjoint then the risk can be improved using another linear estimator. We write the risk as

$$\operatorname{tr} NN^* + \|(N-I)\eta\|^2, \tag{8.3.17}$$

and show that the operator $N_0 = I - ((I-N)(I-N)^*)^{1/2}$ results in an improvement. If the risk is finite for some θ then $\operatorname{tr} NN^* < \infty$. Since $\|(N-I)\eta\| = \|(N_0 - I)\eta\|$ it suffices to show $\operatorname{tr} NN^* \ge \operatorname{tr} N_0 N_0^*$. We suppose N is Hilbert–Schmidt so $N + N^* - NN^*$ is a Hilbert–Schmidt operator and can be diagonalized, hence $(I-N)(I-N)^*$ can be diagonalized by an orthonormal set $e_1, e_2, \ldots$. Let $(I-N)(I-N)^* e_i = \lambda_i^2 e_i > 0$. $(I-N)(I-N)^* e = 0$ if and only if $(I-N)^* e = 0$. Define U by $(I-N)^* = U((I-N)(I-N)^*)^{1/2}$, that is, for the e_i, $\lambda_i^{-1}(I-N)^* e_i = Ue_i$. Then $(Ue_i, Ue_j) = (\lambda_i\lambda_j)^{-1}(e_i, (I-N)(I-N)^* e_j) = 0$ and $(Ue_i, Ue_i) = 1$. Hence U is orthogonal on the closure of the span of $\{e_n, n \ge 1\}$ and we may define $Ue = e$ on the orthogonal complement. Since N_0 is self-adjoint,

$$\begin{aligned}\sum_{i=1}^{n}(N_0e_i, N_0e_i) &= \sum_{i=1}^{n}[(e_i, e_i) - 2(((I-N)(I-N)^*)^{1/2}e_i, e_i) \\ &\quad + ((I-N)(I-N)^*e_i, e_i)] \\ &= 2n - 2\sum_{i=1}^{n}(Ne_i, e_i) + \sum_{i=1}^{n}(NN^*e_i, e_i) \\ &\quad - 2\sum_{i=1}^{n}(((I-N)(I-N)^*)^{1/2}e_i, e_i).\end{aligned} \tag{8.3.18}$$

Therefore to show $\operatorname{tr} N_0N_0^* \le \operatorname{tr} NN^*$ it is sufficient to show that

$$\sum_{i=1}^{n}((I-N)^*e_i, e_i) \le \sum_{i=1}^{n}(((I-N)(I-N)^*)^{1/2}e_i, e_i) = \sum_{i=1}^{n}\lambda_i. \tag{8.3.19}$$

By construction the left side is

$$\sum_{i=1}^{n}(U(I-N)(I-N)^*e_i, e_i) = \sum_{i=1}^{n}\lambda_i(Ue_i, e_i) \le \sum_{i=1}^{n}\lambda_i. \tag{8.3.20}$$

Improvement will be strict unless $Ue_i = e_i$ for all i, hence that $U = I$, in which case N is self-adjoint. This proves the following. □

Theorem 8.3.35. *If M is a bounded operator then relative to the risk function described in* (8.3.16), *if the linear estimator $M\mathbf{X}$ is admissible, then $C^{-1/2}MC^{1/2}$ is Hilbert–Schmidt, positive semidefinite, and self-adjoint.*

Remark 8.3.36. For the Gaussian process $\mathbf{X}$, $C^{1/2}\mathbf{H} = \mathbf{H}_0$ is contained in all subspaces of measure one. See Lemma 8.3.25 and the discussion of Shepp (1965). If $\mathbf{H}_1$ is such a subspace and $M: \mathbf{H}_1 \to \mathbf{H}_0$ is defined and linear, then $C^{-1/2}MC^{1/2}$ is defined as an operator on $\mathbf{H}$. Such transformations M are a special case of *measurable linear transformations* which have been studied by, amoung others, Skorohod (1974) and Rozanov (1971). A source of results on Gaussian measures on Hilbert space is Rozanov, op. cit. Mandelbaum (1983) obtains a complete characterization of the admissible linear estimators in terms of conditions on measurable linear transformations. In particular there exist unbounded M such that $M\mathbf{X}$ has finite risk and is an admissible estimator.

If $\mathbf{H}$ is a Hilbert space, and $\mathfrak{B}$ is the σ-algebra of Borel subsets, ν a probability measure on $\mathfrak{B}$, then Skorohod, op. cit., proves the following.

Theorem 8.3.37. *The following are equivalent.*

(1) *There exists a sequence $L_n: \mathbf{H} \to \mathbf{H}$ of continuous linear transformations such that the strong limit $\lim_{n\to\infty} Lx_n = Lx$ exists almost surely ν.*
(2) *There exists a subspace $\mathbf{H}_1 \subset \mathbf{H}$ such that $\nu(\mathbf{H}_1) = 1$ and $L: \mathbf{H}_1 \to \mathbf{H}$ is defined and is linear.*

8.4. Sets of Zero Measure

Problems 4.4.4 and 4.4.5, needed in the discussion of Wishart's proof, are special cases of more general results presented in this section. The author is indebted to Malley (1982) for calling many of these results to the author's attention. Two problems, 4.4.6 and 4.4.7, deal with multiplicities of eigenvalues of random matrices, and were left as problems in Chapter 4 rather than being presented formally as results in this section. The argument needed for Problems 4.4.4 and 4.4.5 is based on mathematical induction and Fubini's theorem, and is based on the assumption that the measures involved give zero measure to hyperplanes. To make this a useful hypothesis in subsequent discussions we will need to know that products of nonatomic measures do assign zero mass to hyperplanes. Hyperplanes as the boundaries of half spaces are the boundaries of convex sets. We now state a more general result about convex sets. Following the proof of Theorem 8.4.1 some generalizations and other results that were suggested by Malley (1982) are given.

Theorem 8.4.1. *Let* $\mathbf{C}$ *be a closed convex subset of* $\mathbb{R}^n$. *Let* $\mu_1, \ldots, \mu_n$ *be positive* σ*-finite nonatomic measures defined on the Borel subsets of* $\mathbb{R} = \mathbb{R}^1$. *Then*

$$\mu_1 \times \cdots \times \mu_n\,(\textit{boundary of}\ \mathbf{C}) = 0. \tag{8.4.1}$$

PROOF. By induction on the dimension n. If $n = 1$ the boundary of $\mathbf{C}$ contains at most two points and the conclusion follows since μ_1 is a nonatomic measure. We consider sections

$$\mathbf{C}_{x_n} = \{t \mid t \in \mathbb{R}^{n-1} \text{ and } (t, x_n) \in \mathbf{C}\}. \tag{8.4.2}$$

Since $\mathbf{C}$ is a closed convex set there exists $a \le b$ such that $\mathbf{C}_{x_n} \neq \phi$ if and only if $a \le x_n \le b$. If $a = b$ then both a and b are finite so that

$$\mu_1 \times \cdots \times \mu_n(\mathbf{C}) = \mu_1 \times \cdots \times \mu_{n-1}(\mathbf{C}_a)\mu_n(\{a\}) = 0, \tag{8.4.3}$$

so that in particular the boundary of $\mathbf{C}$, being a subset of $\mathbf{C}$, has zero mass. We assume in the sequel that $a < b$. If each section $\mathbf{C}_{x_n}$, $a < x_n < b$, has void interior then each $\mathbf{C}_{x_n}$ is its own boundary and by inductive hypothesis $\mu_1 \times \cdots \times \mu_{n-1}(\mathbf{C}_{x_n}) = 0$. Then

$$\begin{aligned}\mu_1 \times \cdots \times \mu_n(\mathbf{C}) &= \mu_1 \times \cdots \times \mu_{n-1}(\mathbf{C}_a)\mu_n(\{a\}) \\ &\quad + \mu_1 \times \cdots \times \mu_{n-1}(\mathbf{C}_b)\mu_n(\{b\}) \\ &\quad + \int_{a+}^{b-} 0\mu_n(dt) = 0.\end{aligned} \tag{8.4.4}$$

In (8.4.4) it is understood that if $a = -\infty$ or if $b = \infty$ then that term is zero in value, or in other words, is omitted.

In the remaining case, for some x_n, $a < x_n < b$, and $\mathbf{C}_{x_n}$ has nonvoid interior and hence **C** has nonvoid interior. This implies that every section $\mathbf{C}_{x_n}$, $a < x_n < b$, has nonvoid interior. Then using the inductive hypothesis and Fubini's theorem,

$$\begin{aligned}\mu_1 \times \cdots \times \mu_n(\text{boundary } \mathbf{C}) \le \int_{a+}^{b-} &\mu_1 \times \cdots \times \mu_{n-1}(\text{boundary } \mathbf{C}_x)\mu_n(dx)\\ &+ \mu_1 \times \cdots \times \mu_{n-1}(\mathbf{C}_a)\mu_n(\{a\})\\ &+ \mu_1 \times \cdots \times \mu_{n-1}(\mathbf{C}_b)\mu_n(\{b\}) = 0. \qquad (8.4.5)\end{aligned}$$

□

Corollary 8.4.2. *If* **C** *is a convex set in* $\mathbb{R}^n$ *then the boundary of* **C** *has zero Lebesgue measure.*

Remark 8.4.3. Consider the $n \times k$ matrix X as a point in $\mathbb{R}^{nk}$. Let X have row vectors $X_1^t, \ldots, X_n^t$. Then Problem 4.4.4 and Theorem 8.4.1 imply that the set (of first k rows)

$$\mathbf{N} = \{X \mid X_1, \ldots, X_k \text{ are linearly dependent}\} \qquad (8.4.6)$$

has Lebesgue measure zero. If X is not in **N** then the first k rows of X are linearly independent and form a $k \times k$ nonsingular matrix $G^t = (X_1, \ldots, X_k)$. Set

$$Y^t = (I_k, Y_1, \ldots, Y_{n-k}) \quad \text{and} \quad YG = X. \qquad (8.4.7)$$

The matrix Y is uniquely determined.

The null set **N** is invariant under the action of $\mathbf{GL}(k)$ acting as right multipliers of X. If $X = YG$ and $G_1 \in \mathbf{GL}(k)$ then $XG_1 = Y(GG_1)$. Thus Y represents a maximal invariant under the group action.

The remainder of this section is a discussion of generalizations of the idea of Theorem 8.4.1 and the consequences thereof, due to Malley (1982) and others, as named.

Definition 8.4.4. A $p \times 1$ random vector **X** is *flat free* if for every $p \times 1$ vector a and constant c,

$$P(a^t\mathbf{X} = c) = 0. \qquad (8.4.8)$$

Corollary 8.4.5. *If* **X** *is* $p \times 1$ *consisting of independent nonatomic random variables then* **X** *is a flat free random vector.*

A function f of n $p \times 1$ vector variables will be called a *multilinear functional* if f is linear in each variable, the others being held fixed.

Theorem 8.4.6. *If* $\mathbf{X}_1, \ldots, \mathbf{X}_n$ *are independent flat free* $p \times 1$ *random vectors and if* $f \neq 0$ *is a multilinear function of* n *variables then* $P(f(\mathbf{X}_1, \ldots, \mathbf{X}_n) = 0) = 0$.

PROOF. By induction on n. *Case* $n = 1$. If $f \neq 0$ is a function of one variable and $f(\mathbf{X}) = 0$ then for some $p \times 1$ vector a, $a^t\mathbf{X} = 0$. By definition of being flat free, $P(a^t\mathbf{X} = 0) = 0$.

Given $\mathbf{X}_1 = x_1, \ldots, \mathbf{X}_{n-1} = x_{n-1}$, by hypothesis $P(f(x_1, \ldots, x_{n-1}, \mathbf{X}_n) = 0) = 0$. By Fubini's theorem the result follows. □

EXAMPLE 8.4.7. The event that $0 = \det(\mathbf{X}_1, \ldots, \mathbf{X}_p)$ is an event of zero probability. See Problem 4.4.4.

We now consider generalizations to algebraic independence of random variables.

Lemma 8.4.8. *Let* $g(x, y)$ *be a nonzero polynomial in the entries of a* $p \times 1$ *vector* x *and* 1×1 *vector* y. *Then* $\{y|\text{all } x, g(x, y) = 0\}$ *is a finite set.*

PROOF. Write $g(x, y) = \Sigma_{i=0}^n g_i(x)y^i$ where n is the largest degree of a y-factor. Then if for a given x, $g(x, y_i) = 0$ for $n + 1$ distinct $y_0, \ldots, y_n$ the equations $0 = \Sigma_{i=0}^n g_i(x)y_j^i$ have the unique solution $g_0(x) = 0, \ldots, g_n(x) = 0$ since the coefficient matrix is nonsingular. Consequently if $\{y|\text{all } x, g(x, y) = 0\}$ has at least $n + 1$ elements then $g_j(x)$ vanishes identically, hence g is identically zero. □

Definition 8.4.9. Univariate random variables $\mathbf{X}_1, \ldots, \mathbf{X}_n$ are *algebraically independent* if they satisfy nontrivial polynomial equations $f(\mathbf{X}_1, \ldots, \mathbf{X}_n) = 0$ only on sets of measure zero.

Remark 8.4.10. Definition 8.4.9 requires the random variables to be nonatomic.

Theorem 8.4.11. *Stochastically independent nonatomic univariate random variables* $\mathbf{X}_1, \ldots, \mathbf{X}_p$ *are algebraically independent.*

PROOF. By induction on the number of variables p. If f is a polynomial of $p + 1$ variables then the section $f(x_1, \ldots, x_p, \cdot) = 0$ can have at most $p + 1$ elements unless $f(x_1, \ldots, x_p, x_{p+1}) = \Sigma_{i=0}^n f_i(x_1, \ldots, x_p)x_{p+1}^i$ and the coefficient functions vanish. Since $\mathbf{X}_{p+1}$ is nonatomic if $P(f(\mathbf{X}_1, \ldots, \mathbf{X}_{p+1}) = 0) > 0$ by Fubini's theorem there must be a p-dimensional set $\mathbf{A}$ with $P((\mathbf{X}_1, \ldots, \mathbf{X}_p) \in \mathbf{A}) > 0$ on which $f_i(\mathbf{X}_1, \ldots, \mathbf{X}_p) = 0$, $1 \le i \le n$. By inductive hypothesis $P(f_i(\mathbf{X}_1, \ldots, \mathbf{X}_p) = 0) = 0$. This contradiction shows that $\mathbf{A}$ cannot exist. Thus the inductive argument will be complete if we verify the case $p = 1$. Then f is a nonzero polynomial and if $f \neq 0$ then $P(f(\mathbf{X}_1) = 0) > 0$ implies $\mathbf{X}_1$ has an atom. This contradiction completes the argument. □

Theorem 8.4.12 (Okamota (1973)). *Let the* $n \times p$ *random matrix* $\mathbf{X}$ *have a density function relative to* $n \times p$ *Lebesgue measure. If* $f \neq 0$ *is a polynomial of the np entries of the matrix* X *then* $P(f(\mathbf{X}) = 0) = 0$.

PROOF. Because there is a density function, if $P(f(\mathbf{X}) = 0) > 0$ then $\{x \mid f(x) = 0\}$ has positive Lebesgue measure. Consequently if we assign to the component variables of X normal (0, 1) densities the matrix $\mathbf{X}$ becomes a $n \times p$ matrix of independent normal (0, 1) random variables and $P(f(\mathbf{X}) = 0) > 0$. By Theorem 8.4.11, $f = 0$. □

Remark 8.4.13. Random variables which are nonzero polynomial functions of random variables which are independent and nonatomic, such as the sample moments, central moments, and U-statistics, again have nonatomic distributions.

Theorem 8.4.14 (Eaton and Perlman (1973)). *Let* $\tilde{\mathbf{X}} = (\mathbf{X}, A)$ *where*

(1) $\mathbf{X}$ *is* $n \times p$ *and has independent flat-free columns;*
(2) A *is* $n \times r$ *and is a constant matrix of rank* r.
(3) $n \geq p + r$.

Then with probability one, $\tilde{\mathbf{X}}$ *has* rank $p + r$.

PROOF. Permute the rows so that the last r rows of A have rank r. Partition $(\mathbf{X}, A) = \begin{vmatrix} \mathbf{X}_1 & A_1 \\ \mathbf{X}_2 & A_2 \end{vmatrix}$ where $\mathbf{X}_1$ is $(n - r) \times p$, $\mathbf{X}_2$ is $r \times p$, A_1 is $(n - r) \times r$ and A_2 is $r \times r$, A_2 of rank r. Multiply on the left by

$$\begin{vmatrix} I_{n-r} & -A_1 A_2^{-1} \\ 0 & I_r \end{vmatrix} \quad \text{to obtain the matrix} \quad \begin{vmatrix} \mathbf{X}_1 - A_1 A_2^{-1} \mathbf{X}_2 & 0 \\ \mathbf{X}_2 & A_2 \end{vmatrix}.$$

Since A_2 has rank r, $(\mathbf{X}, A)$ is singular if and only if $\mathbf{X}_1 - A_1 A_2^{-1} \mathbf{X}_2$ is singular. $\mathbf{X}_1$ has $n - r \geq p$ independent flat-free rows, hence by Example 8.4.6 the first p rows of $\mathbf{X}_1$ are of full rank almost surely. Therefore conditional on $\mathbf{X}_2 = X_2$, $\mathbf{X}_1 - A_1 A_2^{-1} \mathbf{X}_2$ is of rank p, so by Fubini's theorem, $\mathbf{X}_1 - A_1 A_2^{-1} \mathbf{X}_2$ is of rank p almost surely, hence $(\mathbf{X}, A)$ is of rank $p + r$ almost surely since A_2 is of rank r.

Corollary 8.4.15 (Eaton and Perlman (1973)). *Let the random* $n \times p$ *matrix* $\mathbf{X}$ *have independent flat-free rows. Let* A *be a* $n \times n$ *symmetric positive semidefinite matrix of* rank r *and assume* $r \geq p$. *Then almost surely* $\mathbf{X}^t A \mathbf{X}$ *is positive definite.*

PROOF. Let Γ be $n \times (n - r)$, the columns of Γ orthonormal, each column an eigenvector, for the eigenvalue zero, of A. Let $A^{1/2}$ be the positive semidefinite square root of A. Then $\mathbf{X}^t A \mathbf{X}$ and $X^t A^{1/2}$ have the same rank, so that $\mathbf{X}^t A \mathbf{X}$ has rank $< p$ if and only if there exists a $p \times 1$ vector a with $a^t \mathbf{X}^t A^{1/2} = 0$. Then $a^t \mathbf{X}^t$ is a linear combination of the columns of Γ and there exists a $(n - r) \times 1$ vector b with $a^t \mathbf{X}^t = \Gamma b$. Hence $\mathbf{X}, \Gamma$ is of less than full rank. By Theorem 8.4.14 this is an event of probability zero. □

Corollary 8.4.16 (Okamoto (1973)). *Let the $n \times p$ random matrix $\mathbf{X}$ have independent flat-free rows. Let A be a $n \times n$ symmetric matrix of* rank $r < p$. *Then almost surely* rank $\mathbf{X}^t A \mathbf{X} = r$.

PROOF. Let $A = \Sigma_{i=1}^r \lambda_i u_i u_i^t$ where the $n \times 1$ vectors $u_1, \ldots, u_r$ are orthonormal. Here we allow possibly negative values λ_i. Write $U = (u_1, \ldots, u_r)$ and Λ the $r \times r$ diagonal matrix with $\lambda_1, \ldots, \lambda_r$ on the diagonal so that $A = U\Lambda U^t$. Partition $\mathbf{X} = (\mathbf{X}_1, \mathbf{X}_2)$ with $\mathbf{X}_1$ a $n \times r$ matrix and $\mathbf{X}_2$ a $n \times (p - r)$ matrix. Then

$$\mathbf{X}^t A \mathbf{X} = \begin{vmatrix} \mathbf{X}_1^t U\Lambda U^t \mathbf{X}_1 & \mathbf{X}_1^t U\Lambda U^t \mathbf{X}_2 \\ \mathbf{X}_2^t U\Lambda U^t \mathbf{X}_1 & \mathbf{X}_2^t U\Lambda U^t \mathbf{X}_2 \end{vmatrix}. \tag{8.4.9}$$

Since $\mathbf{X}_1^t U$ is $r \times r$, with real entries and since Λ is nonsingular, with a complex square root $\Lambda^{1/2}$, we have

$$0 = \det(\mathbf{X}_1^t U\Lambda U^t \mathbf{X}_1) = (\det(\mathbf{X}_1^t U\Lambda^{1/2}))^2 = (\det(\mathbf{X}_1^t U))^2(\det \Lambda) \tag{8.4.10}$$

if and only if $\mathbf{X}_1^t U$ is singular. Adjoin to U columns V so that U, V is $n \times n$ orthogonal. Then $a^t \mathbf{X}_1^t U = 0$ implies there exists a vector b with $a^t X_1^t = (Vb)^t$. V is $n \times (n - r)$ and $\mathbf{X}$, V is $n \times n$ singular in contradiction of Theorem 8.4.14. Hence $r \le \operatorname{rank}(\mathbf{X}_1^t U\Lambda U^t \mathbf{X}_1) \le \operatorname{rank}(\mathbf{X}^t A \mathbf{X}) \le \operatorname{rank} A = r$. □

8.5. Problems

PROBLEM 8.5.1 (Gram–Schmidt Orthogonalization). Let X be a $n \times k$ matrix of rank h with $h \le k \le n$ and suppose the first h columns of X are linearly independent. Then there exists a $h \times h$ matrix $T_1 \in \mathbf{T}(h)$, a $(k - h) \times h$ matrix T_2 and a $n \times h$ matrix A such that if $T^t = (T_1^t, T_2^t)$ then

$$A^t A = I_h \quad \text{and} \quad X = AT^t. \tag{8.5.1}$$

PROBLEM 8.5.2. Continue Problem 8.5.1. Assume rank $X = k$. Compute $\bigwedge_{i=1}^n \bigwedge_{j=1}^k dx_{ij}$ in terms of dA and dT, using (8.5.1). Augment A so that the matrix $(A, B) \in \mathbf{O}(n)$ and consider

$$\begin{vmatrix} A^t \\ B^t \end{vmatrix} dX(T^t)^{-1} = \begin{vmatrix} I_k \\ 0 \end{vmatrix} dT(T^t)^{-1} + \begin{vmatrix} A^t \\ B^t \end{vmatrix} dA. \tag{8.5.2}$$

When the indicated join is formed the left side is

$$\bigwedge_{i=1}^n \bigwedge_{j=1}^k \left(\begin{vmatrix} A^t \\ B^t \end{vmatrix} dX(T^t)^{-1} \right)_{ij} = (\det T)^{-n} \bigwedge_{i=1}^n \bigwedge_{j=1}^k dx_{ij}. \tag{8.5.3}$$

On the right side of (8.5.2), $dT^t(T^t)^{-1}$ is an upper triangular matrix. This forces the answer on the right side to be of the form

$$\pm \bigwedge_{1\le i\le j\le k} (dT^t(T^t)^{-1})_{ij} \bigwedge_{1\le j<i\le k} a_i^t\, da_j \bigwedge_{i=1}^{n-k} \bigwedge_{j=1}^{k} b_i^t\, da_j. \tag{8.5.4}$$

This is the product of two invariant differential forms. Use Problems 6.7.6 and 6.7.7.

Problem 8.5.3. Continue Problem 8.5.2. Make the change of variable $S = TT^t$ and obtain the Wishart density function. See Section 4.1, Section 5.2, and Problem 6.7.6.

Problem 8.5.4 (James (1954)). Let X be $n \times p$ and Y be $n \times q$, each of maximal rank, with $p \le q \le n$. Let $H \in \mathbf{GL}(p)$. Then XH makes the same critical angles with Y as does X. Hence the critical angles are an invariant.

Problem 8.5.5 (James (1954)). Continue Problem 8.5.4. Let Y represent a fixed hyperplane Q and X represent a variable hyperplane P. We are going to decompose X into a product of two Stiefel manifolds and a simplex as follows (cf. James, op. cit., for a treatment of the cases not discussed in this problem). Refer to Theorem 8.2.1. Project Xa_i on Q and $Q^\perp$ as in (8.5.5) below. Let

$$\begin{aligned} \tilde{\alpha}_i &= Y(Y^tY)^{-1}Y^tXa_i, \qquad 1 \le i \le p; \\ \tilde{\beta}_i &= (I_n - Y(Y^tY)^{-1}Y^t)Xa_i, \qquad 1 \le i \le p. \end{aligned} \tag{8.5.5}$$

Show that

$$\tilde{\alpha}_i^t\tilde{\beta}_j = 0, \qquad 1 \le i, j \le p \tag{8.5.6}$$

and

$$\|\tilde{\alpha}_i\|^2 + \|\tilde{\beta}_i\|^2 = 1, \qquad 1 \le i \le p;$$

further show that $\tilde{\alpha}_1, \ldots, \tilde{\alpha}_p$ are mutually orthogonal and $\tilde{\beta}_1, \ldots, \tilde{\beta}_p$ are mutually orthogonal.

Write

$$\tilde{\alpha}_i = \alpha_i \cos\theta_i \quad \text{and} \quad \tilde{\beta}_i = \beta_i \sin\theta_i, \tag{8.5.7}$$

where

$$\|\alpha_i\| = \|\beta_i\| = 1, \qquad 1 \le i \le p.$$

Then

$$Xa_i = \tilde{\alpha}_i + \tilde{\beta}_i = \alpha_i\cos\theta_i + \beta_i\sin\theta_i. \tag{8.5.8}$$

If the canonical correlations are distinct real numbers then the vectors $Xa_1, \ldots, Xa_p$, being eigenvectors, are uniquely determined except for sign changes. Fix the signs by requiring

$$\tilde{\tilde{\alpha}}_i = \pm\tilde{\alpha}_i \text{ such that the first component of } \tilde{\tilde{\alpha}}_i \text{ is} > 0. \tag{8.5.9}$$

$$\begin{array}{l}\cos\theta_i > 0 \text{ so } -\pi/2 < \theta_i < \pi/2 \text{ and we eliminate} \\ \text{sign changes on } \beta_i \text{ by requiring } 0 \le \theta_i < \pi/2.\end{array} \tag{8.5.10}$$

Then show

$$\text{each plane of the Grassman manifold } \mathbf{G}_{p,n-p} \tag{8.5.11}$$

whose critical angles with Y are distinct is uniquely representable by $(\tilde{\tilde{\alpha}}_1, \ldots, \tilde{\tilde{\alpha}}_p)$, $(\tilde{\tilde{\beta}}_1, \ldots, \tilde{\tilde{\beta}}_p)$ and $(\theta_1, \ldots, \theta_p)$ with $0 < \theta_1 < \cdots < \theta_p$. Count dimensions and show that the correct dimension results for $\mathbf{G}_{p,n-p}$. That is, $\tilde{\alpha}_1, \ldots, \tilde{\alpha}_p$ generate a p-frame in a q-dimensional space and $\tilde{\beta}_1, \ldots, \tilde{\beta}_p$ generate a p-frame in a $(n-q)$-dimensional space. Compute the dimensions of $\mathbf{V}_{p,q}$ and $\mathbf{V}_{p,n-q}$. Then, use Problem 8.5.4 to write X as a product of $H \in \mathbf{GL}(k)$ and a point of $\mathbf{G}_{p,n-p}$. Check by adding dimensions.

PROBLEM 8.5.6. Let D be a diagonal matrix with pairwise distinct diagonal entries such that D^{-1} exists. If $U \in \mathbf{O}(n)$ such that DU is symmetric then U is also a diagonal matrix.

PROBLEM 8.5.7. Suppose S is a symmetric positive definite $n \times n$ matrix and $U \in \mathbf{O}(n)$ such that $US^2U^t = D^2$ is the square of a diagonal matrix D. If the entries of D are nonnegative then $USU^t = D$.

HINT. Show that $V = D^{-1}USU^t$ is an orthogonal matrix and that DV is a symmetric matrix. Hence $V = I_n$. □

PROBLEM 8.5.8. Suppose S, $T \in \mathbf{S}(n)$ and $S^2 = T^2$. Then $S = T$.

HINT. Associate to each eigenvalue of S the subspace of eigenvectors. These are the eigenspaces for S^2. Therefore S and T have the same eigenspaces and eigenvalues. □

PROBLEM 8.5.9. Let X be a $n \times k$ matrix and $\operatorname{rank} X = k \le n$. Then there exists a $n \times k$ matrix A and a $k \times k$ matrix $S \in \mathbf{S}(k)$ such that

$$X = AS \quad \text{and} \quad A^tA = I_k. \tag{8.5.12}$$

Show that this factorization is unique.

HINT. $X = [X(X^tX)^{-1/2}](X^tX)^{1/2}$. □

PROBLEM 8.5.10. In Theorem 8.1.3 replace (8.1.1) by the hypothesis that

$$A \text{ has rank } p \le n \text{ and } \det \begin{vmatrix} a_{11} & \cdots & a_{1k} \\ \vdots & & \vdots \\ a_{k1} & \cdots & a_{kk} \end{vmatrix} \ge 0 \tag{8.5.13}$$

if $1 \le k \le n$ and is greater than zero if $1 \le k \le p$.

Show that there exists a $n \times p$ matrix $T = (t_{ij})$ such that

$$t_{ij} = 0 \quad \text{if} \quad j > i, \qquad TT^t = A, \qquad \text{and} \quad t_{ii} > 0 \quad \text{if} \quad 1 \le i \le p. \tag{8.5.14}$$

HINT. $A + \varepsilon I_n \in \mathbf{S}(n)$. Find $T_\varepsilon \in \mathbf{T}(n)$ such that $T_\varepsilon T_\varepsilon^t = A + \varepsilon I_n$. As $\varepsilon \to 0$, T_ε is bounded. Take a convergent subsequence convergent to some T_0. Then rank $T_0 T_0^t = \text{rank } T_0 = p$. The minor

$$\begin{vmatrix} t_{11} & 0 & \cdots & 0 \\ t_{21} & t_{22} & \cdots & 0 \\ \vdots & \vdots & & \\ t_{p1} & t_{p2} & \cdots & t_{pp} \end{vmatrix} = S \quad \text{must satisfy} \quad SS^t = \begin{vmatrix} a_{11} & \cdots & a_{1p} \\ \vdots & & \vdots \\ a_{p1} & \cdots & a_{pp} \end{vmatrix} > 0$$

so S is nonsingular. This implies the result. □

PROBLEM 8.5.11 (Uniqueness). Under the hypotheses of Problem 8.5.10, if T_1 and T_2 are $n \times p$ matrices satisfying (8.5.14) then $T_1 = T_2$.

HINT. Write $T_1 = \begin{vmatrix} T_{11} \\ T_{12} \end{vmatrix}$ and $T_2 = \begin{vmatrix} T_{21} \\ T_{22} \end{vmatrix}$ with T_{11} and T_{21} in $\mathbf{T}(p)$. Then $T_{11} = T_{21}$ and these matrices are nonsingular. Examine $T_1 T_1^t = T_2 T_2^t$. □

PROBLEM 8.5.12. Let X be a $n \times p$ matrix, $p \le n$, such that the minor $\begin{vmatrix} x_{11} & \cdots & x_{1p} \\ \vdots & & \vdots \\ x_{p1} & \cdots & x_{pp} \end{vmatrix}$ is nonsingular. There exist matrices $U \in \mathbf{O}(p)$ and $T = (t_{ij})$ such that

$$X = TU \quad \text{and if} \quad j > i \quad \text{then} \quad t_{ij} = 0. \tag{8.5.15}$$

This factorization is unique.

HINT. $X = \begin{vmatrix} X_1 \\ X_2 \end{vmatrix}$ with X_1 a $p \times p$ matrix. Then $X_1 = T_1 U$ with $U \in \mathbf{O}(p)$ and $T_1 \in \mathbf{T}(p)$. T_1 and U are uniquely determined. Required is T_2 solving the equation $T_2 U = X_2$. □

The following result which is stated as a problem has already been used in Problem 4.4.7.

PROBLEM 8.5.13. Let S_1 and S_2 be symmetric positive semidefinite matrices such that $S_1 + S_2 = S > 0$. There exists $G \in \mathbf{GL}(n)$ such that $GSG^t = I_n$ and GS_1G^t is a diagonal matrix. (Uniqueness) If the diagonal entries of GS_1G^t are pairwise distinct and if $G_1 \in \mathbf{GL}(n)$ satisfies

$$G_1 S G_1^t = I_n \quad \text{and} \quad G_1 S_1 G_1^t = G S_1 G^t \tag{8.5.16}$$

then $GG^{-1} \in \mathbf{O}(n)$ and is a diagonal matrix.

PROBLEM 8.5.14. Suppose $\begin{vmatrix} A & B \\ B^t & C \end{vmatrix}$ is a $n \times n$ matrix and that $n = p + q$, A is $p \times p$ and C is $q \times q$. If A^{-1} exists then show

$$\begin{vmatrix} I_p & 0 \\ -B^t A^{-1} & I_q \end{vmatrix} \begin{vmatrix} A & B \\ B^t & C \end{vmatrix} \begin{vmatrix} I_p & -A^{-1}B \\ 0 & I_q \end{vmatrix} = \begin{vmatrix} A & 0 \\ 0 & C - B^t A^{-1} B \end{vmatrix}. \tag{8.5.17}$$

Therefore show

$$\det \begin{vmatrix} A & B \\ B^t & C \end{vmatrix} = (\det A)(\det (C - B^t A^{-1} B)). \tag{8.5.18}$$

PROBLEM 8.5.15. Suppose $\begin{vmatrix} A & B \\ B^t & C \end{vmatrix}$ is as in Problem 8.5.14. Suppose A^{-1} and C^{-1} exist. Then $(\det A)(\det (C - B^t A^{-1} B)) = (\det C)(\det (A - BC^{-1}B^t))$. In addition, if $\begin{vmatrix} D & E \\ E^t & F \end{vmatrix}$ is the inverse of $\begin{vmatrix} A & B \\ B^t & C \end{vmatrix}$ then $F^{-1} = C - B^t A^{-1} B$.

HINT. Pass to $\begin{vmatrix} C & B^t \\ B & A \end{vmatrix}$ by permutation of rows and columns and show the number of interchanges is even. Use Problem 8.5.14. □

PROBLEM 8.5.16. (G. P. H. Styan). If A is a $p \times p$ matrix then $\operatorname{rank}(I_p - A) + \operatorname{rank} A = p + \operatorname{rank}(A - A^2)$.

HINT.

$$\begin{vmatrix} I_p & A \\ A & A \end{vmatrix} = \begin{vmatrix} I_p & I_p \\ 0 & I_p \end{vmatrix} \begin{vmatrix} I_p - A & 0 \\ 0 & A \end{vmatrix} \begin{vmatrix} I_p & 0 \\ I_p & I_p \end{vmatrix}$$

$$= \begin{vmatrix} I_p & 0 \\ A & I_p \end{vmatrix} \begin{vmatrix} I_p & 0 \\ 0 & A - A^2 \end{vmatrix} \begin{vmatrix} I_p & A \\ 0 & I_p \end{vmatrix}.$$ □

PROBLEM 8.5.17. $\operatorname{rank} A + \operatorname{rank}(I_p - A) = p$ if and only if $A^2 = A$.

PROBLEM 8.5.18. Let $\mathbf{X}_1, \ldots, \mathbf{X}_n$ be independently distributed random variables, and $\mathfrak{F}_i$ be the least σ-algebra in which $\mathbf{X}_1, \ldots, \mathbf{X}_{i-1}, \mathbf{X}_{i+1}, \ldots, \mathbf{X}_n$ are measurable. Set $P_i = \mathbf{E}(\ /\mathfrak{F}_i)$. Show that P_i is a projection on the $\mathbf{L}_2$ space of functions f and show that $P_i P_j = P_j P_i$ for all i, j.

PROBLEM 8.5.19. If projections $P_1, \ldots, P_n$ commute then $\prod_{i=1}^n ((I - P_i) + P_i) = \Sigma_s \prod_{i \in s} P_i \prod_{i \notin s} (I - P_i)$ expresses the identity map as a sum of orthogonal projections to which the Cochran type theorem applies. See Theorem 8.1.14.

PROBLEM 8.5.20. In the study of U-statistics the projections of Problem 8.5.18 are used in the decomposition 8.5.19 and applied to a symmetric function u. See Dynkin and Mandelbaum (1982).

PROBLEM 8.5.21 (Karlin (1960)). Write $|A|$ for the determinant of the square matrix A. Show that $|XX^t + S| = |S||I + X^tS^{-1}X|$.

HINT. Expand the matrix products:

$$\begin{vmatrix} I & 0 \\ X^tS^{-1} & -I \end{vmatrix}\begin{vmatrix} S & X \\ X^t & -I \end{vmatrix}\begin{vmatrix} I & -S^{-1}X \\ 0 & I \end{vmatrix}; \begin{vmatrix} -I & 0 \\ -X & I \end{vmatrix}\begin{vmatrix} -I & X^t \\ X & S \end{vmatrix}\begin{vmatrix} -I & X^t \\ 0 & I \end{vmatrix}.$$

Show by permutation of rows and columns that

$$\det\begin{vmatrix} -I & X^t \\ X & S \end{vmatrix} = \det\begin{vmatrix} S & X \\ X^t & -I \end{vmatrix}.$$

□

PROBLEM 8.5.22. If X is $n \times h$, take $S = -\lambda I$ and obtain

$$|XX^t - \lambda I| = \lambda^{n-h}|-I||X^tX - \lambda I|.$$

Thus the nonzero eigenvalues of XX^t and X^tX are the same, multiplicities included.

PROBLEM 8.5.23. Let X, Y be $n \times h$ matrices and D a $n \times n$ diagonal matrix with diagonal entries $d_1, \ldots, d_n$. Then show $\det(X^tDY)$ is a homogeneous polynomial of degree h in $d_1, \ldots, d_n$. By taking subsets $d_{i_1}, \ldots, d_{i_h}$ and setting the others zero show the only terms with nonzero coefficients are $d_{i_1}d_{i_2}\cdots d_{i_h}$, $i_1 < i_2 < \cdots < i_h$.

PROBLEM 8.5.24. Define $X(i_1, \ldots, i_h)$ to be the $h \times h$ principal minor obtained from X by elimination of the other $n - h$ rows. Use Problem 8.5.23, taking subsets of h of the d_i and setting these equal one, to show

$$|X^tY| = \sum_{i_1<\cdots<i_h}\cdots\sum |X(i_1, \ldots, i_h)||Y(i_1, \ldots, i_h)|.$$

This result is part of the Binet–Cauchy theorem. See Marshall and Olkin (1979), page 503.

Definition 8.5.25. From a square $n \times n$ matrix A form the *compound matrix* $A^{(k)}$ which is the $\binom{n}{k} \times \binom{n}{k}$ matrix with $((i_1, \ldots, i_k), (j_1, \ldots, j_k))$-entry the determinant of the corresponding minor of A. The indices $(i_1, \ldots, i_k)$ are to be ordered lexiographically.

PROBLEM 8.5.26. Use Problem 8.5.24 to show

$$(AB)^{(k)} = A^{(k)}B^{(k)}. \tag{8.5.19}$$

That is, the compound matrix formed from AB is the product of $A^{(k)}$ and $B^{(k)}$. Then show

$$I^{(k)} \text{ is the identity matrix};$$

$$(A^*)^{(k)} = (A^{(k)})^*; \quad \text{(transpose or complex conjugate transpose)}$$

$$(A^{-1})^{(k)} = (A^{(k)})^{-1};$$

If A is lower triangular then so is $A^{(k)}$.

(8.5.20)

Problem 8.5.27 (Weyl (1949)). Over the complex numbers the eigenvalues of $A^{(k)}$ are the homogeneous products of degree k of distinct factors $\lambda_{i_1} \cdot \lambda_{i_2} \cdots \lambda_{i_k}$, $\lambda_i = \lambda_i(A)$.

Problem 8.5.28. Over the complex numbers if λ is the eigenvalue of A of largest modulus, and v the largest eigenvalue of $(A^*A)^{1/2}$, then $|\lambda| \le v$.

Hint. Let x be a unit vector with $\lambda x = Ax$. Then $|\lambda|^2 = |(A^*Ax, x)| \le v^2$. □

Problem 8.5.29 (Weyl (1949)). If the eigenvalues $\lambda_1, \ldots, \lambda_n$ of A are ordered so the $|\lambda_1| \ge |\lambda_2| \ge \cdots \ge |\lambda_n|$, and the eigenvalues of $(A^*A)^{1/2}$ are ordered, $v_1 \ge v_2 \ge \cdots \ge v_n$, then for $1 \le k \le n$, $|\lambda_1 \cdots \lambda_k| \le v_1 \cdots v_k$, and equality holds if $k = n$.

CHAPTER 9

Examples Using Differential Forms

9.0. Introduction

In this chapter we calculate probability density functions for the canonical correlations (Section 9.1), Hotelling T^2 (Section 9.2), and the eigenvalues of the sample covariance matrix (Section 9.3). The calculations of Section 9.1 were stated by James (1954) who did the problem in the central case only. The noncentral case was computed by Constantine (1963). Our derivation differs somewhat from that of Constantine in that we place more emphasis on the use of differential forms. The results of Section 9.3 are taken directly from James (1954). The calculations of Section 9.2 are original to the author and are inserted in order to include this important example. In the problems, Section 9.4, several problems present background material. Problems 9.4.7 and 9.4.8 treat the distribution of the correlation coefficients using differential forms. There are other examples of the use of differential forms in the book, notably Section 5.3 on the eigenvalues of the covariance matrix, which should be compared with Section 9.3, Section 10.3 on the noncentral multivariate beta density function, which should be compared with Section 5.7, and Section 11.0.1 on the decomposition $X = AT^t$, $T \in \mathbf{T}(k)$ and A a $n \times k$ matrix such that $A^tA = I_k$.

In reading Chapter 9 keep in mind that we are dealing with examples of method as applied to standard examples. Thus, in Chapter 5 we explored methods of Karlin (1960) which in many problems lead to a complete determination of the noncentral density function by means of an application of the existence and uniqueness theorems for invariant measures, discussed in Section 3.3. All the statistics discussed in Chapter 5 and in this chapter are maximal invariants or are closely related to maximal invariants. As in Chapter 5, although the method is different, the use of differential forms as a method allows determination not only of the form of the density but also the normalizations required to give mass one. In Chapter 10 a totally different method of calculation of the density function of a maximal invariant is developed. As will be seen on reading of Chapter 10 the method developed there is best suited to determination of ratios of density functions in which normalizations cancel.

A basic concept of the computation methods used in this chapter is as follows. Given are manifolds $\mathfrak{M}_1$ and $\mathfrak{M}_2$ with a transformation $f: \mathfrak{M}_1 \to \mathfrak{M}_2$. Unlike the development given in Chapter 6, f is not assumed to be a homeomorphism onto. We suppose f is continuous and onto. We suppose $\mathfrak{G}$ is a transformation group acting on $\mathfrak{M}_1$ such that

$$(g, m) \to g(m) \tag{9.0.1}$$

is jointly continuous. We suppose f satisfies

$$\begin{gathered} \text{if} \quad g \in \mathfrak{G},\ x, y \in \mathfrak{M}_1, \quad \text{and} \quad f(x) = f(y) \\ \text{then} \quad f(g(x)) = f(g(y)). \end{gathered} \tag{9.0.2}$$

Then f induces a transformation group $\bar{\mathfrak{G}}$ on $\mathfrak{M}_2$ by, if $x \in \mathfrak{M}_1$ then

$$\bar{g}f(x) = f(g(x)). \tag{9.0.3}$$

Then clearly by (9.0.2) it follows that $\bar{g}(y)$ is defined uniquely for all $y \in \mathfrak{M}_2$. If μ is a σ-finite measure on the Borel subsets of $\mathfrak{M}_1$ then f induces a measure $\bar{\mu}$ on the Borel subsets of $\mathfrak{M}_2$ by

$$\bar{\mu}(\mathbf{A}) = \mu(f^{-1}(\mathbf{A})). \tag{9.0.4}$$

Then if $g \in \mathfrak{G}$, by (9.0.3) it follows that

$$\bar{\mu}(\bar{g}^{-1}(\mathbf{A})) = \mu(f^{-1}(\bar{g}^{-1}(\mathbf{A}))) = \mu(g^{-1}f^{-1}(\mathbf{A})). \tag{9.0.5}$$

In particular, if the measure μ is an invariant measure for the group $\mathfrak{G}$ then the induced measure $\bar{\mu}$ is invariant for the induced group $\bar{\mathfrak{G}}$. We state this formally.

Lemma 9.0.1. (James (1954)). *If μ is invariant for $\mathfrak{G}$ then the induced measure $\bar{\mu}$ is invariant for the induced group $\bar{\mathfrak{G}}$.*

We illustrate this discussion by the example of $n \times k$ matrices X as points of $\mathbb{R}^{nk}$, $k \le n$, rank $X = k$. We let $\mathfrak{G}$ be $\mathbf{O}(n)$ acting as left multipliers of X, that is,

$$g(X) = GX, \qquad G \in \mathbf{O}(n). \tag{9.0.6}$$

We let $f(X) \in \mathbf{G}_{k,n-k}$ be the hyperplane through 0 spanned by the columns of X. Clearly

$$\begin{gathered} f(X_1) = f(X_2) \quad \text{if and only if} \\ \text{there exists} \quad H \in \mathbf{GL}(k) \quad \text{such that} \quad X_1 H = X_2. \end{gathered} \tag{9.0.7}$$

Then $g(X_1) = GX_1$ and $g(X_2) = GX_2$ and $g(X_1)H = g(X_2)$, that is,

$$f(g(X_1)) = f(g(X_2)). \tag{9.0.8}$$

Therefore the group $\overline{\mathfrak{G}}$ of induced transformations is defined. Lebesgue measure on $\mathbb{R}^{nk}$ is invariant under $\mathfrak{G}$, but since $\mathbb{R}^{nk}$ is being factored by the noncompact subgroup $\mathbf{GL}(k)$ (under the action of f) the measure induced on the compact manifold $\mathbf{G}_{k,n-k}$ is not regular. Given some other finite invariant measure ν on $\mathbb{R}^{nk}$ the induced measure $\bar{\nu}$ will be finite and invariant. Since $\mathbf{G}_{k,n-k}$ is $\mathbf{O}(n)/\mathbf{O}(k) \times \mathbf{O}(n-k)$ the invariant measures are uniquely determined up to normalization. See Theorem 3.3.8. Thus if $\nu(\mathbb{R}^{nk}) = 1$ the induced measure $\bar{\nu}$ has mass one and is given by a differential form, cf. Section 7.5,

$$K^{-1} \bigwedge_{j=1}^{n-k} \bigwedge_{i=1}^{k} b_j^t \, da_i. \tag{9.0.9}$$

This differential form is expected to appear in the integrations as the form of the integrating measure. From Section 7.9,

$$K = A_n \cdots A_{n-k+1}/A_k \cdots A_1. \tag{9.0.10}$$

9.1. Density Function of the Critical Angles

Definitions have been given in Section 8.2. See (8.2.38) for a definition of the critical angles, studied below. The $n \times p$ matrix $X = A_1 T_1^t$ with $T_1 \in \mathbf{T}(p)$ and the $n \times p$ matrix A_1 satisfies $A_1^t A_1 = I_p$. The p-frame A_1 decomposes into a plane $\mathbf{P} \in \mathbf{G}_{p,n-p}$ and an orientation matrix $U_1 \in \mathbf{O}(p)$. We write $U_1 T_1^t = G_1 \in \mathbf{GL}(p)$.

$$X \sim (\mathbf{P}, U_1 T_1^t) = (\mathbf{P}, G_1). \tag{9.1.1}$$

We assume Y is $n \times q$ and that $p \le q$. Our Theorem 9.1.1 will require also $n \ge p + q$. Then for the hyperplane $\mathbf{Q}$ determined by Y,

$$Y \sim (\mathbf{Q}, U_2 T_2^t) = (\mathbf{Q}, G_2), \tag{9.1.2}$$

where $T_2 \in \mathbf{T}(q)$, $U_2 \in \mathbf{O}(q)$, and $G_2 \in \mathbf{GL}(q)$, $G_2 = U_2 T_2^t$, and $\mathbf{Q}$ represents a point of $\mathbf{G}_{q,n-q}$.

In the plane $\mathbf{P}$ we take $a_1, \ldots, a_p$ to be an orthonormal basis. As in Section 7.5 $(a_1, \ldots, a_p)$ represent $\mathbf{P}$ locally analytically in terms of local coordinates. In Section 8.2 these vectors were called $Xa_1, \ldots, Xa_p$ while in this section for brevity we call them $a_1, \ldots, a_p$. As in Section 8.2 and Problem 8.5.5, we let $\alpha_1, \ldots, \alpha_p$ be the normalized projections of $a_1, \ldots, a_p$ on $\mathbf{Q}$ and $\beta_1, \ldots, \beta_p$ be the normalized projections of $a_1, \ldots, a_p$ on $\mathbf{Q}^\perp$. It follows from (8.2.20), (8.2.27) and (8.5.5) that

$$\begin{aligned} &\alpha_1, \ldots, \alpha_p \quad \text{are an orthonormal set;} \\ &\beta_1, \ldots, \beta_p \quad \text{are an orthonormal set.} \end{aligned} \tag{9.1.3}$$

Thus we have the relations

$$\alpha_i^t\alpha_j = \delta_{ij}, \qquad \beta_i^t\beta_j = \delta_{ij}, \qquad \alpha_i^t\beta_j = 0, \qquad 1 \le i,j \le p, \tag{9.1.4}$$

and

$$\text{if} \quad 1 \le i \le p \quad \text{then} \quad a_i = \alpha_i \cos\theta_i + \beta_i \sin\theta_i, \tag{9.1.5}$$

where $\theta_1 \le \theta_2 \le \cdots \le \theta_p$ are the critical angles. We may assume in this construction that $\alpha_1, \ldots, \alpha_p$ have the first component positive thereby uniquely determining $\alpha_1, \ldots, \alpha_p$, and that $0 \le \theta_1 \le \theta_p \le \pi/2$ thereby uniquely fixing $\beta_1, \ldots, \beta_p$. See Problem 8.5.5. Recall that given **P**, **Q** and the critical angles $\theta_1, \ldots, \theta_p$ the vectors $a_1, \ldots, a_p$ in **P**, which are eigenvectors, are uniquely determined except for sign changes.

The plane **Q** is determined by a set of $n - q$ linear restrictions

$$x^t c_i = 0, \qquad i = 1, \ldots, n - q. \tag{9.1.6}$$

In the following remember that these conditions require $\beta_1, \ldots, \beta_p$ to be linear combinations of $c_1, \ldots, c_{n-q}$, while $\alpha_1, \ldots, \alpha_p$ are orthogonal to $c_1, \ldots, c_{n-q}$ so that

$$\alpha_i^t c_j = 0 \quad \text{and} \quad (d\alpha_i)^t c_j = 0, \qquad 1 \le i \le p, \qquad 1 \le j \le n - q. \tag{9.1.7}$$

Therefore the vectors $d\alpha_1, \ldots, d\alpha_p$ also lie in **Q** and

$$(d\alpha_i)^t \beta_j = 0 \quad \text{and} \quad \alpha_i^t(d\beta_j) = 0, \qquad 1 \le i,j \le p. \tag{9.1.8}$$

In addition, from (9.1.4) if follows that

$$\begin{aligned} &(\alpha_i)^t(d\alpha_i) = 0, \quad \text{and} \quad (\beta_i)^t(d\beta_i) = 0, \qquad 1 \le i \le p; \\ &(\alpha_i)^t(d\alpha_j) = -(\alpha_j)^t(d\alpha_i), \\ \text{and} \\ &(\beta_i)^t(d\beta_j) = -(\beta_j)^t(d\beta_i), \qquad 1 \le i \ne j \le p. \end{aligned} \tag{9.1.9}$$

In using the differential form (7.5.1) the orthonormal set of vectors $b_1, \ldots, b_{n-p}$ are chosen mutually orthogonal to $a_1, \ldots, a_p$ but are otherwise arbitrary, so we make a suitable choice

> if $1 \le i \le p$ then $b_i = -\alpha_i \sin\theta_i + \beta_i \cos\theta_i$; choose $b_{p+1}, \ldots, b_q$ in **Q** orthonormal and orthogonal to $\alpha_1, \ldots, \alpha_p$. And in $\mathbf{Q}^\perp$ choose $b_{q+1}, \ldots, b_{n-p}$ orthonormal and orthogonal to $\beta_1, \ldots, \beta_p$. (9.1.10)

Then clearly the orthonormal set of vector $b_1, \ldots, b_{n-p}$ are orthogonal to $a_1, \ldots, a_p$. The following relations hold.

$$da_i = b_i\, d\theta_i + (d\alpha_i \cos\theta_i + d\beta_i \sin\theta_i), \qquad 1 \le i \le p; \tag{9.1.11}$$

$$b_i^t\, da_i = d\theta_i, \qquad 1 \le i \le p; \tag{9.1.12}$$

$$\begin{aligned} &\text{if} \quad i \ne j \quad \text{and} \quad 1 \le i,j \le p \quad \text{then} \\ &b_i^t\, da_j = -\alpha_i^t\, d\alpha_j \sin\theta_i \cos\theta_j + \beta_i^t\, d\beta_j \cos\theta_i \sin\theta_j. \end{aligned} \tag{9.1.13}$$

By use of these relations we may evaluate the join

$$\bigwedge_{j=1}^{n-p} \bigwedge_{i=1}^{p} b_j^t \, da_i. \tag{9.1.14}$$

At the start we use the symmetries. If $1 \le i \ne j \le p$, then

$$-(b_j^t \, da_i) \wedge (b_i^t \, da_j) = (\alpha_j^t \, d\alpha_i) \wedge (\beta_j^t \, d\beta_i)(\cos^2 \theta_j - \cos^2 \theta_i), \tag{9.1.15}$$

which are obtainable from (9.1.13) together with (9.1.9) and

$$b_i^t \, db_j = -b_j^t \, db_i, \qquad 1 \le i,j \le n-p. \tag{9.1.16}$$

Using the orthogonality relations, further

$$\begin{aligned}
&\text{if} \quad i = 1, \ldots, p \quad \text{and} \quad j = p+1, \ldots, q, \quad \text{then} \\
&b_j^t \, da_i = b_j^t \, d\alpha_i \cos \theta_i; \quad \text{and,} \\
&\text{if} \quad i = 1, \ldots, p \quad \text{and} \quad j = q+1, \ldots, n-p, \quad \text{then} \\
&b_j^t \, da_i = b_j^t \, d\beta_i \sin \theta_i.
\end{aligned} \tag{9.1.17}$$

Use of these relations yields

$$\begin{aligned}
\bigwedge_{j=1}^{n-p} \bigwedge_{i=1}^{p} b_j^t \, da_i &= \varepsilon \bigwedge_{j<i} \alpha_j^t \, d\alpha_i \bigwedge_{j<i} \beta_j^t \, d\beta_i \bigwedge_{i=1}^{p} d\theta_i \\
&\times \bigwedge_{j=p+1}^{q} \bigwedge_{i=1}^{p} b_j^t \, d\alpha_i \bigwedge_{j=q+1}^{n-p} \bigwedge_{i=1}^{p} b_j^t \, d\beta_i \\
&\times \Big(\prod_{i=1}^{n} \cos \theta_i\Big)^{q-p} \Big(\prod_{i=1}^{p} \sin \theta_i\Big)^{n-p-q} \prod_{j<i} (\cos^2 \theta_j - \cos^2 \theta_i),
\end{aligned} \tag{9.1.18}$$

with the sign $\varepsilon = \pm 1$ to be determined by the number of interchanges made.

The differential form (9.1.18) is computed for the Grassman manifold containing **P**, **Q** being fixed. Since the computed differential form is invariant under left multiplication by $H \in \mathbf{O}(n)$ we modify the variables as follows. Let $(\tilde{A}, \tilde{B})$ represent **Q** in the sense of Section 7.5. We set

$$\begin{aligned}
\alpha_i &= \tilde{A}\tilde{\alpha}_i, \qquad 1 \le i \le p; \\
\beta_i &= \tilde{B}\tilde{\beta}_i, \qquad 1 \le i \le p; \\
b_i &= \tilde{A}\tilde{b}_i, \qquad p+1 \le i \le q; \\
b_i &= \tilde{B}\tilde{b}_i, \qquad q+1 \le i \le n-p.
\end{aligned} \tag{9.1.19}$$

Then

$$\text{if} \quad 1 \le i \le p, \qquad b_i = -\tilde{A}\tilde{\alpha}_i \sin \theta_i + \tilde{B}\tilde{\beta}_i \cos \theta_i. \tag{9.1.20}$$

We will use the fact that $b_1, \ldots, b_p$ do not appear in the differential form (9.1.18). With $(\tilde{A}, \tilde{B})$ fixed, in the new variables $\tilde{\alpha}$, $\tilde{\beta}$, $\tilde{b}$, the differential form (9.1.18) is unchanged except to replace α, β, b by $\tilde{\alpha}$, $\tilde{\beta}$, $\tilde{b}$. The representation of **P** has become

$$\begin{aligned} a_i &= \tilde{A}\tilde{\alpha}_i \cos\theta_i + \tilde{B}\tilde{\beta}_i \sin\theta_i, \qquad i = 1, \ldots, p,\\ b_i &= -\tilde{A}\tilde{\alpha}_i \sin\theta_i + \tilde{B}\tilde{\beta}_i \cos\theta_i, \qquad i = 1, \ldots, p,\\ b_i &= \tilde{A}\tilde{b}_i, \qquad i = p+1, \ldots, q,\\ b_i &= \tilde{B}\tilde{b}_i, \qquad i = q+1, \ldots, n-p. \end{aligned} \tag{9.1.21}$$

We now compute in the case of a multivariate normal density function. We assume $\mathbf{Z} = (\mathbf{X}, Y)$ has a joint normal density function

$$(2\pi)^{-n(p+q)/2}(\det \Sigma)^{-n/2} \operatorname{etr} - \tfrac{1}{2}\Sigma^{-1} ZZ^t, \tag{9.1.22}$$

where $\operatorname{etr} A = \exp(\operatorname{tr} A)$, exponential of the trace of the matrix A. Using

$$X = AU_1T_1 \quad \text{and} \quad Y = \tilde{A}U_2T_2, \tag{9.1.23}$$

we find

$$\begin{aligned} Z^tZ &= \begin{vmatrix} X^tX & X^tY \\ Y^tX & Y^tY \end{vmatrix}\\ &= \begin{vmatrix} T_1^tU_1^t & 0 \\ 0 & T_2^tU_2^t \end{vmatrix} \begin{vmatrix} I_p & A^t\tilde{A} \\ \tilde{A}^tA & I_q \end{vmatrix} \begin{vmatrix} U_1T_1 & 0 \\ 0 & U_2T_2 \end{vmatrix}. \end{aligned} \tag{9.1.24}$$

Let $\tilde{\alpha}$ be the $q \times p$ matrix with $\tilde{\alpha}_i$ as the i-th column. From (9.1.21) if a_i is the i-th column of A, we find

$$\tilde{A}^tA = \tilde{\alpha}\operatorname{diag}(\cos\theta_1, \ldots, \cos\theta_p) = \tilde{\alpha}E, \tag{9.1.25}$$

where E is the $p \times p$ diagonal matrix in (9.1.25). Then we write

$$D^t = (E, 0) \quad \text{where } D \text{ is } q \times p. \tag{9.1.26}$$

To get back to X and Y we use (9.1.23) and set

$$G = U_1T_1 \quad \text{and} \quad \tilde{G} = U_2T_2. \tag{9.1.27}$$

Then

$$dX = dA\,G + A\,dG \tag{9.1.28}$$

and

$$\begin{aligned} a_i^t\,dx_j &= a_i^t\,dA\,g_j + dg_{ij}, \qquad 1 \le i \le p, \qquad 1 \le j \le p,\\ b_i^t\,dx_j &= b_i^t\,dA\,g_j \qquad 1 \le i \le n-p, \qquad 1 \le j \le p. \end{aligned}$$

This yields

$$\begin{aligned} (\det(A, B))^p \bigwedge_{j=1}^{p}\bigwedge_{i=1}^{n} dx_{ij} &= \bigwedge_{j=1}^{p}\bigwedge_{i=1}^{p} a_i^t\,dx_j \bigwedge_{i=1}^{n-p} b_i^t\,dx_j\\ &= \bigwedge_{j=1}^{p}\bigwedge_{i=1}^{p} dg_{ij} \bigwedge_{i=1}^{n-p} b_i^t\,dA\,g_j. \end{aligned} \tag{9.1.29}$$

On the otherhand

$$\bigwedge_{j=1}^{p} (b_i^t\, dA)g_j = (\det G) \bigwedge_{j=1}^{p} b_i^t\, da_j \tag{9.1.30}$$

so that

$$(\det G)^{n-p} \bigwedge_{j=1}^{p} \bigwedge_{i=1}^{n-p} b_i^t\, da_j = \text{part of (9.1.29)}. \tag{9.1.31}$$

Therefore

$$\bigwedge_{j=1}^{p} \bigwedge_{i=1}^{n} dx_{ij} = (\det G)^{n-p}(\det(A, B))^p \bigwedge_{j=1}^{p} \bigwedge_{i=1}^{p} dg_{ij} \bigwedge_{i=1}^{n-p} b_i^t\, da_j. \tag{9.1.32}$$

In the new variables the differential form $\bigwedge_{i=1}^{n-p} \bigwedge_{j=1}^{p} b_i^t\, da_j$ is given by (9.1.18). Similarly for $\bigwedge_{j-1}^{q} \bigwedge_{i-1}^{n} dy_{ij}$.

In the integration, integrate first over the variables g_{ij} and $\tilde{g}_{ij}$. Then one obtains

$$\begin{aligned} &\int \operatorname{etr} -\tfrac{1}{2}\Sigma^{-1} \begin{vmatrix} G^t & 0 \\ 0 & \tilde{G}^t \end{vmatrix} \begin{vmatrix} I_p & E^t\tilde{\alpha}^t \\ \tilde{\alpha}E & I_q \end{vmatrix} \begin{vmatrix} G & 0 \\ 0 & \tilde{G} \end{vmatrix} (\det G)^{n-p}(\det \tilde{G})^{n-q} \\ &\quad \times \prod_{i=1}^{p} \prod_{j=1}^{p} dg_{ij} \prod_{i=1}^{p} \prod_{j=1}^{p} d\tilde{g}_{ij}. \end{aligned} \tag{9.1.33}$$

Since Haar measure on $\mathbf{GL}(q)$ is orthogonally invariant and since there exists $H \in \mathbf{O}(q)$ such that $H\alpha = \begin{vmatrix} I_p \\ 0 \end{vmatrix}$,

$$\begin{aligned} (9.1.33) = &\int \operatorname{etr} -\tfrac{1}{2}\Sigma^{-1} \begin{vmatrix} G^t & 0 \\ 0 & \tilde{G}^t \end{vmatrix} \begin{vmatrix} I_p & D^t \\ D & I_q \end{vmatrix} \begin{vmatrix} G & 0 \\ 0 & \tilde{G} \end{vmatrix} \\ &\times (\det G)^{n-p}(\det \tilde{G})^{n-q} \times \prod_{i=1}^{p} \prod_{j=1}^{p} dg_{ij} \prod_{i=1}^{p} \prod_{j=1}^{p} d\tilde{g}_{ij}. \end{aligned} \tag{9.1.34}$$

The remaining variables consist of $2^{-p}\mathbf{V}_{p,q}$, the factor 2^{-p} resulting from the choice of signs in $\alpha_1, \ldots, \alpha_p$, of $\mathbf{V}_{p,n-q}$, and the Grassman manifold $\mathbf{G}_{q,n-q}$. See Problem 8.5.5 and refer to Sections 7.8 and 7.9. The resulting normalization from integration over these manifolds is

$$\begin{aligned} K = &(2\pi)^{n(p+q)/2}(\det \Sigma)^{-n/2} 2^{-p} A_q \ldots A_{q-p+1} \\ &\times A_{n-q} \ldots A_{n-q-p+1} \frac{A_n \ldots A_{n-q+1}}{A_q \ldots A_1}. \end{aligned} \tag{9.1.35}$$

Theorem 9.1.1. *Let E be as in* (9.1.25) *and $f(E)$ be the function in* (9.1.34). *Then if $n \geq q \geq p$ and $n \geq p + q$,*

$$K^{-1} f(E)\Big(\prod_{i=1}^{p} \cos\theta_i\Big)^{q-p}\Big(\prod_{i=1}^{p} \sin\theta_i\Big)^{n-p-q} \times \prod_{j<i} (\cos^2\theta_j - \cos^2\theta_i) \tag{9.1.36}$$

is the joint probability density function of the angles $\theta_1 < \cdots < \theta_p$.

The canonical correlations are $\lambda_i^{1/2} = \cos\theta_i$, $i = 1, \ldots, p$. Substitution into (9.1.36) gives the joint probability density function of the correlations. The function $f(E)$ with E as in (9.1.25) may be derived using the methods of Chapter 10 on maximal invariants. However to obtain the normalization K and weight function in $\theta_1, \ldots, \theta_p$, a more precise calculation seems necessary.

In the derivations given we have assumed $q - p \geq 0$. The exponent $n - p - q$ in (9.1.36) cannot be negative. If it is negative then the term $(\prod_{i=1}^{p} \sin\theta_i)^{n-p-q}$ is missing, corresponding to the nonempty intersection of **P** and **Q**. For an analysis of this case, see James (1954).

The squares of the canonical correlations are roots of the equation (8.2.39) and are the eigenvalues of the matrix considered in Section 5.7. Thus the results of Section 5.7 together with the results of Section 5.3 give an alternative method of calculation of the density functions of the angles and correlations.

9.2. Hotelling T^2

In this section we assume **X** is $n \times k$, that **Y** is $k \times 1$, and that **X** and **Y** are stochastically independent random variables. We will suppose the rows of **X** are independently and identically distributed, each normal $(0, \Sigma)$, and that **Y** is normal (a, Σ) with the same covariance matrix Σ. The problem of inference is to test $a = 0$ against the alternative $a \neq 0$. The problem is invariant under transformations

$$(X, Y) \to (XG, YG) \tag{9.2.1}$$

where $G \in \mathbf{GL}(k)$. Under the action of $\mathbf{GL}(k)$ the maximal invariant constructed from the sufficient statistic

$$X^tX + YY^t, Y \tag{9.2.2}$$

is

$$(Y^t(X^tX + YY^t)^{-1}Y)^{1/2} = \lambda^{1/2}. \tag{9.2.3}$$

The number λ is the nonzero root of the equation

$$0 = \det(YY^t - \lambda(X^tX + YY^t)). \tag{9.2.4}$$

This follows from the general rule that the matrices AB and BA have the same eigenvalues. See Problem 8.5.22.

In the sequel we let $\mathbf{T}_1, \mathbf{T} \in \mathbf{T}(k)$ such that

$$\mathbf{T}_1\mathbf{T}_1^t = \mathbf{X}^t\mathbf{X} \quad \text{and} \quad \mathbf{T}\mathbf{T}^t = \mathbf{X}^t\mathbf{X} + \mathbf{Y}\mathbf{Y}^t. \tag{9.2.5}$$

The aim of this section is to calculate the probability density function of $\mathbf{T}^{-1}\mathbf{Y}$ by use of differential forms. Different evaluations result by use of the ideas of Chapter 5 or the ideas of Chapter 10 on maximal invariants.

We note for reference in the sequel that if

$$S = TT^t = (s_{ij}), \tag{9.2.6}$$

then

$$\begin{aligned}\bigwedge_{j\le i} ds_{ij} &= 2^k t_{11}^k \dots t_{kk} \bigwedge_{j\le i} dt_{ij} \\ &= 2^k(\det T)^{k+1}\Big(\prod_{j\le i} dt_{ij}\Big)/(t_{11} \dots t_{kk}^k) \\ &= 2^k(\det T)^{k+1} \prod_{j\le i} (T^{-1}\, dT)_{ij}.\end{aligned} \tag{9.2.7}$$

See Section 7.2 and Problem 6.7.6.

The joint density function of $\mathbf{X}$ and $\mathbf{Y}$ is

$$(2\pi)^{-(n+1)k/2}(\det \Sigma)^{-(n+1)/2} \times \operatorname{etr} - \tfrac{1}{2}\Sigma^{-1}(X^tX + YY^t - 2aY^t + aa^t). \tag{9.2.8}$$

We now begin the process of changes of variables followed by integrations. Write

$$X = AT_1^t \text{ with } A^tA = I_k,\ A \text{ a } n \times k \text{ matrix,} \tag{9.2.9}$$

$$\text{and } T_1 \text{ as in (9.2.5).}$$

Let B be $n \times (n-k)$ satisfying

$$B^tA = 0, \qquad B^tB = I_{n-k}. \tag{9.2.10}$$

We assume B is chosen to be an analytic function of the variables A so the differential form for an invariant measure on the Stiefel manifold is

$$\bigwedge_{j=1}^{n-k} \bigwedge_{i=1}^{k} b_j^t\, da_i \bigwedge_{j<i} a_j^t\, da_i. \tag{9.2.11}$$

See Section 7.6. We compute

$$\begin{aligned} dX &= A\, dT_1^t + (dA)T_1^t; \\ A^t\, dX(T_1^t)^{-1} &= ((dT_1^t)(T_1^t)^{-1}) + A^t\, dA; \\ B^t\, dX(T_1^t)^{-1} &= B^t\, dA. \end{aligned} \tag{9.2.12}$$

These relations give

$$\begin{aligned}&(\det A, B)^k(\det T_1)^{-n} \bigwedge_{j=1}^{k} \bigwedge_{i=1}^{n} dx_{ij} \\ &= \varepsilon \bigwedge_{j=1}^{k} \bigwedge_{i=1}^{n-k} b_i^t\, da_j \bigwedge_{j<i} a_i^t\, da_j \bigwedge_{i\le j} ((dT_1^t)(T_1^t)^{-1})_{ij}.\end{aligned} \tag{9.2.13}$$

$\varepsilon = \pm 1$, depending on the undetermined number of interchanges. Note that

$$\bigwedge_{i\le j} ((dT_1^t)(T_1^t)^{-1})_{ij} = \bigwedge_{i\le j} ((T_1^{-1}\, dT_1)^t)_{ij} = \bigwedge_{j\le i} (T_1^{-1}\, dT_1)_{ij}. \tag{9.2.14}$$

In the density function (9.2.8), $X^tX = T_1T_1^t$ so that (9.2.8) does not depend on points A of the Stiefel manifold. This integration out over the Stiefel manifold gives

$$\begin{aligned}&(2\pi)^{-(n+1)k/2}(\det \Sigma)^{-(n+1)/2}A_n \dots A_{n-k+1}\\ &\times (\det T_1)^n \operatorname{etr} -\tfrac{1}{2}\Sigma^{-1}(T_1T_1^t + YY^t - 2Ya^t + aa^t)\\ &\times \bigwedge_{j\le i}(T_1^{-1}\,dT_1)_{ij}\bigwedge_{i=1}^{k} dy_i.\end{aligned} \tag{9.2.15}$$

Introduce variables S by

$$S = T_1T_1^t + YY^t, \tag{9.2.16}$$

so that

$$\bigwedge_{j\le i} ds_{ij}\bigwedge_{i=1}^{k} dy_i = 2^k(\det T_1)^{k+1}\bigwedge_{j\le i}(T_1^{-1}\,dT_1)_{ij}\bigwedge_{i=1}^{k} dy_i. \tag{9.2.17}$$

Factor S,

$$S = TT^t, \qquad T \in \mathbf{T}(k), \tag{9.2.18}$$

so that

$$\bigwedge_{j\le i} ds_{ij} = 2^k(\det T)^{k+1}\bigwedge_{j\le i}(T^{-1}\,dT)_{ij}. \tag{9.2.19}$$

The ratio $(\det T)/(\det T_1)$ may be evaluated as follows.

$$\begin{aligned}T_1T_1^t &= TT^t - YY^t = T(I_k - (T^{-1}Y)(T^{-1}Y)^t)T^t\\ &= T(I_k - WW^t)T^t,\end{aligned} \tag{9.2.20}$$

with $W = T^{-1}Y$. Therefore $(\det T_1)^2 = (\det T)^2(\det(I_k - WW^t))$. Choose $U \in \mathbf{O}(k)$ such that $(UW)^t = (\|W\|, 0, \dots, 0)$. Then it follows that

$$\det(I_k - WW^t) = 1 - \|W\|^2 \tag{9.2.21}$$

and

$$(\det T_1)/(\det T) = (1 - \|W\|^2)^{1/2}. \tag{9.2.22}$$

Upon substitution, (9.2.15) becomes

$$\begin{aligned}&(2\pi)^{-(n+1)k/2}(\det \Sigma)^{-(n+1)/2} \times A_n \dots A_{n-k+1}(\det T)^{n+1}(1 - \|W\|^2)^{(n-k-1)/2}\\ &\qquad \times \operatorname{etr} -\tfrac{1}{2}\Sigma^{-1}(TT^t - 2TWa^t + aa^t)\\ &\qquad \times \bigwedge_{j\le i}(T^{-1}\,dT)_{ij}\bigwedge_{i=1}^{k} dW_i.\end{aligned} \tag{9.2.23}$$

We use here the relation (think of T as fixed)

$$\bigwedge_{i=1}^{k} dy_i = (\det T)\bigwedge_{i=1}^{k} dW_i. \tag{9.2.24}$$

Integration on the variables T by the left invariant Haar measure $\bigwedge_{j\le i}(T^{-1}\,dT)_{ij}$ gives the density function of $\mathbf{W}$. We let

$$\Sigma = \Sigma_1\Sigma_1^t, \qquad \Sigma_1 \in \mathbf{T}(k). \tag{9.2.25}$$

Then $\Sigma^{-1} = (\Sigma_1^t)^{-1}\Sigma_1^{-1}$ and we use the change of variable $T \to \Sigma_1^{-1}T$. Note that

$$\operatorname{tr}\Sigma^{-1}TWa^t = a^t(\Sigma_1^t)^{-1}\Sigma_1^{-1}TW \tag{9.2.26}$$

and

$$\operatorname{tr}\Sigma^{-1}aa^t = \operatorname{tr}(\Sigma_1^{-1}a)(\Sigma_1^{-1}a)^t. \tag{9.2.27}$$

The density function of $\mathbf{W}$ is then

$$\begin{aligned}&(2\pi)^{-(n+1)k/2}A_n \ldots A_{n-k+1}(1 - \|W\|^2)^{(n-k-1)/2}\\ &\quad\times \int (\det T)^{n+1}\operatorname{etr} -\tfrac{1}{2}(TT^t - 2TWb^t + bb^t)\bigwedge_{j\le i}(T^{-1}\,dT)_{ij}.\end{aligned} \tag{9.2.28}$$

The new parameter is $b = \Sigma_1^{-1}a$.

The statistic often used in inference is $\|W\|^2 = \mathbf{Y}^t(\mathbf{T}\mathbf{T}^t)^{-1}\mathbf{Y} = \mathbf{Y}^t(\mathbf{X}^t\mathbf{X} + \mathbf{Y}\mathbf{Y}^t)^{-1}\mathbf{Y}$. Problems 9.4.1 and 9.4.2 suggest the derivation of the noncentral probability density function of $\|W\|^2$. A different derivation will be obtained in Chapter 11 using random variable techniques. The answer (9.2.28) should be compared with Giri, Kiefer, and Stein (1963).

9.3. Eigenvalues of the Sample Covariance Matrix $\mathbf{X}^t\mathbf{X}$

We assume $\mathbf{X}$ is $n \times k$ and that $n \ge k$. We use the decomposition contained in Theorem 8.1.10. By Problem 4.4.7 together with Section 8.4, that X^tX has two equal nonzero eigenvalues is an event of zero Lebesgue measure, so, except on this set, we write

$$\begin{aligned}X = ADG^t, \qquad &A\ n \times k \ \text{ such that } \ A^tA = I_k, \qquad D \in \mathbf{D}(k),\\ &\text{and} \quad G \in \mathbf{O}(k).\end{aligned} \tag{9.3.1}$$

This decomposition is unique provided the diagonal entries of D are in order of decreasing magnitude, and G is restricted so the entries of the first column of G are all positive. Since

$$X^tX = GD^2G^t, \tag{9.3.2}$$

the entries of D^2 are the eigenvalues of X^tX. We now begin the computation of differential forms. The k-frame A is to be augmented by a $n \times (n-k)$ matrix B such that $(A, B) \in \mathbf{O}(n)$ and $\det(A, B) = 1$. Then

$$dX = (dA)DG^t + A(dD)G^t + AD(dG^t), \tag{9.3.3}$$

and

$$B^t dX = B^t(dA)DG^t.$$

Then,

$$\left(\bigwedge_{j=1}^{n-k}\bigwedge_{i=1}^{k} b_j^t\, dx_i\right)(\det DG^t)^{-(n-k)} = \bigwedge_{j=1}^{n-k}\bigwedge_{i=1}^{k} b_j^t\, da_i. \tag{9.3.4}$$

The remainder of the computation is more complex.

$$\begin{aligned} A^t dX\,G &= A^t(dA)D + dD + D(dG^t)G \\ &= A^t(dA)D + dD - DG^t(dG). \end{aligned} \tag{9.3.5}$$

We use here the fact that $(dG^t)G + G^t(dG) = 0$. From the left side of (9.3.5),

$$\bigwedge_{j=1}^{k}\bigwedge_{i=1}^{k} (A^t dX\,G)_{ij} = (\det G)^k \bigwedge_{j=1}^{k}\bigwedge_{i=1}^{k} (a_j^t\, dx_i). \tag{9.3.6}$$

On the right side of (9.3.5) we write

$$D = \operatorname{diag}(c_1, \ldots, c_k). \tag{9.3.7}$$

In forming the join of terms from the matrix on the right side of (9.3.5), the terms $dc_1 \wedge dc_2 \wedge \ldots \wedge dc_k$ are the $(1,1), (2,2), \ldots, (k,k)$ entries in the product since any other entry from dD is zero. If $i \neq j$ then we form the symmetrical product

$$(A^t dX\,G)_{ij} \wedge (A^t dX\,G)_{ji} \tag{9.3.8}$$

and form the join of these. We use the relations

$$a_i^t\, da_j + (da_i^t)a_j = 0 \quad \text{and} \quad g_i^t\, dg_j + (dg_i^t)g_j = 0. \tag{9.3.9}$$

Note that the j-th column of $(dA)D$ is $c_j\, da_j$ and the i-th row of DG^t is $c_i g_i^t$ where g_i is the i-th column of G. Thus,

$$\text{if} \quad i \neq j \quad \text{then} \quad (A^t\, dX G)_{ij} = c_j a_i^t\, da_j - c_i g_i^t\, dg_j, \tag{9.3.10}$$

and the symmetrical product is, if $i \neq j$, then

$$\begin{aligned} (A^t\, dX\,G)_{ij} \wedge (A^t\, dX\,G)_{ji} &= (c_j a_i^t\, da_j - c_i g_i^t\, dg_j) \wedge (c_i a_j^t\, da_i - c_j g_j^t\, dg_i) \\ &= (c_j^2 - c_i^2)(a_i^t\, da_j) \wedge (g_i^t\, dg_j). \end{aligned} \tag{9.3.11}$$

In this calculation we have used (9.3.9) several times. Putting together (9.3.6) and (9.3.11) we obtain

$$\begin{aligned} (\det(A,B))^k \bigwedge_{j=1}^{n}\bigwedge_{i=1}^{k} dx_{ij} &= \pm(\det DG^t)^{n-k}(\det G)^k \bigwedge_{j=1}^{n-k}\bigwedge_{i=1}^{k} b_j^t\, da_i \bigwedge_{i<j} a_i^t\, da_j \\ &\quad \times \bigwedge_{i<j} g_i^t\, dg_j \prod_{j<i} (c_j^2 - c_i^2) \bigwedge_{i=1}^{k} dc_i. \end{aligned} \tag{9.3.12}$$

Let us assume the rows of **X** are independently and identically distributed normal $(0, \Sigma)$ random vectors. Then the joint probability density function is

$$\begin{aligned} &(2\pi)^{nk/2}(\det\Sigma)^{-n/2}\operatorname{etr} -\tfrac{1}{2}\Sigma^{-1}X^tX \\ &= (2\pi)^{nk/2}(\det\Sigma)^{-n/2}\operatorname{etr} -\tfrac{1}{2}\Sigma^{-1}GD^2G^t, \end{aligned} \tag{9.3.13}$$

where $X = ADG^t$. Let Λ be the diagonal matrix with entries in decreasing order equal to the eigenvalues of Σ. Integration over A and G gives for the probability density function

$$\begin{aligned} &(2\pi)^{nk/2}(\det\Sigma)^{-n/2}\frac{A_n\cdots A_{n-k+1}}{2^k}\times\int \operatorname{etr} -\tfrac{1}{2}\Lambda^{-1}GD^2G^t \bigwedge_{i<j} g_i^t\,dg_j \\ &\qquad\times\Big(\prod_{i=1}^k c_i\Big)^{n-k}\prod_{j<i}(c_j^2 - c_i^2)\bigwedge_{i=1}^k dc_i. \end{aligned} \tag{9.3.14}$$

Here we assume $c_1 \geq c_2 \geq \cdots \geq c_k$.

9.4. Problems

PROBLEM 9.4.1. Let $X^t = (x_1, \ldots, x_n)$ have differential form $\bigwedge_{i=1}^n dx_i$. Let $z = \|X\|$ and $Y^t = (y_1, \ldots, y_n) = X^t/z$. Then show

$$\bigwedge_{i=1}^n dx_i = z^{n-1}\,dz\sum_{i=1}^n(-1)^{i+1}y_i\,dy_1 \wedge \ldots \wedge dy_{i-1} \wedge dy_{i+1} \wedge \ldots \wedge dy_n. \tag{9.4.1}$$

The sum is the differential form for an invariant measure, cf. (6.7.5), on a 1-frame and can be written in a coordinate free form $\bigwedge_{i=1}^{n-1} b_i^t\,dy$. Here $B = (b_1, \ldots, b_{n-1})$ satisfies $B^tB = I_{n-1}$ and $B^ty = 0$.

PROBLEM 9.4.2. Use the results of Section 9.2 together with Problem 9.4.1 to obtain the probability density function of the Hotelling's T^2 statistic, i.e., $\|\mathbf{W}\|^2$.

PROBLEM 9.4.3. Integrate (9.2.28) over a set $\mathbf{A}$ in the W-space which satisfies, if $G \in \mathbf{O}(n)$, then $G\mathbf{A} = \mathbf{A}$. Show the resulting function $\beta(b)$ has the property

$$\text{if} \quad G \in \mathbf{O}(n) \quad \text{then} \quad \beta(Gb) = \beta(b). \tag{9.4.2}$$

PROBLEM 9.4.4. Suppose $\mathbf{X}$ is a $n \times k$ random matrix such that the probability density function of $\mathbf{X}$ depends on X only through the eigenvalues of $(X^tX)^{1/2}$. If these eigenvalues are $c_1 \geq c_2 \geq \cdots \geq c_k$ then the density function of $c_1, \ldots, c_k$ is

$$2^{-k}A_n \cdots A_{n-k+1}A_k \cdots A_1 f(X) \times \Big(\prod_{i=1}^k c_i\Big)^{n-k}\prod_{i<j}(c_i^2 - c_j^2)\bigwedge_{i=1}^k dc_i. \tag{9.4.3}$$

In this formula f is the density function of $\mathbf{X}$.

Problem 9.4.5. Let X be a $n \times k$ matrix, $A \in \mathbf{O}(n)$, $G \in \mathbf{O}(k)$. Then

$$\iint f(AXG) \bigwedge_{j<i} a_i^t\, da_j \bigwedge_{j<i} g_i^t\, dg_j \tag{9.4.4}$$

is a function of the eigenvalues of $(X^tX)^{1/2}$.

Problem 9.4.6 (Refer to Chapter 4). Let $\mathbf{X}$ be a $n \times k$ random matrix with probability density function f. Let $\mathbf{Y}$ be a $n \times k$ random matrix with probability density function (9.4.4). Then the eigenvalues of $\mathbf{X}^t\mathbf{X}$ and the eigenvalues of $\mathbf{Y}^t\mathbf{Y}$ have the same probability density function.

Problem 9.4.7. Continue Problem 9.4.5 and Problem 9.4.6. The eigenvalues of $(\mathbf{X}^t\mathbf{X})^{1/2}$ have probability density function

$$\begin{aligned} &2^{-k}A_n \ldots A_{n-k+1}A_k \ldots A_1\left(\iint f(AXG) \bigwedge_{j<i} a_i^t\, da_j \bigwedge_{j<i} g_i^t\, dg_j)\right) \\ &\quad \times (\prod_{i=1}^k c_i)^{n-k} \prod_{i<j} (c_i^2 - c_j^2) \bigwedge_{i=1}^k dc_i. \end{aligned} \tag{9.4.5}$$

Problem 9.4.8. Let S be a $k \times k$ positive definite matrix. There exists a unique diagonal matrix D such that

$$S = DRD, \tag{9.4.6}$$

and the diagonal entries of R are all equal one. (The off diagonal entries of R are the correlations.)

Problem 9.4.9. Continue Problem 9.4.8. Let

$$D = \operatorname{diag}(c_1, \ldots, c_k), \qquad c_1 \geq \cdots \geq c_k. \tag{9.4.7}$$

Show

$$D^{-1}\, dS\, D^{-1} = D^{-1}(dD)R + R(dD)D^{-1} + dR, \tag{9.4.8}$$

and

$$D^{-1}(dD) = (dD)D^{-1} = \operatorname{diag}(dc_1/c_1, \ldots, dc_k/c_k).$$

Since the diagonal of dR is zero, it follows that

$$\bigwedge_{j\leq i}(D^{-1}\, dS\, D^{-1})_{ij} = \pm 2^k \bigwedge_{i=1}^k (dc_i/c_i) \bigwedge_{j<i}(dR)_{ij}. \tag{9.4.9}$$

By Problem 6.7.7, the left side of (9.4.9) is $(\det D)^{-(k+1)} \bigwedge_{j\leq i}(dS)_{ij}$.
Therefore

$$\bigwedge_{j\leq i}(dS)_{ij} = \pm(\det D)^k 2^k \bigwedge_{i=1}^k dc_i \bigwedge_{j<i}(dR)_{ij}. \tag{9.4.10}$$

PROBLEM 9.4.10. Continue Problem 9.4.9. If S has a Wishart density function

$$2^{-k}(2\pi)^{-nk/2}(\det \Sigma)^{-n/2} A_n \ldots A_{n-k+1} (\det S)^{(n-k-1)/2} \times \operatorname{etr} -\tfrac{1}{2}\Sigma^{-1} S \bigwedge_{j \le i} (dS)_{ij}, \tag{9.4.11}$$

then use the substitutions of Problem 9.4.9 to obtain the joint probability density function of the correlation coefficients.

PROBLEM 9.4.11. The factorization of Problem 8.5.9 does not appear to be useful here since one obtains the density function of $(X^t X)^{1/2}$.

CHAPTER 10

Cross-Sections and Maximal Invariants

10.0. Introduction

In some problems in statistical inference, as in the construction of Bayes tests, what is needed is a ratio of probability density functions. In the formation of these ratios normalizations cancel, and functions of the variables not dependent on the parameters also cancel. In the case of probability density functions of maximal invariants these factors result in part from integration out of unwanted variables which lie in a factor of the manifold. In these cases the detailed answer obtained from use of differential forms may be unnecessary.

The method discussed in this chapter is due to Stein (1956c). Since then several developments related to Stein's ideas have appeared. One approach was developed by Schwartz (1966a), unpublished. In the sequel we develop Schwartz's ideas. Wijsman (1966) develops a theory of cross-sections and uses this as a general tool in obtaining a factorization of an invariant measure. This method was further developed by Koehn (1970). The papers of Koehn and Wijsman require differential geometry and Lie group theory. More recent papers include Bondar (1976), Andersson (1982), and Wijsman (1984).

The approach of Schwartz that is developed in Section 10.1 notes that in many examples, after discarding an obvious null set, the remaining manifold factors. Then the theory of Section 3.4 applies directly. In Section 10.2 the examples of the sample covariance matrix and the general linear hypothesis are examined in detail. In the summary of examples given in this section these examples are listed as Example 10.0.1 and 10.0.4.

To contrast the methods of Chapters 5 and 10, in Chapter 5 it has been assumed that the group $\mathfrak{G}$ acts transitively on $\mathfrak{X}$ such that for each $x \in \mathfrak{X}$, $\{g | gx = x\}$ is compact, and $\mathfrak{X}$ isomorphic to $\mathfrak{G}/\{g | gx_0 = x_0\}$ was used. The invariant under $\mathfrak{G}_0 \subset \mathfrak{G}$ was then studied. In this chapter we do not attempt to enlarge the group to one that acts transitively on $\mathfrak{X}$. Instead the orbits $\{y | y = gx, g \in \mathfrak{G}\}$ become the maximal invariant. In general $\mathfrak{G}$ does not act in a one-to-one way, and a compact subgroup $\mathfrak{H} \subset \mathfrak{G}$ is obtained. The problem of cross-sections is to establish an isomorphism between the Cartesian product of $(\mathfrak{G}/\mathfrak{H}) \times$ (space of orbits) and $\mathfrak{X} - \mathbf{N}$, $\mathbf{N}$ a null set. This is explained in much greater detail in (10.1.11) in the next section.

It should be noted that the method of cross-sections of this chapter or of Wijsman (1966) is a method that can be routinely applied once the hypotheses are verified. In terms of the formulation of Section 10.1, construction of a function K satisfying (10.1.12) and (10.1.20) is the main difficulty. Where the "obvious" choice of K fails to work the effort may sometimes be salvaged as explained in Section 10.4. By contrast, application of the methods of Chapter 5 give the user a better end result but cannot be so routinely applied. Their use may test the ingenuity of the user.

Schwartz's theory is a global type of theory and Example 10.4.5 provides an easy example of the difficulty one may have in verifying the required hypotheses. This difficulty arises in the construction of a suitable maximal invariant, the problem users of cross-sections hope to solve. In the example at hand it seems that one must factor the Stiefel manifold $\mathbf{V}_{p,k}$ by the group $\mathbf{O}(p)$ and from each orbit pick a suitable representative of the point in the Grassman manifold $\mathbf{G}_{p,k-p}$. It should be realized that already in Chapter 7, we have assumed the existence of *local cross-sections*. In this chapter needed is a global cross-section that is measurable in order to obtain an explicit factorization of the manifold and measure.

In earlier chapters of this book a number of examples already examined in detail are examples of maximal invariants to which the theory of this chapter might be applied. We now list these examples.

Example 10.0.1. The sample covariance matrix $\mathbf{X}'\mathbf{X}$, given that $\mathbf{X}$ is $n \times k$, is a maximal invariant under the action of $\mathbf{O}(n)$ as left multipliers of $\mathbf{X}$.

Example 10.0.2. The matrix of correlation coefficients is a maximal invariant under the action $X \to UXD$, where X is $n \times k$, $U \in \mathbf{O}(n)$, and $D \in \mathbf{D}(k)$. See Problems 9.4.8, 9.4.9, and 9.4.10.

Example 10.0.3. The Hotelling T^2-statistic $\mathbf{Y}'(\mathbf{X}'\mathbf{X})^{-1}\mathbf{Y}$ discussed in Section 9.2 is a maximal invariant under the action $(X, Y) \to (UXG, GY)$, where $U \in \mathbf{O}(n)$, $G \in \mathbf{GL}(k)$, and X is $n \times k$ and Y is $k \times 1$.

Example 10.0.4. The generalization of the Hotelling T^2-statistic to the general linear hypothesis is discussed by Lehmann (1959) and will provide the basis of Example 10.1.1 which is further discussed in Section 10.2 as Example 10.2.3 and in Section 10.4 as Example 10.4.5. The problem is to find the distribution of the maximal invariant when X is $p \times k$, Y is $q \times k$, and the group action is $X \to UXG^t$, $Y \to VYG^t$, $U \in \mathbf{O}(p)$, $V \in \mathbf{O}(q)$, and $G \in \mathbf{GL}(k)$. A maximal invariant is the vector of roots of the equation $0 = \det(X^tX - \lambda Y^tY)$. The exact probability density function has been computed by Constantine (1963).

Example 10.0.5. Certain tests relating to canonical correlations and discriminant functions can be based on the maximal invariant of $X \to UXT^t$ and $Y \to VYT^t$ where $U \in \mathbf{O}(p)$, $V \in \mathbf{O}(q)$ and $T \in \mathbf{T}(k)$. The invariant here is the $k \times k$ matrix L satisfying $X^tX = TLT^t$, $Y^tY = T(I_k - L)T^t$, with $L \in \mathbf{S}(k)$ and $T \in \mathbf{T}(k)$. See Kshirsagar (1961). In case $\mathbf{E}\,\mathbf{X} = M$ and $\mathbf{E}\,\mathbf{Y} = 0$ the probability density function of L is the noncentral multivariate beta density function. We further discuss this example in Section 10.3. See also Example 10.0.9, which has a multivariant beta density, and in this respect, Section 5.7.

Example 10.0.6. The eigenvalues of the sample covariance matrix, discussed in Section 9.3, are a maximal invariant of the action $X \to UXV$, $U \in \mathbf{O}(n)$, $V \in \mathbf{O}(k)$ and X an $n \times k$ matrix.

Example 10.0.7. The canonical correlations, discussed in Section 8.2, Problems 8.5.4 and 8.5.5 and Section 9.1, are a maximal invariant under the action $(X, Y) \to (UXG_1, UYG_2)$ with X a $n \times p$ matrix, Y a $n \times q$ matrix, $G_1 \in \mathbf{GL}(p)$, $G_2 \in \mathbf{GL}(q)$, and $U \in \mathbf{O}(n)$, $p + q \leq n$.

Example 10.0.8. Certain of the manifolds listed in these notes may be thought of as maximal invariants. The Stiefel manifold results from the action $X \to XT^t$, X as $n \times k$ matrix and $T \in \mathbf{T}(k)$, and the Grassman manifold results from the action $X \to XG^t$, X an $n \times k$ matrix and $G \in \mathbf{GL}(k)$.

Example 10.0.9. Other invariants are listed in (5.6.2)–(5.6.5). A full discussion of the invariant $S_{22}^{-1/2}S_{21}S_{11}^{-1}S_{12}S_{22}^{-1/2}$ is given in Section 5.7.

10.1. Basic Theory

In order to illustrate the structure needed we choose the example of a restricted version of the general linear hypothesis.

Example 10.1.1. Let the random matrices $\mathbf{X}$ and $\mathbf{Y}$ be $p \times k$ and $q \times k$ respectively with $n = p + q$. We assume $\mathbf{X}$ and $\mathbf{Y}$ are independently distributed and that the rows of $\mathbf{X}$ are independently distributed, that the rows of $\mathbf{Y}$ are independently distributed, that the i-th row of $\mathbf{X}$ is normal (m_i, Σ) and the i-th row of $\mathbf{Y}$ is normal $(0, \Sigma)$. We write

$$M = \mathbf{E}\,\mathbf{X}, \qquad \text{a } p \times k \text{ matrix.} \tag{10.1.1}$$

Classically the hypothesis to be tested is

$$M = 0, \qquad \Sigma > 0 \text{ arbitrary}, \tag{10.1.2}$$

against the alternative

$$M \neq 0, \qquad \Sigma > 0 \text{ arbitrary}. \tag{10.1.3}$$

The problem is left invariant by transformations

$$(X, Y) \to (U_1 X G^t, U_2 Y G^t) \tag{10.1.4}$$

with $U_1 \in \mathbf{O}(p)$, $U_2 \in \mathbf{O}(q)$ and $G \in \mathbf{GL}(k)$. That is to say, the random matrices $U_1 \mathbf{X} G^t$ and $U_2 \mathbf{Y} G^t$ satisfy the conditions given and the hypothesis parameter sets defined by (10.1.2) and (10.1.3) are invariant sets. We may consider the transformation group to be $\mathbf{O}(p) \times \mathbf{O}(q) \times \mathbf{GL}(k)$ since the actions of U_1, U_2 and G commute.

Clearly

$$X^t X + Y^t Y \quad \text{and} \quad X^t X, \qquad Y^t Y \tag{10.1.5}$$

are invariants under the action $(X, Y) \to (U_1 X, U_2 Y)$ with $U_1 \in \mathbf{O}(p)$ and $U_2 \in \mathbf{O}(q)$. The action of $G \in \mathbf{GL}(k)$ on (10.1.5) becomes

$$X^t X + Y^t Y \to G(X^t X + Y^t Y)G^t, \qquad X^t X \to G(X^t X)G^t. \tag{10.1.6}$$

As suggested in Problem 8.5.13, G may be chosen so that

$$I_k = G(X^t X + Y^t Y)G^t \quad \text{and} \quad D = G X^t X G^t, \tag{10.1.7}$$

with $D \in \mathbf{D}(k)$, the diagonal entries of D between 0 and 1. In fact

$$\begin{aligned} &\text{rank } X = k - (\text{number of 0's on the diagonal of } D); \\ &\text{rank } Y = k - (\text{number of 1's on the diagonal of } D). \end{aligned} \tag{10.1.8}$$

The numbers c_i on the diagonal of D are the roots of the equation

$$0 = \det(c(X^t X + Y^t Y) - (X^t X)). \tag{10.1.9}$$

By Problem 4.4.7, that $0 < c_i = c_j < 1$, $i \neq j$, is a set of zero Lebesgue measure which is an invariant set under the group action. In the sequel we let $\mathbf{R}$ be the maximal invariant given by (10.1.7), i.e., $\mathbf{R}$ is the set of such diagonal matrices D in this example.

The action of the group $\mathfrak{G} = \mathbf{O}(p) \times \mathbf{O}(q) \times \mathbf{GL}(k)$ is not in general one-to-one and we analyze this question in complete detail in Section 10.2. It suffices to say that the group $\mathfrak{G}$ has a compact subgroup $\mathbf{H}$ such that the action $\mathfrak{G}/\mathbf{H}$ is one-to-one, provided rank $X = k$ and rank $Y = k$ and the roots of (10.1.9) between 0 and 1 are pairwise distinct. The exceptional set $\mathbf{N}$ consisting of those (X, Y) such that at least one of

$$\text{rank } X < k, \qquad \text{rank } Y < k, \quad \text{or} \quad 0 < c_i = c_j < 1, \qquad i \neq j, \tag{10.1.10}$$

occurs, is an invariant set of Lebesgue measure zero.

We obtain the following diagram, with $\pi\colon \mathfrak{G} \to \mathfrak{G}/\mathbf{H}$ the projection map.

$$\begin{array}{ccc} h\colon \mathfrak{G} \times \mathbf{R} & \rightarrow & \mathfrak{X} - \mathbf{N} \\ \pi \times i \downarrow & \nearrow & K^{-1} \\ (\mathfrak{G}/\mathbf{H}) \times \mathbf{R} & & \end{array} \tag{10.1.11}$$

with the relations

$$K^{-1}(\pi(g), r) = h(g, r); \tag{10.1.12}$$

and

$$\mathbf{H} = \{g \mid g \in \mathfrak{G} \text{ and } gh(e,r) = h(e,r)\} \text{ is a subgroup that does not depend on } r.$$

The thing that makes the factorization work, see Theorem 10.1.2, is the fact that the invariant measures on $\mathfrak{G}/\mathbf{H}$ are a one-dimensional family when $\mathbf{H}$ is a compact subgroup of $\mathfrak{G}$. In (10.1.11) the manifold to be factored is $\mathfrak{X}$. It will appear in our examples that π and h are continuous and K^{-1} is a homeomorphism. In many examples the null set $\mathbf{N}$ is a closed subset of $\mathfrak{X}$ so that $\mathfrak{X} - \mathbf{N}$ is open in $\mathfrak{X}$.

Our discussion above has shown that $\mathbf{R}$ is an invariant. In Section 10.2 we reconstruct X and Y from $D \in \mathbf{R}$ thereby showing $\mathbf{R}$ is a maximal invariant. We now begin a formal statement of the results of Schwartz (1966a).

Theorem 10.1.2. *Assume the following.*

$\mathfrak{G}$ is a locally compact separable group with a compact subgroup $\mathbf{H}$ and factor space $\mathfrak{G}/\mathbf{H} = \bar{\mathfrak{G}}$ with elements $\bar{g} = \pi(g)$. (10.1.13)

$\mathfrak{X}$ is a separable metric space and μ is a σ-finite Borel measure on $\mathfrak{X}$. (10.1.14)

That $\mathfrak{G}$ acts on $\mathfrak{X}$ and the maps $(g, x) \to g(x)$ are jointly measurable. (10.1.15)

That $\mathbf{N} \subset \mathfrak{X}$ is a Borel subset which is $\mathfrak{G}$-invariant and $\mu(\mathbf{N}) = 0$. (10.1.16)

That $K\colon \mathfrak{X} - \mathbf{N} \to (\bar{\mathfrak{G}}/\mathbf{H}) \times \mathbf{R}$ is a one-to-one onto map. (10.1.17)

That $\mathbf{R}$ has a separable locally compact topology. (10.1.18)

That $\mathbf{G}$ acts on $(\bar{\mathfrak{G}}/\mathbf{H}) \times \mathbf{R}$ by the action $g_1(\bar{g}_2, r) = (\overline{g_1 g_2}, r)$. Hereafter the same letters $\mathfrak{G}$, g will be used for the actions of $\mathfrak{G}$ on $\mathfrak{X} - \mathbf{N}$ and $\mathfrak{G}$ on $(\bar{\mathfrak{G}}/\mathbf{H}) \times \mathbf{R}$. (10.1.19)

$Kg = gK$ and hence $K^{-1}g = gK^{-1}$, assuming the notational convention of (10.1.19). (10.1.20)

K^{-1} is continuous. If $\mathbf{A} \subset \mathbf{R}$ and $\bar{\mathbf{B}} \subset \bar{\mathfrak{G}}$ are compact sets then $K^{-1}(\bar{\mathbf{B}} \times \mathbf{A})$ is a compact subset of $\mathfrak{X} - \mathbf{N}$. (10.1.21)

μ is a left invariant regular Borel measure on $\mathfrak{X} - \mathbf{N}$. (10.1.22)

That the σ-algebra of Borel subsets of $\mathbf{R}$ is $\mathfrak{C}$. The σ-algebra of Borel subsets of $\bar{\mathfrak{G}}$ is $\mathfrak{B}(\bar{\mathfrak{G}})$ and of $\mathfrak{X}$ is $\mathfrak{B}(\mathfrak{X})$. (10.1.23)

Then there exist regular measures $\bar{\mu}$ defined on $\mathfrak{B}(\bar{\mathfrak{G}})$ and v defined on $\mathfrak{C}$ such that if $\bar{\mathbf{A}} \in \mathfrak{B}(\bar{\mathfrak{G}})$ and $\mathbf{B} \in \mathfrak{C}$ then

$$\mu K^{-1}(\bar{\mathbf{A}} \times \mathbf{B}) = \bar{\mu}(\bar{\mathbf{A}}) v(\mathbf{B}). \tag{10.1.24}$$

The measure $\bar{\mu}$ is left invariant.

PROOF. The argument is very similar to that of Section 3.4. Let $\mathfrak{C}_0 \subset \mathfrak{C}$ be the set of $\mathbf{B}$ such that for all $\bar{\mathbf{A}} \in \mathfrak{B}(\bar{\mathfrak{G}})$, $K^{-1}(\bar{\mathbf{A}} \times \mathbf{B}) \in \mathfrak{B}(\mathfrak{X})$. We start by showing that if $\mathbf{B}$ is a compact set then $\mathbf{B} \in \mathfrak{C}_0$. Let $\mathfrak{B}(\bar{\mathfrak{G}}, \mathbf{B})$ be the set of those $\bar{\mathbf{A}} \in \mathfrak{B}(\bar{\mathfrak{G}})$ such that $K^{-1}(\bar{\mathbf{A}} \times \mathbf{B}) \in \mathfrak{B}(\mathfrak{X})$. By (10.1.21), since $\mathbf{B}$ is a compact subset, if $\bar{\mathbf{A}}$ is a compact subset then $K^{-1}(\bar{\mathbf{A}} \times \mathbf{B})$ is a compact subset of $\mathfrak{X} - \mathbf{N}$, and it follows that $K^{-1}(\bar{\mathbf{A}} \times \mathbf{B}) \in \mathfrak{B}(\mathfrak{X})$. Further, it is clear that since K^{-1} is a one-to-one function, the set $\mathfrak{B}(\bar{\mathfrak{G}}, \mathbf{B})$ is a monotone class. Since $\bar{\mathfrak{G}}$ has a separable topology a monotone class containing all the compact sets must contain all the Borel subsets. Therefore $\mathfrak{B}(\bar{\mathfrak{G}}, \mathbf{B}) = \mathfrak{B}(\bar{\mathfrak{G}})$.

The argument of the last paragraph shows that $\mathfrak{C}_0$ must contain all compact subsets of $\mathfrak{C}$. Also, $\mathfrak{C}_0$ is a monotone class. By (10.1.18) the topology of $\mathbf{R}$ is separable so it follows that $\mathfrak{C}_0$ contains all the Borel subsets of $\mathbf{R}$, hence $\mathfrak{C}_0 = \mathfrak{C}$. Since $K^{-1}(\bar{\mathbf{A}} \times \mathbf{B})$ is a Borel subset for every rectangle $\bar{\mathbf{A}} \times \mathbf{B}$ with measurable factors, and since K^{-1} is one-to-one, it follows that $K: \mathfrak{X} - \mathbf{N} \to \bar{\mathfrak{G}} \times \mathbf{R}$ is a Borel measurable function.

The measure μK^{-1} is therefore defined for all subsets of the product σ-algebra $\mathfrak{B}(\bar{\mathfrak{G}}) \times \mathfrak{C}$. If $\mu K^{-1} = 0$ then Theorem 10.1.2 follows trivially. Otherwise, since $\bar{\mathfrak{G}} \times \mathbf{R}$ is a countable union of open subsets with compact closure it follows that μK^{-1} is a σ-finite measure that is finite valued on compact subsets of $\bar{\mathfrak{G}} \times \mathbf{R}$. Hence μK^{-1} is a regular Borel measure. In particular there must exist a rectangle $\mathbf{A} \times \mathbf{B}_0$ of compact factors such that $\mu K^{-1}(\bar{\mathbf{A}} \times \mathbf{B}_0) > 0$, and since $K^{-1}(\bar{\mathbf{A}} \times \mathbf{B}_0)$ is a compact subset of $\mathfrak{X} - \mathbf{N}$, $\mu K^{-1}(\bar{\mathbf{A}} \times \mathbf{B}_0) < \infty$. We use $\mathbf{B}_0$ as a reference set and define a measure on $\mathfrak{B}(\bar{\mathfrak{G}})$ by, if $\bar{\mathbf{A}} \in \mathfrak{B}(\bar{\mathfrak{G}})$ then

$$\bar{\mu}(\bar{\mathbf{A}}, \mathbf{B}_0) = \mu K^{-1}(\bar{\mathbf{A}} \times \mathbf{B}_0). \tag{10.1.25}$$

This is clearly a σ-finite measure that is finite in value whenever $\bar{\mathbf{A}}$ is a compact subset of $\bar{\mathfrak{G}}$. Therefore $\bar{\mu}(\ , \mathbf{B}_0)$ is a regular measure. Further,

$$\begin{aligned} \bar{\mu}(\overline{g\mathbf{A}}, \mathbf{B}_0) &= \mu K^{-1}(\overline{g\mathbf{A}} \times \mathbf{B}_0) = \mu K^{-1} g(\bar{\mathbf{A}} \times \mathbf{B}_0) = \mu(g K^{-1}(\bar{\mathbf{A}} \times \mathbf{B}_0)) \\ &= \mu K^{-1}(\bar{\mathbf{A}} \times \mathbf{B}_0) = \bar{\mu}(\bar{\mathbf{A}}, \mathbf{B}_0). \end{aligned} \tag{10.1.26}$$

By uniqueness, Theorem 3.3.8, if $\bar{\mu}_0$ is a nonzero σ-finite regular left invariant measure for $\overline{\mathfrak{G}}$, then there exists a constant $v(\mathbf{B}_0)$ such that

$$\bar{\mu}(\bar{\mathbf{A}}, \mathbf{B}_0) = v(\mathbf{B}_0)\bar{\mu}_0(\bar{\mathbf{A}}). \tag{10.1.27}$$

The equation (10.1.27) holds for all $\bar{\mathbf{A}} \in \mathfrak{B}(\overline{\mathfrak{G}})$ and all compact subsets $\mathbf{B}_0 \in \mathfrak{C}$. Further

$$\mu K^{-1}(\bar{\mathbf{A}} \times \mathbf{B}_0) = \bar{\mu}(\bar{\mathbf{A}}, \mathbf{B}_0) = v(\mathbf{B}_0)\bar{\mu}_0(\mathbf{A}). \tag{10.1.28}$$

If we let $\mathfrak{C}_1 \subset \mathfrak{C}$ be the collection of subsets $\mathbf{B}$ for which, if $\bar{\mathbf{A}} \in \mathfrak{B}(\overline{\mathfrak{G}})$ then

$$\mu K^{-1}(\bar{\mathbf{A}} \times \mathbf{B}) = v(\mathbf{B})\bar{\mu}_0(\bar{\mathbf{A}}), \tag{10.1.29}$$

then $\mathfrak{C}_1$ is clearly a monotone class containing all compact subsets of $\mathbf{R}$, hence $\mathfrak{C}_1 = \mathfrak{C}$. We use here assumption (10.1.23). This implies the result that v is a countably additive measure that is finite valued on compact sets, hence that v is a regular measure. Finally we take $\bar{\mu} = \bar{\mu}_0$ for the statement (10.1.24). □

Remark 10.1.3. It is conceivable that although $\mu \neq 0$, $\mu K^{-1} = 0$. Clearly if K is continuous then $K^{-1}(\mathfrak{B}(\overline{\mathfrak{G}}) \times \mathfrak{C}) = \mathfrak{B}(\mathfrak{X} - \mathbf{N})$ and this problem does not occur. More generally, the argument above shows that K is measurable and it has been assumed that K is one-to-one and onto. Our hypotheses imply that $\overline{\mathfrak{G}} \times \mathbf{R}$ is a complete separable metric space. $\mathfrak{X}$ is a metric space and $\mathfrak{X}$ is a union of compact sets $\bigcup_{n=1}^{\infty} K^{-1}(\mathbf{C}_n)$. On each $K^{-1}(\mathbf{C}_n)$, K^{-1} is a homeomorphism and is a Borel measurable function. Hence K^{-1} is a measurable function, $\mathfrak{B}(\mathfrak{X} - \mathbf{N}) = \mathfrak{B}(\overline{\mathfrak{G}}) \times \mathfrak{C}$, and $\mu K^{-1} \neq 0$.

Lemma 10.1.4. *Let μ be a regular left invariant measure defined on $\mathfrak{B}(\mathfrak{X})$ with the factorization* (10.1.24). *If $f: \mathfrak{X} - \mathbf{N} \to \mathbb{R}$ is a μ-integrable function then*

$$\int_{K^{-1}(\bar{\mathbf{A}} \times \mathbf{B})} f(x)\mu(dx) = \iint_{\bar{\mathbf{A}} \times \mathbf{B}} f(K^{-1}(\bar{g}, r))\bar{\mu}(d\bar{g})v(dr). \tag{10.1.30}$$

Proof. By Theorem 10.1.2, equation (10.1.30) holds if f is the indicator function of the set $K^{-1}(\bar{\mathbf{A}}_1 \times \mathbf{B}_1)$. Since K^{-1} maps $\mathfrak{B}(\overline{\mathfrak{G}}) \times \mathfrak{C}$ onto $\mathfrak{B}(\mathfrak{X} - \mathbf{N})$, the equation (10.1.30) holds for all $\mathfrak{B}(\mathfrak{X} - \mathbf{N})$-measurable simple functions. By taking monotone limits of sequences of simple functions (10.1.30) then follows for all μ-integrable functions f. □

Lemma 10.1.5. *Let the hypotheses of Theorem* 10.1.2 *hold. Let η be a nonzero left invariant Haar measure for $\mathfrak{G}$ and let m be the modular function for η. Let $f: \mathfrak{X} - \mathbf{N} \to [0, \infty)$ be a probability density function relative to the regular left invariant measure μ defined on $\mathfrak{B}(\mathfrak{X} - \mathbf{N})$. Let $\mu = \bar{\eta} \cdot v$ be the factorization of μ where η induces $\bar{\eta}$ on $\mathfrak{G}/\mathbf{H}$. Then* (cf. (10.1.32))

$$m(h) \int_{\mathfrak{G}} f(gx)\eta(dg) \tag{10.1.31}$$

is the probability density function for the maximal invariant $r \in \mathbf{R}$ relative to the measure ν on $\mathfrak{C}$. In (10.1.31), *h is any group element satisfying*

$$x = K^{-1}(\pi(h), r) = hK^{-1}(\bar{e}, r). \tag{10.1.32}$$

PROOF. The modular function is discussed in Section 3.5. This function is a continuous group homomorphism to the multiplicative group on $(0, \infty)$. Since we assume $\mathbf{H}$ is a compact group, $m(\mathbf{H}) = \{1\}$. In the notation of Lemma 10.1.5, if $\pi(x) = \pi(y)$, i.e., $x\mathbf{H} = y\mathbf{H}$, then there exists $h \in \mathbf{H}$ such that $m(y) = m(xh) = m(x)m(h) = m(x)$. Hence $m(x) = \bar{m}(\pi(x))$ defines the function $\bar{m}$ on $\mathfrak{G}/\mathbf{H}$. If $\mathbf{U} \subset \mathbb{R}$ is an open subset then $m^{-1}(\mathbf{U}) = \pi^{-1}(\bar{m}^{-1}(\mathbf{U}))$ so that since π is an open mapping, $\pi(m^{-1}(\mathbf{U})) = \bar{m}^{-1}(\mathbf{U})$ is an open set. Hence $\bar{m}$ is a continuous function.

In our proof $K(x) = (\pi(h), r)$ and if π_1 is the projection on the first coordinate then $\pi_1(K(x)) = \pi(h)$ defines $\pi(h)$ implicitly as a function of x. Thus $m(h) = \bar{m}(\pi(h)) = \bar{m}(\pi_1(K(x)))$ is a measurable function of x. It should be noted that if $K(x_1) = (\pi(h), r_1)$ and $K(x_2) = (\pi(h), r_2)$ then $\bar{m}(\pi_1(K(x_1))) = \bar{m}(\pi_1(K(x_2)))$ independent of r_1 and r_2. Thus the constant $m(h)$ which occurs in (10.1.34) does not depend on the variable of integration r in (10.1.35). Therefore

$$\text{if } x = K^{-1}(\pi(h), r) \quad \text{then} \quad f(gx) = f(ghK^{-1}(\bar{e}, r)). \tag{10.1.33}$$

Using Lemma 3.5.3 and (3.5.2),

$$\begin{aligned}\int_{\mathfrak{G}} f(gx)\eta(dg) &= \int_{\mathfrak{G}} f(ghK^{-1}(\bar{e}, r))\eta(dg) \\ &= (m(h))^{-1} \int_{\mathfrak{G}h} f(K^{-1}(\pi(g), r))\eta(dg) \qquad (10.1.34) \\ &= (m(h))^{-1} \int_{\bar{\mathfrak{G}}} f(K^{-1}(\bar{g}, r)\bar{\eta}(d\bar{g}).\end{aligned}$$

The last step of (10.1.34) uses Lemma 3.3.10 to justify introduction of the induced measure $\bar{\eta}$. For a rectangle $\bar{\mathfrak{G}} \times \mathbf{B} \in \mathfrak{B}(\bar{\mathfrak{G}}) \times \mathfrak{C}$, from Lemma 10.1.4,

$$\int_{\bar{\mathfrak{G}} \times \mathbf{B}} f(K^{-1}(\bar{g}, r))\bar{\eta}(d\bar{g})\nu(dr) = \int_{K^{-1}(\bar{\mathfrak{G}} \times \mathbf{B})} f(x)\mu(dx). \tag{10.1.35}$$

Since the most general invariant subset has the form

$$K^{-1}(\bar{\mathfrak{G}} \times \mathbf{B}), \tag{10.1.36}$$

the right side of (10.1.35) is a measure on the Borel subsets of $\mathbf{R}$ whose density function is

$$\int_{\bar{\mathfrak{G}}} f(K^{-1}(\bar{g}, r)\bar{\eta}(d\bar{g}) = m(h) \int_{\mathfrak{G}} f(gx)\eta(dg). \tag{10.1.37}$$

The result follows. □

Theorem 10.1.6. *Given the hypotheses of Theorem* 10.1.2 *and Lemma* 10.1.5, *if* f_1 *and* f_2 *are probability density functions on* $\mathfrak{X} - \mathbf{N}$ *relative to the regular left invariant measure* μ, *then*

$$\int_{\mathfrak{G}} f_2(gx)\eta(dg) \Big/ \int_{\mathfrak{G}} f_1(gx)\eta(dg) \tag{10.1.38}$$

is the ratio of the probability density functions for the maximal invariant.

10.2. Examples

Example 10.2.1.

Case I. Rank $X = k < n$. *The Sample Covariance Matrix.* This continues the Example 10.0.1. Here $\mathfrak{X}$ is the set of $n \times k$ matrices with real entries and we factor $X \in \mathfrak{X}$ into $X = AT^t$, $T \in \mathbf{T}(k)$ and the k-frame A satisfying $A^tA = I_k$, A an $n \times k$ matrix. Take for $\mathfrak{G}$ the matrix group $\mathbf{O}(n)$. In terms of the diagram (10.1.11), with A $n \times k$, $(A, B) \in \mathbf{O}(n)$,

$$h((A, B), T) = AT^t. \tag{10.2.1}$$

Note that h is an open mapping which implies K is a continuous map. The null set $\mathbf{N}$ is the set of X with rank $X < k$, and this set is clearly closed and is invariant. By uniqueness

$$\begin{aligned} &h((A_1, B_1), T_1) = h((A_2, B_2), T_2) \text{ if and only if } A_1 = A_2, \\ &T_1 = T_2, \text{ and for some } U \in \mathbf{O}(n-k),\ B_1 U = B_2. \end{aligned} \tag{10.2.2}$$

We take $\mathbf{H}$ to be the subgroup of matrices $\begin{vmatrix} I_k & 0 \\ 0 & U \end{vmatrix}$ with $U \in \mathbf{O}(n-k)$ and have that (10.2.2) holds if and only if there exists some $V \in \mathbf{H}$ such that $(A_1, B_1)V = (A_2, B_2)$. Then $\bar{\mathfrak{G}} = \mathfrak{G}/\mathbf{H}$ and

$$(\pi \times i)((A, B), T) = ((A, B)\mathbf{H}, T). \tag{10.2.3}$$

This defines the mapping K^{-1} by

$$K^{-1}((A, B)\mathbf{H}, T) = AT^t, \tag{10.2.4}$$

and then (10.1.12) holds.

To verify the hypotheses of Theorem 10.1.2, $\mathfrak{X} - \mathbf{N}$ is an open subset of $\mathbb{R}^{nk}$ and μ is Lebesgue measure which is a left invariant measure. Relative to Lebesgue measure $\mathbf{N}$ is a null set. K^{-1} is one-to-one onto $\mathfrak{X} - \mathbf{N}$. $\mathbf{R} = \mathbf{T}(k)$ is a locally compact group. If $U \in \mathbf{O}(n)$ the condition $UK(\ ) = K(U(\ ))$ requires $U(A, T) = (UA, T)$ in which case (10.1.19) holds and $UK^{-1}((A, B)\mathbf{H}, T) = U(AT^t) = (UA)T^t = K^{-1}((UA, B)\mathbf{H}, T)$ is correct. Since the mapping K^{-1} is continuous, (10.1.21) holds. (10.1.22) holds for μ as defined and (10.1.23) is a definition.

By Theorem 10.1.2, Lebesgue measure factors into a left invariant measure on the Stiefel manifold $\mathbf{V}_{k,n}$ and a measure v on $\mathbf{T}(k)$. The Haar measures η on $\mathbf{O}(n)$ are unimodular. From (10.1.31) we find

$$\int_{\mathbf{O}(n)} f(gx)\eta(dg) \tag{10.2.5}$$

is the probability density function of the maximal invariant T relative to the measure v. Compare this result with that of Chapter 5. The factored measure v does not, of course, depend on f, and may be computed explicitly using differential forms.

Case II. Rank $X = n < k$. The decomposition of Case I no longer holds. From Theorem 8.1.10 we may write, almost surely,

$$X = ADG, \tag{10.2.6}$$

with $A \in \mathbf{O}(n)$, $D \in \mathbf{D}(n)$, and G an $n \times k$ matrix such that $GG^t = I_n$. Almost surely the diagonal elements of D are pairwise distinct and we assume the diagonal of D is in decreasing order of magnitude, so that (10.2.6) is unique up to sign changes which result from multiplication by diagonal orthogonal matrices. If $U \in \mathbf{O}(n)$ then

$$UX = (UA)DG \tag{10.2.7}$$

so the maximal invariant under $\mathbf{O}(n)$ is DG. In this case

$$X^tX = G^tD^2G \tag{10.2.8}$$

and it follows that the rows of G are eigenvectors for $(X^tX)^{1/2}$ and are (almost surely) uniquely determined except for the sign changes noted. Because of uniqueness, application of Theorem 10.1.2 is immediate.

In the sequel, in order to show K^{-1} to be continuous, the following Lemma is useful.

Lemma 10.2.2. *In the diagram* (10.1.11), *if h is a continuous function and if the function K maps compact sets to compact sets, then K is a homeomorphism of* $\mathfrak{X} - \mathbf{N}$ *to* $\overline{\mathfrak{G}} \times \mathbf{R}$.

Proof. We first show that K^{-1} is a continuous function. If $\mathbf{U} \subset \mathfrak{X} - \mathbf{N}$ is an open set then $h^{-1}(\mathbf{U})$ is an open subset of $\mathfrak{G} \times \mathbf{R}$. Since $\pi \times i$ is an open mapping, $K(\mathbf{U}) = K(h(h^{-1}(\mathbf{U}))) = (\pi \times i)(h^{-1}(\mathbf{U}))$ is an open subset. Thus K is an open mapping and K^{-1} is continuous.

If in addition K maps compact sets to compact sets, then since K is one-to-one, the continuity of K as a function follows since $\mathfrak{X} - \mathbf{N}$ and $\mathfrak{G} \times \mathbf{R}$ are separable locally compact spaces, hence metrizable. □

EXAMPLE 10.2.3. The general linear hypothesis, discussed as Examples 10.0.4 and 10.1.1 is continued here in the case $p \geq k$ and $q \geq k$. The setup as described in Section 10.1 may fail without a suitable choice of the function h. In the discussion here we suppose

$$h(U, V, G, D) = (U_k D^{1/2} G^t, V_k(I_k - D)^{1/2} G^t) \tag{10.2.9}$$

where U_k and V_k are the $p \times k$ and $q \times k$ matrices consisting of the first k columns of U and V respectively. For this choice the condition $h(g_0 g, r) = g_0 h(g, r)$ is readily verified. As in (10.1.12), $K^{-1}(\pi(g), r) = h(g, r)$ defines K^{-1} as a one-to-one function from $\mathfrak{G}/\mathbf{H} \times \mathbf{R}$ to $\mathfrak{X} - \mathbf{N}$. We will verify this in the sequel. It is also easy to see from

$$X^t X + Y^t Y = GG^t \tag{10.2.10}$$

that on a bounded set in $\mathfrak{X} - \mathbf{N}$, $\operatorname{tr} GG^t$ is bounded. Thus, since $0 \leq D \leq I_k$, the maps h^{-1} and $(\pi \times i)^{-1}$ map compact sets to compact sets. Thus both K and K^{-1} map compact sets to compact sets.

To examine the question of being one-to-one, suppose

$$(X, Y) = (U_k D^{1/2} G^t, V_k(I_k - D)^{1/2} G^t) = (\tilde{U}_k \tilde{D}^{1/2} \tilde{G}^t, \tilde{V}_k(I_k - \tilde{D})^{1/2} \tilde{G}^t). \tag{10.2.11}$$

Since $X^t X + Y^t Y = GG^t = \tilde{G}\tilde{G}^t$ it follows that $\tilde{G}^{-1} G \in \mathbf{O}(k)$. Also $X^t X = GDG^t = \tilde{G}\tilde{D}\tilde{G}^t$ so that D and $\tilde{D}$ are the matrices of eigenvalues of the same matrix since $\tilde{D} = (\tilde{G}^{-1}G)D(\tilde{G}^{-1}G)^t$. Since the entries of D are in decreasing order, $D = \tilde{D}$. Since the diagonal of D consists of pairwise distinct numbers, $G = \tilde{G}W$ with W diagonal and orthogonal. Then $U_k W^t = \tilde{U}_k$. Then, in the notation of (10.1.12),

$$\begin{aligned}&\text{if } h(g, r) = h(\tilde{g}, \tilde{r}) \text{ then } r = \tilde{r} \text{ and } \tilde{g}^{-1} g\, h(e, r) = h(e, r).\\ &\text{Thus } \tilde{g}^{-1} g \in \mathbf{H} = \{g \mid gh(e, r) = h(e, r)\}.\end{aligned} \tag{10.2.12}$$

This implies that K and K^{-1} are well defined and are one-to-one.

In terms of the analysis just given, if

$$(U, V, G) \text{ acting on } \begin{vmatrix} I_k \\ 0 \end{vmatrix} D^{1/2}, \qquad \begin{vmatrix} I_k \\ 0 \end{vmatrix} (I - D)^{1/2} \tag{10.2.13}$$

leaves this fixed, then G is a diagonal orthogonal matrix and $U = \begin{vmatrix} G & 0 \\ 0 & U' \end{vmatrix}$, $V = \begin{vmatrix} G & 0 \\ 0 & V' \end{vmatrix}$ with $G \in \mathbf{O}(k) \cap \mathbf{D}(k)$, $U' \in \mathbf{O}(p-k)$ and $V' \in \mathbf{O}(q-k)$. Thus the subgroup $\mathbf{H}$ is compact.

In order to apply Theorem 10.1.2 it is necessary to construct an invariant measure μ on the Borel subsets of $\mathfrak{X} - \mathbf{N}$. Clearly the measure determined by the following differential form is an invariant measure.

$$\frac{\bigwedge_{i=1}^{p} \bigwedge_{j=1}^{k} dx_{ij} \bigwedge_{i=1}^{q} \bigwedge_{j=1}^{k} dy_{ij}}{(\det(X^t X + Y^t Y))^{(p+q)/2}}. \tag{10.2.14}$$

If **X** and **Y** are normally distributed as explained in Example 10.1.1 then the joint density of **X**, **Y** relative to μ is

$$f_{M,\Sigma} = (2\pi)^{-(p+q)k/2}(\det \Sigma)^{-(p+q)/2}(\det(X^tX + Y^tY))^{(p+q)/2} \times \operatorname{etr} -\tfrac{1}{2}\Sigma^{-1}(X^tX + Y^tY - 2X^tM + M^tM). \tag{10.2.15}$$

This corresponds to $\mathbf{E}\mathbf{X} = M$ and $\mathbf{E}\mathbf{Y} = 0$, independent rows each with covariance matrix Σ. Integration over dU, dV, dG for Haar measure on $\mathbf{O}(p) \times \mathbf{O}(q) \times \mathbf{GL}(k)$ gives

$$\int f_{M,\Sigma}(UXG^t, VYG^t)\, dU\, dV\, dG = (2\pi)^{-(p+q)k/2}(\det(X^tX + Y^tY))^{(p+q)/2} \times \int (\det(G^tG))^{-k/2} \operatorname{etr} -\tfrac{1}{2}H(X, U, G)\, dU\, dG, \tag{10.2.16}$$

$$H(X, U, G) = G^t(X^tX + Y^tY)G - 2G^tX^tU^tM\Sigma^{-1/2} + M\Sigma^{-1}M^t.$$

Since the variables of V do not explicitly enter into (10.2.16) the integration dV merely gives a factor of one. By Lemma 10.1.5, after multiplying by the modular function, (10.2.16) becomes the density function of the maximal invariant. In this case the modular function is identically one.

A more general form of the general linear hypothesis was studied by Schwartz (1966a), (1966b), (1967) and (1969).

10.3. Examples: The Noncentral Multivariate Beta Density Function

In this section we describe the problem in random variable terms and in terms of differential forms. Since the quantity in question is a maximal invariant, the application of Sections 10.1 and 10.2 provides a suitable exercise, so stated in the problems of Section 10.5.

We will suppose **X** is $p \times k$ and **Y** is $q \times k$ such that **X** and **Y** have independently normally distributed rows each with covariance matrix I_k. We set

$$S = X^tX, \qquad T = Y^tY, \quad \text{and} \quad CC^t = S + T, \qquad CLC^t = S, \quad \text{with} \quad C \in \mathbf{T}(k) \quad \text{and} \quad L \in \mathbf{S}(k). \tag{10.3.1}$$

The problem discussed here and by Kshirsagar (1961) is the probability density function of **L**. It is at once clear that **L** is a maximal invariant under the group action

$$(X, Y) \to (UXG^t, VYG^t), \qquad U \in \mathbf{O}(p), \qquad V \in \mathbf{O}(q), \qquad G \in \mathbf{T}(k), \tag{10.3.2}$$

provided $p \geq k$ and $q \geq k$.

If we begin with the differential form

$$\bigwedge_{j\le i} ds_{ij} \bigwedge_{j\le i} dt_{ij} \tag{10.3.3}$$

the change of variables involves

$$ds_{ij} + dt_{ij} = (dC\,C^t + C\,dC^t)_{ij}, \tag{10.3.4}$$

and

$$ds_{ij} = (dC\,LC^t + C(dL)C^t + CL(dC^t))_{ij}.$$

Using the fact that if ω is a 1-form then $\omega \wedge \omega = 0$, we obtain at once

$$\begin{aligned}
&\bigwedge_{j\le i} dt_{ij} \bigwedge_{j\le i} ds_{ij}\\
&\quad = \bigwedge_{j\le i}(ds_{ij} + dt_{ij}) \bigwedge_{j\le i} ds_{ij}\\
&\quad = \bigwedge_{j\le i}(dC\,C^t + C\,dC^t)_{ij} \bigwedge_{j\le i}(dC\,LC^t + C(dL)C^t + CL(dC^t))_{ij} \qquad (10.3.5)\\
&\quad = \bigwedge_{j\le i}(dC\,C^t + C\,dC^t)_{ij}(\det C)^{k+1} \bigwedge_{j\le i}(dL)_{ij}\\
&\quad = 2^k c_{11}^k c_{22}^{k-1} \ldots c_{kk}(\det C)^{k+1} \bigwedge_{j\le i}(dC)_{ij} \bigwedge_{j\le i}(dL)_{ij}.
\end{aligned}$$

For notational reasons we write $(C)_{ij}$ and $(L)_{ij}$ for the (i,j)-entries of the indicated matrices. It follows from (10.3.5) that the underlying joint measure for C and L factors, a fact long known. See for example Anderson (1958).

In the central case, $\mathbf{E}\mathbf{X} = 0$ and $\mathbf{E}\mathbf{Y} = 0$, and the joint probability density function of $\mathbf{S}$ and $\mathbf{T}$ is (from the Wishart density)

$$\begin{aligned}
&2^{-(p+q)k/2}\pi^{-k(k-1)/2}(\det S)^{(p-k-1)/2}(\det T)^{(q-k-1)/2}\\
&\qquad \times [\operatorname{etr} - \tfrac{1}{2}(S+T)]/\prod_{i=0}^{k-1}[\Gamma(\tfrac{1}{2}(p-i))\Gamma(\tfrac{1}{2}(q-i))]\\
&\quad = 2^{-(p+q)k/2}\pi^{-k(k-1)/2}(\det CC^t)^{(p+q-2k-2)2}\operatorname{etr} - \tfrac{1}{2}(CC^t) \qquad (10.3.6)\\
&\qquad \times (\det L)^{(p-k-1)/2}(\det(I_k - L))^{(q-k-1)/2}/\prod_{i=1}^{k-1}[\Gamma(\tfrac{1}{2}(p-i))\Gamma(\tfrac{1}{2}(q-i))].
\end{aligned}$$

The function (10.3.6) depends only on $E = CC^t$ so we introduce new variables $(E)_{ij}$. Using Problem 6.7.7 and in particular (6.7.11) we find

$$\bigwedge_{j\le i}(dE)_{ij} = 2^k c_{11}^k c_{22}^{k-1} \ldots c_{kk} \bigwedge_{j\le i}(dC)_{ij}. \tag{10.3.7}$$

Substitution into (10.3.5) and integration of (10.3.6) then yields the central probability density function of the $(L)_{ij}$ to be

$$\begin{aligned}
&\pi^{-k(k-1)/4}\prod_{i=0}^{k-1}\Gamma(\tfrac{1}{2}(p+q-i))/\Gamma(\tfrac{1}{2}(p-i))\Gamma(\tfrac{1}{2}(q-i)) \qquad (10.3.8)\\
&\qquad \times (\det L)^{(p-k-1)/2}(\det(I_k - L))^{(q-k-1)/2} \bigwedge_{j\le i}(dL)_{ij}.
\end{aligned}$$

The differential form (10.3.5) and the function (10.3.6) show the well-known fact that **C** and **L** are stochastically independent in the central case.

We now consider the noncentral case in which it is assumed that $\mathbf{E}\,\mathbf{X} = M \neq 0$ and $\mathbf{E}\,\mathbf{Y} = 0$. In (10.3.6) we must replace the central Wishart probability density function for **S** by the noncentral Wishart density function for **S**, which is

$$2^{-pk/2}\pi^{-k(k-1)/4}(\det S)^{(p-k-1)/2}\operatorname{etr} -\tfrac{1}{2}(S + M^tM)$$
$$\times \int_{\mathbf{O}(k)} \operatorname{etr}((UX)^tM)\,dU \Big/ \prod_{i=0}^{k-1}\Gamma(\tfrac{1}{2}(p-i)). \tag{10.3.9}$$

The integral by Haar measure of unit mass on $\mathbf{O}(k)$ is equal to

$$\int_{\mathbf{O}(k)} \operatorname{etr}(U^t(MX^tXM^t)^{1/2})\,dU = \int_{\mathbf{O}(k)} \operatorname{etr}(U^t(MCLC^tM^t)^{1/2})\,dU. \tag{10.3.10}$$

The integral in (10.3.10) must be substituted into (10.3.9) which in turn is substituted into (10.3.6) for the integration over the variables C. The variables C are to be integrated out, leading to a problem whose answer in closed form is unknown.

The matrix L which is symmetric may be decomposed as

$$L = WW^t, \qquad W \in \mathbf{T}(k). \tag{10.3.11}$$

It is known that the random variables on the diagonal of $\mathbf{W}$, which we call $\mathbf{w}_{11}, \ldots, \mathbf{w}_{kk}$, are independently distributed (in the central case) beta random variables such that the density function of $\mathbf{w}_{ii}$ is

$$w^{(p-1-i)/2}(1-w)^{(q-2)/2}\Gamma(\tfrac{1}{2}(p+q+1-i))/\Gamma(\tfrac{1}{2}(p+1-i))\Gamma(\tfrac{1}{2}q). \tag{10.3.12}$$

This result is usually obtained by calculation of the joint moments of the random variables. See Anderson (1958). In Chapter 11 of this book we will obtain part of this result from the canonical decomposition of the sample covariance matrix. See Problem 11.13.2.

The Wilks criterion for independence of a sample **X** from a sample **Y** uses a statistic which computationally, on the null hypothesis, is equivalent to computing $\det L = (w_{11} \cdots w_{kk})^2$. Therefore the null distribution, that of $\det L$ in the central case, is the distribution of a product of independently distributed beta random variables. The corresponding problem for chi-square random variables was studied by Kullback (1934) and a general formulation of the problem has been made in Mathai and Saxena (1969). The form of the density function is very unpleasant.

We close this section with some observations on moments. If $r \geq 1$ is an integer then

$$\mathbf{E}(\det \mathbf{L})^r = \text{constant} \int (\det L)^{r+(p-k-1)/2} \times (\det(I_k - L))^{(q-k-1)/2} \bigwedge_{j \leq i}(dL)_{ij}. \tag{10.3.13}$$

From (10.3.8), by normalization of (10.3.13), we find

$$\mathbf{E}(\det \mathbf{L})^r = \prod_{i=0}^{k-1} \frac{\Gamma(\frac{1}{2}(2r+p-i))\Gamma(\frac{1}{2}(p+q-i))}{\Gamma(\frac{1}{2}(2r+p+q-i))\Gamma(\frac{1}{2}(p-i))}. \tag{10.3.14}$$

We note that if S_s, T_s, L_s, and C_s are the $s \times s$ principal minors then

$$S_s + T_s = C_s C_s^t \quad \text{and} \quad S_s = C_s L_s C_s^t, \qquad 1 \le s \le k. \tag{10.3.15}$$

Therefore (10.3.14) gives the value of $\mathbf{E}(\det \mathbf{L}_s)^r$ valid for all $p \ge 1$, $q \ge 1$, $r \ge 0$, and $k \ge 1$. More general moment calculations may be found in Anderson, op. cit., Section 9.4.

10.4. Modifications of the Basic Theory

The previous examples have been arranged to fit the hypotheses of Theorem 10.1.2. Here we explore modifications in case obvious choices of the functions h and K do not work. We assume as in Section 10.1 that diagram (10.1.11) is constructible with functions K, h, π, i as described there. Our main hypothesis is

Hypothesis 10.4.1. *If $r \in \mathbf{R}$ and $g_1, g_2 \in \mathfrak{G}$ then*

$g_1 K^{-1}(\bar{e}, r_1) = g_2 K^{-1}(\bar{e}, r_2)$ *if and only if*
$r_1 = r_2$ *and* $g_1^{-1} g_2 \in \mathbf{H}$. *In addition, every* $x \in \mathfrak{X} - \mathbf{N}$ (10.4.1)
must be expressible as $x = gK^{-1}(\bar{e}, r)$.

Note that in the discussion of Example 10.2.2 the Hypothesis 10.4.1 was verified at (10.2.12).

Definition 10.4.2. The function $L: \mathfrak{G} \times \mathbf{R} \to \bar{\mathfrak{G}} \times \mathbf{R}$ is defined by

$$L(g, r) = K(gK^{-1}(\bar{e}, r)). \tag{10.4.2}$$

The group $\mathfrak{G}'$ of transformations of $\bar{\mathfrak{G}} \times \mathbf{R}$ is defined by

$$g_1' L(g, r) = K(g_1 K^{-1}(L(g, r))). \tag{10.4.3}$$

Lemma 10.4.3. *$L(g_1, r_1) = L(g_2, r_2)$ if and only if $r_1 = r_2$ and $g_1^{-1} g_2 \in \mathbf{H}$. Further,*

$$g_1' L(g, r) = L(g_1 g, r). \tag{10.4.4}$$

The action of g' is thus well defined and

$$g_1' L(g, r) = g_2' L(g, r) \quad \text{if and only if } g_1^{-1} g_2 \in \mathbf{H}. \tag{10.4.5}$$

Proof. By (10.4.2), $L(g_1, r_1) = L(g_2, r_2)$ requires $g_1 K^{-1}(\bar{e}, r_1) = g_2 K^{-1}(\bar{e}, r_2)$ which by Hypothesis 10.4.1 requires $r_1 = r_2$ and $g_1^{-1} g_2 \in \mathbf{H}$. Then, by definition,

$$g_1' L(g,r) = K(g_1 K^{-1}(L(g,r))) = K(g_1 K^{-1}(KgK^{-1}(\bar{e},r))) \\ = L(g_1 g, r). \tag{10.4.6}$$

Thus $g_1' L(g,r) = g_2' L(g,r)$ means $L(g_1 g, r) = L(g_2 g, r)$ and $g_1^{-1} g_2 \in \mathbf{H}$. Also, if $L(g_1, r_1) = L(g_2, r_2)$ then $g' L(g_1, r_1) = L(gg_1, r_1)$ and $g' L(g_2, r_2) = L(gg_2, r_2)$ so that $r_1 = r_2$ and $g_1^{-1} g_2 \in \mathbf{H}$ follows and $L(gg_1, r_1) = L(gg_2, r_2)$. □

In order to construct invariant measures we prove the following.

Lemma 10.4.4.

$$K^{-1}(g_1'(\bar{g}, r)) = g_1 K^{-1}(\bar{g}, r). \tag{10.4.7}$$

Proof. First we show that $(\bar{g}, r) = L(g_2, r)$ for some $g_2 \in \mathfrak{G}$. In fact, by definition, $L(g_2, r) = K(g_2 K^{-1}(\bar{e}, r))$ so we wish to solve the equation $K^{-1}(\bar{g}, r) = g_2 K^{-1}(\bar{e}, r)$. By Hypothesis 10.4.1 the element g_2 exists. Then

$$K^{-1}(g_1'(\bar{g}, r)) = K^{-1}(g_1' L(g_2, r)) = K^{-1} L(g_1 g_2, r) \\ = K^{-1} K(g_1 g_2 K^{-1}(\bar{e}, r)) = g_1 K^{-1} K(g_2 K^{-1}(\bar{e}, r)) \\ = g_1 K^{-1} L(g_2, r) = g_1 K^{-1}(\bar{g}, r). \tag{10.4.8}$$

□

The group $\mathfrak{G}'$ will generally be isomorphic to $\mathfrak{G}$. In fact if $g_1' = g_2'$ then $L(g_1 g, r) = L(g_2 g, r)$ for all $g \in \mathfrak{G}$ and $r \in \mathbf{R}$ so $g^{-1} g_1^{-1} g_2 g \in \mathbf{H}$ for all $g \in \mathbf{G}$ or $g_1^{-1} g_2 \in \bigcap_g g\mathbf{H}g^{-1} = \mathbf{H}'$. This is a subgroup of $\mathbf{H}$ which is a normal subgroup of $\mathfrak{G}$. Frequently this is $\{e\}$, which would imply $g_1 = g_2$, and that $\mathfrak{G}$ and $\mathfrak{G}'$ are isomorphic. Otherwise $\mathbf{H}'$ is a proper normal subgroup and

$$\mathfrak{G}' \text{ is isomorphic to } \mathfrak{G}/\mathbf{H}' \text{ under a projection } \pi_0. \tag{10.4.9}$$

We obtain the following diagram.

$$\begin{array}{ccc} (g,r) & \longrightarrow & gK^{-1}(\bar{e},r) \\ \mathfrak{G} \times \mathbf{R} & \longrightarrow & \mathfrak{X} - \mathbf{N} \\ \pi_0 \times i \downarrow & \nearrow h' & \uparrow K^{-1} \\ \mathfrak{G}' \times \mathbf{R} & \longrightarrow & \mathfrak{G}'/\mathbf{H}' \times \{(\bar{e},r) \mid r \in \mathbf{R}\} \\ (g',r) & \longrightarrow & g'L(e,r) = g'(\bar{e},r) \end{array} \tag{10.4.10}$$

where the function h' satisfies

$$h'(g',r) = gK^{-1}(\bar{e},r) = K^{-1}(L(g,r)) = K^{-1}(g'L(e,r)) \\ = K^{-1}(g'(\bar{e},r)). \tag{10.4.11}$$

Theorem 10.1.2 now applies directly to the factoring of an invariant measure.

EXAMPLE 10.4.5. We consider once again the general linear hypothesis with $p \geq k$ and $q \geq k$ using the definition

$$K(X, Y) = (X(X^tX)^{-1/2}, Y(Y^tY)^{-1/2}, G(X, Y), D) \qquad (10.4.12)$$

where $G(X, Y)$ is a coset, any element of which, say G, satisfies

$$X^tX + Y^tY = G^tG \quad \text{and} \quad X^tX = G^tDG. \qquad (10.4.13)$$

As noted in the discussion of Example 10.2.2, $\mathbf{GL}(k)$ must be factored by the group of 2^k diagonal orthogonal matrices which we call $\mathbf{GL}(\pm)$. The inverse function K^{-1} is readily seen to be

$$K^{-1}(U_k, V_k, G, D) = (U_k(G^tDG)^{1/2}, V_k(G^t(I_k - D)G)^{1/2}) \quad (10.4.14)$$

so that in the notation of Hypothesis 10.4.1 we must examine

$$\left(\begin{vmatrix} D^{1/2} \\ 0 \end{vmatrix}, \begin{vmatrix} (I_k - D)^{1/2} \\ 0 \end{vmatrix} \right) \qquad (10.4.15)$$

and determine the result of transformation of this quantity. That is, if

$$U_1 \begin{vmatrix} D_1^{1/2} \\ 0 \end{vmatrix} G_1^t = U_2 \begin{vmatrix} D_2^{1/2} \\ 0 \end{vmatrix} G_2^t, \qquad (10.4.16)$$

and

$$V_1 \begin{vmatrix} (I_k - D_1)^{1/2} \\ 0 \end{vmatrix} G_1^t = V_2 \begin{vmatrix} (I_k - D_2)^{1/2} \\ 0 \end{vmatrix} G_2^t$$

then as previously shown in Example 10.2.2, $G_1 = G_2 W$ with $W \in \mathbf{O}(k) \cap \mathbf{D}(k)$, $D_1 = D_2$, and

$$U_1 = U_2 \begin{vmatrix} W & 0 \\ 0 & U' \end{vmatrix}, \; V_1 = V_2 \begin{vmatrix} W & 0 \\ 0 & V' \end{vmatrix}. \qquad (10.4.17)$$

Thus Hypothesis 10.4.1 is satisfied.

10.5. Problems

PROBLEM 10.5.1. Examples 10.0.2, 10.0.3 and 10.0.5 can be treated using the theory of Sections 10.1 and 10.2. That is, find the density function of the maximal invariants by the integration process described in those sections.

PROBLEM 10.5.2. The F-statistic used in the analysis of variance is the maximal invariant in the case $k = 1$ of the general linear hypothesis. In this case

$\mathbf{GL}(k)$ consists of the multiplicative group on $(0, \infty)$. Write the density function for the maximal invariant, and compare the result with the noncentral F-probability density function.

Remark 10.5.3. Do not forget that the measure μ on the Borel subsets of $\mathfrak{X} - \mathbf{N}$ must be a left invariant measure.

PROBLEM 10.5.4. Use the theory of Sections 10.1 and 10.2 to write the density function of $\mathbf{L}$ which was defined in Section 10.3. As noted in Section 10.3, $\mathbf{L}$ is a maximal invariant. In particular show that an integral similar to (10.3.10) results in the noncentral problem. Use transformations $(X, Y) \to (UXG^t, VYG^t)$.

PROBLEM 10.5.5. Suppose $\mathbf{U}_1, \ldots, \mathbf{U}_k$ are mutually independent real valued random variables such that if $1 \le i \le k$ then $\mathbf{U}_i$ has a gamma density $u^{a_i - 1}e^{-u}/\Gamma(a_i)$. Under the transformation by scale change $(\mathbf{U}_1, \ldots, \mathbf{U}_k) \to (x\mathbf{U}_1, \ldots, x\mathbf{U}_k)$, the maximal invariant is a set of random variables satisfying $\mathbf{S}_i(\mathbf{U}_1 + \cdots + \mathbf{U}_k) = \mathbf{U}_i$, $1 \le i \le k$. Then $\mathbf{S}_1 + \cdots + \mathbf{S}_k \equiv 1$. Find the joint density function of $\mathbf{S}_1, \ldots, \mathbf{S}_{k-1}$. Do this first by applying the theory of Sections 10.1 and 10.2. Then compute the differential form

$$t = u_1 + \cdots + u_k, \quad ts_i = u_i, \tag{10.5.1}$$

so that

$$t^{k-1}\Big(\bigwedge_{i=1}^{k-1} ds_i\Big)\,dt = \bigwedge_{i=1}^{k} du_i,$$

and integrate out the variable t. The answer,

$$s_1^{a_1-1} \ldots s_{k-1}^{a_{k-1}-1}(1 - s_1 - \cdots - s_{k-1})^{a_k-1} \frac{\Gamma(\Sigma a_i)}{\Gamma(a_1)\ldots\Gamma(a_k)} \tag{10.5.2}$$

is sometimes called the multivariate beta density function.

CHAPTER 11

Random Variable Techniques

11.0. Introduction

In some parts of statistical inference it is customary to speak entirely in random variable terms, as in the analysis of variance. One considers his job finished when he writes the ratio of two independently distributed chi-square random variables, the denominator a central chi-square. There are a number of distribution problems in which by manipulation of the random variables involved one reduces the distribution problem to determination of the distribution of a relatively simple function of several independently distributed random variables. The theory of best linear unbiased estimation whereby a noncentral chi-square is obtained that is distributed independently of the sum of squares of error is in the meaning of this chapter a theory using random variable techniques. Another example that can at least partially be treated using random variable techniques is the example of Section 10.3, the multivariate beta density functions. See also Remark 5.7.3.

Aside from the literature of the analysis of variance the literature on random variable techniques seems to be very small. We have found only two papers, Wijsman (1957) and Graybill and Milliken (1969). The treatment of the analysis of variance given in this chapter is my own but other coordinate free treatments exist in the literature, for example Kruskal (1961) and Stein (1959). Basic tools used in this chapter include the generalized inverse of matrices due to Penrose (1955) and discussed in Theorem 8.1.13. Much of the theory can be developed using conditional inverses of matrices as defined by Graybill and Milliken, op. cit., Lemma 2.3. Another tool, the canonical decomposition of the sample covariance matrix, can be derived using random variable techniques as is done in Wijsman (1957) as well as by use of differential forms, illustrated at the end of this section. The decomposition itself seems to be due to Mauldon (1955), but is called the Bartlett decomposition by some authors. See Bartlett (1933). The author learned about the use of random variable techniques from lectures of Wijsman (1958) at the University of Illinois, and some of the results of this chapter are reworkings of my 1958 lecture notes.

Examples treated in this chapter using the canonical decomposition of the sample covariance matrix are (1) the Wishart distribution, (2) Hotelling T^2 statistic, (3) Wilks's generalized variance, (4) the multivariate beta density functions, (5) the sample correlation coefficient, (6) the sample multiple correlation coefficient, and (7) the conditional covariance matrix.

We now state and prove the decomposition theorem. The result stated is a bit more than the Bartlett decomposition since independence of the orthogonal component is of interest also. The orthogonal component is distributed as uniform measure of unit mass on k-frames as follows from invariance of the distribution of $\mathbf{X}$ under orthogonal transformations, while independence can be argued using Basu's lemma, Theorem 3.4.5.

Theorem 11.0.1. *Let* $\mathbf{X}$ *be an* $n \times k$ *random matrix such that the rows of* $\mathbf{X}$ *are mutually independent random vectors each normal* $(0, I_k)$. *Let* $(\mathbf{t}_{ij}) = \mathbf{T} \in \mathbf{T}(k)$, *and* $\mathbf{A}$ *be a random* $n \times k$ *matrix such that*

$$\mathbf{T}\mathbf{T}^t = \mathbf{X}^t\mathbf{X} \quad \textit{and} \quad \mathbf{X} = \mathbf{A}\mathbf{T}^t \tag{11.0.1}$$

Then $\mathbf{A}$ *and* $\mathbf{T}$ *are distributed independently,* $\mathbf{A}$ *has the orthogonally invariant distribution of unit mass on k-frames, and the* $k(k+1)/2$ *elements of* $\mathbf{T}$ *are independently distributed random variables. If* $1 \le i \le k$ *then* $\mathbf{t}_{ii}^2$ *is a central chi-square with* $n - i + 1$ *degrees of freedom, and if* $1 \le j < i \le k$ *then* $\mathbf{t}_{ij}$ *is a normal* $(0, 1)$ *random variable. We assume* $n \ge k$.

PROOF. The proof is based on the use of differential forms; see Chapter 7. Let A be $n \times k$ such that $A^tA = I_k$ and $X = AT^t$. Since almost surely rank $X = \min(k, n) = k$, almost surely A, T are uniquely determined. Then

$$dX = (dA)T^t + A(dT)^t. \tag{11.0.2}$$

Choose B so that $(A, B) \in \mathbf{O}(n)$ and $\det(A, B) = 1$. Then

$$A^t(dX)(T^t)^{-1} = A^t(dA) + (T^{-1}\,dT)^t, \tag{11.0.3}$$

and

$$B^t(dX)(T^t)^{-1} = B^t(dA).$$

As in Chapter 9, forming the join yields

$$(\det T)^{-k} \bigwedge_{i=1}^{k} \bigwedge_{j=1}^{k} (A^t(dX))_{ij} = \bigwedge_{j<i} (A^t\,dA)_{ij} \bigwedge_{j\ge i} (T^{-1}\,dT)^t_{ij}, \tag{11.0.4}$$

and

$$(\det T)^{-(n-k)} \bigwedge_{i=1}^{n-k} \bigwedge_{j=1}^{k} (B^t(dX))_{ij} = \bigwedge_{i=1}^{n-k} \bigwedge_{j=1}^{k} (B^t(dA))_{ij}.$$

Therefore,

$$\begin{aligned}\bigwedge_{i=1}^{n}\bigwedge_{j=1}^{k} dx_{ij} &= (\det(A,B))^k \bigwedge_{i=1}^{n}\bigwedge_{j=1}^{k} dx_{ij} \\ &= \pm(\det T)^n \bigwedge_{i=1}^{n-k}\bigwedge_{j=1}^{k} (B^t(dA))_{ij} \bigwedge_{j<i}(A^t\,dA)_{ij} \bigwedge_{j\le i}(T^{-1}\,dT)_{ij}.\end{aligned} \tag{11.0.5}$$

The variables of A represent a k-frame of the Stiefel manifold and the orthogonally invariant measure $\bigwedge_{i=1}^{n-k}\bigwedge_{j=1}^{k}(B^t(dA))_{ij}\bigwedge_{j<i}(A^t\,dA)_{ij}$ appears as a factor of the integrating measure. Since the variables A do not explicitly enter the density function, after integration over these variables, the density function in the variables of T is

$$(2\pi)^{-nk/2}A_n\cdots A_{n-k+1}(\det T)^n \exp\left(-\tfrac{1}{2}\sum_{j\le i} t_{ij}^2\right)\bigwedge_{j\le i}(T^{-1}\,dT)_{ij}. \tag{11.0.6}$$

The notations $A_n, \ldots, A_{n-k+1}$ are as defined in the statement of Theorem 7.8.1, Section 7.8. From (7.2.1) and Problem 7.10.1, the expression (11.0.6) becomes

$$(2\pi)^{-nk/2}A_n\cdots A_{n-k+1}t_{11}^{n-1}\cdots t_{kk}^{n-k} \times \exp\left(-\tfrac{1}{2}\sum_{j\le i} t_{ij}^2\right)\bigwedge_{j\le i} dt_{ij}. \tag{11.0.7}$$

Substitution of

$$s_{ii} = t_{ii}^2, \qquad ds_{ii} = 2t_{ii}\,dt_{ii}, \qquad i = 1, \ldots, k, \tag{11.0.8}$$

into

$$t_{ii}^{n-i}\exp(-\tfrac{1}{2}t_{ii}^2)\,dt_{ii} \tag{11.0.9}$$

yields

$$\tfrac{1}{2}s_{ii}^{(n-i+1)/2-1}\exp(-\tfrac{1}{2}s_{ii})\,ds_{ii}, \tag{11.0.10}$$

which is an unnormalized form of the χ_{n-i+1}^2 density function. Therefore (11.0.7) clearly factors into

$$\prod_{j\le i} f_{ij}(s_{ij})\,ds_{ij}, \qquad s_{ij} = t_{ij}, \qquad j < i, \tag{11.0.11}$$

and

$$s_{ii} = t_{ii}^2, \qquad 1 \le i \le k,$$

where if

$$1 \le j < i \le k \tag{11.0.12}$$

then

$$f_{ij}(s_{ij}) = (2\pi)^{-1/2}\exp(-\tfrac{1}{2}s_{ij}^2);$$

if

$$1 \le i \le k$$

then

$$a_i f_{ii}(s_{ii}) = \frac{s_{ii}^{(n-i+1)/2-1} \exp(-\frac{1}{2}s_{ii})}{\Gamma(\frac{1}{2}(n-i+1))2^{(n-i+1)/2}}.$$

Since (11.0.7) is a density function it follows from (11.0.12) that the product $a_1 \cdots a_k = 1$. That completes the proof. □

11.1. Random Orthogonal Matrices

Expressed in the language of measure theory the basic lemma is as follows.

Lemma 11.1.1. *Let* $\mathbf{X}$ *and* $\mathbf{Y}$ *be random variables with ranges* $\mathbf{R}_\mathbf{X}$ *and* $\mathbf{R}_\mathbf{Y}$. *We suppose* $\mathbf{X}$ *and* $\mathbf{Y}$ *are independently distributed. Let* $\mathbf{X}$ *induce the probability measure* μ *on* $\mathbf{R}_\mathbf{X}$ *and let* $\mathbf{Y}$ *induce the probability measure* ν *on* $\mathbf{R}$ *so that the joint distribution of* $(\mathbf{X}, \mathbf{Y})$ *on* $\mathbf{R}_\mathbf{X} \times \mathbf{R}_\mathbf{Y}$ *is* $\mu \times \nu$.

Suppose $T: \mathbf{R}_\mathbf{X} \times \mathbf{R}_\mathbf{Y} \to \mathbf{R}_\mathbf{Y}$ *is jointly measurable and if* $X \in \mathbf{R}_\mathbf{X}$ *then* $\nu(\{Y | T(X, Y) \in \mathbf{B}\}) = \nu(\mathbf{B})$. *Then* $(\mu \times \nu)(\{X, Y) | T(X, Y) \in \mathbf{B}\}) = \nu(\mathbf{B})$.

Proof. Let $\mathbf{A} = \{(X, Y) | T(X, Y) \in \mathbf{B}\}$ and let $\mathbf{A}_X = \{Y | T(X, Y) \in \mathbf{B}\}$. Since $\mu(\mathbf{R}_\mathbf{X}) = 1$, by Fubini's theorem,

$$(\mu \times \nu)(\mathbf{A}) = \int \nu(\mathbf{A}_X)\mu(dX) = \nu(\mathbf{B}). \tag{11.1.1}$$

□

We will apply Lemma 11.1.1 in the specific context that $\mathbf{Y}$ is an $n \times k$ random matrix with independently distributed rows each normal $(0, I_k)$. We suppose $U: \mathbf{R}_\mathbf{X} \to \mathbf{O}(n)$ is measurable, and let

$$T(X, Y) = U(X)Y. \tag{11.1.2}$$

Suppose $\mathbf{Y}$ has columns $\mathbf{Y}_1, \ldots, \mathbf{Y}_k$ so that the columns of $U(\mathbf{X})\mathbf{Y}$ are $U(\mathbf{X})\mathbf{Y}_1, \ldots, U(\mathbf{X})\mathbf{Y}_k$. Since for each X, the random variables $U(X)\mathbf{Y}_1, \ldots, U(X)\mathbf{Y}_k$ are independently distributed, $\mathbf{E}\, U(X)\mathbf{Y}_i = 0$, $\mathbf{Cov}(U(X)\mathbf{Y}_i) = I_n$, so it follows that for each X, the random variables $\mathbf{Y}_i$ and $U(X)\mathbf{Y}_i$ have the same probability density function, $i = 1, \ldots, k$. By Lemma 11.1.1, the random variable

$$T(\mathbf{X}, \mathbf{Y}) = U(\mathbf{X})\mathbf{Y} \tag{11.1.3}$$

has independently distributed rows, each normal $(0, I_k)$. In applications of this idea the matrix $U(\mathbf{X})$ will be defined with some specific property such as, for example, $(U(\mathbf{X})\mathbf{Y}_1)^t = (\|\mathbf{Y}_1\|, 0, \ldots, 0)$.

Lemma 11.1.2. *If* $\mathbf{R}_\mathbf{X} = \mathbb{R}^n$ *and* x *is an* $n \times 1$ *vector there exists a matrix valued function* $U(\ )$ *with values in* $\mathbf{O}(n)$ *such that* U *is measurable and if*

$$x \in \mathbb{R}^n \quad \textit{then} \quad (U(x)x)^t = (0, \ldots, 0, \|x\|). \tag{11.1.4}$$

PROOF. The proof is left as an exercise; see Problem 11.11.1. □

11.2. Decomposition of the Sample Covariance Matrix Using Random Variable Techniques. The Bartlett Decomposition

We assume $\mathbf{X}$ is an $n \times k$ matrix consisting of nk independently and identically distributed entries, each a normal (0, 1) random variable. We write X_i and $\mathbf{X}_i$ for the i-th columns of X and $\mathbf{X}$ respectively. Define U measurable, see Lemma 11.1.2, so that

$$(U(X_1)X_1)^t = (0, \ldots, 0, \|X_1\|) = Z_1^t, \tag{11.2.1}$$

and define

$$Z_1 = U(X_1)X_1, \qquad \text{and} \quad \mathbf{Z}_i = U(\mathbf{X}_1)\mathbf{X}_i, \qquad 2 \le i \le k. \tag{11.2.2}$$

By the lemmas of Section 11.1, the $n \times (k-1)$ random matrix $\mathbf{Z}_2, \ldots, \mathbf{Z}_k$ consists of $n(k-1)$ independently distributed normal (0, 1) random variables. Then if we write

$$Z_1^t = (z_{1i}, \ldots, z_{ni}), \qquad 1 \le i \le k, \tag{11.2.3}$$

it follows that

$$z_{n1}^2 = Z_1^t Z_1 = X_1^t X_1. \tag{11.2.4}$$

We further write

$$Y_{j-1}^t = (z_{1j}, \ldots, z_{n-1\,j}), \qquad 2 \le j \le k, \tag{11.2.5}$$

so that

$$Y_{i-1}^t Y_{j-1} + z_{ni} z_{nj} = Z_i^t Z_j, \qquad 2 \le i, \qquad j \le k. \tag{11.2.6}$$

Further let

$$z^t = (z_{n2}, \ldots, z_{nk}) \quad \text{and} \quad Y = (Y_1, \ldots, Y_{k-1}). \tag{11.2.7}$$

Then the following is a matrix identity.

$$X^t X = \begin{vmatrix} z_{n1} & 0 \\ z & I_{k-1} \end{vmatrix} \begin{vmatrix} 1 & 0 \\ 0 & Y^t Y \end{vmatrix} \begin{vmatrix} z_{n1} & z^t \\ 0 & I_{k-1} \end{vmatrix}. \tag{11.2.8}$$

In random variable terms, $\mathbf{Y}$ is an $(n-1) \times (k-1)$ random matrix with $(n-1)(k-1)$ independently distributed entries each normal (0, 1) which are stochastically independent of $\mathbf{z}_{n1}, \ldots, \mathbf{z}_{nk}$. Further

$$\mathbf{z}_{n1}^2 = \mathbf{X}_1^t \mathbf{X}_1 \text{ is } \chi_n^2 \text{ and } \mathbf{z}_{n2}, \ldots, \mathbf{z}_{nk} \text{ are independent normal (0, 1) random variables, each independent of } \mathbf{z}_{n1}. \tag{11.2.9}$$

The proof now proceeds by mathematical induction on the number of columns k. In the inductive step the requirement that $k \le n$ appears. The decomposition is clearly correct if $k = 1$. By induction hypothesis there exists a $(k-1) \times (k-1)$ random matrix **S** with values in $\mathbf{T}(k-1)$ such that

$$\mathbf{S}\mathbf{S}^t = \mathbf{Y}^t\mathbf{Y} \tag{11.2.10}$$

and such that the joint distribution of the entries of **S** are as stated in the Theorem 11.0.1. We use (11.2.8) and set

$$T = \begin{vmatrix} z_{n1} & 0 \\ z & I_{k-1} \end{vmatrix} \begin{vmatrix} 1 & 0 \\ 0 & S \end{vmatrix} = \begin{vmatrix} z_{n1} & 0 \\ z & S \end{vmatrix}. \tag{11.2.11}$$

The conclusions of Theorem 11.0.1 are clearly satisfied except possibly for the distribution of the diagonal elements of **T**. If $2 \le i \le k$ then $(\mathbf{T}_{ii})^2 = (\mathbf{S}_{i-1\,i-1})^2$ which is a $\chi^2_{(n-1)-(i-1)+1} = \chi^2_{n-i+1}$ random variable. □

Remark 11.2.1. If instead of the $k \times k$ matrix $(X_i^t X_j)$, $1 \le i, j \le k$, we consider the $p \times p$ matrix $(X_i^t X_j)$, $1 \le i, j \le p$, then the decomposition (11.2.8) now reads

$$\begin{vmatrix} z_{n1} & 0 & \cdots & 0 \\ z_{n2} & 1 & \cdots & 0 \\ \vdots & & \ddots & \\ z_{np} & 0 & \cdots & 1 \end{vmatrix} \begin{vmatrix} 1 & 0 & & 0 \\ 0 & Y_1^t Y_1 & \cdots & Y_1^t Y_{p-1} \\ & & \ddots & \\ 0 & Y_{p-1}^t Y_1 & \cdots & Y_{p-1}^t Y_{p-1} \end{vmatrix} \begin{vmatrix} z_{n1} & z_{n2} & \cdots & z_{np} \\ 0 & 1 & \cdots & 0 \\ & & \ddots & \\ 0 & 0 & \cdots & 1 \end{vmatrix} \tag{11.2.12}$$

That is, the same vectors $Y_1, \ldots, Y_{p-1}$ occur. By induction it is then clear that if $X^tX = TT^t$ and if $(T)_p$ is the principal $p \times p$ minor of T then

$$\begin{vmatrix} X_1^t X_1 & \cdots & X_1^t X_p \\ & \ddots & \\ X_p^t X_1 & \cdots & X_p^t X_p \end{vmatrix} = (T)_p (T)_p^t. \tag{11.2.13}$$

In particular

Theorem 11.2.2.

$$\frac{\det((T)_{p+1}(T)_{p+1}^t)}{\det((T)_p(T)_p^t)} = t_{p+1\,p+1}^2, \tag{11.2.14}$$

where t_{ii} is the (i, i)-element of T. Also from (11.2.12),

$$\det X^tX = t_{11}^2 t_{22}^2 \cdots t_{ii}^2 \det Y^tY \tag{11.2.15}$$

where the $(k-i) \times (k-i)$ matrix $\mathbf{Y}^t\mathbf{Y}$ *has a Wishart distribution with $n - i$ degrees of freedom.*

A complex valued random variable $\mathbf{z} = \mathbf{x} + i\mathbf{y}$ will be said to be complex normal (0, 2) if $\mathbf{x}$ and $\mathbf{y}$ are independently and identically distributed normal (0, 1) random variables. In this case $\|z\|^2$ is χ_2^2. The complex form of the Bartlett decomposition is as follows.

Theorem 11.2.3. *Let* $\mathbf{Z}$ *be a* $n \times k$ *matrix of independently and identically distributed complex normal* (0, 2) *random variables. There exists a random lower triangular matrix* $\mathbf{T} \in \mathbf{CT}(k)$ *such that* $\mathbf{TT}^* = \mathbf{Z}^*\mathbf{Z}$ *and such that the elements of* $\mathbf{T}$ *are mutually independent. The diagonal elements of* $\mathbf{T}$, $\mathbf{t}_{ii}$, *are positive and real valued such that* $\mathbf{t}_{ii}^2$ *is* $\chi_{2(n-i+1)}^2$ *and the off diagonal elements* $\mathbf{t}_{ij}$ *are complex normal* (0, 2) *random variables.*

Proof. Using a random unitary transformation $\mathbf{U}$ which is a function of the first column $\mathbf{z}_1$ of $\mathbf{Z}$, reduce $\mathbf{Z}$ to

$$\mathbf{UZ} = \begin{vmatrix} 0 \\ \vdots \\ 0 \\ \|\mathbf{z}_1\| \end{vmatrix} \mathbf{W}. \tag{11.2.16}$$

By Lemma 11.1.1; the elements of $\mathbf{W}$ are complex normal (0, 2) random variables and are independent of $\|\mathbf{z}_1\|$. As before the inductive step is

$$\begin{vmatrix} \|z_1\| & 0 \\ z & I \end{vmatrix} \begin{vmatrix} I & 0 \\ 0 & Z_1^*Z_1 \end{vmatrix} \begin{vmatrix} \|z_1\| & z^* \\ 0 & I \end{vmatrix} = \begin{vmatrix} \|z_1\|^2 & \|z_1\|z^* \\ \|z_1\|z & zz^* + Z_1^*Z_1 \end{vmatrix} = Z^*Z. \tag{11.2.17}$$

Z_1 is the matrix of the first $n-1$ rows of W and z^* is the last row of W so that $zz^* + Z_1^*Z_1 = W^*W$. Thus $\|\mathbf{z}_1\|^2$ is χ_{2n}^2 and the entries of z are complex normal (0, 2). The result now follows by an inductive argument. □

Remark 11.2.4. An $n \times k$ matrix Z of rank k over the complex numbers factors to $Z(Z^*Z)^{-1/2}(Z^*Z)^{1/2}$ and $U = Z(Z^*Z)^{-1/2}$ satisfies $U^*U = I_k$. If $\mathbf{Z}$ is a matrix of independently and identically distributed complex normal random variables then the density function of the $2nk$ variables of Z is a function of $\operatorname{tr} Z^*Z$. Since Haar measure on $\mathbf{CGL}(k)$, the complex general linear group, will factor to Haar measure on the unitary matrices times a measure on $(Z^*Z)^{1/2}$, independence of the random variables follows.

11.3. The Generalized Variance, Zero Means

In this section we use (11.2.15) but require a more expressive notation. So we let

$\mathbf{S}(k, n)$ be a $k \times k$ random matrix such that
$\mathbf{S}(k, n)$ has a central Wishart probability density
function with n degrees of freedom. (11.3.1)

Then (11.2.15) may be phrased as

$$\begin{aligned}&\det S(k,n) = t_{11}^2 \cdots t_{ii}^2 \det S(k-i, n-i);\\ &\mathbf{t}_{11}, \ldots, \mathbf{t}_{ii} \text{ and } \mathbf{S}(k-i, n-i) \text{ are mutually}\\ &\text{independent random variables.}\end{aligned} \tag{11.3.2}$$

Consider the decomposition of the $n \times k$ matrix $X = AT^t$, with $T \in \mathbf{T}(k)$ and the $n \times k$ matrix A satisfying $A^tA = I_k$. This decomposition arises from use of the Gram–Schmidt orthogonalization process. If the columns of X are $X_1, \ldots, X_k$ and those of A are $A_1, \ldots, A_k$ then from the Gram–Schmidt process it follows that

$$(X_1, \ldots, X_{k-1}) = \tilde{X} = (A_1, \ldots, A_{k-1})\tilde{T}^t \tag{11.3.3}$$

where $\tilde{T} \in \mathbf{T}(k-1)$ such that $(\tilde{T})_{ij} - (T)_{ij}$, $1 \le j \le i \le k-1$. Note that the (k,k)-entry of $(X^tX)^{-1}$ is given by

$$\det(\tilde{X}^t\tilde{X})/\det(X^tX) = ((X^tX)^{-1})_{kk}. \tag{11.3.4}$$

We may now state and prove the following theorem.

Theorem 11.3.1. *The* (k,k)*-entry of* $(\mathbf{X}^t\mathbf{X})^{-1}$ *is distributed as* $1/\chi^2_{n-k+1}$, *where* $\mathbf{X}$ *is an* $n \times k$ *matrix consisting of nk mutually independent random variables, each normal* $(0,1)$, $k \le n$.

Proof. $((\mathbf{X}^t\mathbf{X})^{-1})_{kk} = \det(\tilde{\mathbf{X}}^t\tilde{\mathbf{X}})/\det(\mathbf{X}^t\mathbf{X}) = 1/\mathbf{t}_{kk}^2$. We use here Theorem 11.2.2 followed by Theorem 11.0.1. □

Definition 11.3.2. If $\mathbf{X}$ is an $n \times k$ random matrix and $\mathbf{E}\mathbf{X} = 0$ then the quantity $\det(\mathbf{X}^t\mathbf{X})$ is called the *generalized variance*.

Theorem 11.3.3. *Let* $\mathbf{X}$ *be as in Theorem* 11.3.1. *Then the trace of* $\mathbf{X}^t\mathbf{X}$ *is distributed as an* χ^2_{nk} *random variable, and* $\det(\mathbf{X}^t\mathbf{X})$, *that is, the generalized variance, is* $\chi^2_n\chi^2_{n-1} \cdots \chi^2_{n-k+1}$, *a product of k mutually independent chi-square random variables.*

11.4. Noncentral Wishart, Rank One Means

We let f_n be the $n \times 1$ vector having all entries equal one. Suppose the random $n \times k$ matrix $\mathbf{X}$ satisfies

$$\mathbf{E}\mathbf{X} = f_n a^t \tag{11.4.1}$$

where $a \in \mathbb{R}^k$. We further assume that $\mathbf{X} - f_n a^t$ consists of nk mutually independent normal $(0,1)$ entries. Suppose $U \in \mathbf{O}(n)$ and $(Uf_n)^t = (0, \ldots, 0, \sqrt{n})$. Then $\mathbf{X}^t\mathbf{X} = (U\mathbf{X})^t(U\mathbf{X})$. Let $\mathbf{Y}$ be the $(n-1) \times k$ matrix of the first $n-1$ rows of $U\mathbf{X}$ and $\mathbf{Z}^t$ be the last row of $U\mathbf{X}$. Then

$$\mathbf{X}^t\mathbf{X} = \mathbf{Y}^t\mathbf{Y} + \mathbf{Z}\mathbf{Z}^t. \tag{11.4.2}$$

Then $\mathbf{X}^t\mathbf{X}$ is the sum of two independent Wishart random variables such that $\mathbf{Y}^t\mathbf{Y}$ is a central Wishart $n - 1$, k random variable and $\mathbf{Z}\mathbf{Z}^t$ is a noncentral Wishart 1, k random variable, with $\mathbf{E}\,\mathbf{Z}^t = \sqrt{n}a^t$.

As an alternative, let $V \in \mathbf{O}(k)$ and $a^t V = (\|a\|, 0, \ldots, 0)$. Set

$$\mathbf{X}' = \mathbf{X}V \quad \text{so that} \quad \mathbf{E}\,\mathbf{X}' = f_n(\|a\|, 0, \ldots, 0). \tag{11.4.3}$$

Then the columns of $\mathbf{X}'$ satisfy

$$\mathbf{E}\,(\mathbf{X}')_i = 0, \qquad 2 \le i \le k, \qquad \mathbf{E}\,(\mathbf{X}')_1 = \|a\| f_n. \tag{11.4.4}$$

We make the same decomposition as in Section 11.2. The random variables $\mathbf{z}_{n1}, \ldots, \mathbf{z}_{nk}$ obtained at the first step, see (11.2.7), are mutually independent, $\mathbf{z}_{n1}^2$ is a noncentral chi-square, n degrees of freedom, and noncentrality parameter $na^t a/2$. $\mathbf{z}_{n2}, \ldots, \mathbf{z}_{nk}$ are normal (0, 1) random variables and the entries of $\mathbf{Y}$ (not the $\mathbf{Y}$ of the preceding paragraph), see (11.2.5), are $(n - 1)(k - 1)$ independent normal (0, 1) random variables. Hence, after the first step the decomposition proceeds as before. This process gives a canonical decomposition of $V^t\mathbf{X}^t\mathbf{X}V$ as $\mathbf{T}\mathbf{T}^t$ which satisfies,

> $\mathbf{t}_{11}^2$ is a noncentral χ_n^2 random variable with noncentrality parameter $na^t a/2$, and the other entries of $\mathbf{T}$ are distributed as described in Theorem 11.0.1. (11.4.5)

11.5. Hotelling T^2 Statistic, Noncentral Case

We continue the notation of Section 11.4 and let $e = f_n$ be the $n \times 1$ vector having all entries one. We note that

$$ee^t/n \quad \text{and} \quad I_n \quad ee^t/n \tag{11.5.1}$$

are $n \times n$ orthogonal projection matrices.

Theorem 11.5.1. *Let* $\mathbf{X}$ *be a random* $n \times k$ *matrix with independently distributed rows, each normal* (a, Σ). *Then*

$$n(n - k)k^{-1}(n^{-1}e^t\mathbf{X} - a_0^t)(\mathbf{X}^t(I_n - ee^t/n)\mathbf{X})^{-1}(n^{-1}\mathbf{X}^t e - a_0) \tag{11.5.2}$$

is distributed as a noncentral F *statistic* $(n - k)\chi_k^2/k\chi_{n-k}^2$ *with noncentrality parameter*

$$\tfrac{1}{2}n(a - a_0)^t\Sigma^{-1}(a - a_0). \tag{11.5.3}$$

Proof. $\mathbf{X}^t e/n$ is the $k \times 1$ vector of sample means. We show that $\mathbf{X}^t e/n$ and $(I_n - ee^t/n)\mathbf{X}$ are mutually independent random variables by computing covariances.

$$\mathbf{E}(I_n - ee^t/n)\mathbf{X} = (I_n - ee^t/n)ea^t = 0, \tag{11.5.4}$$

and therefore the matrix of covariances is

$$\mathbf{E}\,\mathbf{X}^t e(I_n - ee^t/n)\mathbf{X} = 0. \tag{11.5.5}$$

That is, the sum of squares has been partitioned into

$$X^tX = X^t(ee^t/n)X + X^t(I_n - ee^t/n)X. \tag{11.5.6}$$

Using random variable techniques, the following transformations leave the problem invariant. Let $\Omega \in \mathbf{O}(n)$ satisfy

$$e^t\Omega/\sqrt{n} = (0, 0, \ldots, 1). \tag{11.5.7}$$

Then

$$\mathbf{E}\,\mathbf{X}^t\Omega = ae^t\Omega = (0, \ldots, 0, \sqrt{n}a), \tag{11.5.8}$$

a $k \times n$ matrix in which 0 means the $k \times 1$ vector of zeros. Then

$$\mathbf{E}(\Omega^t\mathbf{X} - \Omega^t ea^t)_i(\Omega^t\mathbf{X} - \Omega^t ea^t)_j^t = \sigma_{ij}I_n \tag{11.5.9}$$

so the covariances remain unchanged.

Write $\tilde{\mathbf{X}}$ to be the $(n-1) \times k$ matrix obtained by deletion of the n-th row from $\Omega^t\mathbf{X}$ so that $\mathbf{E}\tilde{\mathbf{X}} = 0$. Then

$$\tilde{\mathbf{X}}^t\tilde{\mathbf{X}} = (\Omega^tX)^t\Omega^t(I_n - ee^t/n)\Omega(\Omega^tX) = X^t(I_n - ee^t/n)X, \tag{11.5.10}$$

so the random variables $\tilde{\mathbf{X}}^t\tilde{\mathbf{X}}$ and $e^t\mathbf{X}$ are stochastically independent. From (11.5.2) we seek the distribution of

$$\mathbf{S} = n(n-k)k^{-1}(n^{-1}e^t\mathbf{X} - a_0^t)(\tilde{\mathbf{X}}^t\tilde{\mathbf{X}})^{-1}(n^{-1}\mathbf{X}^te - a_0), \tag{11.5.11}$$

and transform this problem to

$$n(n-k)k^{-1}(n^{-1}e^t\mathbf{X} - a_0^t)\Sigma^{-1/2}\Sigma^{1/2}(\tilde{\mathbf{X}}^t\tilde{\mathbf{X}})^{-1}\Sigma^{1/2}\Sigma^{-1/2}(n^{-1}\mathbf{X}^te - a_0). \tag{11.5.12}$$

The $(n-1) \times k$ random matrix $\tilde{\mathbf{X}}\Sigma^{-1/2}$ has $(n-1)k$ independent normal $(0, 1)$ entries and

$$\mathbf{E}(n^{-1}e^t\mathbf{X} - a_0^t)\Sigma^{-1/2} = (a - a_0)^t\Sigma^{-1/2}, \tag{11.5.13}$$

and

$$\mathbf{Cov}\left[(n^{-1}e^t\mathbf{X} - a_0^t)\Sigma^{-1/2}\right] = n^{-1}I_k.$$

Therefore without loss of generality in the sequel we assume $\Sigma = I_k$ and that the vector of means is $\Sigma^{-1/2}a$. We require the distribution of (11.5.11) under this assumption and the assumption that $n^{-1}e^t\mathbf{X}$ is independent of $\tilde{\mathbf{X}}$.

Let Ω be a random orthogonal matrix such that $\Omega(e^t\mathbf{X}) \in \mathbf{O}(k)$ and

$$(\Omega(e^t\mathbf{X})(n^{-1}e^t\mathbf{X} - a_0^t\Sigma^{-1/2}))^t = (0, \ldots, 0, \|n^{-1}e^t\mathbf{X} - a_0^t\Sigma^{-1/2}\|)^t. \tag{11.5.14}$$

By Section 11.1, $\tilde{\mathbf{X}}\Omega(e^t\mathbf{X})$ is distributed as a matrix of $(n-1)k$ independent normal (0, 1) random variables. Therefore, if $\mathbf{S}$ is as in (11.5.11),

$$n^{-1}(n-k)^{-1}k\mathbf{S} = \|n^{-1}e^t\mathbf{X} - a_0^t\Sigma^{-1/2}\|^2\,((\Omega\mathbf{X}^t\mathbf{X}\Omega^t)^{-1})_{kk}. \tag{11.5.15}$$

Apply Theorem 11.3.1. The random variable

$$n\|n^{-1}e^t\mathbf{X} - a_0^t\Sigma^{-1/2}\|^2 = \|n^{-1/2}e^t\mathbf{X} - n^{1/2}a_0^t\Sigma^{-1/2}\|^2 = \mathbf{Y}_1 \tag{11.5.16}$$

is a noncentral chi-square with k degrees of freedom and noncentrality parameter

$$\tfrac{1}{2}n(a-a_0)^t\Sigma^{-1}(a-a_0). \tag{11.5.17}$$

Using Theorem 11.3.1, it follows that the distribution of $\mathbf{S}$ is

$$k^{-1}\mathbf{Y}_1/(n-k)^{-1}Y_2, \tag{11.5.18}$$

with $\mathbf{Y}_1$ defined in (11.5.16) and $\mathbf{Y}_2$ distributed as a central chi-square with $n-k$ degrees of freedom, $\mathbf{Y}_1$ and $\mathbf{Y}_2$ stochastically independent, and the noncentrality parameter being the expression in (11.5.17). □

11.6. Generalized Variance, Nonzero Means

In the situation of Section 11.5, where $\mathbf{E}\mathbf{X} = ea^t$, the sample covariance matrix is

$$(n-1)^{-1}\mathbf{X}^t(I_n - ee^t/n)\mathbf{X}. \tag{11.6.1}$$

Definition 11.6.1. If the $n \times k$ random matrix $\mathbf{X}$ has independently distributed rows each normal (a, Σ) then $\det(\mathbf{X}^t(I_n - ee^t/n)\mathbf{X})/\det\Sigma$ is the generalized variance.

Theorem 11.6.2. *The generalized variance is distributed as a product $\chi^2_{n-1}\cdots\chi^2_{n-k}$ of k independently distributed central chi-square random variables.*

Proof. Define $\tilde{\mathbf{X}}$ as in (11.5.10) so that the generalized variance is $\det(\tilde{\mathbf{X}}^t\tilde{\mathbf{X}})$. By Theorem 11.3.3 the result now follows. □

Remark 11.6.3. The distribution of products of independently distributed chi-square random variables has been studied by Kullback (1934). In his paper the Fourier transform of a product is computed. Inversion of the transform is done in several cases by use of the residue theorem of complex variables together with knowledge of residues of the gamma function. The answers are not simple and are expressed as finite series of n-th derivatives (uncomputed).

11.7. Distribution of the Sample Correlation Coefficient

In Problems 9.4.8, 9.4.9 and 9.4.10 the joint probability density function of the sample correlation coefficients has been described using differential forms. In this section we give a random variable description for a single correlation coefficient. Thus we may assume $\mathbf{X}$ is $n \times k$ with $k = 2$. Let $\mathbf{X}$ have first column $\mathbf{X}_1$ and second column $\mathbf{X}_2$. The sample quantity is defined to be

$$r = \frac{n^{-1}X_1^tX_2 - (n^{-1}e^tX_1)(n^{-1}e^tX_2)}{(n^{-1}X_1^tX_1 - (n^{-1}e^tX_1)^2)^{1/2}(n^{-1}X_2^tX_2 - (n^{-1}(e^tX_2)^2)^{1/2}}. \tag{11.7.1}$$

The function (11.7.1) is homogeneous of degree 0, so we may assume at the onset by scale changes that

$$\Sigma = \begin{vmatrix} 1 & \rho \\ \rho & 1 \end{vmatrix} = \begin{vmatrix} 1 & 0 \\ \rho & (1-\rho^2)^{1/2} \end{vmatrix} \begin{vmatrix} 1 & \rho \\ 0 & (1-\rho^2)^{1/2} \end{vmatrix}. \tag{11.7.2}$$

Write

$$C = \begin{vmatrix} 1 & 0 \\ \rho & (1-\rho^2)^{1/2} \end{vmatrix} \quad \text{and} \quad Z = X(C^t)^{-1}. \tag{11.7.3}$$

Then the random variable $\mathbf{Z} = \mathbf{X}(C^t)^{-1}$ has I_2 as covariance matrix.

Apply the canonical decomposition Theorem 11.0.1 to the 2×2 matrix

$$(C^{-1}\mathbf{X}^t)(I_n - ee^t/n)(C^{-1}\mathbf{X}^t)^t = \mathbf{T}\mathbf{T}^t \tag{11.7.4}$$

with the 2×2 matrix T given by

$$T = \begin{vmatrix} t_{11} & 0 \\ t_{21} & t_{22} \end{vmatrix}. \tag{11.7.5}$$

Then

$$S = \begin{vmatrix} s_{11} & s_{12} \\ s_{21} & s_{22} \end{vmatrix} \underset{\text{def}}{=} X^t(I_n - ee^t/n)X = (CT)(CT)^t, \tag{11.7.6}$$

and

$$(CT) = \begin{vmatrix} t_{11} & 0 \\ \rho t_{11} + (1-\rho^2)^{1/2} t_{21} & (1-\rho^2)^{1/2} t_{22} \end{vmatrix}.$$

From (11.7.1) we find the sample quantity to be

$$r = s_{12}/(s_{11}s_{22})^{1/2}. \tag{11.7.7}$$

The distribution of $\mathbf{r}/(1-\mathbf{r}^2)^{1/2}$ is easier to describe in random variable terms so the distribution of this quantity is computed below. From (11.7.7)

$$r/(1-r^2)^{1/2} = s_{12}/(s_{11}s_{22} - s_{12}^2)^{1/2}. \tag{11.7.8}$$

From (11.7.6) we compute

$$S = (CT)(CT)^t, \tag{11.7.9}$$

with

$$s_{11} = t_{11}^2;$$
$$s_{21} = s_{12} = \rho t_{11}^2 + (1-\rho^2)^{1/2} t_{11} t_{21};$$
$$s_{22} = (1-\rho^2) t_{22}^2 + (\rho t_{11} + (1-\rho^2)^{1/2} t_{21})^2.$$

Then

$$s_{11}s_{22} - s_{12}^2 = (1-\rho^2) t_{11}^2 t_{22}^2, \tag{11.7.10}$$

and

$$r/(1-r^2)^{1/2} = \frac{(\rho/(1-\rho^2)^{1/2})t_{11} + t_{21}}{t_{22}}. \tag{11.7.11}$$

Theorem 11.7.1. *If* $\mathbf{X}$ *is a random* $n \times 2$ *matrix with independently distributed rows such normal* (a, Σ), *then the distribution of* $\mathbf{r}/(1-\mathbf{r}^2)^{1/2}$ *with* $\mathbf{r}$ *defined* by (11.7.1) *is the same as that of*

$$[(\rho/(1-\rho^2)^{1/2})\mathbf{t}_{11} + \mathbf{t}_{21}]/\mathbf{t}_{22} \tag{11.7.12}$$

where $\mathbf{t}_{11}$, $\mathbf{t}_{21}$ *and* $\mathbf{t}_{22}$ *are mutually independent random variables,* $\mathbf{t}_{11}^2$ *is* χ_{n-1}^2, $\mathbf{t}_{22}^2$ *is* χ_{n-2}^2, *and* $\mathbf{t}_{21}$ *is normal* (0, 1). *The variables* t_{11}, t_{21}, t_{22} *are those of* T *defined in* (11.7.4). *The population parameter* ρ *is defined by* (11.7.2). *The chi-squares are central chi-squares.*

11.8. Multiple Correlation, Algebraic Manipulations

Multiple correlation, as are the canonical correlations, is defined in terms of a maximization. The multiple correlation is in fact a canonical correlation when one of the matrices has a single column. See the discussion of Section 8.2.

We suppose X is $n \times k$ and that $k = p + q$. X is partitioned into $X = (X_1, X_2)$ such that X_1 is $n \times p$ and X_2 is $n \times q$. We assume the rows of $\mathbf{X}$ are independently distributed normal $(0, \Sigma)$ random vectors and

$$\Sigma = \begin{vmatrix} \Sigma_{11} & \Sigma_{12} \\ \Sigma_{21} & \Sigma_{22} \end{vmatrix}, \quad \text{where } \Sigma_{11} \text{ is a } p \times p \text{ matrix.} \tag{11.8.1}$$

Then the rows of $\mathbf{X}_1$ have Σ_{11} as covariance matrix.

The multiple correlation problem is to choose a $q \times 1$ vector a such that the correlation between $(\mathbf{X}_1)_{1i}$ and $((\mathbf{X}_2)_{11}, \ldots, (\mathbf{X}_2)_{1q})a$ is maximized. Call the maximum value of the correlation $\mathbf{R}_{i \cdot p+1, \ldots, p+q}$. Here $(\mathbf{X}_1)_{1i}$ is the first component of the i-th column of $\mathbf{X}_1$ and $(\mathbf{X}_2)_{1i}$ is the first component of the i-th column of $\mathbf{X}_2$.

We let, in this section, e_i be the $p \times 1$ vector of norm one with the i-th position of e_i equal one. Then, since it has been assumed that $\mathbf{E}\,\mathbf{X} = 0$,

$$\mathbf{E}\, a^t \mathbf{X}_2^t \mathbf{X}_2 a = n a^t \Sigma_{22} a, \tag{11.8.2}$$

and

$$\mathbf{E}\, e_i^t \mathbf{X}_1^t \mathbf{X}_2 a = n e_i^t \Sigma_{12} a. \tag{11.8.3}$$

The population parameter, the correlation, is

$$e_i^t \Sigma_{12} a / \sigma_{1ii} (a^t \Sigma_{22} a)^{1/2}, \tag{11.8.4}$$

and this is to be maximized through choice of a. As in Section 8.2, the problem may be solved by use of the Cauchy–Schwarz inequality. The maximum value is

$$R_{i \cdot p+1, \ldots, p+q} = \sigma_{1ii}^{-1} (e_i^t \Sigma_{12} \Sigma_{22}^{-1} \Sigma_{21} e_i)^{1/2}. \tag{11.8.5}$$

The maximization just described is related to the following minimization problem.

Lemma 11.8.1. *Let* $\begin{vmatrix} P & Q \\ R & S \end{vmatrix}$ *be a symmetric positive definite matrix in which P and S are square matrices. Suppose P is $p \times p$ and S is $q \times q$ and that A is a $p \times q$ matrix. Then using the partial ordering of semidefinite matrices,*

$$P - QS^{-1}R \le P - AR - QA^t + ASA^t. \tag{11.8.6}$$

Equality holds if and only if $A = QS^{-1}$.

Proof. Let

$$A = QS^{-1} + C. \tag{11.8.7}$$

Substitution into the right side of (11.8.6) yields the result that

$$P - AR - QA^t - ASA^t = P - QS^{-1}R + CSC^t. \tag{11.8.8}$$

Since $CSC^t \ge 0$ in the ordering of semidefinite matrices, the result follows. □

Apply Lemma 11.8.1 to the vector $\mathbf{X}_i e_i - \mathbf{X}_2 a$. The components of this vector are independently and identically distributed, each with variance

$$e_i^t \Sigma_{11} e_i - e_i^t \Sigma_{12} a - a^t \Sigma_{21} e_i + a^t \Sigma_{22} a. \tag{11.8.9}$$

We set $P = e_i^t \Sigma_{11} e_i = \sigma_{1ii}^2$, $Q = e_i^t \Sigma_{12}$, and $S = \Sigma_{22}$, and obtain

$$e_i^t (\Sigma_{11} - \Sigma_{12} \Sigma_{22}^{-1} \Sigma_{21}) e_i = (\sigma_{1ii}^2 - e_i^t \Sigma_{12} \Sigma_{22}^{-1} \Sigma_{21} e_i) \le (11.8.9). \tag{11.8.10}$$

From (11.8.5) and the lemma we obtain

Lemma 11.8.2. *The vector a which maximizes the correlation between $\mathbf{X}_1 e_i$ and $\mathbf{X}_2 a$ also minimizes the variance of $\mathbf{X}_1 e_i - \mathbf{X}_2 a$ and the minimum variance is*

$$\sigma_{1ii}^2(1 - R_{i\cdot p+1,\ldots,p+q}^2) = e_i^t(\Sigma_{11} - \Sigma_{12}\Sigma_{22}^{-1}\Sigma_{21})e_i. \tag{11.8.11}$$

Remark 11.8.3. The number $e_i^t(\Sigma_{11} - \Sigma_{12}\Sigma_{22}^{-1}\Sigma_{21})e_i$ is the (i, i)-entry of the conditional covariance matrix of a row of $\mathbf{X}_1$ given the same row of $\mathbf{X}_2$.

By Problem 8.5.14, in the notation of Lemma 11.8.1,

$$\det\begin{vmatrix} P & Q \\ R & S \end{vmatrix} = (\det S)(\det(P - QS^{-1}R)). \tag{11.8.12}$$

With the interpretation $P = e_i^t\Sigma_{11}e_i$, $Q = e_i^t\Sigma_{12}$ and $S = \Sigma_{22}$, from (11.8.12) it follows that $\det(P - QS^{-1}R) = e_i^t(\Sigma_{11} - \Sigma_{12}\Sigma_{22}^{-1}\Sigma_{21})e_i$ and the following Lemma has been verified.

Lemma 11.8.4. *The following identity holds.*

$$((e_i^t\Sigma_{11}e_i)\det\Sigma_{22})(1 - R_{i\cdot p+1,\ldots,p+q}^2) = \det\begin{vmatrix} e_i^t\Sigma_{11}e_i & e_i^t\Sigma_{12} \\ \Sigma_{21}e_i & \Sigma_{22} \end{vmatrix}. \tag{11.8.13}$$

11.9. Distribution of the Multiple Correlation Coefficient

Start with $\mathbf{X} = (\mathbf{X}_1, \mathbf{X}_2)$ a random $(n + 1) \times k$ matrix with independently and identically distributed rows, each normal (a, Σ). Eliminate the means by considering the "sample" matrix

$$\begin{vmatrix} S_{11} & S_{12} \\ S_{21} & S_{22} \end{vmatrix} = \begin{vmatrix} X_1^t \\ X_2^t \end{vmatrix}(I_n - ee^t/n)(X_1, X_2). \tag{11.9.1}$$

Then the sample multiple correlation coefficient may be defined by the equation

$$(1 - R^2)(e_i^tS_{11}e_i)(\det S_{22}) = \det\begin{vmatrix} e_i^tS_{11}e_i & e_i^tS_{12} \\ S_{21}e_i & S_{22} \end{vmatrix}. \tag{11.9.2}$$

The computation is done with formula (11.9.2). As in Section 11.7, calculate the distribution of $R/(1 - R^2)^{1/2}$. The general technique is the same as in Section 11.7.

The population and sample multiple correlation coefficients are invariant under addition of an arbitrary $k \times 1$ vector to the mean vector a, together with transformation of the random variables. That is, if the transformation is $a \to a + a_1$ and $(\mathbf{X}_1, \mathbf{X}_2) \to (\mathbf{X}_1, \mathbf{X}_2) + ea_1^t$ then the correlations are unchanged. Thus for simplicity one may suppose $a = 0$. Other transformations that leave the correlations invariant are

$$(\mathbf{X}_1, \mathbf{X}_2) \to (a\mathbf{X}_1, \mathbf{X}_2 E), \tag{11.9.3}$$

where a is a constant and E is a nonsingular $q \times q$ matrix. For, by (11.9.1), S_{11}, S_{21}, S_{22} transform to $a^2 S_{11}, aS_{12}E$, and $E^t S_{22} E$ so that (11.9.2) becomes

$$\frac{\det\begin{vmatrix} a & 0 \\ 0 & E \end{vmatrix} \det\begin{vmatrix} e_i^t S_{11} e_i & e_i^t S_{12} \\ S_{21} e_i & S_{22} \end{vmatrix} \det\begin{vmatrix} a & 0 \\ 0 & E \end{vmatrix}}{a^2 (e_i^t S_{11} e_i)(\det E^t S_{22} E)}. \tag{11.9.4}$$

Similarly for (11.8.13). In order to find the distribution of $\mathbf{R}$ choose a and E so that $\sigma_{1ii} = 1$ and $\Sigma_{22} = I_q$, and (11.8.13) now reads

$$(1 - R^2_{i \cdot p+1,\ldots,p+q}) = \det\begin{vmatrix} 1 & e_i^t \Sigma_{12} \\ \Sigma_{21} e_i & I_q \end{vmatrix}. \tag{11.9.5}$$

The problem may be further transformed by matrices of the form $\begin{vmatrix} 1 & 0 \\ 0 & U \end{vmatrix}$ with $U \in \mathbf{O}(q)$ and the result that is obtained is

$$(1 - R^2_{i \cdot p+1,\ldots,p+q}) = \det\begin{vmatrix} 1 & 0 & \cdots & 0 & z \\ 0 & 1 & \cdots & 0 & 0 \\ \vdots & \vdots & \ddots & \vdots & \vdots \\ z & 0 & \cdots & 0 & 1 \end{vmatrix} = 1 - z^2, \tag{11.9.6}$$

where $z = \|e_i^t \Sigma_{12}\|$. The orthogonal matrix U would be chosen to satisfy $e_i^t \Sigma_{12} U = (0, \ldots, 0, z)$. The transformed covariance matrix is then

$$\begin{vmatrix} 1 & b^t \\ b & I_q \end{vmatrix}, \quad b^t = (0, \ldots, 0, R), \quad \text{and} \quad R = R_{i \cdot p+1,\ldots,p+q} \geq 0. \tag{11.9.7}$$

In order to simplify notations, relabel the variables, with $\mathbf{X} = (\mathbf{X}_1, \mathbf{X}_2)$, $\mathbf{X}_1$ an $n \times q$ matrix, $\mathbf{X}_2$ an $n \times 1$ matrix, such that the covariance matrix of a row of $\mathbf{X}$ is the matrix in (11.9.7). In the new notation wanted is the distribution of $\mathbf{R}_{q+1 \cdot 1,\ldots,q}$. Let

$$\begin{vmatrix} I_q & 0 & \cdots & 0 & 0 \\ \cdot & \cdot & \cdots & \cdot & \cdot \\ R & 0 & \cdots & 0 & (1 - R^2)^{1/2} \end{vmatrix} = C, \tag{11.9.8}$$

a lower triangular matrix, so that $CC^t = (11.9.7)$. Apply the canonical decomposition to $(\mathbf{X}(C^{-1})^t)^t(\mathbf{X}(C^{-1})^t$ so that

$$\mathbf{T}\mathbf{T}^t = C^{-1}\mathbf{X}^t\mathbf{X}(C^{-1})^t. \tag{11.9.9}$$

Needed are the entries of CT, which are the same as the entries of T except for the last row. The last row of CT is

$$\bar{t}_{q+1\,1} = Rt_{11} + (1 - R^2)^{1/2} t_{q+1\,1}, \quad \text{and if } 2 \le i \le q+1,$$
$$\bar{t}_{q+1\,i} = (1 - R^2)^{1/2} t_{q+1\,i}. \tag{11.9.10}$$

Write

$$CT = \begin{vmatrix} \bar{T} & 0 \\ \bar{t}^t & \bar{t}_{q+1\,q+1} \end{vmatrix}. \tag{11.9.11}$$

Then

$$(CT)(CT)^t = \begin{vmatrix} \bar{T}\bar{T}^t & \bar{T}\bar{t} \\ \bar{t}^t\bar{T}^t & \bar{t}^t\bar{t} + \bar{t}^2_{q+1\,q+1} \end{vmatrix}. \tag{11.9.12}$$

Then, the sample quantity satisfies

$$1 - R^2 = \frac{(\det \bar{T})^2(\bar{t}_{q+1\,q+1})^2}{(\det \bar{T})^2(\bar{t}^t\bar{t} + \bar{t}^2_{q+1\,q+1})} \tag{11.9.13}$$

and

$$\mathbf{R}^2/(1 - \mathbf{R}^2) = \frac{(R\mathbf{t}_{11} + (1 - R^2)^{1/2}\mathbf{t}_{q+1\,1})^2 + (1 - R^2)\sum_{i=2}^{q} \mathbf{t}^2_{q+1\,i}}{(1 - R^2)\mathbf{t}^2_{q+1\,q+1}}. \tag{11.9.14}$$

Theorem 11.9.1. *If* $\mathbf{X}$ *is an* $(n + 1) \times k$ *random matrix of independently and identically distributed rows such that each row is normal* (a, Σ), *then the distribution of* $\mathbf{R}_{i\cdot p+1,\ldots,p+q}$, *transformed to* (11.9.14), *is the same as the distribution of the right side of* (11.9.14), *where* $\mathbf{t}_{11}, \mathbf{t}_{q+1\,1}, \ldots, \mathbf{t}_{q+1\,q+1}$ *are independently distributed random variables,* $\mathbf{t}^2_{11}$ *is a central* χ^2_n, $\mathbf{t}^2_{q+1\,q+1}$ *is a central* χ^2_{n-q}, *and* $\mathbf{t}_{q+1\,1}, \ldots, \mathbf{t}_{q+1\,q}$ *are normal* $(0, 1)$ *random variables.*

11.10. BLUE: Best Linear Unbiased Estimation, an Algebraic Theory

In this section, $\mathbf{X}$ is an $n \times 1$ random vector satisfying

$$\mathbf{E}\mathbf{X} = B\phi \quad \text{and} \quad \mathbf{E}(\mathbf{X} - B\phi)(\mathbf{X} - B\phi)^t = \sigma^2 I_n, \tag{11.10.1}$$

where B is an $n \times k$ matrix assumed known to the statistician and ϕ, σ are unknown parameters, ϕ a $k \times 1$ vector.

Definition 11.10.1. An estimator which is a function of $\mathbf{X}$ is a *linear estimator* if and only if there exists $a \in \mathbb{R}^n$ such that $a^t\mathbf{X}$ is the estimator.

Remark 11.10.2. A linear estimator $a^t\mathbf{X}$ is considered to estimate its expectation, that is, $a^t\mathbf{X}$ estimates $\mathbf{E}\,a^t\mathbf{X} = a^tB\phi$.

Definition 11.10.3. A linear estimator $a^t\mathbf{X}$ is a *best linear unbiased estimator* (BLUE) of $a_0^t B\phi$ if and only if

$$a_0^t B\phi = a^t B\phi \text{ for all } \phi \in \mathbb{R}^k, \text{ and} \tag{11.10.2}$$

$a^t\mathbf{X}$ has minimum variance amoung all linear unbiased estimators of $a_0^t B\phi$.

The questions dealt with in this chapter are (i) necessary and sufficient conditions that $a_0^t B\phi$ have a best linear unbiased estimator, (ii) if a best linear unbiased estimator does exist, how does one compute it, and (iii) sums of squares.

Lemma 11.10.4. *If a and b are in* $\mathbb{R}^n$ *then*

$$\mathbf{cov}(a^t\mathbf{X}, b^t\mathbf{X}) = \sigma^2 a^t b. \tag{11.10.3}$$

PROOF. $\mathbf{E}(a^t\mathbf{X} - a^t B\phi)(b^t\mathbf{X} - b^t B\phi) = \text{trace}\,\mathbf{E}(a^t\mathbf{X} - a^t B\phi)(b^t\mathbf{X} - b^t B\phi)^t = a^t\,\mathbf{E}((\mathbf{X} - B\phi)(\mathbf{X} - B\phi)^t)b = a^t(\sigma^2 I_n)b = \sigma^2 a^t b.$ □

Lemma 11.10.5. *If* $c \in \mathbb{R}^k$ *then* $c^t\phi$ *is estimable by a linear estimator if and only if* c^t *is in the row space of B, i.e.,* c^t *is a linear combination of the rows of B.*

PROOF. If $c^t\phi$ is estimable then for some $a \in \mathbb{R}^n$,

$$\mathbf{E}\,a^t\mathbf{X} = a^t B\phi = c^t\phi. \tag{11.10.4}$$

This is to hold for all $\phi \in \mathbb{R}^k$ so that $c^t = a^t B$ follows. That is, c^t is a linear combination of the rows of B. Conversely, if $a \in \mathbb{R}^n$ and $a^t B = c^t$ then $\mathbf{E}\,a^t\mathbf{X} = a^t B\phi = c^t\phi$. □

Lemma 11.10.6. *In order that every linear combination* $c^t\phi$ *be estimable (by linear unbaised estimators),* $c \in \mathbb{R}^k$*, it is necessary and sufficient that* rank $B = k$.

PROOF. rank B = dimension of the row space of B. Use Lemmas 11.10.5 and 11.10.7. □

Lemma 11.10.7. *If B is an* $n \times k$ *matrix with real entries then the following are equal:*

(a) rank $B \underset{\text{def}}{=}$ *dimension of the row space of* $B =$ *row* rank *of B.*
(b) *dimension of the column space of* $B =$ *column* rank *of B.*
(c) rank $B^t B$.
(d) rank BB^t.

PROOF. The row space of B is the set of vectors $\{B^t a \mid a \in \mathbb{R}^n\}$. Two vectors a_1 and a_2 in $\mathbb{R}^n$ satisfy $B^t a_1 = B^t a_2$ if and only if $B^t(a_1 - a_2) = 0$. Therefore

$$\dim \text{row space } B = n - \dim(\text{column space } B)^{\perp} = n - (n - \text{column rank } B). \tag{11.10.5}$$

Further, if $c \in \mathbb{R}^k$ and

$$B^tBc = 0, \quad \text{then} \quad c^tB^tBc = 0, \quad \text{then} \quad (Bc)^t(Bc) = 0, \quad \text{then} \quad Bc = 0. \tag{11.10.6}$$

Therefore,

$$(\text{row space } B)^{\perp} = (\text{row space } B^tB)^{\perp}. \tag{11.10.7}$$

Similarly,

$$(\text{column space } B)^{\perp} = \text{column space } BB^t)^{\perp}. \tag{11.10.8}$$

□

Lemma 11.10.8. *If $a \in \mathbb{R}^n$ there exists $c \in \mathbb{R}^k$ such that*

$$a^tB = c^tB^tB. \tag{11.10.9}$$

Therefore $d \in \mathbb{R}^k$ and $d^t\phi$ is estimable if and only if there exists $c \in \mathbb{R}^k$ with $d^t = c^tB^tB$.

PROOF. By Lemma 11.10.7, B and B^tB have the same row spaces. Therefore (11.10.9) is correct. By Lemma 11.10.5, $d^t\phi$ is estimable if and only if d^t is in the row space of B if and only if there exists $c \in \mathbb{R}^k$ such that $d^t = c^tB^tB$.

□

Definition 11.10.9.

$$\mathbf{H} = \{a | a \in \mathbb{R}^n \text{ and for some } c \in \mathbb{R}^k, a = Bc\};$$

$$\mathbf{N}(B) = \{a | a \subset \mathbb{R}^n \text{ and } a^tB = 0\}. \tag{11.10.10}$$

In words, $\mathbf{H}$ is the column space of B.

Lemma 11.10.10. *The following hold.*

$$\mathbf{H}^{\perp} = \mathbf{N}(B) \text{ and the dimension of } \mathbf{H} = \text{rank of } B;$$

$$\mathbf{H} \cap \mathbf{N}(B) = \{0\}; \tag{11.10.11}$$

$$\mathbb{R}^n = \mathbf{H} \oplus \mathbf{N}(B).$$

PROOF. Obvious. □

Lemma 11.10.11. *If $c \in \mathbb{R}^k$ and $c^t\phi$ is estimable then there is a uniquely determined $a \in \mathbf{H}$ such that $a^t\mathbf{X}$ is* the *best linear unbiased estimator of $c^t\phi$. Conversely, if $a \in \mathbf{H}$ then $a^t\mathbf{X}$ is* the *best linear unbiased estimator of $a^tB\phi$.*

PROOF. If $c^t\phi$ is estimable then there exists $a \in \mathbb{R}^n$ such that $a^tB = c^t$, and by Lemma 11.10.8, there exists $r \in \mathbb{R}^k$ such that $c^t = a^tB = r^tB^tB$. Then $Br \in \mathbf{H}$ and $\mathbf{E}(Br)^t\mathbf{X} = r^tB^tB\phi = c^t\phi$.

To prove uniqueness, if a_1 and a_2 are in $\mathbf{H}$ and if $\mathbf{E}a_1^t\mathbf{X} = \mathbf{E}a_2^t\mathbf{X}$ then $(a_1 - a_2)^tB\phi = 0$ for all ϕ in $\mathbb{R}^k$. Therefore $a_1 - a_2 \in \mathbf{H} \cap \mathbf{N}(B)$ and by Lemma 11.10.10, $a_1 = a_2$.

To prove minimum variance, suppose $a \in \mathbb{R}^n$ and $a = a_1 + a_2$ with $a_1 \in \mathbf{H}$ and $a_2 \in \mathbf{N}(B)$. Then $a^tB\phi = \mathbf{E}a^t\mathbf{X} = a_1^tB\phi + a_2^tB\phi = a_1^tB\phi$, and by Lemma 11.10.4, $\mathbf{var}(a^t\mathbf{X}) = \sigma^2(a_1 + a_2)^t(a_1 + a_2) = \sigma^2(a_1^ta_1 + a_2^ta_2)$, since $a_1^ta_2 = 0$. Clearly, the variance is then minimized by the choice $a_2 = 0$, i.e., by the choice $a \in \mathbf{H}$. In particular the best linear unbiased estimator is uniquely determined. □

11.11. The Gauss–Markov Equations and Their Solution

The preceding results can be summarized in a theorem.

Theorem 11.11.1. *If $c \in \mathbb{R}^k$ then $c^t\phi$ is estimable if and only if c is in the row space of B. In this case, there exists $r \in \mathbb{R}^k$ such that $c = B^tBr$ and the best linear unbiased estimator of $c^t\phi$ is $(Br)^t\mathbf{X}$. The set of vectors $\{a \mid a \in \mathbb{R}^n$ and $a^t\mathbf{X}$ is a BLUE$\}$ = column space of $B = \mathbf{H}$.*

If $a \in \mathbf{R}^n$ then a^tB is in the row space of B and $a^tB\phi$ is estimable, so that there exists $r \in \mathbb{R}^k$ such that

$$r^tB^tB = a^tB, \tag{11.11.1}$$

and

$$r^tB^t\mathbf{X} \text{ is the best linear unbiased estimator of } a^tB\phi. \tag{11.11.2}$$

Closely related to the equations (11.11.1) are the Gauss–Markov equations (11.11.3).

$$B^tB\mathbf{t} = B^t\mathbf{X}. \tag{11.11.3}$$

Clearly if $(B^tB)^{-1}$ exists then $\mathbf{t} = (B^tB)^{-1}B^t\mathbf{X}$ solves the equations (11.11.3) and if $c \in \mathbb{R}^k$ then $B(B^tB)^{-1}c$ is in the column space of B so that $(B(B^tB)^{-1}c)^t\mathbf{X}$ is the best linear unbiased estimator of

$$\mathbf{E}(B(B^tB)^{-1}c)^t\mathbf{X} = c^t\phi. \tag{11.11.4}$$

In the sequel we will study in detail the relation between being a solution of the Gauss–Markov equations (11.11.3) and being a best linear unbiased estimator.

Theorem 11.11.2. *If* $\mathbf{X}$ *has a joint normal density function and if* $\mathbf{E}\,\mathbf{X} = B\phi$, $\mathbf{E}(\mathbf{X} - B\phi)(\mathbf{X} - B\phi)^t = \sigma^2 I_n$, *then the maximum likelihood estimators* $\hat{\phi}$ *of* ϕ *are the solutions of the equations* (11.11.3).

PROOF. The joint density function is

$$(2\pi)^{-n/2}\alpha^{-n}\exp -\tfrac{1}{2}(X - B\phi)^t(X - B\phi)\sigma^{-2}. \qquad (11.11.5)$$

Therefore $\hat{\phi}$ satisfies

$$(X - B\hat{\phi})^t(X - B\hat{\phi}) = \inf_{\phi\in\mathbb{R}^k}(X - B\phi)^t(X - B\phi). \qquad (11.11.6)$$

If s solves (11.11.3) then

$$\begin{aligned}((X - Bs) + B(s - \phi))^t((X - Bs) + B(s - \phi))\\ = (X - Bs)^t(X - Bs) + (s - \phi)^t(B^tB)(s - \phi) \qquad (11.11.7)\\ + (X^t - s^tB^t)B(s - \phi) + (s - \phi)^tB^t(X - Bs).\end{aligned}$$

By (11.11.3) the last two terms of (11.11.7) vanish and the matrix B^tB being positive semi-definite the minimum is attained if $\hat{\phi} = s$ is used. Conversely if $\hat{\phi}$ is a maximum likelihood estimator then, since s solves (11.11.3),

$$(s - \hat{\phi})^tB^tB(s - \hat{\phi}) = 0; \qquad (11.11.8)$$

therefore

$$B(s - \hat{\phi}) = 0;$$

therefore

$$B^tB\hat{\phi} = B^tBs = B^t\mathbf{X}. \qquad \square$$

Remark 11.11.3. The minimization of Theorem 11.11.2 implies equations (11.11.3) always have a solution. This fact may be seen dirctly as follows. $B^t\mathbf{X}$ is an element of the row space of B, while by Lemma 11.10.8 the row space of B is the set of vectors $\{c \mid \text{for some } t\in\mathbb{R}^k, c = B^tBt\}$. Hence a vector $t\in\mathbb{R}^k$ exists solving the equations.

Theorem 11.11.4 (Gauss–Markov). *Suppose* $\mathbf{t}$ *solves the Gauss–Markov equations. Let* $c\in\mathbb{R}^k$ *and suppose* $c^t\phi$ *is estimable. Then* $c^t\mathbf{t}$ *is the best linear unbiased estimator of* $c^t\phi$.

PROOF. We write $c^t = r^tB^tB$. Then $c^t\mathbf{t} = r^tB^tB\mathbf{t} = r^tB^t\mathbf{X}$. By Theorem 11.11.1 the result now follows. $\square$

The solutions of the Gauss–Markov equations have an elegant expression in terms of the generalized inverse $(B^tB)^+$. We now develop this theory and refer the reader to Theorem 8.1.11 for the preliminary mathematics, Definition 8.1.12 for the definition of the generalized inverse, and Theorem 8.1.13

for the statement and proof of its properties. To begin, consider (11.11.4). Here we may write $c^t = r^t(B^tB)$ so that (11.11.4) becomes

$$r^tB^t(B(B^tB)^{-1}B^t)\mathbf{X}. \tag{11.11.9}$$

The essential things here are the projection matrix $B(B^tB)^{-1}B^t = P$, which is a symmetric matrix satisfying $P = P^2$, and that $\mathbf{E}\, r^tB^tP\mathbf{X} = c^t\phi$. In the general theory B^tB may be a singular matrix, i.e., rank $B^tB < k$, in which case we consider the projection matrix (as follows directly from Theorem 8.1.13)

$$P = B(B^tB)^+B^t. \tag{11.11.10}$$

The superscript "+" is used to denote the generalized inverse, and one readily verifies that P is symmetric and $P = P^2$. The basic computational lemma is the following.

Lemma 11.11.5. *The $n \times n$ projection matrix* (11.11.10) *has as its range the column space of B, and hence*

$$B = (B(B^tB)^+B^t)B. \tag{11.11.11}$$

PROOF. Clearly the columns of P in (11.11.10) are in the column space of B. Further

$$\begin{aligned} \operatorname{rank} P &\geq \operatorname{rank} B^tPB = \operatorname{rank}\left[(B^tB)(B^tB)^+(B^tB)\right] \\ &= \operatorname{rank} B^tB = \operatorname{rank} B = \text{the column dimension.} \end{aligned} \tag{11.11.12}$$

Therefore linear combinations of the columns of P generate the entire column space of B. □

Lemma 11.11.6. *The following projection matrices are equal.*

$$B(B^tB)^+B^t = (BB^t)(BB^t)^+. \tag{11.11.13}$$

PROOF. By Theorem 8.1.13, $(BB^t)(BB^t)^+$ is an $n \times n$ symmetric projection matrix. By the proof of Lemma 11.11.5, the column space of B is representable as BB^tc. Thus the column space comes from

$$(BB^t)(BB^t)^+(BB^t)c = BB^tc. \tag{11.11.14}$$

Since dimension of the column space $= \operatorname{rank} B \geq \operatorname{rank}(BB^t)(BB^t)^+$, it follows that both orthogonal projections of (11.11.13) have the same range, hence are equal as matrices. □

Theorem 11.11.7. *The $k \times 1$ vector*

$$s = (B^tB)^+B^tX \tag{11.11.15}$$

solves the Gauss–Markov equations (11.11.3). *If s_1 and s_2 are solutions of* (11.11.3) *then $s_1 - s_2$ is an element of* (*row space of B*)$^\perp$. *Therefore the*

Gauss–Markov equations have a unique solution in the row space of B which is the vector (11.11.15).

Proof. Substitution of (11.11.15) into (11.11.3) gives

$$B^tB(B^tB)^+(B^tX) = B^tX. \tag{11.11.16}$$

The computation uses Lemma 11.11.5. Given two solutions s_1 and s_2,

$$B^tB(s_1 - s_2) = 0 \tag{11.11.17}$$

so that

$$B(s_1 - s_2) = B(B^tB)^+(B^tB)(s_1 - s_2) = 0.$$

Thus $s_1 - s_2$ is orthogonal to the rows of B. The theorem now follows. □

Theorem 11.11.8. *The matrix*

$$Q = I_n - P = I_n - B(B^tB)^+B^t \tag{11.11.18}$$

is an $n \times n$ symmetric projection matrix. In addition,

$$\mathbf{E}\,Q\mathbf{X} = 0 \quad \text{and} \quad \mathbf{E}\,\mathbf{X}^tQ\mathbf{X} = (n - \operatorname{rank} B)\sigma^2. \tag{11.11.19}$$

Proof. From Lemma 11.11.6, $QB = 0$. Thus $\mathbf{E}\,Q\mathbf{X} = 0$ and $\mathbf{E}\,\mathbf{X}^tQ\mathbf{X} = \mathbf{E}(\operatorname{tr} Q(\mathbf{X} - B\phi)(\mathbf{X} - B\phi)^t) = \operatorname{tr}(\sigma^2 QI_n) = (n - \operatorname{rank} B)\sigma^2$. □

Theorem 11.11.9. *Let $a \in \mathbb{R}^n$, and let $c \in \mathbb{R}^k$ such that $c^t\phi$ is estimable. Let $\mathbf{t}$ be a solution of the Gauss–Markov equations and let Q be the projection matrix* (11.11.18). *Then*

$$\mathbf{cov}(c^t\mathbf{t}, a^tQ\mathbf{X}) = 0. \tag{11.11.20}$$

Proof. Write $c^t = r^tB^tB$ so that $c^t\mathbf{t} = r^tB^t\mathbf{X}$. Then (11.11.20) becomes

$$\mathbf{E}\,r^tB^t\mathbf{X}\mathbf{X}^tQa = \sigma^2 r^tB^tQ = 0. \tag{11.11.21}$$

□

Remark 11.11.10. In the analysis of variance the quantity $\mathbf{X}^t(I_n - B(B^tB)^+B^t)\mathbf{X} = \mathbf{X}^tQ\mathbf{X}$ is known as *sum of squares of error* and $\sigma^{-2}\mathbf{X}^tQ\mathbf{X}$ is the central chi-square statistic used in the denominator of the F-statistic.

11.12. Normal Theory. Idempotents and Chi-Squares

Throughout this section we assume that $\mathbf{X}$ has a joint normal density function such that $\mathbf{E}\,\mathbf{X} = 0$ and $\mathbf{E}\,\mathbf{X}\mathbf{X}^t = \sigma^2 I_n$.

Lemma 11.12.1. *Let P be an $n \times n$ symmetric projection matrix of* rank r. *Then $\mathbf{X}^tP\mathbf{X}$ is a (central) χ_r^2 random variable.*

PROOF. Take $U \in \mathbf{O}(n)$ such that $UPU^t = \begin{vmatrix} I_r & 0 \\ 0 & 0 \end{vmatrix}$ and set $\mathbf{Y} = U\mathbf{X}$. Then $\mathbf{Y}$ has a joint normal density function and $\mathbf{E}\,\mathbf{Y} = 0$, $\mathbf{E}\,\mathbf{Y}\mathbf{Y}^t = \sigma^2 I_n$. We have that

$$\mathbf{X}^t P \mathbf{X} = \mathbf{Y}^t \begin{vmatrix} I_r & 0 \\ 0 & 0 \end{vmatrix} \mathbf{Y} = \mathbf{Y}_1^2 + \cdots + \mathbf{Y}_r^2. \tag{11.12.1}$$

□

Lemma 11.12.2. *If P is an $n \times n$ symmetric matrix and $\mathbf{X}^t P \mathbf{X}$ has a chi-square probability density function then $P^2 = P$.*

PROOF. As in the preceding lemma, take $U \in \mathbf{O}(n)$ so that UPU^t is a diagonal matrix with nonzero diagonal entries $p_1, \ldots, p_r$. If $\mathbf{Y} = U\mathbf{X}$ and $\mathbf{Y}^t = (\mathbf{Y}_1, \ldots, \mathbf{Y}_n)$ then

$$\mathbf{X}^t P \mathbf{X} = \sum_{i=1}^{r} p_i \mathbf{Y}_i^2, \tag{11.12.2}$$

which is a sum of independently distributed random variables. Therefore the Laplace transform of the random variable $\mathbf{X}^t P \mathbf{X}$ is

$$\prod_{i=1}^{r} 1/(1 - 2p_i s)^{1/2}. \tag{11.12.3}$$

The hypothesis of the lemma is that for some integer $r' > 0$,

$$\prod_{i=1}^{r} 1/(1 - 2p_i s)^{1/2} = 1/(1 - 2s)^{r'/2}. \tag{11.12.4}$$

Since both sides are analytic functions of s, their singularities must be the same. Therefore $p_1 = \cdots = p_r = 1$, and $r = r'$ follows. □

Lemma 11.12.3. *The random variable*

$$\sigma^{-2}\mathbf{X}^t Q \mathbf{X} = \sigma^{-2}\mathbf{X}^t(I_n - B(B^tB)^+ B^t)\mathbf{X} \tag{11.12.5}$$

has a central χ^2_{n-r} probability density function, where $r =$ rank of B.

PROOF. Since Q is an $n \times n$ symmetric projection matrix of rank $n - r$, Lemma 11.12.1 applies directly to the random variable $\mathbf{X} - B\phi$. Clearly by Lemma 11.11.5

$$(\mathbf{X} - B\phi)^t Q(\mathbf{X} - B\phi) = \mathbf{X}^t Q \mathbf{X}. \tag{11.12.6}$$

□

Remark 11.12.4. In the sequel we will call $\mathbf{X}^t Q \mathbf{X}$ the *sum of squares of error, SSE*, which appears in the denominators of the F-statistics below.

Theorem 11.12.5. *In the normal theory* BLUE *and* SSE *are stochastically independent random variables.*

PROOF. If $\mathbf{t}$ is a solution of the Gauss–Markov equations then by Theorem 11.11.9, $c^t\mathbf{t}$ is stochastically independent of $Q\mathbf{X}$, provided c is in the row space of B. □

Theorem 11.12.6. *If c is in the row space of B and $\mathbf{t}$ is a solution of the Gauss–Markov equation then*

$$(n-r)(c^t(B^tB)^+c)^{-1}(c^t\mathbf{t})^2/(\mathrm{SSE}) \tag{11.12.7}$$

is distributed as a noncentral F-statistic $(n-r)\chi_1^2/\chi_{n-r}^2$ with noncentrality parameter

$$\tfrac{1}{2}(c^t\phi)^2(\sigma^2(c^t(B^tB)^+c))^{-1}. \tag{11.12.8}$$

PROOF. We may write $c^t = r^t(B^tB)$. Of the possible choices of r we want r in the row space of B. To see that this is possible, since c is in the row space of B,

$$c^t = c^t(B^tB)^+(B^tB) = r^t(B^tB)(B^tB)^+(B^tB), \tag{11.12.9}$$

and $(B^tB)^+(B^tB)r$ is in the row space of B by virtue of Lemmas 11.11.5 and 11.11.6. Then, using Lemmas 11.10.4 and 11.11.5,

$$c^t\mathbf{t} = r^tB^t\mathbf{X}, \tag{11.12.10}$$

and

$$\mathbf{var}\,(c^t\mathbf{t}) = \sigma^2 r^t(B^tB)r. \tag{11.12.11}$$

Since r is in the row space of B we have that

$$c^t(B^tB)^+ = r^t, \tag{11.12.12}$$

and thus

$$\mathbf{var}\,(c^t\mathbf{t}) - \sigma^2 c^t(B^tB)^+c. \tag{11.12.13}$$

Thus the numerator of (11.12.7) has variance σ^2 and (11.12.8) is one-half of the expectation squared. By Lemma 11.12.3 and Theorem 11.12.5, the result now follows. □

Lemma 11.12.7. *If c_1 and c_2 are in the row space of B and $\mathbf{t}$ is a solution of the Gauss–Markov equations then*

$$\mathbf{cov}\,(c_1^t\mathbf{t}, c_2^t\mathbf{t}) = \sigma^2 c_1^t(B^tB)^+c_2. \tag{11.12.14}$$

PROOF. As in the proof of Theorem 11.12.6. □

The general setup of the analysis of variance is expressed in the next theorem. As follows from Lemma 11.12.7 the best linear unbiased estimators indicated in the following statement are stochastically independent and the theorem follows trivially.

Theorem 11.12.8. *Let* **S** *be a subspace of the row space of* B *and let* $s = \dim \mathbf{S}$. *Choose* $c_1, \ldots, c_s$ *in* **S** *such that*

$$\begin{aligned} c_i^t(B^tB)^+c_j &= 1, \qquad 1 \le i = j \le s, \\ &= 0, \qquad 1 \le i \neq j \le s. \end{aligned} \tag{11.12.15}$$

Then

$$(n - r)\sum_{i=1}^{s} (c_i^t\mathbf{t})^2/s(\text{SSE}) \tag{11.12.16}$$

is a noncentral F-statistic $(n - r)\chi_s^2/s\chi_{n-r}^2$ *with noncentrality parameter*

$$\tfrac{1}{2}\sigma^{-2}\sum_{i=1}^{s} (c_i^t\phi)^2. \tag{11.12.17}$$

Remark 11.12.9. Normally the experimenter has $k \times 1$ vectors $c_1, \ldots, c_s$ which are a basis of **S** and needs to find a new basis of **S** to satisfy the condition of Theorem 11.12.8. Rather than solve this problem the needed result can be obtained as follows. Let $C = (c_1, \ldots, c_s)$ be the indicated $k \times s$ matrix. With $\mathbf{t} = (B^tB)^+B^t\mathbf{X}$ the $s \times 1$ vector $C^t\mathbf{t}$ has expectation

$$\mathbf{E}\, C^t\mathbf{t} = C^t(B^tB)^+B^tB\phi = C^t\phi, \tag{11.12.18}$$

where because the rows of C^t are in the row space of B, Lemma 11.11.5 applies. Also

$$\mathbf{Cov}\, C^t\mathbf{t} = \sigma^2 C^t(B^tB)^+C, \tag{11.12.19}$$

and by Lemma 11.10.7

$$\operatorname{rank} C^t(B^tB)^+B^t \ge \operatorname{rank} C^t(B^tB)^+(B^tB) = \operatorname{rank} C, \tag{11.12.20}$$

and

$$\operatorname{rank} C^t(B^tB)^+C = \operatorname{rank} C^t(B^tB)^+B^t.$$

Thus the $s \times s$ matrix $C^t(B^tB)^+C$ is nonsingular and

$$[C^t(B^tB)^+C]^{-1/2}C^t(B^tB)^+B^t\mathbf{X} \tag{11.12.21}$$

has $\sigma^2 I_s$ as covariance matrix. Then the corresponding chi-square has the noncentrality parameter

$$\tfrac{1}{2}\sigma^{-2}\|[C^t(B^tB)^+C]^{-1/2}C^t\phi\|^2 = \tfrac{1}{2}\sigma^{-2}\phi^tC[C^t(B^tB)^+C]^{-1}C^t\phi. \tag{11.12.22}$$

We summarize this discussion in a theorem.

Theorem 11.12.10. *Let* **S** *be an s-dimensional subspace of the row space of* B. *Let* C *be a* $k \times s$ *matrix of* rank s *such that the columns of* C *span* **S**. *Then there exists a noncentral F-statistic* $(n - r)\chi_s^2/s\chi_{n-r}^2$ *with noncentrality parameter* (11.12.22). *This parameter is independent of the choice of* C *as basis of* **S**.

Remark 11.12.11. That the noncentrality parameter vanishes means ϕ is orthogonal to $\mathbf{S}$. Therefore the F-test constructed using (11.12.21) is a test of the hypothesis that $\phi \in \mathbf{S}^{\perp}$ against the alternative that $\phi \notin \mathbf{S}^{\perp}$.

Remark 11.12.12. The covariance matrix of $C^t\mathbf{t}$ is $\sigma^2 C^t(B^tB)^+C$ while the matrix of the quadratic form in the noncentrality parameter is $C[C^t(B^tB)^+C]^{-1}C^t$, which is not the inverse of the covariance matrix. Early discussions of optimality of Latin square designs by Wald and Ehrenfeld defined D- and E-optimality in terms of $\det(C^t(B^tB)^+C)$ and the largest eigenvalue of $C^t(B^tB)^+C$. Kiefer (1958) discussed these definitions and also defined optimality in terms of the power functions of tests. The user who is interested only in the comparision of F-tests need look only at the matrix in (11.12.23) to make a design decision based on the power functions of tests.

11.13. Problems

Problem 11.13.1. Let X be an $n \times 1$ vector, $X^t = (x_1, \ldots, x_n)$. Define $U(X)$ to be an $n \times n$ matrix such that if $x_1 = 0$ then $U(X) = I_n$ and if $x_1 \neq 0$ then

$$U(X) = \begin{vmatrix} x_2(x_1^2 + x_2^2)^{-1/2} & -x_1(x_1^2 + x_2^2)^{-1/2} & 0, \ldots, 0 \\ x_1(x_1 + x_2)^{-1/2} & x_2(x_1 + x_2)^{-1/2} & 0, \ldots, 0 \\ 0 & 0 & \\ \cdots\cdots & \cdots\cdots & \cdot I_{n-2} \\ 0 & 0 & \end{vmatrix} \tag{11.13.1}$$

In each case the positive square root is to be used. Show that $U(X) \in \mathbf{O}(n)$, that the entries of $U(X)$ are measurable functions of $x_1, \ldots, x_n$, and that $(U(X)X)^t = (0, (x_1^2 + x_2^2)^{1/2}, x_3, \ldots, x_n)$. By induction construct a matrix valued function $U(X)$ with values in $\mathbf{O}(n)$ such that $(U(X)X)^t = (0, \ldots, 0, \|X\|)$.

Problem 11.13.2. This problem is about the multivariate beta density function. Refer to Section 10.3 for notations. In this problem, $k \times k$ symmetric random matrices $\mathbf{S}$ and $\mathbf{T}$ are stochastically independent central Wishart random matrices with p and q degrees of freedom respectively. Write $(S)_h$ and $(T)_h$ for the principal minors with elements s_{ij}, t_{ij}, $1 \le i, j \le h$. By (10.3.1), $CC^t = S + T$, so that $\mathbf{C}$ has the canonical distribution specified by Theorem 11.0.1. Determine the parameter values.

Again by (10.3.1), $L = C^{-1}S(C^t)^{-1}$, so that if $D \in \mathbf{T}(k)$, and $DD^t = S$ then $L = (C^{-1}D)(C^{-1}D)^t$. Show that for the $h \times h$ principal minor the same relations hold, namely,

$$(L)_h = (C)_h^{-1}(S)_h(C)_h^{-1} = ((C)_h^{-1}(D)_h))((C)_h^{-1}(D)_h)^t, \tag{11.13.2}$$

and

$$(S + T)_h = (C)_h (C)_h^t.$$

Let the diagonal elements of $W = C^{-1}D$ (see (10.3.11)) be $w_{11}, \ldots, w_{kk}$. Then show

$$w_{h+1\,h+1}^2 = \det(L)_{h+1}/\det(L)_h = \frac{(\det(D)_{h+1}\det(C)_h)^2}{(\det(C)_{h+1}\det(D)_h)^2}. \qquad (11.13.3)$$

Show by use of Theorem 11.2.2 that $\mathbf{w}_{hh}$ is stochastically independent of $\mathbf{w}_{11}, \ldots, \mathbf{w}_{h-1\,h-1}$.

PROBLEM 11.13.3. Let the nonnegative real valued random variables $\mathbf{X}$ and $\mathbf{Y}$ have probability density functions f and g respectively. Then the random variable $\mathbf{W} = \mathbf{X}/(\mathbf{X} + \mathbf{Y})$ has density function

$$(1 - w)^{-2}\int_0^\infty yf(wy/(1 - w))g(y)\,dy. \qquad (11.13.4)$$

PROBLEM 11.13.4. Continue Problem 11.13.3. Suppose

$$f(x) = x^{n-1}e^{-x}/\Gamma(n) \quad \text{and} \quad g(y) = y^{m-1}e^{-y}/\Gamma(m). \qquad (11.13.5)$$

Then the density function of $\mathbf{X}/(\mathbf{X} + \mathbf{Y})$ is the beta density function

$$h(w) = w^{n-1}(1 - w)^{m-1}\Gamma(m + n)/\Gamma(m)\Gamma(n). \qquad (11.13.6)$$

PROBLEM 11.13.5. Continue Problem 11.11.4 Calculate the r-th moment of the density function h and show that the answer is

$$\frac{\Gamma(m + n)\Gamma(n + r)}{\Gamma(m + n + r)\Gamma(n)}. \qquad (11.13.7)$$

PROBLEM 11.13.6. Using (10.3.14) and (10.3.15), together with (11.13.7), show that

$$\frac{\mathbf{E}(\det(\mathbf{L})_{h+1})^r}{\mathbf{E}(\det(\mathbf{L})_h)^r} = \frac{\Gamma(r + \frac{1}{2}(p - h))\Gamma(\frac{1}{2}(p + q - h))}{\Gamma(r + \frac{1}{2}(p + q - h))\Gamma(\frac{1}{2}(p - h))}. \qquad (11.13.8)$$

In (11.13.8) make the identification

$$n = \tfrac{1}{2}(p - h) \quad \text{and} \quad m = \tfrac{1}{2}q. \qquad (11.13.9)$$

Refer to (10.3.11) and conclude that $\mathbf{w}_{h+1\,h+1}$ has a beta probability density function. What theorem about moments are you using?

PROBLEM 11.13.7 (unsolved by me). Obtain the result of Problem 11.13.6 that $\mathbf{w}_{h+1\,h+1}$ has a beta probability density function by random variable techniques.

Problems on Conditional Distributions

PROBLEM 11.13.8. Suppose $\mathbf{X} = (\mathbf{X}_1, \mathbf{X}_2)$ is an $n \times k$ random matrix such that the rows of $\mathbf{X}$ are independent normal $(0, \Sigma)$ random vectors. Let $\mathbf{X}_1$ be $n \times p$ and $\mathbf{X}_2$ be $n \times q$, and partition the covariance matrix

$$\Sigma = \begin{vmatrix} \Sigma_{11} & \Sigma_{12} \\ \Sigma_{21} & \Sigma_{22} \end{vmatrix}, \quad \Sigma_{11} \text{ a } p \times p \text{ matrix.} \tag{11.13.10}$$

Let

$$C = \begin{vmatrix} I_p & -\Sigma_{12}\Sigma_{22}^{-1} \\ 0 & I_q \end{vmatrix}. \tag{11.13.11}$$

Show that the covariance matrix of a row of $\mathbf{X}C^t$ is

$$\begin{vmatrix} \Sigma_{11} - \Sigma_{12}\Sigma_{22}^{-1}\Sigma_{21} & 0 \\ 0 & \Sigma_{22} \end{vmatrix}. \tag{11.13.12}$$

Therefore $\mathbf{X}_2$ and $\mathbf{X}_1 - \mathbf{X}_2\Sigma_{22}^{-1}\Sigma_{21}$ have zero covariance.

PROBLEM 11.13.9. Continue Problem 11.13.8. The conditional distribution of the first row of $\mathbf{X}_1$ given the first row of $\mathbf{X}_2$ is normal $(\mathbf{X}_2\Sigma_{22}^{-1}\Sigma_{21}, \Sigma_{11} - \Sigma_{12}\Sigma_{22}^{-1}\Sigma_{21})$.

PROBLEM 11.13.10. Continue Problem 11.13.9. We have assumed that $\mathbf{E}\,\mathbf{X} = 0$. Let $\mathbf{S}_{ij} = \mathbf{X}_i^t\mathbf{X}_j$, $1 \le i, j \le 2$. The $p \times p$ random matrix

$$\mathbf{S}_{11} - \mathbf{S}_{12}\mathbf{S}_{22}^{-1}\mathbf{S}_{21} \tag{11.13.13}$$

has a Wishart density with parameters $n - q, p, \Sigma_{11} - \Sigma_{12}\Sigma_{22}^{-1}\Sigma_{21}$.

HINT. The quantity (11.13.13) is

$$\mathbf{X}_1^t\mathbf{X}_1 - \mathbf{X}_1^t\mathbf{X}_2(\mathbf{X}_2^t\mathbf{X}_2)^{-1}\mathbf{X}_2^t\mathbf{X}_1 = \mathbf{X}_1^t(I_n - \mathbf{X}_2(\mathbf{X}_2^t\mathbf{X}_2)^{-1}\mathbf{X}_2^t)\mathbf{X}_1. \tag{11.13.14}$$

The $n \times n$ matrix in the parentheses is a $n \times n$ orthogonal projection. Choose $U = U(\mathbf{X}_2)$ a random orthogonal $n \times n$ matrix such that

$$U(I_n - \mathbf{X}_2(\mathbf{X}_2^t\mathbf{X}_2)^{-1}\mathbf{X}_2^t)U^t = \begin{vmatrix} I_{n-q} & 0 \\ 0 & 0 \end{vmatrix}. \tag{11.13.15}$$

The random matrices $\mathbf{X}_1$ and $U(\mathbf{X}_2)\mathbf{X}_1$ have the same distribution. Therefore (11.13.13) has the same distribution as

$$\mathbf{X}_1^t \begin{vmatrix} I_{n-q} & 0 \\ 0 & 0 \end{vmatrix} \mathbf{X}_1. \tag{11.13.16}$$

□

PROBLEM 11.13.11. Continue Problem 11.13.10. The conditional distribution of $\mathbf{S}_{11} - \mathbf{S}_{12}\mathbf{S}_{22}^{-1}\mathbf{S}_{21}$ given $\mathbf{X}_2$ is a Wishart density with parameters $n - q$, p, $\Sigma_{11} - \Sigma_{12}\Sigma_{22}^{-1}\Sigma_{21}$.

HINT. Use the choice of U in (11.13.15). As in (11.13.16),

$$\mathbf{S}_{11} - \mathbf{S}_{12}\mathbf{S}_{22}^{-1}\mathbf{S}_{21} = (U(\mathbf{X}_2)\mathbf{X}_1)^t \begin{vmatrix} I_{n-q} & 0 \\ 0 & 0 \end{vmatrix} (U(\mathbf{X}_2)\mathbf{X}_1).$$

Show that the conditional means of $U(\mathbf{X}_2)\mathbf{X}_1$ are zero by using Problem 11.13.9 and (11.13.15). □

PROBLEM 11.13.12. Continue Problem 11.13.10. If $\Sigma_{12} = 0$ then the conditional distribution of $\mathbf{S}_{12}\mathbf{S}_{22}^{-1}\mathbf{S}_{21}$ given $\mathbf{X}_2$ is Wishart with parameters p, q, Σ_{11}. The quantities $\mathbf{S}_{11} - \mathbf{S}_{12}\mathbf{S}_{22}^{-1}\mathbf{S}_{21}$ and $\mathbf{S}_{12}\mathbf{S}_{22}^{-1}\mathbf{S}_{21}$ are stochastically independent.

HINT. In terms of $U(\mathbf{X}_2)$ the random variables are

$$(U(\mathbf{X}_2)\mathbf{X}_1)^t \begin{vmatrix} 0 & 0 \\ 0 & I_q \end{vmatrix} (U(\mathbf{X}_2)\mathbf{X}_1) \quad \text{and} \quad (U(\mathbf{X}_2)\mathbf{X}_1)^t \begin{vmatrix} I_{n-q} & 0 \\ 0 & 0 \end{vmatrix} (U(\mathbf{X}_2)\mathbf{X}_1).$$

Since $\Sigma_{12} = 0$ the conditional means are zero. The result now follows since $U(\mathbf{X}_2)\mathbf{X}_1$ and $\mathbf{X}_1$ have the same conditional distribution. □

Problems on the Analysis of Variance

PROBLEM 11.13.13. Suppose A is an $n \times s$ matrix and $\sigma\ {}^{2}\mathbf{X}^t A A^t \mathbf{X}$ is a noncentral χ_s^2 random variable. Suppose $\mathbf{X}$ has a joint normal density function with $\mathbf{E}\mathbf{X} = B\phi$ and $\mathbf{E}(\mathbf{X} - B\phi)(\mathbf{X} - B\phi)^t = \sigma^2 I_n$. Then show that AA^t is an $n \times n$ projection matrix.

HINT. See Lemma 11.12.2 and Chapter 2. □

PROBLEM 11.13.14. Continue Problem 11.13.13. Show the noncentrality parameter is

$$\tfrac{1}{2}\sigma^{-2}\|A^t B\phi\|^2 = \tfrac{1}{2}\sigma^{-2}\phi^t B^t (AA^t) B\phi. \tag{11.13.17}$$

More generally, if P is an $n \times n$ orthogonal projection matrix then the noncentral chi-square $\sigma^{-2}\mathbf{X}^t P\mathbf{X}$ has noncentrality parameter $\tfrac{1}{2}\sigma^{-2}(B\phi)^t P(B\phi)$.

PROBLEM 11.13.15. Let A_1 be an $n \times s$ matrix and A_2 an $n \times r$ matrix and suppose the columns of A_1 and A_2 are in the column space of B. Let

$\sigma^{-2}\mathbf{X}^t A_1 A_1^t \mathbf{X}$ and $\sigma^{-2}\mathbf{X}^t A_2 A_2^t \mathbf{X}$ be noncentral chi-square random variables. Suppose $0 = \phi^t B^t (A_1 A_1^t) B\phi$ implies $0 = \phi^t B^t (A_2 A_2^t) B\phi$. Then show $A_1 A_1^t \geq A_2^t A_2$ in the partial ordering of semidefinite matrices.

PROBLEM 11.13.16. Continuation of Problem 11.13.15. The power function of an analysis of variance test is a strictly increasing function of the noncentrality parameter. Let two analyses of variance tests have power functions β_1 and β_2 and assume each F-statistic is a function of the $n \times 1$ random vector $\mathbf{X}$. Let the numerators of the two F-statistics be $\sigma^{-2}\mathbf{X}^t A_1 A_1^t \mathbf{X}$ and $\sigma^{-2}\mathbf{X}_2^t A_2 A_2^t \mathbf{X}$ with noncentrality parameters $\|A_1^t B\phi\|^2/2\sigma^2$ and $\|A_2^t B\phi\|^2/2\sigma^2$ respectively. Let $\mathbf{S}$ be an s-dimensional subspace of the row space of B and suppose $\phi \in \mathbf{S}^\perp$ if and only if $\|A_1^t B\phi\| = 0$. See Remark 11.12.11. If both tests are similar size α for the null hypothesis $\phi \in \mathbf{S}^\perp$, show that if $\phi \in \mathbb{R}^k$ then

$$\|A_1^t B\phi\| \geq \|A_2^t B\phi\|. \tag{11.13.18}$$

PROBLEM 11.13.17 (Kiefer (1958)). Let $n_1 \leq n_1'$ and $n_1 + n_2 \geq n_1' + n_2'$ with at least one strict inequality. Let noncentral F-tests with power functions $\beta_{n_1 n_2}(\lambda, \alpha)$ and $\beta_{n_1' n_2'}(\lambda, \alpha)$ be given with size α and noncentrality parameter λ. Then if $\lambda > 0$ and $0 < \alpha < 1$,

$$\beta_{n_1 n_2}(\lambda, \alpha) > \beta_{n_1' n_2'}(\lambda, \alpha). \tag{11.13.19}$$

HINT. Given four stochastically independent chi-squares $\chi^2_{m_1}$, $\chi^2_{m_2}$, $\chi^2_{m_1'}$, $\chi^2_{m_2'}$ with the interpretations $m_1 = n_1$, $m_2 = n_2'$, $m_1' = n_1' - n_1$ and $m_2' = n_1 + n_2 - n_1' - n_2'$, construct test statistics which give the two power functions. The test based on one statistic is a *UMP* unbiased size α test, which is a uniquely determined test. Hence the inequality. □

PROBLEM 11.13.18. Continue Problems 11.13.16 and 11.13.17. Show that under the hypotheses of these two problems, if $\phi \in \mathbb{R}^k$ then

$$\beta_1(\|A_1^t B\phi\|^2/2\sigma^2, \alpha) \geq \beta_2(\|A_2^t B\phi\|^2/2\sigma^2, \alpha). \tag{11.13.20}$$

PROBLEM 11.13.19. Let $\mathbf{X}$ be an $n \times 1$ random vector with a joint normal density function such that

$$\mathbf{E}\mathbf{X} = M \quad \text{and} \quad \mathbf{E}(\mathbf{X} - M)(\mathbf{X} - M)^t = \Sigma. \tag{11.13.21}$$

Suppose A is an $n \times n$ symmetric matrix such that $\mathbf{X}^t A\mathbf{X}$ has a noncentral chi-square probability density function. This can hold if and only if $(A\Sigma)^2 = (A\Sigma)$.

PROBLEM 11.13.20. Suppose A and B are $n \times n$ symmetric matrices and $AB = 0$. Show that $BA = 0$ and that there exists $U \in \mathbf{O}(n)$ such that UAU^t and UBU^t are both diagonal matrices.

PROBLEM 11.13.21. If $\mathbf{X}$ is as in Problem 11.13.19 and if A and B are $n \times n$ symmetric matrices such that $A\Sigma B = 0$ then $\mathbf{X}^t A\mathbf{X}$ and $\mathbf{X}^t B\mathbf{X}$ are stochastically independent.

HINT. Show that $B\Sigma A = 0$. Thus use Problem 11.13.20 and simultaneously diagonalize $\Sigma^{1/2} A \Sigma^{1/2}$ and $\Sigma^{1/2} B \Sigma^{1/2}$. □

PROBLEM 11.13.22 (Graybill and Milliken (1969)). Let $\mathbf{X}$ be an $n \times 1$ random vector having a joint normal probability density function such that $\mathbf{E}\mathbf{X} = M$ and $\mathbf{E}(\mathbf{X} - M)(\mathbf{X} - M)^t = I_n$. Let K be an $r \times n$ matrix and L an $n \times n$ matrix such that $KL^t = 0$. Let $\mathbf{A}$ be a random symmetric $n \times n$ matrix whose entries are measurable functions of $K\mathbf{X}$. Suppose

(i) $\mathbf{A}^2 = \mathbf{A}$;
(ii) $\mathbf{A} = L^t \mathbf{A} L$;
(iii) $\operatorname{tr} \mathbf{A} = m$, a nonrandom constant;
(iv) $M^t \mathbf{A} M = 2\lambda$, a nonrandom constant.

Then $\mathbf{X}^t \mathbf{A} \mathbf{X}$ has a noncentral χ_m^2 probability density function with noncentrality parameter λ. If m and λ are random, what mixture of probability distributions results?

HINT. The random variables $L\mathbf{X}$ and $K\mathbf{X}$ are independent and $\mathbf{X}^t\mathbf{A}\mathbf{X} = \mathbf{X}^t L^t \mathbf{A} L \mathbf{X}$. Use Problem 11.13.1. □

PROBLEM 11.13.23 (Graybill and Milliken (1969)). Continue Problem 11.13.22. Suppose $\mathbf{A}$ and $\mathbf{B}$ are $n \times n$ random symmetric matrices such that the entries of $\mathbf{A}$ and $\mathbf{B}$ are functions of $K\mathbf{X}$. Assume

(i) $\mathbf{A}^2 = \mathbf{A}$ and $\mathbf{B}^2 = \mathbf{B}$;
(ii) $\mathbf{A} = L^t \mathbf{A} L$ and $\mathbf{B} = L^t \mathbf{B} L$;
(iii) $\mathbf{A}\mathbf{B} = 0$;
(iv) $\operatorname{tr} \mathbf{A}$, $\operatorname{tr} \mathbf{B}$, $M^t \mathbf{A} M$, and $M^t \mathbf{B} M$ are nonrandom constants.

Then the random variables $\mathbf{X}^t \mathbf{A} \mathbf{X}$ and $\mathbf{X}^t \mathbf{B} \mathbf{X}$ are stochastically independent.

HINT. By Section 11.1, since conditional on $K\mathbf{X}$ the random variables $\mathbf{A}L\mathbf{X}$ and $\mathbf{B}L\mathbf{X}$ have a joint normal probability density function with zero covariances, conditional independence follows. Since the joint distribution is independent of $K\mathbf{X}$, unconditional independence follows. □

PROBLEM 11.13.24 (Graybill and Milliken (1969)). Continue Problem 11.13.23. In the analysis of variance model, $\mathbf{E}\mathbf{X} = B\phi$. In the notations of Problem 11.13.23 let

$$
\begin{aligned}
L &= I_n - B(B^t B)^+ B^t; \\
K &= B^t; \\
M &= B\phi.
\end{aligned}
\tag{11.13.22}
$$

In these expressions "+" means generalized inverse. Graybill and Milliken, op. cit., use conditional inverses instead of generalized inverses. See Sections 8.1 and 11.10, and the paper by Graybill and Milliken.

Let $\mathbf{Q}$ be a random $n \times r$ matrix whose entries are functions of $K\mathbf{X} = B^t\mathbf{X}$ such that

$$\operatorname{tr} L\mathbf{Q} = m, \quad \text{a nonrandom constant.} \tag{11.13.23}$$

Define

$$\mathbf{A} = (L\mathbf{Q})((L\mathbf{Q})^t(L\mathbf{Q}))^+(L\mathbf{Q})^t, \tag{11.13.24}$$

and

$$\mathbf{C} = L - \mathbf{A}.$$

Show in sequence that

(i) $\mathbf{A}^2 = \mathbf{A}$;
(ii) $\mathbf{A} \le L$;
(iii) $L\mathbf{A} = \mathbf{A}L = \mathbf{A}$ and $L^2 = L$;
(iv) $\mathbf{C}^2 = \mathbf{C}$ and $\mathbf{C}L = L\mathbf{C} = \mathbf{C}$;
(v) $\mathbf{AC} = 0$;
(vi) $\operatorname{tr} \mathbf{A} = m$, a nonrandom constant;
(vii) $\operatorname{tr} \mathbf{C} = n - \operatorname{rank} B - m$, a nonrandom constant;
(viii) $M^t\mathbf{A}M = M^t\mathbf{C}M = 0$;
(ix) $\mathbf{X}^t\mathbf{AX}$ and $\mathbf{X}^t\mathbf{CX}$ are independently distributed noncentral chi-square random variables.

PROBLEM 11.13.25 (Graybill and Milliken (1969)). Continue Problem 11.13.24. In the two-way classification we let $n = IJ$ and speak of the ij-component of $\mathbf{X}$. Our model is

$$\mathbf{E}\mathbf{X} = \{\phi_{ij}\}, \qquad 1 \le i \le I, 1 \le j \le J. \tag{11.13.25}$$

subject to the side conditions

$$I^{-1}\sum_{i=1}^{I} \phi_{ij} = J^{-1}\sum_{j=1}^{J} \phi_{ij}, \quad \text{i.e., } \phi_{i.} = \phi_{.j}. \tag{11.13.26}$$

A test due to Tukey is a test of whether the interactions $\phi_{ij} - \phi_{i.} - \phi_{.j} + \phi_{..}$ all vanish. We assume that they do all vanish so that $\mathbf{E}\mathbf{X} = \phi_{i.} + \phi_{.j} - \phi_{..}$. The statistic is described as follows.

$$s_1^2 = \frac{\left(\sum_{i=1}^{I}\sum_{j=1}^{J}(x_{ij} - x_{i.} - x_{.j} + x_{..})(x_{i.} - x_{..})(x_{.j} - x_{..})\right)^2}{\sum_{i=1}^{I}(x_{i.} - x_{..})^2 \sum_{j=1}^{J}(x_{.j} - x_{..})^2}, \tag{11.13.27}$$

and

$$s_2^2 = \sum_{i=1}^{I} \sum_{j=1}^{J} (x_{ij} - x_{i.} - x_{.j} + x_{..})^2.$$

The numerator of the F-test is $s_1^2 = \mathbf{X}^t\mathbf{AX}$ and the denominator is $s_2^2 - s_1^2 = \mathbf{X}^t\mathbf{CX}$. The problem is to obtain descriptions of these statistics in the notations of Problem 11.13.24. Let P_1 and P_2 be the $IJ \times IJ$ orthogonal projection matrices such that

$$(P_1X)_{ij} = X_{.j} \quad \text{and} \quad (P_2X)_{ij} = X_{i.}. \tag{11.13.28}$$

Then $P = P_1P_2 = P_2P_1$ is a rank one projection matrix. Then

$$I_n - B(B^tB)^+B^t = L = I_n - P_1 - P_2 + P = (I_n - P_1)(I_n - P_2). \tag{11.13.29}$$

The matrix Q is an $n \times 1$ matrix of rank 1 defined to have ij-entry equal $x_{i.}x_{.j}$. This is a function of B^tX. Then

$$LQ \text{ has } ij\text{-entry } (x_{i.} - x_{..})(x_{.j} - x_{..}). \tag{11.13.30}$$

Find $((LQ)^t(LQ))^+$ and the matrix $A = LAL = L^tAL$. Since

the ij-entry of LX is $x_{ij} - x_{i.} - x_{.j} + x_{..}$

it follows that $s_1^2 = \mathbf{X}^t\mathbf{AX}$ and $s_2^2 - s_1^2 = \mathbf{X}^t\mathbf{BX}$. (11.13.31)

What is the rank of $\mathbf{A}$?

By Problem 11.13.24 the statistic

$$s_1^2/(s_2^2 - s_1^2) = \chi_1^2/\chi_{(I-1)(J-1)-1}^2 \tag{11.13.32}$$

is an unnormalized central F-statistic. In case the second order interactions do not vanish then $B = I_{IJ}$ and the above analysis no longer applies.

CHAPTER 12

The Construction of Zonal Polynomials

12.0. Introduction

The discussion of previous chapters has shown that in many of the noncentral problems the answer involves an integral that cannot be evaluated in closed form, typically an integral over a locally compact topological group with respect to a Haar measure. In this book the groups have been matrix groups. For example the discussion of James (1955a) given in Chapter 4 obtains the probability density function of the noncentral Wishart distribution in terms of an integral $\int_{\mathbf{O}(n)} \exp(\operatorname{tr} X^t HM)\, dH$, where dH means the Haar measure of unit mass on $\mathbf{O}(n)$. The possibility of deriving noncentral density functions by integration with respect to Haar measures is further discussed in Chapters 5, 9 and 10 where also examples of integrals over $\mathbf{T}(n)$ and $\mathbf{GL}(n)$ have been given.

The theory of zonal polynomials allows the evaluation of some of these integrals in terms of infinite series whose summands are a coefficient multiplying a zonal polynomial. The general theory was presented in a series of papers in the years 1960, 1961, 1962, 1963, 1964, and 1968 by James and the papers by Constantine (1963, 1966). The second paper by Constantine is really about a related topic of Laguerre polynomials of a matrix argument. In the 1963 paper Constantine produced some very basic computational formulas without which the series representation using zonal polynomials would not have succeeded. Constantine's paper showed that the series being obtained in the multivariate problems were in fact series representations of hypergeometric functions as defined by Herz (1955). Herz also defined Laguerre polynomials and the 1966 paper by Constantine is a further development of this theory.

Roughly speaking the substance of this chapter is the necessary algebra needed for the existence and uniqueness theory of the polynomials. The chapter provides sufficient background for the reading of Constantine (1963) and James (1964) but the reader should refer to these papers for the actual calculation of multivariate examples. The original definition of a zonal polynomial in James (1960) defined the polynomials implicitly using group representations. The main theoretical paper by James was James (1961b) in which enough combinatorial analysis was done to explicitly calculate the polynomials of low degree and to give an algorithm for the calculation of polynomials of higher degree. James (1964) is a survey paper giving a complete summary of all results known to Constantine and James at that time. This remarkable survey paper is apparently completely without error but contains a number of unproven assertions. James has told this author that some results and proofs about zonal polynomials of complex matrix argument known to James were never submitted as James did not feel these results were sufficiently useful for publication. Proofs of some of the unproved results of James (1964) as well as results on polynomials of complex matrix argument may be found in this chapter. Algorithms for the computation of zonal polynomials are discussed in Sections 12.12, 12.13 and again in Section 13.5. The results of Section 12.12 derive from the ideas of Saw (1977), those of Section 12.13 derive from the Takemura (1982) development of Saw's idea into a complete development. Over the years James (1961b, 1964, 1968) developed three essentially different computational algorithms which have produced the same polynomials, which is evidence for the correctness of the existing tables and methods. Section 13.5 discusses the 1961b algorithm. In the survey paper, James (1964), he makes reference to Helgason (1962) which if followed through implies that the zonal polynomials of real matrix argument are spherical functions in the meaning of Helgason. This fact is used implicitly in James (1968) where the Laplace–Beltrami operator is used to derive a differential equation for the zonal polynomials. Solution of the equation, which was the third algorithm of James, represented the polynomials in terms of the monomial symmetric functions and provided the first proof that the coefficients were all nonnegative. By contrast the first two algorithms of James in the earlier papers were combinatorial in character. To this author it has not seemed possible to pull these results into a coherent readable form based on group representations. However if one takes the algebra as developed by Weyl (1946) and makes a direct algebraic attack on the subject a unified self-contained presentation does result. The subject is very deep and not rewarding to any but the brightest students. However it should be noted that parts of the tensor algebra discussed in the sequel can be taken out of the present context and used in subjects like the analysis of variance.

The subject as originally discussed by James was about polynomials of matrix argument that had two-sided orthogonal invariance. The result can then be reduced to a discussion of polynomials of a symmetric positive semidefinite matrix invariant under transformations $A \to UAU^t$, $U \in \mathbf{O}(k)$, and such polynomials are in one-to-one correspondence with the symmetric homogeneous polynomials in the eigenvalues of A. The more recent treatments of Saw (1977) and Takemura (1982) ignore group representations and the subject of polynomials of matrix argument and work directly with symmetric functions of real and/or complex variables. This approach produces a usable theory for invariant functions of a single matrix argument but apparently fails to obtain all the formulae that James obtained through group representation considerations. Also, when one considers invariant polynomials of two matrix arguments, as has been done by Davis (1981), the use of group representations produces a theory whereas the algebraic approach of this chapter does not seem to easily extend. Alternative expositions are available in Muirhead (1982), who has made extensive use of the partial differential equations satisfied by the hypergeometric functions, and in Takemura (1982).

James in his development of the subject depended heavily on Littlewood (1940, 1950) for the theory of algebras which decompose into a direct sum of closed minimal ideals. A more modern presentation using normed algebras may be found in Loomis (1953) which helps in getting more directly to the essential things.

An integral

$$\int_{\mathbf{O}(n)} \exp(\operatorname{tr} HX)\, dH = \sum_{k=0}^{\infty} (k!)^{-1} \int_{\mathbf{O}(n)} (\operatorname{tr} HX)^k\, dH \tag{12.0.1}$$

is an infinite series of homogeneous polynomials of even degree. Our theory is therefore a theory of homogeneous polynomials of degree $2k$ in the variables of a matrix X. Note that if k is odd

$$\int_{\mathbf{O}(n)} (\operatorname{tr} HX)^k\, dH = \int_{\mathbf{O}(n)} (\operatorname{tr} -HX)^k\, dH = -\int_{\mathbf{O}(n)} (\operatorname{tr} HX)^k\, dH, \tag{12.0.2}$$

so that the odd degree terms vanish. The theory which follows establishes an isomorphism between *bi*-symmetric linear transformations and the homogeneous polynomials. The *bi*-symmetric linear transformations are an H^* algebra in the meaning of Loomis (1953) and the space of *bi*-symmetric transformations is a direct sum of its minimal closed ideals.

The *bi*-symmetric linear transformations act on $\mathbf{M}(\mathbf{E}^k, \mathbb{C})$, the space of *bi*-linear k-forms over the complex numbers $\mathbb{C}$. Within the algebra of endomorphisms of $\mathbf{M}(\mathbf{E}^k, \mathbb{C})$ the commutator algebra with the *bi*-symmetric transformations has a very special form which plays a key role in the theory. This algebra is a representation of the group algebra of the symmetric group. A complete description may be obtained from Weyl (1946). We use Weyl as our source.

The idempotents in the center of the commutator algebra are directly identifiable with the zonal polynomials of complex transpose symmetric matrices. This fills in a part of the theory mentioned without proof by James (1964). When X is an $n \times n$ matrix and we are considering polynomials of degree k the idempotents in the center of the commutator algebra are in one-to-one correspondence with partitions of the integer k into not more than n parts. This correspondence is established using Young's diagrams, discussed below.

Each polynomial $\int_{\mathbf{O}(n)} (\operatorname{tr} HX)^k \, dH = f(X)$ satisfies, if $G_1, G_2 \in \mathbf{O}(n)$ then

$$f(X) = f(G_1 X G_2) = f(G_2^t (X^t X)^{1/2} G_2) = f(\operatorname{diag}(x_1, \ldots, x_n)), \tag{12.0.3}$$

where we write $x_1, \ldots, x_n$ for the eigenvalues of $(X^t X)^{1/2}$. From this it is easily seen that f is a homogeneous symmetric polynomial of degree k in $x_1, \ldots, x_n$, which vanishes if k is odd. In fact (see the problems of Chapter 4) for even exponents $2k$, with $H = (h_{ij})$,

$$f(X) = \int_{\mathbf{O}(n)} (h_{11} x_{\sigma(1)} + \cdots + h_{nn} x_{\sigma(n)})^{2k} \, dH \tag{12.0.4}$$

valid for all permutations σ of $1, \ldots, n$, so that f is a polynomial in the symmetric functions

$$\operatorname{tr} X^t X, \operatorname{tr} (X^t X)^2, \ldots, \operatorname{tr} (X^t X)^n. \tag{12.0.5}$$

Therefore part of the algebraic theory is a theory of polynomials which are symmetric functions. Our sources were Littlewood (1940) and Weyl (1946), with a more recent book Macdonald (1979).

Zonal polynomials as defined by James have the invariance property (12.0.3) and form a basis for the symmetric functions. Therefore, given the degree $2k$, the number of zonal polynomials of that degree is the number of linearly independent symmetric polynomials of degree k in n variables. As a function of k this number grows rapidly. James and Parkhurst (1974) have published tables of the polynomials of degree less than or equal twelve. The number of polynomials of degree twelve is almost prohibitive to the table maker, and because of loss of digits of accuracy through subtractions, integer coefficients had to be computed with multiprecision accuracy, which, the authors explain, was unexpected and delayed the original publication of the tables.

Zonal polynomials of a complex matrix argument are more easily defined than are the zonal polynomials of a real matrix argument. The definition is given in Section 12.7, James knew, and Farrell (1980) showed that these are the polynomial group characters of $\mathbf{GL}(n)$. Section 12.10 gives an algebraic definition of the zonal polynomials of real matrix argument along the lines of James (1961b). Section 12.11 gives a definition in terms of spherical functions and integrals of group characters as suggested by James (1964). Sections 12.12 and 13.6 explore the consequence of an idea of Saw (1977) while Saw's idea is perfected by Takemura (1982), as discussed in Section 12.13, resulting in a definition and version of the theory that is more easily taught to the uninitiated. Kates (1980), in his Princeton Ph.D. thesis, obtained integral formulas giving the zonal polynomials. These formulas were discussed by Takemura, and we adapt Takemura's development in Section

12.14. The relationship of the group algebra of the symmetric group on n letters and the various natural bases for the symmetric polynomials of degree k in n variables is given an elegant treatment in Garsia and Remmel (1981) where the use of quadratic forms in the establishment of an isomorphism is fully exploited.

Zonal polynomials as special functions are one way of expressing, in infinite series, the hypergeometric functions. James (1964) gave an extensive description of various noncentral multivariate density functions in terms of the hypergeometric functions. Muirhead (1982) gives a contemporary presentation and in his statement of results gives the description of the relevant hypergeometric function that applies to each noncentral problem.

12.1. Kronecker Products and Homogeneous Polynomials

We assume $\mathbf{E}$ is an n-dimensional vector space over the complex numbers $\mathbb{C}$ with fixed basis $e_1, \ldots, e_n$ and let $u_1, \ldots, u_n$ be the canonical basis of the dual space. As in Chapter 6 we let $\mathbf{M}(\mathbf{E}^k, \mathbb{C})$ be the space of multilinear k-forms with coefficients in the complex numbers. Linear transformations $X_1, \ldots, X_k$ of $\mathbf{E}$ induce a linear transformation $X_1 \otimes X_2 \otimes \cdots \otimes X_k$ of $\mathbf{M}(\mathbf{E}^k, \mathbb{C})$. Here, relative to the fixed basis, linear transformations will be represented by the transformation's matrix which allows us to make the definition.

Definition 12.1.1. If $f \in \mathbf{M}(\mathbf{E}^k, \mathbb{C})$ then

$$\begin{aligned}(X_1 \otimes \cdots \otimes X_k)(f)(e_{i_1}, \ldots, e_{i_k}) &= f(X_1^t e_{i_1}, \ldots, X_k^t e_{i_k}),\\ 1 \le i_1, \ldots, i_k \le n.&\end{aligned} \tag{12.1.1}$$

We will think of this definition as defining a matrix $X_1 \otimes \cdots \otimes X_k$, and below we compute the entries of this matrix. In (12.1.1) the use of the transpose enters because of the bracket product

$$[Xu_i, e_j] = [u_i, X^t e_j]. \tag{12.1.2}$$

If

$$X_p^t e_i = \sum_{j=1}^{n} (X_p^t)_{ij} e_j, \tag{12.1.3}$$

then one obtains

$$\begin{aligned}&(X_1 \otimes \cdots \otimes X_k)(u_{j_1} \cdots u_{j_k})(e_{i_1}, \ldots, e_{i_k})\\ &= (u_{j_1} \cdots u_{j_k})\left(\sum_{h=1}^{n} (X_1^t)_{i_1 h} e_h, \ldots, \sum_{h=1}^{n} (X_k^t)_{i_k h} e_h\right)\\ &= (X_1^t)_{i_1 j_1} (X_2^t)_{i_2 j_2} \cdots (X_k^t)_{i_k j_k},\ 1 \le i_1, \ldots, i_k, j_1, \ldots, j_k \le n.\end{aligned} \tag{12.1.4}$$

The identity (12.1.4) gives the $((i_1, \ldots, i_k), (j_1, \ldots, j_k))$-entry of the Kronecker product matrix $X_1 \otimes \cdots \otimes X_k$. Following are some useful lemmas.

Lemma 12.1.2.

$$\operatorname{tr} X_1 \otimes \cdots \otimes X_k = \prod_{i=1}^{k} (\operatorname{tr} X_i).$$

Lemma 12.1.3.

$$X_1^t \otimes \cdots \otimes X_k^t = (X_1 \otimes \cdots \otimes X_k)^t.$$

Lemma 12.1.4.

$$(X_1 \otimes \cdots \otimes X_k)(Y_1 \otimes \cdots \otimes Y_k) = (X_1 Y_1) \otimes \cdots \otimes (X_k Y_k).$$

Note that the use of transposition is needed in Definition 12.1.1 *in order that Lemma* 12.1.4 *be correct. In the sequel, if* $c \in \mathbb{C}$ *then* $\bar{c}$ *is the complex conjugate of* c.

Lemma 12.1.5.

$$\bar{X}_1 \otimes \cdots \otimes \bar{X}_k = \overline{X_1 \otimes \cdots \otimes X_k}.$$

Lemma 12.1.6. *If A is an $n_1 \times n_1$ matrix and B is an $n_2 \times n_2$ matrix then*

$$\det A \otimes B = (\det A)^{n_2} (\det B)^{n_1}. \tag{12.1.5}$$

PROOF. It is easily verified that $\det I_{n_1} \otimes B = (\det B)^{n_1}$, and that, using Lemma 12.1.4, $A \otimes B = (I_{n_1} \otimes B)(A \otimes I_{n_2})$. □

We will be interested mostly in the case that $X_1 = X_2 = \cdots = X_n = X$. The $((i_1, \ldots, i_k), (j_1, \ldots, j_k))$-entry of $X \otimes \cdots \otimes X$ is $(X)_{i_1 j_1} \cdots (X)_{i_k j_k}$ which is a homogeneous polynomial of degree k in the entries of X. The arbitrary homogeneous polynomial of degree k is

$$\Sigma \cdots \Sigma a_{i_1 j_1 \cdots i_k j_k} (X)_{i_1 j_1} \cdots (X)_{i_k j_k} = \operatorname{tr} A^t (X \otimes \cdots \otimes X), \tag{12.1.6}$$

where we use the observation that if $B = (b_{ij})$ and $C = (c_{ij})$ then

$$\operatorname{tr} B^t C = \Sigma_i \Sigma_j b_{ij} c_{ij}. \tag{12.1.7}$$

Usually there will be several different coefficient matrices A resulting in the same homogeneous polynomial. Uniqueness is introduced by requiring A to be a *bi*-symmetric matrix. To define this property we consider the action of permutations σ of $1, 2, \ldots, k$. We let σ act on $\mathbf{M}(\mathbf{E}^k, \mathbb{C})$ by means of the definition of P_σ, to be extended by linearity, that

$$(P_\sigma f)(e_{i_1}, \ldots, e_{i_k}) = f(e_{i_{\sigma(1)}}, \ldots, e_{i_{\sigma(k)}}), \tag{12.1.8}$$
$$1 \le i_1, \ldots, i_k \le n.$$

Note that $P_\sigma(u_{j_1} \cdots u_{j_k})(e_{i_1}, \ldots, e_{i_k}) = u_{j_1}(e_{i_{\sigma(1)}}) \cdots u_{j_k}(e_{i_{\sigma(k)}})$, and that this is zero unless

$$j_1 = i_{\sigma(1)}, \ldots, j_k = i_{\sigma(k)}. \tag{12.1.9}$$

Therefore P_σ is a permutation of the canonical basis of $\mathbf{M}(\mathbf{E}^k, \mathbb{C})$ and P_σ is therefore an orthogonal matrix. In particular, $P_\sigma P_\sigma^t$ is the identity matrix.

Lemma 12.1.7.

$$P_\sigma(X \otimes \cdots \otimes X) = (X \otimes \cdots \otimes X)P_\sigma.$$

PROOF. If $f \in \mathbf{M}(\mathbf{E}^k, \mathbb{C})$ then, with $g = (X \otimes \cdots \otimes X)(f)$,

$$\begin{aligned} P_\sigma(X \otimes \cdots \otimes X)(f)(e_{i_1}, \ldots, e_{i_k}) &= P_\sigma(g)(e_{i_1}, \ldots, e_{i_k}) \\ &= g(e_{i_{\sigma(1)}}, \ldots, e_{i_{\sigma(k)}}) \\ &= f(X^t e_{i_{\sigma(1)}}, \ldots, X^t e_{i_{\sigma(k)}}) \\ &= (P_\sigma f)(X^t e_{i_1}, \ldots, X^t e_{i_k}) \\ &= (X \otimes \cdots \otimes X)(P_\sigma f)(e_{i_1}, \ldots, e_{i_k}), \end{aligned} \tag{12.1.10}$$

and this holds if $1 \le i_1, \ldots, i_k \le n$. □

Definition 12.1.8. A matrix A representing an element of the endomorphisms End $\mathbf{M}(\mathbf{E}^k, \mathbb{C})$, in the basis $u_{j_1}, \ldots, u_{j_k}$, $1 \le j_1, \ldots, j_k \le n$, is said to be a *bi*-symmetric matrix if and only if for all permutations σ of $1, \ldots, k$,

$$P_\sigma A = A P_\sigma. \tag{12.1.11}$$

Lemma 12.1.9. *Let $B = (k!)^{-1}\Sigma_\sigma P_\sigma A P_\sigma^t$. Then as polynomials in the variables of the matrix X the polynomials*

$$\operatorname{tr} AX \otimes \cdots \otimes X \quad \text{and} \quad \operatorname{tr} BX \otimes \cdots \otimes X \tag{12.1.12}$$

are the same polynomial. The matrix B is bi-symmetric.

Lemma 12.1.10. *If a homogeneous polynomial of degree k in n variables is identically zero then all coefficients are zero (coefficients are complex numbers).*

Lemma 12.1.11. *The set of bi-symmetric matrices is the linear span of the matrices of the form $X \otimes \cdots \otimes X$.*

PROOF. If not, there is a linear functional on End $\mathbf{M}(\mathbf{E}^k, \mathbb{C})$, which we represent by a matrix B, and a bi-symmetric matrix A, such that if X is $n \times n$ then

$$\operatorname{tr} B^t X \otimes \cdots \otimes X = 0 \quad \text{and} \quad \operatorname{tr} B^t A \neq 0. \tag{12.1.13}$$

Since

$$0 \neq \operatorname{tr} B^t A = \operatorname{tr} P_\sigma B^t A P_\sigma^t = \operatorname{tr} P_\sigma B^t P_\sigma^t A, \tag{12.1.14}$$

it follows that

$$0 \neq \operatorname{tr}((k!)^{-1}\Sigma_\sigma P_\sigma B P_\sigma^t)^t A, \tag{12.1.15}$$

and therefore we may assume the matrix B to be *bi*-symmetric. By hypothesis and Lemma 12.1.10, all coefficients of the polynomial $\operatorname{tr} B^t X \otimes \cdots \otimes X$ are zero. The term $x_{i_1 j_1} \cdots x_{i_k j_k}$ occurs also as $x_{i_{\sigma(1)} j_{\sigma(1)}} \cdots x_{i_{\sigma(k)} j_{\sigma(k)}}$ and these are the only occurrences, so the coefficient involved is

$$\begin{aligned} 0 &= \Sigma_\sigma b_{i_{\sigma(1)} \cdots i_{\sigma(k)} j_{\sigma(1)} \cdots j_{\sigma(k)}} \\ &= (k!) b_{i_1 \cdots i_k j_1 \cdots j_k}, \end{aligned} \tag{12.1.16}$$

since B is *bi*-symmetric. Therefore $B = 0$. This contradiction shows that the conclusion of the lemma must hold. □

We summarize the results above in a theorem.

Theorem 12.1.12. *The vector space of homogeneous polynomials of degree k in the variables of X is isomorphic to the vector space of bi-symmetric matrices under the representation* (12.1.6).

Lemma 12.1.13. *The product of two bi-symmetric matrices is a bi-symmetric matrix. Hence, under the matrix operations of addition, scalar multiplication and matrix multiplication, the bi-symmetric matrices form an algebra over the complex numbers which is closed under conjugation and transposition.*

We now show that the algebra of *bi*-symmetric matrices is an H^* algebra in the sense of Loomis (1953). The required inner product is defined by

Definition 12.1.14.

$$(A, B) = \operatorname{tr}(\bar{B})^t A = \operatorname{tr} A(\bar{B})^t.$$

Theorem 12.1.15. *The bilinear functional (,) is an inner product under which the bi-symmetric matrices are an H^* algebra.*

PROOF. The involution $A^* = (\bar{A})^t$, i.e., the conjugate transpose, clearly satisfies

$$A^{**} = A; \tag{12.1.17a}$$

$$(A + B)^* = A^* + B^*; \tag{12.1.17b}$$

$$\text{if } c \in \mathbb{C} \text{ then } (cA)^* = \bar{c}A^*; \tag{12.1.17c}$$

$$(AB)^* = B^* A^*. \tag{12.1.17d}$$

Further,

$$(AB, C) = \mathrm{tr}\,(AB)C^* = \mathrm{tr}\,B(C^*A) = \mathrm{tr}\,B(A^*C)^* = (B, A^*C), \tag{12.1.18}$$

and

$$\|A\|^2 = \mathrm{tr}\,AA^* = \mathrm{tr}\,A^*A^{**} = \|A^*\|^2. \tag{12.1.19}$$

Also, $A^*A = 0$ implies $\mathrm{tr}\,AA^* = 0$, that is $\|A\| = 0$ and $A = 0$. Therefore the defining conditions given by Loomis are satisfied.

By Lemma 12.1.13, the inner product (,) is defined on the algebra of *bi*-symmetric matrices, so this algebra is also an H^* algebra. □

Remark 12.1.16. The algebra of *bi*-symmetric matrices is finite dimensional. By the results on H^* algebras in Loomis, op. cit., it follows that the algebra of *bi*-symmetric matrices is a direct sum of its minimal closed ideals. Each minimal closed ideal is isomorphic to a full matrix algebra with unit. Each ideal, being a finite dimensional subspace, is closed topologically.

12.2. Symmetric Polynomials in *n* Variables

The material of this section is taken from Littlewood (1950) and Weyl (1946). Following the notations of Littlewood, given $c_1, \ldots, c_n \in \mathbb{C}$ we define a polynomial

$$x^n f(x^{-1}) = \prod_{i=1}^{n} (x - c_i) = x^n - a_1 x^{n-1} + \cdots + (-1)^n a_n. \tag{12.2.1}$$

As is well known the coefficients $a_1, \ldots, a_n$ are the elementary symmetric functions of $c_1, \ldots, c_n$. We note that

$$\begin{aligned} F(x) = 1/f(x) &= \prod_{i=1}^{n} (1/(1 - c_i x)) \\ &= \prod_{i=1}^{n} (1 + c_i x + c_i^2 x^2 + \cdots) \\ &= 1 + h_1 x + h_2 x^2 + \cdots + h_m x^m + \cdots. \end{aligned} \tag{12.2.2}$$

The coefficient h_m is a symmetric polynomial of degree m which is the sum of the homogeneous products of degree m of $c_1, \ldots, c_n$. Also,

$$\ln f(x) = \sum_{i=1}^{n} \ln (1 - c_i x), \tag{12.2.3}$$

so that by taking derivatives

$$\begin{aligned} f'(x)/f(x) &= \sum_{i=1}^{n} -c_i/(1 - c_i x) \\ &= -\sum_{i=1}^{n} (c_i + c_i^2 x + \cdots + c_i^{m+1} x^m + \cdots) \qquad (12.2.4) \\ &= -(S_1 + S_2 x + \cdots + S_{m+1} x^m + \cdots). \end{aligned}$$

In (12.2.4) the term $S_m = \Sigma_{i=1}^n c_i^m$. From the relation (12.2.4) upon substitution of (12.2.1) for f and f', we obtain

$$(a_1 - 2a_2 x + 3a_3 x^2 - \cdots) = (1 - a_1 x + a_2 x^2 - \cdots)(S_1 + S_2 x + \cdots). \qquad (12.2.5)$$

By matching coefficients we obtain the Newton identities, which are stated in the next lemma.

Lemma 12.2.1.

$$ma_m = S_1 a_{m-1} - S_2 a_{m-2} + S_3 a_{m-3} - \cdots + (-1)^{m+1} S_m. \qquad (12.2.6)$$

It follows from the identities (12.2.6) *that* $a_1, \ldots, a_m$ *are polynomials of the variables* $S_1, \ldots, S_m$ *and that by solving the equations,* $S_1, \ldots, S_m$ *are polynomials of the variables* $a_1, \ldots, a_m$. *In particular every term* $S_{i_1} \cdots S_{i_p}$ *entering into* a_m *is homogeneous of degree* m, *i.e.,* $i_1 + \cdots + i_p = m$, *and similarly when expressing* S_m *as a polynomial of the variables* $a_1, \ldots, a_m$.

Lemma 12.2.2. *The symmetric functions* $a_1, \ldots, a_n$ *of the variables* $c_1, \ldots, c_n$ *are functionally independent in the sense that, if* F *is a polynomial of* n *variables over* $\mathbb{C}$ *such that* $F(a_1, \ldots, a_n)$ *vanishes identically in the variables* $c_1, \ldots, c_n$, *then* $F = 0$.

Proof. Given complex numbers $a_1, \ldots, a_n$ the polynomial $x^n - a_1 x^{n-1} + \cdots + (-1)^n a_n$ has n complex roots $c_1, \ldots, c_n$. Therefore the mapping $(c_1, \ldots, c_n) \to (a_1, \ldots, a_n)$ is onto $\mathbb{C}^n$. If F is a polynomial of n variables such that

$$F(a_1(c_1, \ldots, c_n), \ldots, a_n(c_1, \ldots, c_n)) \equiv 0, \qquad (12.2.7)$$

then $F(a_1, \ldots, a_n) = 0$ for every n-tuple of complex numbers. Hence F is the zero function, i.e., $F = 0$. □

Lemma 12.2.3. *The symmetric functions* $S_1, \ldots, S_n$, *as functions of* $c_1, \ldots, c_n$, *are functionally independent. As functions of* $a_1, \ldots, a_n$, *they are functionally independent.*

Proof. The same as above after noting from the Newton identities (12.2.6) that the mapping $(a_1, \ldots, a_n) \to (S_1, \ldots, S_n)$ is onto $\mathbb{C}^n$. □

If an integer $k > 0$ is a sum of nonnegative integers $k_1, \ldots, k_n$ such that $k_1 \geq k_2 \geq \cdots \geq k_n$ then we write κ for the ordered set $k_1, \ldots, k_n$ which is called a partition of k into not more than n parts. Some authors ask that $k_n > 0$ but our results are more easily stated if $k_n \geq 0$ is assumed.

Partitions κ_1 and κ_2 of k given by $k_{11} \geq \cdots \geq k_{1n}$ and $k_{21} \geq \cdots \geq k_{2n}$ respectively may be linearly ordered as follows. Let i be the least integer such that $k_{1i} \neq k_{2i}$. Then $\kappa_1 > \kappa_2$ if and only if $k_{1i} > k_{2i}$. This is the *lexicographic* ordering of partitions. In using this definition, note that if κ_1 has m nonzero terms and $\kappa_1 \geq \kappa_2$ then either $k_{2\,m+1} = 0$ or else $\kappa_1 > \kappa_2$. With this understanding it will generally not be necessary to state the number of nonzero terms in a partition.

From a partition κ we may define

$$m_n = k_n \quad \text{and} \quad m_i = k_i - k_{i+1}, \qquad 1 \leq i \leq n-1; \tag{12.2.8}$$

$$\pi_i = \text{the number of } j \text{ such that } k_j = i,\ 1 \leq i \leq k.$$

Then

$$k = m_1 + 2m_2 + \cdots + nm_n, \tag{12.2.9}$$

and

$$k_i = m_i + \cdots + m_n,\ 1 \leq i \leq n.$$

In the sequel we begin to eliminate subscripts in favor of vector and matrix representation and will write $c^t = (c_1, \ldots, c_n)$.

Definition 12.2.4. The monomial symmetric function M_κ associated with the partition κ of k is computed by

$$M_\kappa(c) = \Sigma \cdots \Sigma c_{i_1}^{k_1} c_{i_2}^{k_2} \cdots c_{i_n}^{k_n} \tag{12.2.10}$$

taken over all distinct monomial terms. Note that if κ has only m nonzero parts then the sum in (12.2.10) is interpreted as being over all products of m factors, not of n factors.

From (12.2.1) the elementary symmetric functions $a_1, \ldots, a_n$ are functions of the variables c.

Definition 12.2.5. For the partition κ with $m_1, \ldots, m_n$ defined in (12.2.8), define

$$A_\kappa(c) = a_1^{m_1} a_2^{m_2} \cdots a_n^{m_n}. \tag{12.2.11}$$

Here if κ has m nonzero parts then A_κ depends only on $a_1, \ldots, a_m$ and is a homogeneous polynomial of degree k in the variables c.

Definition 12.2.6. For the partition κ define

$$S_\kappa(c) = S_1^{m_1} \cdots S_n^{m_n} \quad \text{where} \quad S_j = \sum_{i=1}^{n} c_i^j, \qquad 1 \leq j \leq n. \tag{12.2.12}$$

Definition 12.2.7. For the partition κ define

$$J_\kappa(c) = S_1^{\pi_1} S_2^{\pi_2} \cdots S_k^{\pi_k} = S_{k_1} S_{k_2} \cdots S_{k_n}. \tag{12.2.13}$$

Remark 12.2.8. The functions M_κ, A_κ, S_κ, and J_κ are symmetric functions of $c_1, \ldots, c_n$, each a homogeneous function of degree k.

Remark 12.2.9. Partitions may be added. If γ and κ are given by $g_1 \geq \cdots \geq g_m$ and $k_1 \geq \cdots \geq k_n$ and if $m \geq n$ then the parts of $\gamma + \kappa$ are $g_i + k_i$, $1 \leq i \leq m$, where $k_i = 0$ if $i > n$. With this interpretation we may also define $\gamma \vee \kappa$ as the partition obtained by ordering $g_1, \ldots, g_m, k_1, \ldots, k_n$.

The following lemma is easy to establish.

Lemma 12.2.10.

$$\begin{aligned} A_\gamma A_\kappa &= A_{\gamma+\kappa}; \\ S_\gamma S_\kappa &= S_{\gamma+\kappa}; \\ J_\gamma J_\kappa &= J_{\gamma\vee\kappa}. \end{aligned} \tag{12.2.14}$$

Proofs of the "classical" basis theorems are made easier by the following observation.

Lemma 12.2.11. *If $\gamma_1 \leq \gamma_2$ and $\kappa_1 \leq \kappa_2$ then $\gamma_1 + \kappa_1 \leq \gamma_2 + \kappa_2$ and $\gamma_1 \vee \kappa_1 \leq \gamma_2 \vee \kappa_2$.*

In terms of the linear ordering of partitions into less than or equal n parts we may speak of the number of distinct partitions and will write r for the number of distinct partitions of k into not more than n parts.

Definition 12.2.12.

$$\begin{aligned} M^t &= (M_{\kappa_1}, \ldots, M_{\kappa_r}); \\ A^t &= (A_{\kappa_1}, \ldots, A_{\kappa_r}); \\ S^t &= (S_{\kappa_1}, \ldots, S_{\kappa_r}); \\ J^t &= (J_{\kappa_1}, \ldots, J_{\kappa_r}); \end{aligned} \tag{12.2.15}$$

where $\kappa_1 > \kappa_2 > \cdots > \kappa_r$ are the partitions ordered in decreasing order.

We now state three theorems, then give proofs.

Theorem 12.2.13. *There exists a nonsingular lower triangular $r \times r$ matrix T_A with integer entries, diagonal entries equal one, such that*

$$A^t = M^t T_A. \tag{12.2.16}$$

The symmetric functions $A_{\kappa_1}, \ldots, A_{\kappa_r}$ are a basis of the space of homogeneous symmetric polynomials of degree k in $c_1, \ldots, c_n$.

Theorem 12.2.14. *There exists a nonsingular lower triangular* $r \times r$ *matrix* T_S *with integer entries such that*

$$S^t = A^t T_S^t. \tag{12.2.17}$$

Thus the entries of S *are a basis for the homogeneous symmetric polynomials of degree* k. *The diagonal entry of* T_S *in the* κ, κ *position is*

$$(-1)^{k+k_1} 1^{m_1} 2^{m_2} \cdots n^{m_n}. \tag{12.2.18}$$

Theorem 12.2.15. *There exists a nonsingular lower triangular* $r \times r$ *matrix* T_J *with integer entries such that*

$$J^t = M^t T_J^t. \tag{12.2.19}$$

Remark 12.2.16. The numbers $\pi_1, \ldots, \pi_k$ may be viewed as giving a conjugate partition γ of k into k parts none of which exceed n in value, and defined by $g_i = \pi_i + \cdots + \pi_k$. Theorem 12.2.14 may then be applied to express J in terms of A with a triangular coefficient matrix.

PROOF OF 12.2.13. For the functions A_κ, if on expansion the monomial term $c_1^{g_1} c_2^{g_2} \cdots c_n^{g_n}$ results then $g_1 \le k_1$. If $g_1 = k_1$ then $g_2 \le k_2$, and so on. Thus the partition γ is less than or equal κ, which in matrix terms is (12.2.16). In the expansion of A_κ using the binomial theorem the coefficients are seen to be nonnegative integers. The lead term $c_1^{k_1} c_2^{k_2} \cdots c_n^{k_n}$ occurs once, hence the diagonal entries are as stated. □

PROOF OF 12.2.14. A partition $g_1 = \cdots = g_k = 1$, $g_{k+1} = \cdots = g_n = 0$ of k is minimal and cannot be expressed as a sum of two partitions. In this case $m_1 = m_2 = \cdots = m_{k-1} = 0$, $m_k = 1$, and $m_{k+1} = \cdots = m_n = 0$. Thus $S_\gamma = S_k = c_1^k + \cdots + c_n^k$. By the Newton identities (12.2.6) S_k is a polynomial in $a_1, \ldots, a_k$ homogeneous of degree k, and each monomial of degree k in $a_1, \ldots, a_k$ corresponds to a partition of k greater than or equal 1^k, the minimal partition.

Suppose then that $k_1 \ge 2$ and $\kappa = \gamma + \rho$. By inductive hypothesis $S_\gamma = \Sigma_{\gamma' \ge \gamma} c'_{\gamma'} A_{\gamma'}$ and $S_\rho = \Sigma_{\rho' \ge \rho} c''_{\rho'} A_{\rho'}$ so that by Lemma 12.2.10, $S_\kappa = S_\gamma S_\rho = \Sigma\, c'_{\gamma'} c''_{\rho'} A_{\gamma' + \rho'}$ and $\gamma' + \rho' \ge \gamma + \rho = \kappa$.

In the Newton identities, solving for $S_1, \ldots, S_k$ as polynomial functions of $a_1, \ldots, a_k$, the coefficients are integers and the coefficient of a_k, corresponding to the partition 1^k, is $(-1)^{k+1} k$. Thus by induction the coefficient of A_κ in the expansion of S_κ is $(-1)^{(\Sigma k m_k) + (\Sigma m_k)} 1^{m_1} 2^{m_2} \cdots n^{m_n} =$ (12.2.18). □

PROOF OF 12.2.15. For the J-functions, let $S_0 = 1$ and suppose $S_{k_1} S_{k_2} \cdots S_{k_n}$ contains the monomial term $c_{i_1}^{k_1} \cdots c_{i_n}^{k_n}$. Other monomial terms from

$S_{k_1} \cdots S_{k_n}$ not resulting from a permutation of subscripts will contain fewer variables. Thus if $c_{i_1}^{g_1} \cdots c_{i_n}^{g_n}$ is such a term and $g_1 = k_1, \ldots, g_h = k_h$, we may, after permutation, think of c_{i_j} as coming from S_{k_j}, $1 \le j \le h$, so that $g_{h+1} \ge k_{h+1}$. Hence there exists a constant c such that $cJ_\kappa - M_\kappa = \Sigma_{\kappa' > \kappa} c'_{\kappa'} M_{\kappa'}$. Then by an induction on the linear ordering of partitions it follows that $M_\kappa = \Sigma_{\kappa' \ge \kappa} c''_{\kappa'} J_{\kappa'}$. Since the inverse of an upper triangular matrix is upper triangular it follows that $J^t = M^t T_J^t$, with T_J lower triangular. Upon expansion the terms of $S_{k_1} \cdots S_{k_n}$ are sums of monomials with integer coefficients. The monomial term $c_1^{k_1} c_2^{k_2} \cdots c_n^{k_n}$ occurs $\pi_1! \cdots \pi_n!$ times in $S_{k_1} \cdots S_{k_n}$. □

Remark 12.2.17 (On the Duality). From $\pi_1, \ldots, \pi_k$, since $\pi_1 + 2\pi_2 + \cdots + k\pi_k = k$, we may construct the partition γ of k into k parts by the definition $g_i = \pi_i + \cdots + \pi_k$. Write $\tau(\kappa) = \gamma$. Then the functions S_κ and J_κ are dual via these relations. By Theorem 12.2.14 there exists a lower triangular matrix T such that $(J_{\tau(\kappa_1)}, \ldots, J_{\tau(\kappa_r)}) = (A_{\tau(\kappa_1)}, \ldots, A_{\tau(\kappa_r)})T^t$.

12.3. The Symmetric Group Algebra

We begin this section with definitions and lemmas that apply to group algebras of finite groups. Then, towards the end of the section, results are specialized to the needed results about the symmetric group on m letters.

We let $\mathfrak{G}$ be a finite group with m elements $g_1, \ldots, g_m$. By considering $g_1, \ldots, g_m$ as linearly independent elements over the complex numbers $\mathbb{C}$ we may form the m-dimensional vector space with elements

$$\sum_{g \in \mathfrak{G}} a(g)g, \tag{12.3.1}$$

where $a\colon \mathfrak{G} \to \mathbb{C}$ is a coefficient function. As elements of a vector space addition and scalar multiplication are done coordinatewise. As elements of an algebra products are defined by

$$\left(\sum_{g \in \mathfrak{G}} a(g)g\right)\left(\sum_{g \in \mathfrak{G}} b(g)g\right) = \sum_{k \in \mathbf{G}} \left(\sum_{gh=k} a(g)b(h)\right) k. \tag{12.3.2}$$

Since each element of the algebra has a unique representation this bilinear functional is well defined. It is easy to show that this product is associative and distributive.

An involution $*$ is defined by

$$\left(\sum_{g \in \mathfrak{G}} a(g)g\right)^* = \sum_{g \in \mathfrak{G}} \overline{a(g^{-1})g} = \sum_{g \in \mathfrak{G}} \overline{a(g)}g^{-1}. \tag{12.3.3}$$

An inner product is defined by

$$\left(\sum_g a(g)g, \sum_g b(g)g\right) = \sum_g a(g)\overline{b(g)}. \tag{12.3.4}$$

It is clear that under this inner product $g_1, \ldots, g_m$ form an orthonormal basis of the additive group of the algebra. Also, if e is multiplicative unit of $\mathfrak{G}$ then e is the unit of the ring of the algebra.

Lemma 12.3.1. *The group algebra of a finite group $\mathfrak{G}$ is an H^* algebra.*

PROOF. The involution $*$ clearly satisfies (i) $a^{**} = a$, (ii) $(a+b)^* = a^* + b^*$, and (iii) $(ca)^* = \bar{c}a^*$, $c \in \mathbb{C}$. We now verify (iv) $(ab)^* = b^*a^*$. We find that

$$(ab)^* = \sum_k \left(\sum_{gh=k} \overline{a(g)b(h)} \right) k^{-1}, \tag{12.3.5}$$

and

$$\begin{aligned} b^*a^* &= \left(\sum_h \overline{b(h)} h^{-1} \right) \left(\sum_g \overline{a(g)} g^{-1} \right) \\ &= \sum_k \left(\sum_{(gh)^{-1}=k^{-1}} \overline{b(h)a(g)} \right) k^{-1} \\ &= \sum_k \left(\sum_{gh=k} \overline{a(g)b(h)} \right) k^{-1}. \end{aligned} \tag{12.3.6}$$

□

The property $\|a\| = \|a^*\|$ is obvious. We compute

$$\begin{aligned} a^*a &= \left(\sum_g \overline{a(g)} g^{-1} \right) \left(\sum_h a(h)h \right) \\ &= \sum_k \left(\sum_{g^{-1}h=k} \overline{a(g)}a(h) \right) k. \end{aligned} \tag{12.3.7}$$

The coefficient of e in the expression for a^*a is $\Sigma_g a(g)\overline{a(g)} = \|a\|^2$. Thus $a^*a = 0$ implies $\|a\| = 0$ or $a = 0$. Last, we establish that $(ab, c) = (b, a^*c)$. Note that

$$(ab, c) = \sum_k \left(\sum_{gh=k} a(g)b(h) \right) \overline{c(k)}, \tag{12.3.8}$$

and

$$\overline{(b, a^*c)} = \sum_h \left(\sum_{gk=h} \overline{a(g^{-1})} c(k) \right) \overline{b(h)},$$

so that

$$(b, a^*c) = \sum_k \left(\sum_{g^{-1}h=k} a(g^{-1})b(h) \right) \overline{c(k)} = (ab, c).$$

□

Corollary 12.3.2. *The group algebra of a finite group $\mathfrak{G}$ is a direct sum of its minimal closed ideals. Each minimal closed ideal is isomorphic to a full matrix algebra and thus each minimal closed ideal contains an idempotent that acts as multiplicative identity in the ideal.*

PROOF. Directly from Loomis (1953) on H^* algebras. □

Definition 12.3.3. The class of an element $h \in \mathfrak{G}$ is the set $[h]$ of all elements conjugate to h.

Lemma 12.3.4. *The set $\mathfrak{G}_h$ of all elements of $\mathfrak{G}$ that commute with h is a subgroup of $\mathfrak{G}$.*

Lemma 12.3.5. *If $g_1, g_2, h \in \mathfrak{G}$ and $g_1 h g_1^{-1} = g_2 h g_2^{-1}$ then $g_1^{-1} g_2 \in \mathfrak{G}_h$ and conversely, so that g_1 and g_2 are in the same coset of $\mathfrak{G}_h$. Hence the number of elements in the class $[h]$ divides the order of $\mathfrak{G}$.*

Lemma 12.3.6. *If $\mathfrak{G}$ has m elements and $[h]$ has p elements then the sequence $g_1 h g_1^{-1}, \ldots, g_m h g_m^{-1}$ repeats each element of $[h]$ exactly m/p times.*

Theorem 12.3.7. *Let the classes of $\mathfrak{G}$ be $C_0 = [e], C_1, \ldots, C_{p-1}$ containing 1, $m_1, \ldots, m_{p-1}$ elements respectively. An element a of the group algebra is in the center of the algebra if and only if there exists coefficients ϕ_i with*

$$a = \sum_{i=0}^{p-1} \phi_i \left(\sum_{g \in C_i} g \right). \tag{12.3.9}$$

PROOF. If a is in the center and $h \in \mathfrak{G}$ then $a = hah^{-1}$ so

$$\begin{aligned} ma &= \sum_h hah^{-1} = \sum_g a(g) \sum_h hgh^{-1} \\ &= \sum_{i=0}^{p-1} \left(\sum_{g \in C_i} a(g) \right) \left((m/m_i) \sum_{k \in C_i} k \right) \end{aligned} \tag{12.3.10}$$

and the coefficient ϕ_i of (12.3.9) is thus $\phi_i = \Sigma_{g \in C_i} a(g)/m_i$.

The converse is obvious after observing that $g(\Sigma_{k \in C_i} k)g^{-1} = \Sigma_{k \in C_i} k$. □

Theorem 12.3.8. *If the group algebra of $\mathfrak{G}$ has p classes then the group algebra is the direct sum of p minimal closed ideals, hence, of p simple matrix algebras.*

PROOF. Let the minimal closed ideals be $\mathbf{N}_1, \ldots, \mathbf{N}_q$ so that $\mathbf{N}_1 \oplus \cdots \oplus \mathbf{N}_q$ is the group algebra. $\mathbf{N}_i$ has an idempotent e_i which is the multiplicative unit in $\mathbf{N}_i$, $1 \le i \le q$. If $a = a_1 + \cdots + a_q$, $a_i \in \mathbf{N}_i$, $1 \le i \le q$, then a being in the center requires that $e_i a = a e_i = a_i e_i$, $1 \le i \le q$. Since $\mathbf{N}_i$ is a full matrix algebra, and since a_i must commute with all elements of $\mathbf{N}_i$, it follows that $a_i = \phi_i e_i$ for some $\phi_i \in \mathbb{C}$, $1 \le i \le q$. Conversely, every element $\phi_i e_i$ is in the center of the group algebra. By Theorem 12.3.7, the center is p-dimensional, so that $p = q$ follows. □

Lemma 12.3.9. *In the symmetric group on m letters two permutations are conjugate if and only if*

(a) *each is the product of the same number of cycles, and,*
(b) *after ordering in decreasing order the sequences of cycle lengths are the same partition of m.*

PROOF. A permutation f of $1, \ldots, r$ is said to be a cycle if no proper nonempty subset of $1, \ldots, r$ is invariant under f. In this case $f(1), f^2(1) = f(f(1)), \ldots, f^r(1) = f(f^{r-1}(1))$ are pairwise distinct. Conversely, if $f(1), f^2(1), \ldots, f^r(1)$ are pairwise distinct then f is a cycle.

Given a permutation f of $1, \ldots, m$ the minimal invariant subsets are uniquely determined and f is a cycle on each of the minimal invariant subsets, giving the cyclic decomposition of f. Also, if g is a permutation of $1, \ldots, m$ then $g^{-1}fg$ has the same number of invariant subsets and of the same ordinalities as those of f.

Conversely, given the invariant subsets $\{a_{11}, \ldots, a_{1n_1}\}$, $\{a_{21}, \ldots, a_{2n_2}\}$, etc., of f and invariant subsets $\{b_{11}, \ldots, b_{n_1}\}$, $\{b_{21}, \ldots, b_{2n_2}\}, \ldots$ of g such that

$$\begin{aligned} f &= (a_{11}, \ldots, a_{1n_1})(a_{21}, \ldots, a_{2n_2}) \cdots \\ g &= (b_{11}, \ldots, b_{1n_1})(b_{21}, \ldots, b_{2n_2}) \cdots \end{aligned} \tag{12.3.11}$$

then the permutation h such that

$$h(b_{ij}) = a_{ij} \tag{12.3.12}$$

satisfies

$$h^{-1}fh = g. \tag{12.3.13}$$

□

Remark 12.3.10. We switch notation. Here and in the sequel k is an integer to be partitioned, not a group element. The group algebra of the symmetric group on $1, \ldots, k$ becomes an algebra of linear transformations of k-forms by means of the mapping

$$\sum_{g \in \mathfrak{G}} a(g)g \to \sum_{g \in \mathfrak{G}} a(g)P_g. \tag{12.3.14}$$

The permutation matrices P_g were defined in (12.1.8). As we will see below, this algebra homomorphism is in general not one-to-one if $k > n$, corresponding to the fact that some k-forms in $n > k$ variables vanish. In particular for the alternating operator A defined in Chapter 6,

$$A = \sum_g \varepsilon(g)(k!)^{-1}g, \tag{12.3.15}$$

where $\varepsilon(g)$ is the sign of the permutation. In case $k > n$ and $f \in \mathbf{M}(\mathbf{E}^k, \mathbb{C})$, then

$$Af = 0. \tag{12.3.16}$$

In the discussion to follow the relationship between alternation and being a zero transformation will become clear.

12.4. Young's Symmetrizers

The terminology and results of this section are taken from Weyl (1946). In our statement and proofs we have found it necessary to interchange "p" and "q" throughout and we are thus in partial disagreement with the results in Weyl, op. cit.

Associated with a partition $k_1 \geq k_2 \geq \cdots \geq k_p$ of k, call it κ, is a diagram $T(\kappa)$ consisting of p rows and k_1 columns. Each row is divided into 1×1 cells, the i-th row has k_i cells, and cells are numbered from left to right,

$$k_1 + \cdots + k_{i-1} + 1, \ldots, k_1 + \cdots + k_{i-1} + k_i. \tag{12.4.1}$$

The i-th and $(i+1)$st rows, $1 \leq i \leq p-1$, are in the relation shown in

$k_1 + \cdots + k_{i-1} + 1$	$k_1 + \cdots + k_{i-1} + 2$	$, \ldots,$	$k_1 + \cdots + k_{i-1} + k_{i+1}$	$, \ldots,$	$k_1 + \cdots k_i$
$k_1 + \cdots + k_i + 1$	$k_1 + \cdots + k_i + 2$	$, \ldots,$	$k_1 + \cdots + k_{i+1}$		

(12.4.2)

A permutation of $1, \ldots, k$ is said to be of type p if the permutation leaves the row sets of $T(\kappa)$ fixed; it is of type q if the permutation leaves the column sets of $T(\kappa)$ fixed.

Definition 12.4.1. The Young's symmetrizer of the diagram $T(\kappa)$ is

$$\sum_p \sum_q \varepsilon(q) pq, \tag{12.4.3}$$

the sum being taken over all permutations of type p and q relative to the diagram $T(\kappa)$.

Lemma 12.4.2. *Relative to a fixed diagram the permutations of type p form a group and those of type q form a group. A permutation that is of both type p and type q is the identity permutation. Therefore, $p_1 q_1 = p_2 q_2$ implies $p_1 = p_2$ and $q_1 = q_2$.*

PROOF. The first two statements are obvious. Suppose a permutation σ is of type p and of type q. If i is in row j_1 and column j_2, then it follows that $\sigma(i)$ is also in row j_1 and column j_2. Since the row number and column number as a pair uniquely determine the cell of the diagram, $\sigma(i) = i$. □

Young's diagrams are partially ordered using the partial ordering of partitions described in Section 12.2. Thus $T(\kappa) \geq T(\kappa')$ if and only if $\kappa \geq \kappa'$. Given a Young's diagram $T(\kappa)$ we may define configurations $\sigma T(\kappa)$ obtained from T by replacing the entry i by $\sigma(i)$ throughout, $1 \leq i \leq k$.

Lemma 12.4.3. *Let $T(\kappa) \geq T(\kappa')$ and let σ be a permutation of $1, \ldots, k$. Then either*

(a) *there are two numbers* i_1 *and* i_2 *occurring in the same row of* $T(\kappa)$ *and the same column of* $\sigma T(\kappa')$, *or,*
(b) $T(\kappa) = T(\kappa')$ *and the permutation* σ *is a product* pq.

PROOF. We have $\kappa \geq \kappa'$. The k_1 numbers in the first row of $T(\kappa)$ must be spread amoung the k_1' columns of $\sigma T(\kappa')$. If $k_1 > k_1'$ then two numbers in the first row of $T(\kappa)$ must occur in some column of $\sigma T(\kappa')$.

For the remainder of the argument we assume $k_1 = k_1'$ and that conclusion (a) is false. Then there exists a permutation τ_1 column sets of $\sigma T(\kappa')$ such that the first row of $(\tau_1\sigma)T(\kappa')$ is the same as the first row of $T(\kappa)$. Delete the first row of $T(\kappa)$ and of $(\tau_1\sigma)T(\kappa')$ and make an induction on the reduced diagrams, with the induction being on the number of rows. By inductive hypothesis the partitions determined by the reduced diagrams are the same, so that $\kappa = \kappa'$ and $T(\kappa) = T(\kappa')$ and there exists a permutation τ_2 of the column sets of $(\tau_1\sigma)T(\kappa')$ such that $(\tau_2\tau_1\sigma)T(\kappa')$ has the same row sets as does $T(\kappa) = T(\kappa')$. Therefore there is a permutation π of the row sets of $(\tau_2\tau_1\sigma)T(\kappa)$ such that $(\pi\tau_2\tau_1\sigma)T(\kappa) = T(\kappa)$. Therefore with $\tau_3 = \tau_2\tau_1$,

$$\pi\tau_3\sigma = \text{identity}, \quad \text{and} \quad \sigma = \tau_3^{-1}\pi^{-1}. \tag{12.4.4}$$

Recall that τ_3 is a permutation of the column sets of $\sigma T(\kappa)$. It is clear that there exists a permutation τ_4 of the column sets of $T(\kappa)$ such that

$$\tau_3\sigma = \sigma\tau_4. \tag{12.4.5}$$

Therefore

$$\pi\tau_3\sigma = \pi\sigma\tau_4 = \text{identity}, \quad \text{and} \quad \sigma = \pi^{-1}\tau_4^{-1}. \tag{12.4.6}$$

This is a permutation of type pq.

In the case not included in the previous paragraphs $k_1 = k_1'$ and conclusion (a) holds on the reduced diagrams. Then (a) holds for the original diagrams. □

Lemma 12.4.4. *Let* $T(\kappa)$ *be a Young's diagram and let* s *be a permutation of* $1, \ldots, k$ *which is not of type* pq. *Then there exists a transposition* u *of type* p *and a transposition* v *of type* q *such that*

$$us = sv. \tag{12.4.7}$$

PROOF. By Lemma 12.4.3 with $\kappa = \kappa'$, $\sigma = s$, and configuration $sT(\kappa)$, by conclusion (a), there exists a row and entries in the row i_1 and i_2 which occur in the same column of $sT(\kappa)$. We let v be the transposition interchanging $s^{-1}(i_1)$ and $s^{-1}(i_2)$, and u be the transposition interchanging i_1 and i_2. Then u is of type p and v is of type q since $s^{-1}(i_1)$ and $s^{-1}(i_2)$ are in the same column of $T(\kappa)$. Further $(sv)T(\kappa) = (us)T(\kappa)$ and thus $sv = us$. □

Lemma 12.4.5. *If* $\kappa \geq \kappa'$ *and* $\kappa \neq \kappa'$ *and* s *is a permutation of* $1, \ldots, k$ *then there exist transpositions* u *of type* p *and* v *of type* q' *(i.e., for the diagram*

$T(\kappa')$) *such that*

$$us = sv. \tag{12.4.8}$$

PROOF. In Lemma 12.4.3 with $T(\kappa) \geq T(\kappa')$ conclusion (a) holds and in some row of $T(\kappa)$ exist i_1 and i_2 which occur in the same column of $sT(\kappa')$. As above, let u transpose i_1 and i_2 and v transpose $s^{-1}(i_1)$ and $s^{-1}(i_2)$. Then clearly $us = sv$, u is of type p and v is of type q' since $s^{-1}(i_1)$ and $s^{-1}(i_2)$ occur in the same column of $T(\kappa')$. □

Lemmas 12.4.3–12.4.5 give the combinatorial basis for the study of Young's symmetrizers defined in (12.4.3). For the purpose of studying the symmetrizers we introduce coefficient functions $c(g)$, $g \in \mathfrak{G}$, defined by (relative to a partition κ)

$$\begin{aligned} &\text{if } g \text{ is of type } pq \text{ and } g = pq \text{ then } c(g) = \varepsilon(q); \\ &\text{otherwise, } c(g) = 0. \end{aligned} \tag{12.4.9}$$

Then relative to a partition κ the Young's symmetrizer is

$$c = \sum_g c(g)g. \tag{12.4.10}$$

Lemma 12.4.6. *The coefficient function defined in* (12.4.9) *satisfies*

$$\begin{aligned} c(pg) &= c(g), \qquad g \in \mathfrak{G}, \qquad \text{and } p \text{ of type } p; \\ c(gq) &= \varepsilon(q)c(g), \qquad g \in \mathfrak{G}, \qquad \text{and } q \text{ of type } q. \end{aligned} \tag{12.4.11}$$

PROOF. $c(g) = 0$ unless $g = p_1 q_1$. Then $pg = (pp_1)q_1$ and $c(pg) = c(q_1) = c(g)$ by definition. Likewise if g is not of type pq then gq is not of type pq and $c(g) = c(gq) = 0$. Otherwise $c(gq) = c(p_1(q_1 q)) = \varepsilon(q_1 q) = \varepsilon(q_1)\varepsilon(q) = \varepsilon(q)c(g)$. □

Lemma 12.4.7. *Let* $d = \Sigma_g d(g)g$ *and assume the coefficient function of* d *relative to the partition* κ *satisfies*

$$\begin{aligned} &d(pg) = d(g) \quad \text{and} \quad d(gq) = \varepsilon(q)d(g) \quad \text{for } g \in \mathfrak{G}, \\ &\qquad p \text{ of type } p \quad \text{and} \quad q \text{ of type } q. \end{aligned} \tag{12.4.12}$$

Then the complex number $d(e)$ *satisfies*

$$\text{if } g \in \mathfrak{G} \quad \text{then} \quad d(e)c(g) = d(g), \tag{12.4.13}$$

where c *is the coefficient function of the Young's symmetrizer.*

PROOF. If e is the unit of $\mathfrak{G}$ then e is both a p and a q. Therefore

$$d(e) = d(pe) = d(p), \tag{12.4.14}$$

and

$$d(e) = \varepsilon(q)d(eq) = \varepsilon(q)d(q).$$

If the permutation g is not of type pq then there exist transpositions u of type p and v of type q such that $ug = gv$. Then

$$d(g) = d(ug) = d(gv) = \varepsilon(v)d(g) = -d(g). \tag{12.4.15}$$

Therefore $d(g) = 0$. Together (12.4.14) and (12.4.15) say

$$d(g) = d(e)c(g),\ g \in \mathfrak{G}. \tag{12.4.16}$$

□

Lemma 12.4.8. *Let $\kappa \geq \kappa'$ and $\kappa \neq \kappa'$. Let the coefficient function d satisfy*

$$d(pg) = d(g) \quad \text{and} \quad d(gq') = \varepsilon(q')d(g) \text{ for all } g \in \mathfrak{G}, \tag{12.4.17}$$
$$\text{and permutations } p, q' \text{ of types } p, q'.$$

Then $d(g) = 0$ for all $g \in \mathfrak{G}$.

Proof. By Lemma 12.4.5 there exist transpositions u of type p and v of type q' such that $ug = gv$. Then

$$d(g) = d(ug) = d(gv) = -d(g). \tag{12.4.18}$$

□

Lemma 12.4.9. *Let κ be a partition of k with Young's diagram $T(\kappa)$ and Young's symmetrizer $c = \Sigma_{g \in \mathfrak{G}} c(g)g$. If $g_0 \in \mathfrak{G}$ there exists a complex number $\phi(g_0)$ such that*

$$cg_0c = \phi(g_0)c. \tag{12.4.19}$$

Proof. Let $d = cg_0c$ and compute the coefficient function of d. It is

$$d(g) = \sum_{rst=g} c(r)g_0(s)c(t). \tag{12.4.20}$$

For a permutation p of type p,

$$\begin{aligned} d(pg) &= \sum_{rst=pg} c(r)g_0(s)c(t) \\ &= \sum_{rst=g} c(pr)g_0(s)c(t) = d(g). \end{aligned} \tag{12.4.21}$$

Similarly for a permutation q of type q, $d(gq) = \varepsilon(q)d(g)$. By Lemma 12.4.7, the number $\phi(g_0)$ exists and its value is

$$\phi(g_0) = \sum_{rst=e} c(r)g_0(s)c(t). \tag{12.4.22}$$

□

Lemma 12.4.10. *Let $\kappa \geq \kappa'$ and $\kappa \neq \kappa'$, with respective Young's symmetrizers c and c'. Then*

$$\text{if } g \in \mathfrak{G}, \qquad cgc' = 0. \tag{12.4.23}$$

Proof. We let $d = cgc'$ and compute the coefficient function of d. This function satisfies (12.4.17) and hence $d = 0$. See Lemma 12.4.8 and the proof of Lemma 12.4.9. □

In the group algebra of the symmetric group $\mathfrak{G}$ on k letters the partition κ and diagram $T(\kappa)$ generate a left ideal $\mathbf{I}_\kappa = \{x | x = yc, y \in \text{group algebra}\}$, and c the Young's symmetrizer for κ. We let the dimension of $\mathbf{I}_\kappa$ be n_κ.

Lemma 12.4.11. *If c is the Young's symmetrizer for κ then*

$$c^2 = \phi c, \qquad \textit{and} \quad \phi = (k!)/n_\kappa \neq 0. \tag{12.4.24}$$

Proof. Define a linear transformation of the group algebra by

$$L(x) = xc. \tag{12.4.25}$$

The range of L is $\mathbf{I}_\kappa$ so the matrix of L has rank n_κ. Using as basis the elements $h \in \mathfrak{G}$, we compute

$$L(h) = hc = \sum_g c(g)hg = \sum_g c(h^{-1}g)g. \tag{12.4.26}$$

Thus the trace of the matrix is

$$c(e) + \cdots + c(e) = (k!)c(e). \tag{12.4.27}$$

On the other hand the action of L on $\mathbf{I}_\kappa$ is

$$L(xc) = (xc)c = \phi xc, \tag{12.4.28}$$

where we have used (12.4.24). If we take a new basis, consisting of a basis $h_1, \ldots, h_{n_\kappa}$ of $\mathbf{I}_\kappa$ together with other elements outside the ideal then the matrix of L is

$$\begin{vmatrix} \phi I_{n_\kappa} & 0 \\ A & 0 \end{vmatrix}, \tag{12.4.29}$$

which has trace

$$\phi n_\kappa = (k!)c(e) = k!, \tag{12.4.30}$$

since $c(e) = 1$. Therefore ϕ has the value (12.4.24) and $c^2 \neq 0$. □

Lemma 12.4.12. *If $e = c/\phi$ with c the Young's symmetrizer relative to κ and ϕ as in* (12.4.24), *then e is a primitive idempotent.*

Proof. $e^2 = c^2/\phi^2 = c/\phi = e$, by (12.4.24). Suppose $e = e_1 + e_2$ such that

$$e_1^2 = e_1,\ e_2^2 = e_2, \qquad \text{and} \quad e_1e_2 = e_2e_1 = 0. \tag{12.4.31}$$

Then $ee_1 = e_1$ and $ee_1e = e_1e = e_1$. By Lemma 12.4.9,

$$e_1 = ce_1c/\phi^2 = \phi(e_1)c/\phi^2 = (\phi(e_1)/\phi)e \tag{12.4.32}$$

and $\phi(e_1)/\phi = 1$ or 0 since $e_1^2 = e_1$ and $e = e^2$. Hence $e_1 = e$ or $e_2 = e$. □

Lemma 12.4.13. *If κ and κ' are distinct partitions of k with Young's symmetrizers c and c' respectively then the subspaces $\mathbf{I}_\kappa$ and $\mathbf{I}_{\kappa'}$ are invariant irreducible subspaces and they are inequivalent.*

PROOF. Assume $\kappa > \kappa'$. The left ideals $\mathbf{I}_\kappa$ and $\mathbf{I}_{\kappa'}$ are irreducible since the corresponding idempotents c/ϕ and c'/ϕ' are irreducible. We are to show there cannot exist x in the group ring such that

$$x(yc)x^{-1} = y'c' \tag{12.4.33}$$

maps $\mathbf{I}_\kappa$ onto $\mathbf{I}_\kappa$, in a one-to-one fashion. To show this, if (12.4.33) holds, then

$$ycx^{-1} = x^{-1}y'c' \quad \text{and} \quad cycx^{-1} = cx^{-1}y'c' = 0 \tag{12.4.34}$$

by (12.4.23). By (12.4.19) there exists a complex number $\phi(y)$ such that $\phi(y)c = cyc$ so that

$$0 = \phi(y)cx^{-1}. \tag{12.4.35}$$

Since y is arbitrary, the choice $y = c$ gives $\phi(y) \neq 0$ and therefore $cx^{-1} = 0$. Therefore, if y' is in the group ring, $y'c' = 0$, so that in particular $c'c' = 0$. Contradiction. Therefore an invertible x satisfying (12.4.33) cannot exist. □

From the Young's symmetrizer c for the partition κ we construct the element in the center of the group ring

$$\phi^{-2}\sum_g gcg^{-1} = \varepsilon_1, \tag{12.4.36}$$

where $\phi c = c^2$. We show $\varepsilon_1^2 = \varepsilon_1$ and that the principal ideal determined by ε_1 is a minimal closed ideal. To do this we consider something that seems different. Take the coefficient function

$$\varepsilon(g) = \phi^{-2}\sum_h c(h^{-1}gh) \tag{12.4.37}$$

and define

$$\varepsilon = \sum_g \varepsilon(g)g. \tag{12.4.38}$$

Relative to a second partition κ' we similarly define ε'.

Lemma 12.4.14. $\varepsilon^2 = \varepsilon$. *If $\kappa \neq \kappa'$ then $\varepsilon\varepsilon' = 0$. Both ε and ε' are in the center and $\varepsilon = \varepsilon_1$ (see (12.4.36)).*

PROOF. Refer to (12.4.14). $\Sigma_t tct^{-1} = \Sigma_t \Sigma_g c(g)tgt^{-1} \Sigma_t \Sigma_g c(t^{-1}(tgt^{-1})t)(tgt^{-1}) = \Sigma_t \Sigma_g c(t^{-1}gt)g = \phi^2\varepsilon$. Therefore $\varepsilon_1 = \varepsilon$ and ε is in the center. Given

$\kappa > \kappa'$ we have by (12.4.23) that $\phi^2(\phi')^2\varepsilon\varepsilon' = \Sigma_t\Sigma_s tct^{-1}sc's^{-1} = 0$. Since both ε and ε' are in the center it follows that $\varepsilon'\varepsilon = 0$.

It remains to show that $\varepsilon^2 = \varepsilon$. The first computation shows that $\varepsilon c = c$. Using the formulas from the preceding paragraph,

$$\varepsilon c = \phi^{-2}\sum_t tct^{-1}c = \phi^{-2}\sum_t \phi(t^{-1})tc, \tag{12.4.39}$$

where $ct^{-1}c = \phi(t^{-1})c$. We compute the value of $\phi(t^{-1})$.

$$ct^{-1}c = \sum_g\sum_h c(g)c(h)gt^{-1}h = \phi(t^{-1})\sum_h c(h)h. \tag{12.4.40}$$

Since $c(e) = 1$ we have on the coefficient of e that

$$\phi(t^{-1}) = \sum_{gt^{-1}h=e} c(g)c(h) = \sum_h c(h^{-1}t)c(h). \tag{12.4.41}$$

Also

$$\begin{aligned} c^2 &= \sum_g\sum_h c(g)c(h)gh = \sum_g\sum_h c(g)c(g^{-1}gh)gh \\ &= \sum_k\sum_g c(g)c(g^{-1}k)k. \end{aligned} \tag{12.4.42}$$

Therefore

$$\phi(t^{-1}) = c^2(t) = \phi c(t). \tag{12.4.43}$$

Resume the calculation of (12.4.39).

$$\varepsilon c = \phi^{-2}\sum_t \phi c(t)tc = \phi^{-1}c^2 = c. \tag{12.4.44}$$

Since ε is in the center,

$$tct^{-1} = t\varepsilon ct^{-1} = \varepsilon tct^{-1}. \tag{12.4.45}$$

The sum of (12.4.45) over t gives

$$\phi^2\varepsilon = \phi^2\varepsilon^2 \text{ or } \varepsilon = \varepsilon^2. \tag{12.4.46}$$

□

Theorem 12.4.15. *To each partition κ of k and diagram $T(\kappa)$ let ε_κ be the corresponding idempotent in the center of the group ring. If $\kappa \neq \kappa'$ then $\varepsilon_\kappa\varepsilon_{\kappa'} = 0$. The sum $\Sigma_\kappa \varepsilon_\kappa$ gives a complete decomposition of the identity of the group algebra. The principal ideals are minimal closed ideals.*

PROOF. The idempotents ε_κ are linearly independent and the dimension of their linear span is the number of partitions of k. By Lemma 12.3.9 the number of partitions of k is the number of conjugate classes, and by Theorem 12.3.8 this is the dimension of the center of the group algebra. Therefore each ε_κ is irreducible in the center. And the identity being an element of the center clearly is $\Sigma_\kappa \varepsilon_\kappa$. It follows that the principal ideals determined by the ε_κ are the minimal closed ideals. □

12.5. Realization of the Group Algebra as Linear Transformations

We have defined permutation matrices P_σ by, if $f \in \mathbf{M}(\mathbf{E}^k, \mathbb{C})$ then $(P_\sigma f)(x_1, \ldots, x_k) = f(x_{\sigma(1)}, \ldots, x_{\sigma(k)})$. Then $(P_\tau(P_\sigma f))(x_1, \ldots, x_k) = (P_\sigma f)(y_1, \ldots, y_k) = f(y_{\sigma(1)}, \ldots, y_{\sigma(k)})$ where $y_j = x_{\tau(j)}$. Thus $(P_\tau(P_\sigma f))(x_1, \ldots, x_k) = f(x_{\tau\sigma(1)}, \ldots, x_{\tau\sigma(k)}) = (P_{\tau\sigma} f)(x_1, \ldots, x_k)$. That is,

$$P_\tau P_\sigma = P_{\tau\sigma}. \tag{12.5.1}$$

Consequently corresponding to a Young's diagram and symmetrizer $c = \Sigma_p \Sigma_q \varepsilon(q)pq$ is the linear transformation

$$\sum_p \sum_q \varepsilon(q) P_{pq} = \sum_p \sum_q \varepsilon(q) P_p P_q. \tag{12.5.2}$$

The action on a multilinear form f is

$$\sum_p P_p \left(\sum_q \varepsilon(q) P_q f \right). \tag{12.5.3}$$

The subgroup $\mathfrak{G}_q$ of permutations leaving the column sets of $T(\kappa)$ invariant is a direct sum $\mathfrak{G}_q = \mathfrak{G}_{q1} \oplus \cdots \oplus \mathfrak{G}_{qk_1}$, where $\mathfrak{G}_{qi}$ is the group of permutations of the letters in the i-th column of $T(\kappa)$. The first column of $T(\kappa)$ is the longest containing some number $r \geq 1$ cells corresponding to a partition of k into r nonzero parts. We write

$$q = q_r q_{r-1} \cdots q_1, \qquad q_i \in \mathfrak{G}_{qi}, \qquad 1 \leq i \leq r. \tag{12.5.4}$$

The factors $q_1, \ldots, q_r$ commute, of course. Then

$$\sum_q \varepsilon(q) P_q = \sum_{q_1} \cdots \sum_{q_r} \varepsilon(q_1) \cdots \varepsilon(q_r) P_{q_1} \cdots P_{q_r} \tag{12.5.5}$$

and if $r > n$ then

$$\sum_{q_1} \varepsilon(q_1) P_{q_1} f = 0, \tag{12.5.6}$$

as was shown in Chapter 6, since the maximal degree of a nonzero alternating k-form in n variables is n.

Lemma 12.5.1. *Let ε be the idempotent in the center of the group algebra corresponding to the partition $k_1 \geq k_2 \geq \cdots \geq k_r \geq 1$ of k. Let $\varepsilon = \Sigma_q \varepsilon(q) q$ and assume $r > n$. Then the linear operator $\Sigma_q \varepsilon(q) P_q$ is the zero operator on $\mathbf{M}(\mathbf{E}^k, \mathbb{C})$.*

PROOF. By (12.4.36) and the proof of Lemma 12.4.14, $\phi^2 = \Sigma_t tct^{-1}$ so that the corresponding linear operator is

$$\left(\sum_t P_t \left(\sum_g c(g) P_g \right) P_{t^{-1}} \right) f = \sum_t P_t \left(\sum_g c(g) P_g \right) (P_{t^{-1}} f) = 0. \tag{12.5.7}$$

□

Lemma 12.5.2. *The map $d \to P_d$ of the group algebra into the endomorphisms of $\mathbf{M}(\mathbf{E}^k, \mathbb{C})$ defined by*

$$\Sigma d(g)g \to \Sigma d(g)P_g = P_d \tag{12.5.8}$$

is an algebra homomorphism. The kernel is an ideal which is a direct sum of minimal ideals contained in the kernel. If ε is the idempotent in the center corresponding to the partition κ and if $P_\varepsilon \neq 0$, then the map $d\varepsilon \to P_{d\varepsilon}$ is one-to-one from the principal ideal $\{d\varepsilon\}$ to the endomorphisms of $\mathbf{M}(\mathbf{E}^k, \mathbb{C})$.

PROOF. We use the results of Loomis (1953) as applied to the group algebra which is an H^* algebra, as stated in Lemma 12.3.1. It follows from Loomis that every ideal in the group algebra is a direct sum of the minimal closed ideals contained in it.

The mapping $d\varepsilon \to P_{d\varepsilon}$ is an algebra homomorphism of the ideal $\{d\varepsilon\}$ into End $\mathbf{M}(\mathbf{E}^k, \mathbb{C})$ and is either one-to-one or has as kernel a proper ideal in $\{d\varepsilon\}$. However $\{d\varepsilon\}$ being a full matrix algebra has no proper nonzero ideals. The lemma now follows. □

Lemma 12.5.3. *If κ is a partition of k into $\leq n$ parts and ε_κ is the idempotent in the center corresponding to κ, then $P_{\varepsilon_\kappa} \neq 0$.*

PROOF. For brevity we write $\varepsilon = \varepsilon_\kappa$. If $P_\varepsilon = 0$ then $P_{d\varepsilon} = 0$ for all d in the group algebra. From (12.4.44) we have $c\varepsilon = \varepsilon c = c$ so that it is sufficient to show $P_c \neq 0$. In turn, $c = \Sigma_p \Sigma_q \varepsilon(q)pq = (\Sigma_p p)(\Sigma_q \varepsilon(q)q)$ so that it is sufficient to show that $\Sigma_q \varepsilon(q)P_q \neq 0$.

Choose σ so that the configuration $\sigma T(\kappa)$ has as columns $1, \ldots, s_1$; $s_1 + 1, \ldots, s_2; \ldots; s_{r-1} + 1, \ldots, s_r = k$. We show $\Sigma_q \varepsilon(q)P_{\sigma q\sigma^{-1}} \neq 0$. The permutations $\sigma q\sigma^{-1}$ leave the column sets of $\sigma T(\kappa)$ fixed. If we factor $\Sigma_q \varepsilon(q)P_{\sigma q\sigma^{-1}} = \prod_{i=1}^r (\Sigma_{q_i} \varepsilon(q_i)P_{\sigma q_i\sigma^{-1}})$ as in (12.5.5) and apply this operator to $f = u_1 u_2 \cdots u_k$ then

$$\begin{aligned} g &= \sum_q \varepsilon(q)P_{\sigma q\sigma^{-1}} f \\ &= (u_1 \wedge \cdots \wedge u_{s_1})(u_{s_1+1} \wedge \cdots \wedge u_{s_2}) \cdots (u_{s_{r-1}+1} \wedge \cdots \wedge u_k). \end{aligned} \tag{12.5.9}$$

If we evaluate $P_{\sigma p\sigma^{-1}} g$ at $e_{i(1)}, \ldots, e_{i(k)}$ then

$$\begin{aligned} &(P_{\sigma p\sigma^{-1}} g)(e_{i(1)}, \ldots, e_{i(k)}) \\ &\quad = \prod_{j=0}^{r-1} (u_{s_j+1} \wedge \cdots \wedge u_{s_{j+1}})(e_{i(\sigma p\sigma^{-1}(s_j+1))}, \ldots, e_{i(\sigma p\sigma^{-1}(s_{j+1}))}). \end{aligned} \tag{12.5.10}$$

We choose the subscript function i of (12.5.10) to be the permutation $\sigma p_0 \sigma^{-1}$ of $1, \ldots, k$. Then $i(\sigma p\sigma^{-1}(j)) = \sigma p_0 \sigma^{-1} \sigma p\sigma^{-1}(j) = \sigma p_0 p\sigma^{-1}(j)$. If p and p_0 are permutations of type p then $\sigma p_0 p\sigma^{-1}$ permutes the row sets of the configuration $\sigma T(\kappa)$. However (12.5.10) vanishes unless $\sigma p_0 p\sigma^{-1}(s_j + 1)$, $\ldots, \sigma p_0 p\sigma^{-1}(s_{j+1})$ is a permutation of the column set $s_j + 1, \ldots, s_{j+1}$. This requires $p_0 p$ to be the identity permutation and $p = p_0^{-1}$. Therefore if $i = \sigma p_0 \sigma^{-1}$ then (12.5.10) vanishes for all permutations p of type p except for $p = p_0^{-1}$. This clearly means $\Sigma_p P_{\sigma p \sigma^{-1}} \Sigma_q \varepsilon(q) P_{\sigma q \sigma^{-1}} \neq 0$, as was to be shown.

Lemma 12.5.4. *The idempotents ε_κ of the center of the group algebra have matrices P_{ε_κ} which are bi-symmetric. The matrices P_{ε_κ} commute with all bi-symmetric matrices.*

PROOF. $P_\sigma P_{\varepsilon_\kappa} = P_{\sigma\varepsilon_\kappa} = P_{\varepsilon_\kappa \sigma} = P_{\varepsilon_\kappa} P_\sigma$, where σ is a permutation of $1, \ldots, k$. If A is any *bi*-symmetric matrix then $AP_{\varepsilon_\kappa} = A \Sigma_p \Sigma_q \varepsilon(q) P_p P_q = \Sigma_p \Sigma_q \varepsilon(q) P_p P_q A = P_{\varepsilon_\kappa} A$. □

12.6. The Center of the Bi-Symmetric Matrices, as an Algebra

It is the purpose of this section to show that the matrices P_{ε_κ} form the center of the algebra of *bi*-symmetric matrices. In order to obtain this conclusion some abstract algebraic argument is needed. We introduce names for the algebras under study.

Let $\mathfrak{S}$ be the representation of the symmetric group algebra in End $\mathbf{M}(\mathbf{E}^k, \mathbb{C})$, as described in Section 12.5. Within End $\mathbf{M}(\mathbf{E}^k, \mathbb{C})$ we let $\mathfrak{A}_k$ be the algebra of *bi*-symmetric matrices of n^k rows and columns. By definition, $\mathfrak{A}_k$ is the set of matrices which commute with all of the matrices in $\mathfrak{S}$. Since the inverse of P_σ is P_σ^t, σ a permutation of $1, \ldots, k$, it follows that $\mathfrak{A}_k$ is closed under transpose and conjugation and hence is an H^* algebra. The same clearly holds of $\mathfrak{S}$.

Lemma 12.6.1 below is a result of Wedderburn which we take from Albert (1939), Page 19. See also Weyl (1946), page 93, Theorem (3.5.A)

Lemma 12.6.1 (Wedderburn). *Let $\mathfrak{M}$ be a matrix algebra and $\mathfrak{A}$ be a matrix subalgebra of $\mathfrak{M}$ such that the unit of $\mathfrak{A}$ is the identity matrix and $\mathfrak{A}$ is a total matrix algebra. Then $\mathfrak{M} = \mathfrak{A} \otimes \mathfrak{C}$, the tensor product, where $\mathfrak{C}$ is the commutator algebra of $\mathfrak{A}$ in $\mathfrak{M}$.*

PROOF. $\mathfrak{A}$ contains elements e_{ij}, $1 \leq i, j \leq p$, such that $e_{11}, e_{22}, \ldots, e_{pp}$ are orthogonal idempotents, $I = \Sigma_{i=1}^p e_{ii}$, and if $a \in \mathfrak{A}$ then $a = \Sigma_{i=1}^p \Sigma_{j=1}^p \bar{a}_{ij} e_{ij}$, where if $1 \leq i, j \leq p$ then $\bar{a}_{ij}$ is a complex number, i.e., $\bar{a}_{ij} \in \mathbb{C}$.

Given a, let

$$a_{ij} = \sum_{k=1}^{p} e_{ki} a e_{jk}. \tag{12.6.1}$$

Then from (12.6.1),

$$a_{ij}e_{rs} = e_{ri}ae_{js} \quad \text{and} \quad e_{rs}a_{ij} = e_{ri}ae_{js}. \tag{12.6.2}$$

Therefore a_{ij} commutes with all the e_{ij} for $\mathfrak{A}$ and the a_{ij} are contained in $\mathfrak{C}$. Using $I = e_{11} + \cdots + e_{pp}$ and (12.6.1),

$$\begin{aligned}\sum_{i=1}^{p}\sum_{j=1}^{p} a_{ij}e_{ij} &= \sum_{i=1}^{p}\sum_{j=1}^{p}\sum_{k=1}^{p} e_{ki}ae_{jk}e_{ij} \\ &= \sum_{i=1}^{p}\sum_{j=1}^{p} e_{ii}ae_{jj} = a.\end{aligned} \tag{12.6.3}$$

Finally, let a_{ij}, $1 \le i, j \le p$ be p^2 elements in $\mathfrak{C}$. Assume

$$a = \sum_{i=1}^{p}\sum_{j=1}^{p} a_{ij}e_{ij} = 0. \tag{12.6.4}$$

Then

$$\begin{aligned}0 &= \sum_{k=1} e_{ki_0}\left(\sum_{i=1}^{p}\sum_{j=1}^{p} a_{ij}e_{ij}\right)e_{j_0k} \\ &= \sum_{k=1}^{p}\sum_{i=1}^{p}\sum_{j=1}^{p} a_{ij}e_{ki_0}e_{ij}e_{j_0k} \\ &= a_{i_0j_0}\sum_{k=1}^{p} e_{kk} = a_{i_0j_0}.\end{aligned} \tag{12.6.5}$$

□

Lemma 12.6.2. *Let $\mathfrak{M}$ be the algebra of all endomorphisms of a finite dimensional vector space $\mathfrak{X}$ over $\mathbb{C}$. Let $\mathfrak{A}$ be a subalgebra of $\mathfrak{M}$ which is an H^* algebra and $\mathfrak{C}$ be the commutator algebra of $\mathfrak{A}$ in $\mathfrak{M}$. Then $\mathfrak{A}$ and $\mathfrak{C}$ have the same center which is the linear span of the idempotents $e_1, \ldots, e_r$. $\mathfrak{A}$ is the commutator algebra of $\mathfrak{C}$ in $\mathfrak{M}$.*

Proof. Since $\mathfrak{M}$ is a finite dimensional H^* algebra $\mathfrak{A}$ is completely reducible. We let $e_1, \ldots, e_r$ be the irreducible idempotents of the center of $\mathfrak{A}$. It is easily verified that $e_i\mathfrak{M}e_i$ is the algebra of endomorphisms of the vector space $e_i\mathfrak{X}$, $1 \le i \le r$. Therefore the center of $e_i\mathfrak{M}e_i$ is one-dimensional. We use this fact below.

An element e_ime_i commutes with all the elements of $\mathfrak{A}e_i$ if and only if

$$\begin{aligned}(e_ime_i)a &= (e_ime_i)(e_ia) = (e_ime_i)(ae_i) \\ &= (ae_i)(e_ime_i) = a(e_ime_i).\end{aligned} \tag{12.6.6}$$

It follows that (e_ime_i) is an element of $\mathfrak{C}$, which implies $(e_ime_i) \in \mathfrak{C}e_i$. Since e_i is in the center of $\mathfrak{C}$, it follows that

$$\mathfrak{C}e_i = e_i\mathfrak{C}e_i \subset e_i\mathfrak{M}e_i \tag{12.6.7}$$

and that $\mathfrak{C}e_i$ is the commutator algebra of $\mathfrak{A}e_i$ in $e_i\mathfrak{M}e_i$. Therefore by Wedderburn, see Lemma 12.6.1, since $\mathfrak{A}e_i$ is a total matrix algebra (see Loomis on H^* algebras) and e_i is the identity of $e_i\mathfrak{M}e_i$,

$$e_i\mathfrak{M}e_i = (\mathfrak{A}e_i) \otimes (\mathfrak{C}e_i). \tag{12.6.8}$$

The centers of $e_i\mathfrak{M}e_i$ and $\mathfrak{A}e_i$ are one-dimensional. It is easy to see that if the center of $\mathfrak{C}e_i$ is more than one-dimensional then so is the center of $e_i\mathfrak{M}e_i$. Hence the center of $\mathfrak{C}e_i$ is one-dimensional and is generated by e_i.

Within the algebra $e_i\mathfrak{M}e_i$ every element is representable as (cf. Lemma 12.6.1)

$$\sum_p \sum_q e_{pq} c_{pq} \tag{12.6.9}$$

where the e_{pq} are fixed elements of the total matrix algebra $\mathfrak{A}e_i$. An element of the form (12.6.9) commutes with every element $c \in \mathfrak{C}e_i$ if and only if

$$\begin{aligned} c\sum_p \sum_q e_{pq}c_{pq} &= \sum_p \sum_q (ce_{pq})c_{pq} = \sum_p \sum_q (e_{pq}c)c_{pq} \\ &= \sum_p \sum_q e_{pq}(cc_{pq}) = \sum_p \sum_q e_{pq}(c_{pq}c). \end{aligned} \tag{12.6.10}$$

Since this representation is unique, for all p, q

$$cc_{pq} = c_{pq}c. \tag{12.6.11}$$

In words, the c_{pq} are in the center of $\mathfrak{C}e_i$ and there exist complex numbers ϕ_{pq} such that

$$c_{pq} = \phi_{pq}e_i \quad \text{and} \quad \sum_p \sum_q e_{pq}c_{pq} = \sum_p \sum_q \phi_{pq}e_{pq} \in \mathfrak{A}e_i. \tag{12.6.12}$$

Hence $\mathfrak{A}e_i$ is the commutator of $\mathfrak{C}e_i$ in $e_i\mathfrak{M}e_i$. Since

$$\mathfrak{C} = \bigotimes_{i=1}^r \mathfrak{C}e_i, \tag{12.6.13}$$

it follows that the center of $\mathfrak{C}$ is the linear span of the e_i and equals the center of $\mathfrak{A}$. And, since

$$\mathfrak{A} = \bigotimes_{i=1}^r \mathfrak{A}e_i, \tag{12.6.14}$$

it follows that $\mathfrak{A}$ is the commutator algebra of $\mathfrak{C}$ in $\mathfrak{M}$. □

Corollary 12.6.3. *The center of $\mathfrak{A}_k$, the algebra of bi-symmetric matrices, is generated by the idempotents P_{ε_κ}, as κ runs over all partitions of k into $\leq n$ parts.*

12.7. Homogeneous Polynomials II. Two-Sided Unitary Invariance

Theorem 12.7.1. *The matrices P_{ε_κ} are symmetric matrices with real number entries.*

PROOF. The elements of the center are obtained as $\phi^2 \varepsilon_\kappa = \Sigma_g gcg^{-1} = \Sigma_p \Sigma_q \varepsilon(q) \Sigma_g gpqg^{-1}$. In the permutation groups permutations σ and σ^{-1} have the same cycle decomposition, hence are conjugate. Thus $\Sigma_g gpqg^{-1} = \Sigma_g g(pq)^{-1}g^{-1}$. Since P_σ is a permutation matrix, $P_{\sigma^{-1}} = P_\sigma^t$, so that $P_{\varepsilon_\kappa} = P_{\varepsilon_\kappa}^t = \Sigma_p \Sigma_q \Sigma_g \varepsilon(q) P_{gpqg^{-1}}$ is a symmetric matrix with real number entries, i.e., the coefficients $\varepsilon(q)$ are real numbers. □

The matrices P_{ε_κ} are directly related to unitarily invariant polynomials. For, if X is an $n \times n$ matrix and U is an $n \times n$ unitary matrix, then, with A^* the conjugate transpose of A and $\bigotimes_{i=1}^k X = X \otimes \cdots \otimes X$,

$$\begin{aligned} \operatorname{tr} P_{\varepsilon_\kappa} \bigotimes_{i=1}^k (UXU^*) &= \operatorname{tr}\left(\bigotimes_{i=1}^k U^* P_{\varepsilon_\kappa} \bigotimes_{i=1}^k U \right) \bigotimes_{i=1}^k X \\ &= \operatorname{tr} P_{\varepsilon_\kappa} \bigotimes_{i=1}^k X. \end{aligned} \tag{12.7.1}$$

We use here the fact that P_{ε_κ} is in the center of the algebra $\mathfrak{A}_k$. By the results of Section 12.6, the center of $\mathfrak{A}_k$ is spanned by the matrices P_{ε_κ} taken over all partitions κ of k into n or fewer parts. Being orthogonal idempotents, these matrices are clearly linearly independent. As polynomials, suppose

$$\operatorname{tr}\left(\sum_\kappa a_\kappa P_{\varepsilon_\kappa} \right) \bigotimes_{i=1}^k X = 0, \tag{12.7.2}$$

identically in X. Since the coefficient matrix is *bi*-symmetric, by Lemma 12.1.11, it follows that

$$\operatorname{tr}\left(\sum_\kappa a_\kappa P_{\varepsilon_\kappa} \right) \left(\overline{\sum_\kappa a_\kappa P_{\varepsilon_\kappa}} \right) = 0, \tag{12.7.3}$$

and hence that the coefficient matrix is zero. As noted this implies $a_\kappa = 0$ for all partitions κ.

Conversely, suppose a homogeneous polynomial has the invariance property that if U is unitary then

$$\operatorname{tr} A \bigotimes_{i=1}^k (UXU^*) = \operatorname{tr} A \bigotimes_{i=1}^k X. \tag{12.7.4}$$

The matrix A is assumed to be a *bi*-symmetric matrix as explained in Theorem 12.1.12. The relation (12.7.4) clearly implies that if U is unitary then for all $n \times n$ matrices X

$$\operatorname{tr} A \bigotimes_{i=1}^k (UX) - \operatorname{tr} A \bigotimes_{i=1}^k (XU) = 0. \tag{12.7.5}$$

If we fix A and X and define a polynomial by

$$f(Y) = \operatorname{tr} A \bigotimes_{i=1}^k (YX) - \operatorname{tr} A \bigotimes_{i=1}^k (XY), \tag{12.7.6}$$

then (12.7.5) says that $f(U) = 0$ for all unitary matrices U. By Weyl (1946) Lemma (7.1.A), page 177, it follows that $f(Y)$ vanishes for all $n \times n$ matrices Y over the complex numbers. Therefore the *bi*-symmetric coefficient matrix

$$A\left(\bigotimes_{i=1}^{k} X\right) - \left(\bigotimes_{i=1}^{k} X\right)A = 0, \tag{12.7.7}$$

and this holds for all X. Therefore by Lemma 12.1.11, it follows that A is in the center of $\mathfrak{A}_k$. Thus we have proven

Theorem 12.7.2. *The polynomial* $\operatorname{tr} A\bigotimes_{i=1}^{k} X$, *with* $A \in \mathfrak{A}_k$, *has the invariance property* (12.7.4) *if and only if A is in the center of* $\mathfrak{A}_k$.

If the polynomial $\operatorname{tr} A\bigotimes_{i=1}^{k} X$ satisfies (12.7.4) then for Hermitian X there exists unitary U such that UXU^* is a diagonal matrix. Therefore the value of $\operatorname{tr} A\bigotimes_{i=1}^{k} X$ depends only on the eigenvalues of Hermitian X. In the sequel we shall want to know that the polynomials

$$\operatorname{tr} P_{\varepsilon_\kappa} \bigotimes_{i=1}^{k} X \tag{12.7.8}$$

are linearly independent when X is restricted to be a real diagonal matrix. This assertion says something about the diagonal elements of P_{ε_κ}. See Sections 12.8, 12.9 and also see Theorem 12.7.7. It will appear on reading Constantine (1963) that this same theory is crucial in the evaluation of important integrals.

To continue with the results of this section, we first prove the important reproducing property. We then use this result to examine group algebras when multiplication is defined by convolution in the unitary group.

Theorem 12.7.3.

$$\int \operatorname{tr} P_{\varepsilon_\kappa} \bigotimes_{i=1}^{k} (UXU^*Y)\, dU = \left(\operatorname{tr} P_{\varepsilon_\kappa} \bigotimes_{i=1}^{k} X\right)\left(\operatorname{tr} P_{\varepsilon_\kappa} \bigotimes_{i=1}^{k} Y\right) / \operatorname{tr} P_{\varepsilon_\kappa}, \tag{12.7.9}$$

where the integral is with respect to the Haar measure of unit mass on the unitary group.

Proof. The polynomial

$$\operatorname{tr} P_{\varepsilon_\kappa} \bigotimes_{i=1}^{k} (UXU^*Y), \tag{12.7.10}$$

as a polynomial in X, has coefficient matrix

$$\left(\bigotimes_{i=1}^{k} U^*Y\right) P_{\varepsilon_\kappa} \bigotimes_{i=1}^{k} U, \tag{12.7.11}$$

which, since P_{ε_κ} is in the center of $\mathfrak{A}_k$, is a matrix in the (closed) ideal $\mathfrak{A}_k P_{\varepsilon_\kappa}$.

Therefore if we view (12.7.9) as an integral of matrices in $\mathfrak{A}_k P_{\varepsilon_\kappa}$, then the coefficient matrix lies in this ideal. The polynomial (12.7.9) is invariant in the sense of (12.7.4) so the coefficient matrix is, by Theorem 12.7.2, in the center of $\mathfrak{A}_k$, and by Lemma 12.6.2 as applied to $\mathfrak{A}$ and $\mathfrak{C}$, the only elements of the center of $\mathfrak{A}_k$ in $\mathfrak{A}_k P_{\varepsilon_\kappa}$ are $\phi P_{\varepsilon_\kappa}$, ϕ a complex number. Therefore

$$P_{\varepsilon_\kappa}\int\left(\bigotimes_{i=1}^{k} U^*YU\right)dU = \phi(Y)P_{\varepsilon_\kappa} \tag{12.7.12}$$

and (12.7.9) is given by

$$\phi(Y)\,\mathrm{tr}\,P_{\varepsilon_\kappa}\bigotimes_{i=1}^{k} X. \tag{12.7.13}$$

A similar argument shows that (12.7.9) is given by

$$\phi'(X)\,\mathrm{tr}\,P_{\varepsilon_\kappa}\bigotimes_{i=1}^{k} Y, \tag{12.7.14}$$

where

$$\phi'(X) = \int\left(\bigotimes_{i=1}^{k} UXU^*\right)dU. \tag{12.7.15}$$

Set $Y =$ identity matrix and find $\phi(I_n) = 1$ and

$$\phi'(X)\,\mathrm{tr}\,P_{\varepsilon_\kappa} = \mathrm{tr}\,P_{\varepsilon_\kappa}\bigotimes_{i=1}^{k} X. \tag{12.7.16}$$

Substitution of (12.7.16) into (12.7.14) yields (12.7.9). □

Theorem 12.7.4. *If A is in the center of $\mathfrak{A}_k$ then there exists a polynomial g of n variables such that*

$$\mathrm{tr}\,A\bigotimes_{i=1}^{k} X = g(\mathrm{tr}\,X, \mathrm{tr}\,X^2, \ldots, \mathrm{tr}\,X^n) \tag{12.7.17}$$

is a homogeneous polynomial of degree k in the entries of X.

Proof. In the proof we use Theorem 8.1.1 (4). Since A is in the center of $\mathfrak{A}_k$, by invariance, if X is unitary, then (12.7.17) is a symmetric function of the eigenvalues of X. By Theorem 12.2.14 there exists a polynomial function g of n variables such that (12.7.17) holds for all unitary matrices X. By Weyl (1946), Lemma (7.1.A), it then follows that (12.7.17) holds for all $n \times n$ matrices X with complex number entries. □

Given polynomials f and g the convolution

$$f * g(X) = \int f(XU^{-1})g(U)\,dU, \tag{12.7.18}$$

with the integral over the unitary group with respect to Haar measure of unit mass, defines a polynomial in the variables X. In case f is a homogeneous polynomial then so is $f*g$ as defined. Consequently with this definition of $*$ the homogeneous polynomials of degree k become an algebra. We show below that the coefficient matrices of the polynomials in the center of the group algebra are the matrices of the center of the algebra $\mathfrak{A}_k$ of bi-symmetric matrices. This discussion should be compared with the discussion of central elements of convolution group algebras as given by Loomis (1953).

Lemma 12.7.5. *Let g be a polynomial of matrix argument. The following are equivalent.*

If U is unitary then $g(UXU^*) = g(X)$ for all $n \times n$ matrices X with complex number entries. (12.7.19)

$g(XY) = g(YX)$ for all X, Y $n \times n$ with complex number entries. (12.7.20)

PROOF. As shown in Theorem 12.7.2, $g(UXU^{-1}) = g(X)$ implies the coefficient matrix of g is in the center, from which (12.7.20) follows. If (12.7.20) holds then $g((UX)U^{-1}) = g(U^{-1}(UX)) = g(X)$ and (12.7.19) holds. □

Loomis (1953), p. 157, calls functions g satisfying (12.7.20) "central" in his discussion of the group algebras of compact groups. It is easily verified that central functions are in the center of the group algebra.

Theorem 12.7.6. *If κ is a partition of k into not more than n parts then define g_κ by, if $X \in \mathbf{GL}(n)$ then*

$$g_\kappa(X) = \operatorname{tr} P_{\varepsilon_\kappa} \bigotimes_{i=1}^{k} X. \tag{12.7.21}$$

Then with integration by Haar measure of unit mass on the unitary group,

If $\kappa \neq \kappa'$ then $\int g_\kappa(U) g_{\kappa'}(XU^{-1})\, dU = 0$. (12.7.22)

If g is a unitarily invariant homogeneous polynomial of degree k in the variables X then there exists a partition κ of k such that $\int g(U) g_\kappa(U^{-1})\, dU \neq 0$. (12.7.23)

If $\gamma \neq 0$ and if g has the splitting property that $\int g(XUYU^{-1})\, dU = g(X)g(Y)/\gamma$, then g is unitarily invariant and there exists a number γ' and partition κ such that $g = \gamma' g_\kappa$. (12.7.24)

If g and h are central in the sense of Loomis then the convolution is central. (12.7.25)

$g_\kappa * g_\kappa = \omega g_\kappa$ *where* $\omega = \|g_\kappa\|_2^2 / g_\kappa(I_n)$. (12.7.26)

PROOF OF (12.7.22). The functions $g_{\kappa'}$ by Theorem 12.7.3, have the splitting property. Thus

$$\begin{aligned} g_g(Y)g_{\kappa'}(I) \int g_\kappa(U)g_{\kappa'}(XU^{-1})\,dU &= g_\kappa(I)g_{\kappa'}(I) \iint g_\kappa(YVUV^{-1})g_{\kappa'}(XU^{-1})\,dV\,dU \\ &= g_\kappa(I)g_{\kappa'}(I) \iint g_\kappa(U)g_{\kappa'}(XU^{-1}V^{-1}YV)\,dU\,dV \\ &= g_\kappa(I)g_{\kappa'}(Y) \iint g_\kappa(U)g_{\kappa'}(XU^{-1})\,dU. \end{aligned} \tag{12.7.27}$$

This identity holds for all unitary matrices Y. Since g_κ and $g_{\kappa'}$ are polynomials in the entries of Y, by Weyl, op. cit., Lemma (7.1.A), the identity holds for all $Y \in \mathbf{GL}(n)$. Since g_κ and $g_{\kappa'}$ are distinct polynomials and since $g_\kappa(I) \neq 0$, $g_{\kappa'}(I) \neq 0$, the integral must be zero. Note that $g_\kappa(I) = \operatorname{tr} P_{\varepsilon_\kappa} > 0$. □

PROOF OF (12.7.23). The coefficient matrix of g_κ is real and symmetric, see Theorem 12.7.1, so that

$$g_\kappa(Y^{-1}) = \overline{g_\kappa(Y)}, \quad Y \text{ unitary.} \tag{12.7.28}$$

Hence the convolution is

$$\begin{aligned} g_\kappa * g_\kappa(I) &= \int g_\kappa(U)g_\kappa(U^{-1})\,dU = \int |g_\kappa(U)|^2\,dU \\ &= \|g_\kappa\|_2^2 > 0. \end{aligned} \tag{12.7.29}$$

By Theorem 12.7.2 an invariant polynomial $g = \Sigma_\kappa a_\kappa g_\kappa$ and if $g \neq 0$ then $a_\kappa \neq 0$ for some κ since the functions g_κ are a basis. Then by (12.7.22) and the preceding, $\int g(U)g_\kappa(U^{-1})\,dU \neq 0$. □

PROOF OF (12.7.24). From

$$g(X)g(VYV^{-1}) = \gamma \int g(XU(VYV^{-1})U^{-1})\,dU = g(X)g(Y) \tag{12.7.30}$$

it follows that if $g \neq 0$ then g is invariant. By (12.7.23) there exists a partition κ such that $g * g_\kappa \neq 0$. Then for unitary matrices X

$$\begin{aligned} g(X) \int g(Y)g_\kappa(Y^{-1})\,dY &= \gamma \iint g(XUYU^{-1})g_\kappa(Y^{-1})\,dU\,dY \\ &= \gamma \iint g(Y)g_\kappa(Y^{-1}U^{-1}XU)\,dU\,dY \\ &= \gamma(g_\kappa(X)/g_\kappa(I)) \int g(Y)g_\kappa(Y^{-1})\,dY. \end{aligned} \tag{12.7.31}$$

Since the integral, a convolution, is by hypothesis, not zero, we obtain $g(X) = \gamma g_\kappa(X)/g_\kappa(I)$. By Weyl, op. cit., since both sides are polynomials in X holding for all unitary X, the identity holds for all $X \in \mathbf{GL}(n)$. □

PROOF OF (12.7.25). Let e and f be central in the sense of Loomis, op. cit., and let X and Y be unitary matrices. Then

$$\begin{aligned}(e*f)(XY) &= \int e(XYU^{-1})f(U)\,dU \\ &= \int e(XU^{-1})f(UY)\,dU = \int e(XU^{-1})f(YU)\,dU \\ &= \int e(XU^{-1}Y)f(U)\,dU \\ &= \int e(YXU^{-1})f(U)\,dU = (e*f)(YX).\end{aligned} \tag{12.7.32}$$

Since this holds for all unitary matrices X and Y it holds for all X, $Y \in \mathbf{GL}(n)$. □

PROOF OF (12.7.26). By (12.7.23) $g_\kappa * g_\kappa(I) > 0$ and by part (12.7.25) $g_\kappa * g_\kappa$ is an invariant function. Since $g_\kappa * g_\kappa(X)$ is homogeneous of degree k in the entries of X,

$$g_\kappa * g_\kappa = \sum_\kappa a_\kappa g_\kappa. \tag{12.7.33}$$

By orthogonality, see (12.7.22), and associativity of convolutions,

$$0 = (g_{\kappa'} * g_\kappa) * g_\kappa = \sum_\kappa a_\kappa g_{\kappa'} * g_\kappa = a_{\kappa'} g_{\kappa'} * g_{\kappa'}. \tag{12.7.34}$$

Hence if $\kappa \neq \kappa'$ then by (12.7.23), $a_{\kappa'} = 0$. Since $g_\kappa * g_\kappa \neq 0$ it follows that $g_\kappa * g_\kappa = a_\kappa g_\kappa \neq 0$ for all $X \in \mathbf{GL}(n)$. From (12.7.29) the value of a_κ may then be shown to be (12.7.26). □

Theorem 12.7.7. *If A is in the center of $\mathfrak{A}_k$ and the polynomial $\operatorname{tr} A \bigotimes_{i=1}^k X$ vanishes for all real diagonal $n \times n$ matrices X then $A = 0$. Consequently the functions defined by*

$$h_\kappa(\lambda_1, \ldots, \lambda_n) = \operatorname{tr} P_{\varepsilon_\kappa} \bigotimes_{i=1}^{k} \operatorname{diag}(\lambda_1, \ldots, \lambda_n) \tag{12.7.35}$$

are a basis of the space of homogeneous symmetric functions of degree k in variables $\lambda_1, \ldots, \lambda_n$.

PROOF. By linearity it follows at once that $\operatorname{tr} A \bigotimes_{i=1}^k X = 0$ for all complex diagonal X. Since A is in the center, if Y is diagonal and U is unitary then $\operatorname{tr} A \bigotimes_{i=1}^k (UYU^{-1}) = \operatorname{tr} A \bigotimes_{i=1}^k Y = 0$. By part 4 of Theorem 8.1.1 every unitary matrix X is representable as $X = UYU^{-1}$, so that $\operatorname{tr} A \bigotimes_{i=1}^{k} X = 0$ for all unitary matrices. Then, by Weyl, op. cit., it follows that the polynomial is identically zero in $X \in \mathbf{GL}(n)$. By Lemma 12.1.11 it then follows that $\operatorname{tr} AA^* = 0$, hence that $A = 0$.

Since the dimension of the space of symmetric functions is r, the number of partitions of k into n or fewer parts, a dimensionality argument completes the proof. □

Remark 12.7.8. The mapping $X \to P_{\varepsilon_\kappa} \bigotimes_{i=1}^{k} X$ is a representation of $\mathbf{GL}(n)$ in the algebra $\mathfrak{A}_k$ of bi-symmetric matrices. Thus the polynomial $\operatorname{tr} P_{\varepsilon_\kappa} \bigotimes_{i=1}^{k} X$ is a group character. This provides a rationale for James (1964), page 487, who defined the zonal polynomials of Hermitian matrix argument X as

$$\tilde{C}_\kappa(X) = \chi_\kappa(1)\chi_\kappa(X). \tag{12.7.36}$$

In this formula $\chi_\kappa(1)$ is the dimension of the representation of the symmetric group on k letters in the ideal of its group algebra determined by the Young's diagram for the partition κ. $\chi_\kappa(X)$ is the irreducible polynomial character of $\mathbf{GL}(n)$ of signature κ. In terms of Section 12.11 the substitution $X \to g(YX)$ changes the coefficient matrix of g from E to $\bigotimes_{i=1}^{k} XE$. Then in the algebra of bi-symmetric matrices the irreducible units of an ideal determine the conjugate representations and the ideal determined by κ has $\chi_\kappa(1)$ irreducible units. Hence $\operatorname{tr} P_{\varepsilon_\kappa} \bigotimes_{i=1}^{k} X = (12.7.36)$. This argument is only a sketch; a bit more detail is given in Farrell (1980), Theorem 2.1.

Remark 12.7.9. The characters $g_\kappa(X)$ are known, are called Schur functions, see McDonald (1979), but were called primative characters by Weyl (1946). Explicit formulas may be found in either source. Certain computation properties are discussed in Farrell (1980) and a much more extensive development is given in Takemura (1982) who establishes fully the parallel between zonal polynomials of a real matrix argument and those of complex matrix argument. Both Farrell, op. cit., and Takemura, op. cit., show that Saw's concept, Saw (1977), can be used to give a development of the properties of the zonal polynomials of complex matrix argument.

The Schur functions are one of several natural bases for the space of homogeneous symmetric polynomials. A combinatorial study of the relationships of the various bases may be found in Garsia and Remmel (1981).

12.8. Diagonal Matrices

The implication of Theorem 12.7.4 is that the polynomials $\operatorname{tr} A \bigotimes_{i=1}^{k} X$, A in the center, depend for their value only on the eigenvalues of X. Since we may take YXY^{-1} upper triangular and have for A in the center that $\operatorname{tr} A \bigotimes_{i=1}^{k} X =$

$\operatorname{tr} A \bigotimes_{i=1}^{k} (YXY^{-1})$, the value of the polynomial is the dot product of the diagonal of A with the eigenvalues of X. The approach of Saw (1977), Farrell (1980) and Takemura (1982) is to treat in the abstract homogeneous symmetric functions of degree k and obtain various properties this way. In this section and in Section 12.9 the development of functions of matrix argument is continued.

If the partition κ is $k_1 \geq k_2 \geq \cdots \geq k_p \geq 1$, then we will say the index set $i_1, \ldots, i_k$ *belongs to* κ if and only if there exists a permutation π of $1, \ldots, k$ such that $i_{j_1} = i_{j_2}$ if and only if $\pi(j_1)$ and $\pi(j_2)$ are in the same row set of the Young's diagram $T(\kappa)$.

In this section we work with the canonical basis elements $e_1, \ldots, e_n$ of $\mathbf{E}$ and define

$$\mathbf{E}_\kappa = \{(e_{i_1}, \ldots, e_{i_k}) | \text{such that } i_1, \ldots, i_k \text{ belongs to } \kappa\}. \qquad (12.8.1)$$

In the following recall that if $f \in \mathbf{M}(\mathbf{E}^k, \mathbb{C})$ then $P_\sigma f$ is defined by $(P_\sigma f)(e_{i_1}, \ldots, e_{i_k}) = f(e_{i_{\sigma(1)}}, \ldots, e_{i_{\sigma(k)}})$, so that $P_\sigma P_\tau = P_{\sigma\tau}$.

It is easy to see that $(e_{i_1}, \ldots, e_{i_k})$ belongs to κ if and only if $(e_{i_{\sigma(1)}}, \ldots, e_{i_{\sigma(k)}})$ belongs to κ, so the action of σ on $\mathbf{E}_\kappa$ is $\sigma\mathbf{E}_\kappa \subset \mathbf{E}_\kappa$. Likewise if G is an $n \times n$ permutation matrix (in the basis $e_1, \ldots, e_n$) then $(\bigotimes_{i=1}^{k} G)\, \mathbf{E}_\kappa \subset \mathbf{E}_\kappa$ since $Ge_{i_1}, \ldots, Ge_{i_k}$ belongs to κ if and only if $e_{i_1}, \ldots, e_{i_k}$ does.

In the sequel we use the same notation P_σ as an operator on k-tuples $(e_{i_1}, \ldots, e_{i_k})$ so that $(P_\sigma f)(e_{i_1}, \ldots, e_{i_k}) = f(P_\sigma(e_{i_1}, \ldots, e_{i_k}))$.

Lemma 12.8.1. *Let $\kappa > \kappa'$. Let $i_1, \ldots, i_k$ belong to κ. Let $c' = \Sigma_{p'} \Sigma_{q'} \varepsilon(q')p'q'$ be the Young's symmetrizer of κ'. Define a function d by*

$$\sum_{p'} \sum_{q'} \varepsilon(q') P_{p'q'}(e_{i_1}, \ldots, e_{i_k}) = \sum_{\sigma} d(\sigma) P_\sigma(e_{i_1}, \ldots, e_{i_k}). \qquad (12.8.2)$$

Then $d \equiv 0$.

Proof. By linear independence of the basis elements in $\mathbf{M}(\mathbf{E}^k, \mathbb{C})$, it follows that

$$d(\sigma) = \sum_{p'} \sum_{q'} \varepsilon(q'),$$

$$\text{taken over those } p', q' \text{ such that} \qquad (12.8.3)$$

$$P_{p'q'}(e_{i_1}, \ldots, e_{i_k}) = P_\sigma(e_{i_1}, \ldots, e_{i_k}).$$

The condition in (12.8.3) is equivalent to the condition

$$i_{p'q'(j)} = i_{\sigma(j)}, \qquad 1 \leq j \leq k. \qquad (12.8.4)$$

Since $i_1, \ldots, i_k$ belongs to κ there exists a permutation π of $1, \ldots, k$ depending only on κ and $i_1, \ldots, i_k$ such that if $1 \leq j \leq k$,

$$\pi p'q'(j) \text{ and } \pi\sigma(j) \text{ are in the same row set of } T(\kappa). \qquad (12.8.5)$$

Thus there exists a permutation p of type p such that

$$p\pi p'q' = \pi\sigma \quad \text{and} \quad \sigma = (\pi^{-1}p\pi)(p'q'). \tag{12.8.6}$$

Therefore the condition (12.8.3) is

$$\begin{aligned} d(\sigma) &= \Sigma\, \varepsilon(q'), \text{ the sum being over} \\ &\qquad \{p', q' \mid \text{for some } p,\ \sigma = (\pi^{-1}p\pi)(p'q')\}. \end{aligned} \tag{12.8.7}$$

Then

$$\begin{aligned} d(\pi^{-1}p_0\pi\sigma) &= \Sigma\, \varepsilon(q'), \text{ the sum being over} \\ &\qquad \{p', q' \mid \text{for some } p,\ \pi^{-1}p_0\pi\sigma = \pi^{-1}p\pi(p'q')\} \\ &= \Sigma\, \varepsilon(q'), \text{ the sum being over} \\ &\qquad \{p', q' \mid \text{for some } p,\ \sigma = \pi^{-1}p_0^{-1}p\pi(p'q')\} \\ &= d(\sigma). \end{aligned} \tag{12.8.8}$$

Also,

$$\begin{aligned} d(\sigma q_0') &= \Sigma\, \varepsilon(q'), \text{ the sum being over} \\ &\qquad \{p', q' \mid \text{for some } p,\ \sigma q_0' = \pi^{-1}p\pi(p'q')\} \\ &= \Sigma\, \varepsilon(q_0')\varepsilon(q'q_0'^{-1}), \text{ the sum being over} \\ &\qquad \{p', q' \mid \text{for some } p,\ \sigma = \pi^{-1}p\pi p'(q'q_0'^{-1})\} \\ &= \varepsilon(q_0')d(\sigma). \end{aligned} \tag{12.8.9}$$

By Lemma 12.4.8 and its proof, $d = 0$. □

Corollary 12.8.2. *If $\kappa > \kappa'$ then $P_{\varepsilon_{\kappa'}}$ is the zero operator on $\mathbf{E}_\kappa$.*

Proof. $\phi^2\varepsilon_{\kappa'} = \Sigma_g g c' g^{-1}$, as shown in the proof of Lemma 12.4.14, first line. As noted above, $P_g\mathbf{E}_\kappa \subset \mathbf{E}_\kappa$ for all permutations g of $1, \ldots, k$, so that since $P_{c'}\mathbf{E}_\kappa = 0$, the result follows. □

Corollary 12.8.3. *Let*

$$P_{\varepsilon_{\kappa'}} u_{i_1} \cdots u_{i_m} = \sum_{j_1, \ldots, j_k} a_{j_1, \ldots, j_k} u_{j_1} \cdots u_{j_k}. \tag{12.8.10}$$

If $\kappa > \kappa'$ and $j_1, \ldots, j_k$ belong to κ then $a_{j_1, \ldots, j_k} = 0$.

Proof.

$$\begin{aligned} a_{j_1, \ldots, j_k} &= (P_{\varepsilon_{\kappa'}} u_{i_1} \cdots u_{i_k})(e_{j_1}, \ldots, e_{j_k}) \\ &= u_{i_1} \cdots u_{i_k}(P_{\varepsilon_{\kappa'}}(e_{j_1}, \ldots, e_{j_m})) = 0. \end{aligned}$$ □

Remark 12.8.4. The canonical basis elements of $\mathbf{M}(\mathbf{E}^k, \mathbb{C})$ are the k-forms $u_{i_1} \cdots u_{i_k}$, so the effect of Corollary 12.8.3 is to identify certain basis elements that map to zero.

Corollary 12.8.5. *If the* $((i_1, \ldots, i_k), (j_1, \ldots, j_k))$*-entry of the matrix* P_{ε_κ} *is nonzero then* $i_1, \ldots, i_k$ *belongs to* κ' *and* $j_1, \ldots, j_k$ *belongs to* κ'' *such that* $\kappa' \le \kappa$ *and* $\kappa'' \le \kappa$.

Lemma 12.8.6. *The diagonal of* P_{ε_κ} *is constant for those indices* $i_1, \ldots, i_k$ *which belong to* κ'.

Proof. The maps $i_1, \ldots, i_k \to \tau(i_{\sigma(1)}), \ldots, \tau(i_{\sigma(k)})$ are transitive on the index set that belongs to κ' and all these maps are realizable by matrix multiplication, which when applied says

$$P_{\varepsilon_\kappa} = \left(\bigotimes_{i=1}^{k} G\right) P_{\varepsilon_\kappa} \left(\bigotimes_{i=1}^{k} G^t\right), \quad \text{and} \quad P_{\varepsilon_\kappa} = P_\sigma P_{\varepsilon_\kappa} P_\kappa^t, \tag{12.8.11}$$

with σ a permutation of $1, \ldots, k$ and τ a permutation of $1, \ldots, n$. The lemma then follows. □

Lemma 12.8.7. *If c is the Young's symmetrizer for the partition* κ *then* P_c *is not zero on* $\mathbf{E}_\kappa$.

Proof. Let $i_1, \ldots, i_k$ belong to κ such that $i_{j_1} = i_{j_2}$ if and only if j_1 and j_2 are in the same row set of $T(\kappa)$. Then

$$\begin{aligned} &(e_{i_{pq(1)}}, \ldots, e_{i_{pq(k)}}) = (e_{i_1}, \ldots, e_{i_k}) \\ &\text{if and only if } i_{pq(j)} = i_j, \\ &\text{if and only if } pq(j) \text{ and } j \text{ are in the same row set,} \\ &\text{if and only if } q(j) \text{ and } j \text{ are in the same row set,} \\ &\text{which implies } q(j) = j \text{ for all } j. \end{aligned} \tag{12.8.12}$$

Therefore $\Sigma_p \Sigma_q \varepsilon(q)(e_{i_{pq(1)}}, \ldots, e_{i_{pq(k)}}) = (\#$ permutations of type $p) \times (e_{i_1}, \ldots, e_{i_k}) +$ other terms. By linear independence this is not zero. □

Lemma 12.8.8. *Let d be in the group algebra of the symmetric group. If* $i_1, \ldots, i_k$ *belongs to* κ *and if* $P_d u_{i_1} \cdots u_{i_k} = \Sigma a_{j_1 \cdots j_k} u_{j_1} \cdots u_{j_k}$*, then* $a_{j_1 \cdots j_k} \neq 0$ *implies* $j_1, \ldots, j_k$ *belongs to* κ.

Proof. It suffices to prove the lemma for d which are permutations. Then

$$\begin{aligned} P_\sigma u_{i_1} \cdots u_{i_k}(e_{h_1}, \ldots, e_{h_k}) &= u_{i_1} \cdots u_{i_k}(e_{h_{\sigma(1)}}, \ldots, e_{h_{\sigma(k)}}) \\ &= \Sigma a_{j_1 \cdots j_k} u_{j_1} \cdots u_{j_k}(e_{h_1}, \ldots, e_{h_k}). \end{aligned} \tag{12.8.13}$$

The left side vanishes except for the single case $i_1 = h_{\sigma(1)}, \ldots, i_k = h_{\sigma(k)}$. Hence the only nonzero term on the right is $u_{i_{\sigma^{-1}(1)}} \cdots u_{i_{\sigma^{-1}(k)}}$ for which the index set $i_{\sigma^{-1}(1)}, \ldots, i_{\sigma^{-1}(k)}$ belongs to κ. □

Lemma 12.8.9. *Let* $\mathbf{M}(\mathbf{E}^k, \mathbb{C}, \kappa)$ *be the linear span of those basis elements* $u_{i_1} \cdots u_{i_k}$ *such that* $i_1, \ldots, i_k$ *belongs to* κ*. Then if d is in the group algebra* *of the symmetric group,* $P_d\mathbf{M}(\mathbf{E}^k, \mathbb{C}, \kappa) \subset \mathbf{M}(\mathbf{E}^k, \mathbb{C}, \kappa)$. *Then* P_{ε_κ} *is not zero on* $\mathbf{M}(\mathbf{E}^k, \mathbb{C}, \kappa)$.

PROOF. By Lemma 12.8.8, the operator P_d is reduced by $\mathbf{M}(\mathbf{E}^k, \mathbb{C}, \kappa)$. As in the proof of Lemma 12.8.7 we compute that

$$\begin{aligned} &P_c u_{j_1} \cdots u_{j_k}(e_{i_1}, \ldots, e_{i_k}) \\ &\quad = u_{j_1} \cdots u_{j_k}((\#\text{ of permutations of type } p)(e_{i_1}, \ldots, e_{i_k})) + \text{other terms} \\ &\quad = (\#\text{ permutations of type } p) \text{ if } (i_1, \ldots, i_k) = (j_1, \ldots, j_k), \text{ and} \\ &\quad = 0 \text{ if } (j_1, \ldots, j_k) \text{ does not belong to } \kappa. \end{aligned} \tag{12.8.14}$$

Since $P_{\varepsilon_\kappa} P_c = P_c$, it follows that $P_{\varepsilon_\kappa}(P_c u_{i_1} \cdots u_{i_k}) \neq 0$, and since $P_c u_{i_1} \cdots u_{i_k}$ is an element of $\mathbf{M}(\mathbf{E}^k, \mathbb{C}, \kappa)$, the result follows. □

Theorem 12.8.10. *If* $i_1, \ldots, i_k$ *belongs to* κ *then*

$$\begin{gathered} P_{\varepsilon_\kappa} u_{i_1} \cdots u_{i_k} = \phi u_{i_1} \cdots u_{i_k} + \textit{other terms, and} \\ \phi \textit{ is a nonzero rational number.} \end{gathered} \tag{12.8.15}$$

PROOF. ϕ is the diagonal entry of P_{ε_κ}. By Lemma 12.8.6, P_{ε_κ} has the same diagonal entry for all $u_{i_1} \cdots u_{i_k}$ such that $i_1, \ldots, i_k$ belongs to κ. P_{ε_κ} is an orthogonal projection, see Theorem 12.7.1, which maps $\mathbf{M}(\mathbf{E}^k, \mathbb{C}, \kappa)$ into itself, see Lemma 12.8.9, with the matrix given in terms of an orthogonal basis. Since P_{ε_κ} is not the zero operator on $\mathbf{M}(\mathbf{E}^k, \mathbb{C}, \kappa)$ it follows that $\operatorname{tr} P_{\varepsilon_\kappa} \neq 0$ and that $\phi \neq 0$. Since $\operatorname{tr} P_{\varepsilon_\kappa} = \phi(\text{dimension } \mathbf{M}(\mathbf{E}^k, \mathbb{C}, \kappa))$ it follows that ϕ is rational. □

12.9. Polynomials of Diagonal Matrices X

We let $A \in \mathfrak{A}_k$, and $B \in \mathfrak{A}_k$ be a diagonal matrix, $B = \bigotimes_{i=1}^k X$, X a diagonal matrix. The polynomial

$$\operatorname{tr} AB = \operatorname{tr} A \bigotimes_{i=1}^{k} X \tag{12.9.1}$$

depends only on the diagonal of A for its value.

Lemma 12.9.1. *Let* X *be an* $n \times n$ *diagonal matrix and* $A \in \mathfrak{A}_k$ *be a diagonal matrix. If*

$$\operatorname{tr} A \bigotimes_{i=1}^{k} X = 0 \textit{ for all } n \times n \textit{ diagonal } X, \tag{12.9.2}$$

then $A = 0$.

PROOF. (12.9.2) is a homogeneous polynomial of n variables $x_1, \ldots, x_n$, and the coefficients of the polynomial are all zero. Clearly $x_{i_1} \cdots x_{i_k} = x_{j_1} \cdots x_{j_k}$ if and only if there exists a permutation σ of $1, \ldots, k$ such that $i_{\sigma(1)} = j_1, \ldots, i_{\sigma(k)} = j_k$. Since A is *bi*-symmetric $a_{(i_1,\ldots,i_k),(i_1,\ldots,i_k)} = a_{(i_{\sigma(1)},\ldots,i_{\sigma(k)}),(i_{\sigma(1)},\ldots,i_{\sigma(k)})}$. Hence the diagonal terms all vanish and $A = 0$. □

Lemma 12.9.2. *The polynomials* $\operatorname{tr} P_{\varepsilon_\kappa} \bigotimes_{i=1}^k Y$, *$Y$ a diagonal matrix, are linearly independent.*

SECOND PROOF. This result was obtained by a different argument in Section 12.7. See Theorem 12.7.6. The argument given here is needed as part of the proof for Theorem 12.9.3.

If $\operatorname{tr}(\Sigma_\kappa a_\kappa P_{\varepsilon_\kappa}) \bigotimes_{i=1}^k Y = 0$ identically for diagonal matrices Y then by Lemma 12.9.1, the diagonal of $\Sigma_\kappa a_\kappa P_{\varepsilon_\kappa}$ is identically zero. Order the partitions as $\kappa_1 > \kappa_2 > \cdots > \kappa_r$. By Lemma 12.8.2 the matrices $P_{\varepsilon_{\kappa_2}}, \ldots, P_{\varepsilon_{\kappa_r}}$ map $u_{i_1} \cdots u_{i_k}$ to zero for those index sets $i_1, \ldots, i_k$ belonging to κ_1. Since the diagonal of $P_{\varepsilon_{\kappa_1}}$ is not zero for those $u_{i_1} \cdots u_{i_k}$ such that $i_1, \ldots, i_k$ belongs to κ_1, it follows that $a_{\kappa_1} = 0$. Then the last term of the sum has zero coefficient. Using the obvious backward induction together with the nested character of the matrices P_{ε_κ} as described in Corollary 12.8.5 and Theorem 12.8.10, the result that $a_{\kappa_i} = 0$, $1 \le i \le r$, follows. □

If G is an $n \times n$ permutation matrix then since P_{ε_κ} is in the center of $\mathfrak{A}_k$,

$$\operatorname{tr} P_{\varepsilon_\kappa} \bigotimes_{i=1}^k (GYG^t) = \operatorname{tr}\left(\left(\bigotimes_{i=1}^k G^t\right) P_{\varepsilon_\kappa} \left(\bigotimes_{i=1}^k G\right)\right) \bigotimes_{i=1}^k Y$$
$$= \operatorname{tr} P_{\varepsilon_\kappa} \bigotimes_{i=1}^k Y. \tag{12.9.3}$$

If Y is a diagonal matrix with diagonal entries $y_1, \ldots, y_n$ then the polynomial (12.9.3) can be expressed as

$$f(y_1, \ldots, y_n) = \operatorname{tr} P_{\varepsilon_\kappa} \bigotimes_{i=1}^k Y, \tag{12.9.4}$$

and (12.9.3) says that f is a symmetric function in the variables $y_1, \ldots, y_n$, homogeneous of degree k. The space of these polynomials has dimension r = number of partitions of k into not more than n parts, which is also the number of idempotents P_{ε_κ} in the center of $\mathfrak{A}_k$. The action

$$\begin{aligned} P_{\varepsilon_\kappa} \bigotimes_{i=1}^k Y u_{i_1} \cdots u_{i_k}(e_{j_1}, \ldots, e_{j_k}) &= P_{\varepsilon_\kappa} u_{i_1} \cdots u_{i_k}(y_{j_1} e_{j_1}, \ldots, y_{j_k} e_{j_k}) \\ &= y_{j_1} \cdots y_{j_k} (P_{\varepsilon_\kappa} u_{i_1} \cdots u_{i_k})(e_{j_1}, \ldots, e_{j_k}) \\ &= 0 \text{ if } i_1, \ldots, i_k \text{ belongs to } \kappa' > \kappa. \end{aligned} \tag{12.9.5}$$

Therefore

Theorem 12.9.3. *The space of polynomials*

$$\operatorname{tr}\left(\sum_{\kappa} a_\kappa P_{\varepsilon_\kappa}\right) \bigotimes_{i=1}^{k} Y, \tag{12.9.6}$$

$Y = \operatorname{diag}(y_1, \ldots, y_n)$, *is the space of homogeneous symmetric polynomials of degree k in the variables* $y_1, \ldots, y_n$. *The polynomial* $P_{\varepsilon_\kappa} \bigotimes_{i=1}^{k} Y$ *expressed as a sum of monomial symmetric functions* (12.2.10) *is*

$$\begin{aligned} P_{\varepsilon_{\kappa'}} \bigotimes_{i=1}^{k} \operatorname{diag}(y_1, \ldots, y_n) &= \sum_{\kappa \le \kappa'} a_\kappa M_\kappa(y), \\ y^t &= (y_1, \ldots, y_n). \end{aligned} \tag{12.9.7}$$

Remark 12.9.4. The result stated in Theorem 12.9.3 is that the coefficient matrix of the (complex) zonal polynomials as linear combinations of the monomial symmetric functions is triangular, upper triangular if partitions are in decreasing order. That the expression of polynomial characters as linear combinations of symmetric functions would result in triangular coefficient matrices was known to Weyl (1946), see Theorem 7.6D. Triangularity of the coefficient matrix was used by Saw (1977) to define real zonal polynomials although Saw's development had a gap in it. An elegant treatment using the triangularity of the matrices has now been given by Takemura (1982) in both the case of real zonal polynomials and in the case of complex zonal polynomials, i.e., the Schur functions. Saw's idea is explored in Section 12.12 and Takemura's development is summarized in Section 12.13.

12.10. Zonal Polynomials of Real Matrices

The theory of the preceding sections has assumed complex numbers as coefficients. In the process we have obtained a theory in Section 12.7 of polynomials f with the unitary invariance

$$f(UXU^*) = f(X), \qquad U \text{ unitary.} \tag{12.10.1}$$

In this section we will study polynomials which satisfy the orthogonal invariance

$$f(UXU^t) = f(X), \qquad U \in \mathbf{O}(n), \qquad X \in \mathbf{S}(n). \tag{12.10.2}$$

The resulting function being a symmetric function of the eigenvalues $\lambda_1(X), \ldots, \lambda_n(X)$ is naturally definable for matrices X having n real eigenvalues, and this has been done by James, et. al. The object is to obtain a representation of f as a linear combination of zonal polynomials. The polynomials will enjoy a reproducing property similar to (12.7.9) but in which integration is over the group $\mathbf{O}(n)$ of $n \times n$ matrices. The polynomials however are not those of Section 12.7 and a more elaborate construction is required.

Related to the construction given are polynomials f with a two-sided invariance property

$$f(UXV) = f(X), \qquad U, V \in \mathbf{O}(n), \qquad X \in \mathbf{GL}(n). \tag{12.10.3}$$

An example of (12.10.3) is the integral

$$\int_{\mathbf{O}(n)} f(\operatorname{tr} UX)\, dU \tag{12.10.4}$$

which does not vanish if f is a positive function. A function satisfying (12.10.3) clearly satisfies

$$f(X) = f(((X^tX)^{-1/2}X^t)X) = f((X^tX)^{1/2}) = f(D) \tag{12.10.5}$$

with D an $n \times n$ diagonal matrix whose entries are the eigenvalues of $(X^tX)^{1/2}$. If f is a homogeneous polynomial of degree m then

$$f(X) = f(-X) = (-1)^m f(X). \tag{12.10.6}$$

Thus if m is odd the polynomial vanishes. In the sequel we write $2k = m$ for the degree. By the theory of orthogonal invariants in Weyl (1946), $f((X^tX)^{1/2})$ is a polynomial $f_1(S)$ of degree k in the entries of $S = X^tX$. Thus

Lemma 12.10.1. *If f is a homogeneous polynomial of degree $2k$ in the entries of X such that* (12.10.3) *holds then there is a polynomial f_1 of n variables such that*

$$f(X) = f_1(\operatorname{tr} X^tX, \operatorname{tr}(X^tX)^2, \ldots, \operatorname{tr}(X^tX)^n) \tag{12.10.7}$$

and such that every term is homogeneous of degree $2k$.

PROOF. $f(X) = f_2(S)$ and f_2 satisfies (12.10.2), so that $f_2(S) = f_3(D^2)$, with D as in (12.10.5). Therefore the value of $f(X)$ is a symmetric function of the eigenvalues of X^tX, hence the form (12.10.7). □

It is clear that if f satisfies (12.10.2) then the polynomial $f(X^tX)$ satisfies (12.10.3). Lemma 12.10.1 states the converse, and implies the dimension of the space of polynomials in question is again r, the number of partitions of k into not more than n parts. See Theorem 12.2.13 and (12.10.7).

A homogeneous polynomial of degree $2k$ satisfying (12.10.3) must have the form

$$\operatorname{tr} A \bigotimes_{i=1}^{2k} X = \operatorname{tr} A \bigotimes_{i=1}^{2k} U \bigotimes_{i=1}^{2k} X \bigotimes_{i=1}^{2k} V, \tag{12.10.8}$$

with $U, V \in \mathbf{O}(n)$. Thus we are led to define an operator

$$E = \int_{\mathbf{O}(n)} \left(\bigotimes_{i=1}^{2k} U\right) dU. \tag{12.10.9}$$

The coefficient matrix A must satisfy, A is *bi*-symmetric, i.e., $A \in \mathfrak{A}_{2k}$, and

$$A = EAE. \tag{12.10.10}$$

The main result of this section is that the polynomials satisfying (12.10.3) are given uniquely by *bi*-symmetric matrices A satisfying (12.10.10). One of the main results of James (1961b) is that the set of matrices EAE is a commutative algebra of dimension r containing r mutually orthogonal idempotents. These idempotents provide the definition of the zonal polynomials of real matrix argument.

Lemma 12.10.2.

$$E^2 = E,\ E^t = E \text{ and } \bar{E} = E. \text{ If } U \in \mathbf{O}(n) \text{ then}$$
$$\left(\bigotimes_{i=1}^{2k} U\right) E = E = E\left(\bigotimes_{i=1}^{2k} U\right). \tag{12.10.11}$$

Proof. The entries of E are real numbers so that $\bar{E} = E$ follows automatically. Since the Haar measure of unit mass on $\mathbf{O}(n)$ is invariant,

$$\begin{aligned} E\left(\bigotimes_{i=1}^{2k} V\right) &= \int_{\mathbf{O}(n)} \left(\bigotimes_{i=1}^{2k} U\right)\left(\bigotimes_{i=1}^{2k} V\right) dU \\ &= \int_{\mathbf{O}(n)} \left(\bigotimes_{i=1}^{2k} UV\right) dU = E. \end{aligned} \tag{12.10.12}$$

Therefore

$$E = \int_{\mathbf{O}(n)} E\, dV = \int_{\mathbf{O}(n)} E\left(\bigotimes_{i=1}^{2k} V\right) dV = E^2. \tag{12.10.13}$$

Last, the mapping $U \to U^t$ is a measure preserving map of Haar measure on $\mathbf{O}(n)$ so that

$$\int_{\mathbf{O}(n)} f(U)\, dU = \int_{\mathbf{O}(n)} f(U^t)\, dU, \tag{12.10.14}$$

f an arbitrary function. In particular $E^t = E$. □

We now formalize the opening discussion in a Lemma.

Lemma 12.10.3. *The two-sided invariant polynomials which are homogeneous of degree* $2k$ *are the polynomials*

$$\operatorname{tr}(EAE)\left(\bigotimes_{i=1}^{2k} X\right), \qquad A \in \mathfrak{A}_{2k}. \tag{12.10.15}$$

The coefficient matrices EAE clearly form an algebra $E\mathfrak{A}_{2k}E$. Since $(EAE)^t = E^tA^tE^t = EA^tE \in E\mathfrak{A}_{2k}E$, and $\overline{EAE} = E\bar{A}E \in E\mathfrak{A}_{2k}E$, the algebra is an H^* algebra. As noted above, if X is $n \times n$ nonsingular and f satisfies

(12.10.3) then $f(X) = f_1(X^t X)$. If X is singular then take $\varepsilon \neq 0$ so that $X + \varepsilon I_n$ is nonsingular. Then $f(X + \varepsilon I_n) = f_1((X + \varepsilon I_n)^t (X + \varepsilon I_n))$. By continuity as $\varepsilon \to 0$, $f(X) = f_1(X^t X)$ follows for all $n \times n$ matrices X. In particular if $(X^t X)^{1/2}$ is a positive semidefinite square root then $f((X^t X)^{1/2}) = f_1((X^t X)^{1/2}(X^t X)^{1/2}) = f_1(X^t X) = f(X)$. The matrix $A \in \mathfrak{A}_k$ can be written as

$$A = \Sigma_i a_i \left(\bigotimes_{j=1}^{2k} X_i \right), \tag{12.10.16}$$

so that

$$\begin{aligned} EAE &= \Sigma_i a_i \left(E \left(\bigotimes_{j=1}^{2k} X_i \right) E \right) = \Sigma_i a_i \left(E \left(\bigotimes_{j=1}^{2k} (X_i^t X_i)^{1/2} \right) E \right) \\ &= \Sigma_i a_i \left(E \left(\bigotimes_{j=1}^{2k} D_i \right) E \right) = E \left(\Sigma_i a_i \bigotimes_{j=1}^{2k} D_i \right) E \end{aligned} \tag{12.10.17}$$

In this expression the matrices D_i are diagonal matrices. Therefore to every $A \in \mathfrak{A}_{2k}$ there is a diagonal matrix $D \in \mathfrak{A}_{2k}$ such that $EAE = EDE$. All the entries of D with index $i_1, \ldots, i_k$ belonging to κ has the same value. See Section 12.8. Therefore D is a sum $D = \Sigma_{2\kappa} D_{2\kappa} I_{2\kappa}$ in which $I_{2\kappa}$ is the identity on $\mathbf{M}(\mathbf{E}^{2k}, \mathbb{C}, 2\kappa)$ defined in Lemma 12.8.10, and $I_{2\kappa}$ is zero otherwise. The notation here is, if κ is given by $k_1 \geq k_2 \geq \cdots \geq k_n$ then the partition 2κ is given by $2k_1 \geq \cdots \geq 2k_n$.

Theorem 12.10.4. *The algebra $E\mathfrak{A}_{2k}E$ is commutative. The irreducible idempotents are those elements $P_{\varepsilon_{2\kappa}}E$ (which are not zero) and $E\mathfrak{A}_{2k}E$ is the linear span of these idempotents. The number of idempotents is the number of partitions of k into not more than n parts. The algebra $E\mathfrak{A}_{2k}E$ is also the linear span of the matrices $EI_\gamma E$, some of which are zero.*

Proof. As seen above the elements of $E\mathfrak{A}_{2k}E$ are expressible as EDE with D a diagonal matrix. Therefore

$$(EDE)^t = E^t D^t E^t = EDE. \tag{12.10.18}$$

If in a matrix algebra $X^t = X$ for all X, then

$$XY = (XY)^t = Y^t X^t = YX, \tag{12.10.19}$$

and the algebra is commutative. □

As noted above the algebra $E\mathfrak{A}_{2k}E$ is an H^* algebra and is therefore a direct sum of its minimal ideals each of which is a total matrix algebra. Thus the minimal ideals are all one-dimensional, and $E\mathfrak{A}_{2k}E$ is the linear span of its irreducible idempotents. We know the dimension of $E\mathfrak{A}_{2k}E$ is the dimension of the space of homogeneous polynomials of degree k in n variables, which we have been calling r, so that $E\mathfrak{A}_{2k}E$ is the linear span of r idempotents.

We now identify the irreducible idempotents. In the process we shall find that the representation $a \to P_{\varepsilon a}E$ of the symmetric group algebra is irreducible. Let $\mathbf{A}$ be the space of $n^{2k} \times n^{2k}$ matrices acting as linear transformations of $\mathbf{M}(\mathbf{E}^{2k}, \mathbb{C})$ in the canonical basis of this space. First we find the endomorphisms of the subspace $E\mathbf{M}(\mathbf{E}^{2k}, \mathbb{C})$. A transformation T of $E\mathbf{M}(\mathbf{E}^{2k}, \mathbb{C})$ may be extended to a transformation T of $\mathbf{M}(\mathbf{E}^{2k}, \mathbb{C})$ by defining $T(f) = 0$ on the orthogonal complement $(I_{n^{2k}} - E)\mathbf{M}(\mathbf{E}^{2k}, \mathbb{C})$. Then T extended has matrix A in the canonical basis, and since

$$T(f) = T(Ef) + T((I - E)f) = T(Ef) = E(T(Ef)), \quad (12.10.20)$$

it follows that $A = EAE$.

Within the matrix set $E\mathbf{A}E$ we seek the commutator algebra of the matrix set $\{EP_aE, a \text{ in the symmetric group algebra}\}$. If g is a permutation of $1, \ldots, 2k$ and (EAE) is in the commutator algebra then

$$(EP_gE)(EAE) = (EAE)(EP_gE). \quad (12.10.21)$$

The matrix E is bi-symmetric so that by definition of the bi-symmetric matrices

$$P_aE = EP_a, \quad a \in \text{the symmetric group algebra}. \quad (12.10.22)$$

It follows that

$$P_g(EAE) = (EAE)P_g, \quad g \text{ a permutation of } 1, \ldots, 2k. \quad (12.10.23)$$

Therefore EAE is a *bi*-symmetric matrix and

$$EAE = E(EAE)E. \quad (12.10.24)$$

Conversely, if A is a *bi*-symmetric matrix and g is a permutation of $1, \ldots, 2k$ then

$$(EP_gE)(EAE) = (EAE)(EP_gE). \quad (12.10.25)$$

Therefore the matrix set EAE with A *bi*-symmetric is the commutator algebra of

$$\{EP_aE \mid a \in \text{symmetric group algebra}\}$$
$$= \{P_aE \mid a \in \text{symmetric group algebra}\}. \quad (12.10.26)$$

The centers of the algebras are the same, by Lemma 12.6.2. Clearly $\{a \mid P_aE = 0\}$ is an ideal of the symmetric group algebra. Thus either $P_{\varepsilon_\gamma}E = 0$ or the map $a\varepsilon_\gamma \to P_aP_{\varepsilon_\gamma}$ is one-to-one into. Therefore the center of the matrix set (12.10.26) is the linear span of the idempotents $P_{\varepsilon_\gamma}E$ that are not zero. These are the irreducible idempotents of center.

Since the commutator algebra is a commutative algebra, the representation $g \to P_gP_{\varepsilon_\gamma}E$ as an endomorphism on the vector space $E\mathbf{M}(\mathbf{E}^{2k}, \mathbb{C})$ is an irreducible representation of the group of permutations of $1, \ldots, 2k$.

$$\operatorname{tr} P_{\varepsilon_\gamma}E = 0, \quad (12.10.27)$$

or

$$\operatorname{tr} P_{\varepsilon_\gamma} E = \text{dimension of the representation} = \chi_\gamma(1).$$

See Lemma 12.11.4 and Problem 13.3.11.

The question remains, which $P_{\varepsilon_\gamma} E = 0$? This question is resolved as the last result of this section by showing that if γ as a partition of $2k$ has only even summands then $P_{\varepsilon_\gamma} E \neq 0$. A dimensionality argument will then show that if γ has an odd summand then $P_{\varepsilon_\gamma} E = 0$. This leads to the natural definition, given above, of 2κ as the partition $2k_1 \geq 2k_2 \geq \cdots \geq 2k_n$ where κ is the partition $k_1 \geq k_2 \geq \cdots \geq k_n$. □

We first prove the splitting properly.

Theorem 12.10.5.

$$\int_{O(n)} \operatorname{tr} P_{\varepsilon_\gamma} E \bigotimes_{i=1}^{2k} (UXU^t Y)\, dU$$
$$= \left(\operatorname{tr} P_{\varepsilon_\gamma} E \bigotimes_{i=1}^{2k} X\right)\left(\operatorname{tr} P_{\varepsilon_\gamma} E \bigotimes_{i=1}^{2k} Y\right) \Big/ (\operatorname{tr} P_{\varepsilon_\gamma} E). \tag{12.10.28}$$

Proof. We assume $P_{\varepsilon_\gamma} E \neq 0$ and write $f(X) = \operatorname{tr} P_{\varepsilon_\gamma} E \otimes_{i=1}^{2k} X$. Then

$$\int_{O(n)} f(UXU^t Y)\, dU \tag{12.10.29}$$

has coefficient matrix in $\mathfrak{A}_{2k} P_{\varepsilon_\gamma} E$, and, as a polynomial in X, is two-sided invariant in the sense of (12.10.3). Consequently, since the only coefficient matrices of two-sided invariant polynomials in $\mathfrak{A}_{2k} P_{\varepsilon_\gamma} E$ are the matrices $\phi P_{\varepsilon_\gamma} E$, ϕ a complex number, it follows that the integral in (12.10.29) has value

$$\phi_1(Y) f(X) = \phi_2(X) f(Y). \tag{12.10.30}$$

Take $X =$ identity to obtain

$$\phi_1(Y)(\operatorname{tr} P_{\varepsilon_\gamma} E) = \phi_2(I) f(Y) = f(Y). \tag{12.10.31}$$

□

Lemma 12.10.6. *If the partition γ of $2k$ involves only even summands then $P_{\varepsilon_\gamma} E \neq 0$. If this partition involves at least one odd summand then $P_{\varepsilon_\gamma} E = 0$.*

Proof. We let $\delta(i,j) = 1$ if $i = j$ and $\delta(i,j) = 0$ if $i \neq j$. We examine the two-sided invariant polynomial

$$(\operatorname{tr} X^t X)^k = \sum_{i_1 \ldots i_{2k}} \sum_{h_1 \ldots h_{2k}} \sum \sum a_{i_1 \ldots i_{2k} h_1 \ldots h_{2k}} x_{i_1 h_1} \cdots x_{i_{2k} h_{2k}}, \tag{12.10.32}$$

with coefficients

$$a_{i_1 \cdots i_{2k} h_1 \cdots h_{2k}} = \delta(i_1, i_2) \cdots \delta(i_{2k-1}, i_{2k}) \delta(h_1, h_2) \cdots \delta(h_{2k-1}, h_{2k}).$$

The coefficient matrix A in (12.10.32) is symmetric but not bi-symmetric.

However it is easy to see that A is positive semidefinite. Consequently if B is an $n^{2k} \times n^{2k}$ matrix of real entries it follows that $AB = 0$ if and only if $B^tAB = 0$. We let $\tilde{A} = (k!)^{-1}\Sigma_g P_g A P_g^t$ so that the bi-symmetric matrix $\tilde{A}$ is the coefficient matrix of the same polynomial as (12.10.32). Further it is easy to see that $(k!)\tilde{A} \geq A$ and $(k!)P_{\varepsilon_\gamma}\tilde{A}P_{\varepsilon_\gamma} \geq P_{\varepsilon_\gamma}AP_{\varepsilon_\gamma}$. Since $P_{\varepsilon_\gamma} = P_{\varepsilon_\gamma}^t$, it follows that if $c = \Sigma_p \Sigma_q \varepsilon(q)pq$ is the Young's symmetrizer for γ then $P_c P_{\varepsilon_\gamma} = P_c$ and $P_c A \neq 0$ implies $P_{\varepsilon_\gamma} A \neq 0$ implies $P_{\varepsilon_\gamma} A P_{\varepsilon_\gamma} \neq 0$ implies $P_{\varepsilon_\gamma}\tilde{A}P_{\varepsilon_\gamma} \neq 0$. Since $\tilde{A}$ is bi-symmetric and P_{ε_γ} is in the center of $\mathfrak{A}_{2k}$, the above implies $P_{\varepsilon_\gamma}\tilde{A}P_{\varepsilon_\gamma} = \tilde{A}P_{\varepsilon_\gamma} \neq 0$.

To show $P_c A \neq 0$ is equivalent to showing $AP_c^t = (P_c A)^t \neq 0$. Further $P_c^t = \Sigma_p \Sigma_q \varepsilon(q)(P_p P_q)^t = \Sigma_p \Sigma_q \varepsilon(q) P_{q^{-1}} P_{p^{-1}}$. Then

$$P_c^t(u_{i_1} \cdots u_{i_{2k}}) = \sum_p \sum_q \varepsilon(q) u_{i_{pq(1)}} \cdots u_{i_{pq(2k)}}. \tag{12.10.33}$$

When A is applied to a term of (12.10.33) then the entries of the row of A are zero except in the case

$$1 = \delta(i_{pq(1)}, i_{pq(2)}) \cdots \delta(i_{pq(2k-1)}, i_{pq(2k)}). \tag{12.10.34}$$

Choose the index set $i_1, \ldots, i_{2k}$ to satisfy (see (12.4.2))

$$\text{if } 1 \leq j \leq 2k \text{ then } i_j = \text{the row number of the row of } T(\gamma) \text{ containing } j. \tag{12.10.35}$$

Then (12.10.34) requires $pq(2j+1)$ and $pq(2j+2)$ to be in the same row of $T(\gamma)$, hence $q(2j+1)$ and $q(2j+2)$ are in the same row. Since $2j+1$ and $2j+2$ are necessarily in the same row set of $T(\gamma)$, and since q maps each column of $T(\gamma)$ into itself, it follows that

$$q(2j+1) + 1 = q(2j+2), \qquad 1 \leq j \leq k-1. \tag{12.10.36}$$

This is because a column of $T(\gamma)$ contains only even integers or only odd integers. Therefore it follows that q is an even permutation and $\varepsilon(q) = 1$. In particular all coefficients of $AP_c^t u_{i_1} \cdots u_{i_{2k}}$ are nonnegative and $AP_c^t u_{i_1} \cdots u_{i_{2k}} \neq 0$.

Thus $\tilde{A}P_{\varepsilon_\gamma} \neq 0$. Since $\tilde{A}$ is the bi-symmetric coefficient matrix of a two-sided invariant polynomial, it follows from Theorem 12.10.4 that $\tilde{A} = \Sigma_{\gamma'} a_{\gamma'} P_{\varepsilon_{\gamma'}} E$. This clearly implies $P_{\varepsilon_\gamma} E \neq 0$. The space $E\mathfrak{A}_{2k}E$ has dimension r, the number of partitions of k into not more than n parts. This is the number of partitions γ of $2k$ into even summands. Since the nonzero matrices $P_{\varepsilon_\gamma}E$ are linearly independent, it follows that $P_{\varepsilon_\gamma}E = 0$ if γ involves an odd summand. □

We now establish the connection between the theory of this section and zonal polynomials as described by James (1964). Since the matrices $P_{\varepsilon_{2\kappa}}$ are in the center of the algebra of bi-symmetric matrices, $P_{\varepsilon_{2\kappa}}E = P_{\varepsilon_{2\kappa}}EP_{\varepsilon_{2\kappa}}$. Then by Lemma 12.8.8 and its proof, the matrix of $P_{\varepsilon_{2\kappa}}E$ in the canonical basis is zero for those $u_{i_1} \cdots i_{i_{2k}}$ such that $i_1, \ldots, i_{2k}$ belongs to $\gamma' > 2\kappa$.

The matrix $P_{\varepsilon_\gamma} E \bigotimes_{i=1}^{2k} X$ with $X = \operatorname{diag}(x_1, \ldots, x_n)$ applied to $u_{i_1} \cdots u_{i_{2k}}$ evaluated at $e_{j_1}, \ldots, e_{j_{2k}}$ is

$$\begin{aligned}(P_{\varepsilon_\gamma} E)&(u_{i_1} \cdots u_{i_{2k}})(x_{j_1} e_{j_1}, \ldots, x_{j_{2k}} e_{j_{2k}}) \\ &= \left(\prod_{h=1}^{2k} x_{j_h}\right)(P_{\varepsilon_\gamma} E)(u_{i_1} \cdots u_{i_{2k}})(e_{j_1}, \ldots, e_{j_{2k}}) \\ &= \left(\prod_{h=1}^{2k} x_{j_h}\right)(\phi u_{i_1} \cdots u_{i_{2k}})(e_{j_1}, \ldots, e_{j_{2k}}) + \text{other terms.}\end{aligned} \tag{12.10.37}$$

Take $e_{j_1}, \ldots, e_{j_{2k}} = e_{i_1}, \ldots, e_{i_{2k}}$ and obtain the value $\phi \prod_{h=1}^{2k} X_{i_h}$.

Thus the $(i_1, \ldots, i_{2k}), (i_1, \ldots, i_{2k})$ diagonal term of $P_{\varepsilon_\gamma} E \bigotimes_{i=1}^{2k} X$ with $X = \operatorname{diag}(x_1, \ldots, x_n)$ consists of the monomial term

$$\phi \prod_{h=1}^{2k} x_{i_h} = \phi \prod_{h=1}^{n} x_h^{f_h} \tag{12.10.38}$$

where $\phi \geq 0$ and $\phi = 0$ if f_h is an odd integer for some $1 \leq h \leq n$.

This discussion may be summarized in the following theorem.

Theorem 12.10.7. *The polynomial* $\operatorname{tr} P_{\varepsilon_{2\kappa}} E \bigotimes_{i=1}^{2k} X$, *with* $X = \operatorname{diag}(x_1, \ldots, x_n)$ *is a sum of monomial symmetric functions* M_γ *homogeneous of degree* $2k$ *such that* $\gamma \leq 2\kappa$. *The coefficients of the monomial terms are nonnegative and* $M_{2\kappa}$ *has positive coefficient in the expansion.*

Remark 12.10.8. We have defined zonal polynomials of a real matrix argument X by the polynomials $P_{\varepsilon_{2\kappa}} E \bigotimes_{i=1}^{2k} X = P_{\varepsilon_{2\kappa}} E \bigotimes_{i=1}^{2k} (X^t X)^{1/2}$. Because of the orthogonal invariance the definition introduced by James, op. cit., without consideration of normalizations, are homogeneous polynomials of the entries of symmetric matrices $A^t A$ defined by

$$f_\kappa(A^t A) = P_{\varepsilon_{2\kappa}} E \bigotimes_{i=1}^{2k} A. \tag{12.10.39}$$

We note that if $A^t A = B^t B$ and both A and B are $n \times n$ matrices then there exists $U \in \mathbf{O}(n)$ with $UA = B$. Then

$$\begin{aligned}f_\kappa(B^t B) &= P_{\varepsilon_{2\kappa}} E \bigotimes_{i=1}^{2k} (UA) = P_{\varepsilon_{2\kappa}} \left(E \bigotimes_{i=1}^{2k} U\right)\left(\bigotimes_{i=1}^{2k} A\right) \\ &= f_\kappa(A^t A).\end{aligned} \tag{12.10.40}$$

Because of orthogonal invariance, $f_\kappa(U^t A^t A U) = f_\kappa(A^t A)$, and f_κ is a symmetric function of the n eigenvalues of $A^t A$. We use the notation $\lambda(A^t A)$ for the vector of eigenvalues. For any matrix with real entries and a complete set of eigenvectors with positive real eigenvalues we may define

$$f_\kappa(A) = P_{\varepsilon_{2\kappa}} E \bigotimes_{i=1}^{2k} \operatorname{diag}(\lambda^{1/2}(A)). \tag{12.10.41}$$

With this extension of the definition, necessary for the theory developed by James, op. cit.,

$$f_\kappa(AB) = f_\kappa(BA). \tag{12.10.42}$$

The reproducing property for positive definite matrices A and B takes the form

$$\begin{aligned} f_\kappa(AUBU^t) &= f_\kappa(A^{1/2}A^{1/2}UB^{1/2}U^tUB^{1/2}U^t) \\ &= f_\kappa((UB^{1/2}U^tA^{1/2})(A^{1/2}UB^{1/2}U^t)) \\ &= P_{\varepsilon_{2\kappa}} E \bigotimes_{i=1}^{2k} (A^{1/2}UB^{1/2}U^t). \end{aligned} \tag{12.10.43}$$

Then by Theorem 12.10.5,

$$\int_{\mathbf{O}(n)} f_\kappa(AUBU^t)\,dU = f_\kappa(A)f_\kappa(B)/f_\kappa(I_n). \tag{12.10.44}$$

Depending on the normalization used and upon the symmetric functions used in the expansion, letters C_κ and Z_κ have been used by James to describe sets of zonal polynomials. See the survey paper, James (1964). This is discussed in Section 12.11. See definitions 12.11.2 for the functions Z_κ and 12.11.3 for the functions C_κ. Takemura (1982) gives a very complete discussion of the normalizations.

In Sections 12.12 and 12.13 we further develop the properties of the symmetric functions f_κ.

12.11. Alternative Definitions of Zonal Polynomials. Group Characters

James (1960) defined zonal polynomials in terms of group representations as follows. A homogeneous polynomial of degree k in the entries of an $n \times n$ symmetric matrix S, say $f(S)$, allows substitutions

$$A \to f(A^tSA) \tag{12.11.1}$$

of the real general linear group. Since $f(A^tSA)$ is again a homogeneous polynomial of degree k, this substitution is a representation of $\mathbf{GL}(n)$. The space of such polynomials, $\mathbf{V}_k$, decomposes into a direct sum $\mathbf{V}_k = \bigotimes_{i=1}^r \mathbf{V}_{ki}$ of invariant subspaces, $r =$ the number of partitions of k into not more than n parts, and each $\mathbf{V}_{ki}$ contains a one-dimensional subspace generated by an orthogonally invariant polynomial ϕ_i, that is, $\phi_i(U^tSU) = \phi_i(S)$ for all $U \in \mathbf{O}(n)$. The polynomial ϕ_i, normalized in the manner described below, is the zonal polynomial for the representation of $\mathbf{GL}(n)$ in $\mathbf{V}_{ki}$.

A polynomial f of degree k determines a polynomial g of degree $2k$ by

$$g(X) = f(X^tX). \tag{12.11.2}$$

Clearly if $U \in \mathbf{O}(n)$ then $g(UX) = g(X)$. If

$$g(X) = \operatorname{tr} A \bigotimes_{i=1}^{2k} X, \tag{12.11.3}$$

with $A \in \operatorname{Re} \mathfrak{A}_{2k}$, the real bi-symmetric matrices, then clearly

$$g(X) = g(UX) = tr\left(A \bigotimes_{i=1}^{2k} U\right)\left(\bigotimes_{i=1}^{2k} X\right), \tag{12.11.4}$$

so on integration over $\mathbf{O}(n)$ by Haar measure of unit mass, it follows that

$$AE = A. \tag{12.11.5}$$

The converse is obvious, modulo Weyl (1946), since by the theory of orthogonal invariants, if $A \in \operatorname{Re} \mathfrak{A}_{2k} E$ there exists a polynomial f such that

$$f(X^t X) = \operatorname{tr} A \bigotimes_{i=1}^{2k} X. \tag{12.11.6}$$

Hence $\mathbf{V}_k$ is isomorphic to $\operatorname{Re} \mathfrak{A}_{2k} E$ and the invariant subspaces of $\mathbf{V}_k$ are clearly given by the ideals

$$\operatorname{Re} \mathfrak{A}_{2k} P_{\varepsilon_{2\kappa}} E \tag{12.11.7}$$

such that $P_{\varepsilon_{2\kappa}} E \neq 0$. Theorem 12.10.4 clearly implies the invariant subspace corresponding to the ideal $\operatorname{Re} \mathfrak{A}_{2k} P_{\varepsilon_{2\kappa}} E$ has a one-dimensional subspace generated by $P_{\varepsilon_{2\kappa}} E$ which is the coefficient matrix of the uniquely determined two-sided invariant polynomial. Therefore the following theorem holds.

Theorem 12.11.1. *If f is a zonal polynomial of degree k in the entries of the real positive definite matrices S, then there exists $\phi \in \mathbb{C}$ and a partition 2κ of even summands such that if X is an $n \times n$ matrix,*

$$f(X^t X) = \phi \operatorname{tr} (P_{\varepsilon_{2\kappa}} E)\left(\bigotimes_{i=1}^{2k} X\right). \tag{12.11.8}$$

Conversely, every such polynomial, except for normalization, is a zonal polynomial.

In tables of zonal polynomials given by James (1960, 1961a, 1964, 1968) and by James and Parkhurst (1974) the polynomials Z_κ are expressed as linear combinations of the functions J. See (12.2.13).

Definition 12.11.2. Let the partition κ be $k_1 \geq \cdots \geq k_n$. The zonal polynomial Z_κ of real positive semidefinite matrices $X^t X$ is the polynomial (12.11.8) with ϕ chosen to make the coefficient of $(\operatorname{tr} X^t X)^k$ equal one when expressed in the basis functions S.

The fact that the term $(\operatorname{tr} X^t X)^k$ always has a nonzero coefficient follows from the proof of Lemma 12.10.6. The polynomials are unknown in explicit closed form, but have been tabled for degrees $k \le 12$. See James and Parkhurse (1974). In James (1960, 1961a, 1964) the polynomials are expressed as polynomials of the symmetric functions $\operatorname{tr}(X^tX), \operatorname{tr}(X^tX)^2, \ldots, \operatorname{tr}(X^tX)^k$. In James (1964) the zonal polynomials were also given as polynomial functions of the elementary symmetric functions. In James (1968) a recurrence relation is derived from the Laplace–Beltrami operator. Use of the recurrence relation leads naturally to expression of the zonal polynomials in terms of the monomial symmetric functions

$$\sum_{\sigma} x_{\sigma(1)} \cdots x_{\sigma(i)}, \qquad 1 \le i \le n, \tag{12.11.9}$$

and James (1968) tables the zonal polynomials for degrees ≤ 5 as functions of the monomial symmetric functions. The recursion established that the coefficients of the monomials would be nonnegative. We have restated this result, with different proofs, in Theorem 12.8.10. Kates (1980) obtained explicit formulas for the zonal polynomials expressed as expectations of products of functions of Wishart matrix argument. A different proof of the Kates formula may be found in Takemura (1982). Takemura's proof of Kates's formulas is sketched in Section 12.14.

It was realized by Constantine (1963) and James (1964) that more elegantly expressed formulas would result if a different normalization were used.

Definition 12.11.3. The zonal polynomial C_κ is the polynomial $\phi_\kappa Z_\kappa$, with $\phi_\kappa \in \mathbb{C}$, such that

$$\sum_{\kappa} C_\kappa(S) = (\operatorname{tr} S)^k, \tag{12.11.10}$$

S a real positive semidefinite matrix. By Theorem 12.10.4 the expansion (12.11.10) is unique.

The numbers ϕ_κ used to pass from one definition to the other are not easy to compute but were computed by Constantine (1963) and were restated by James (1964). We state here without proof the following lemma.

Lemma 12.11.4. *The numbers ϕ_κ of Definition* 12.11.3 *have the value*

$$\chi_{2\kappa}(1)2^k(k!)/(2k)!, \tag{12.11.11}$$

and $\chi_{2\kappa}(1)$ is the dimension of the representation 2κ of the symmetric group on $2k$ symbols, given by

$$\chi_{2\kappa}(1) = (2k)! \prod_{i<j}^{p} (2k_i - 2k_j - i + j) \prod_{i=1}^{p} (2k_i + p - i)!, \tag{12.11.12}$$

the partition being of p nonzero parts $2k_1 \ge \cdots \ge 2k_p \ge 2$.

In regard (12.11.12), cf. Weyl (1946), page 213, Theorem (7.7.B). The formula (12.11.11) involves a quantity obtained from the theory of group representations. The treatments of Saw (1977) and Takemura (1982) discussed in Sections 12.12 and 12.13 do not seem able to obtain these formulas but present a much simpler derivation of basics and avoid the use of group representations in their derivations.

Aside from the three methods of definition mentioned, namely the algebraic approach of this book, group representations, and the methods of Saw and Takemura, another definition arises from the fact that zonal polynomials are spherical functions in the sense of Helgason (1962). This provided the basis of James' (1968) use of the Laplace–Beltrami differential equation to obtain a differential equation solved by the zonal polynomials. See Takemura, op. cit., and Muirhead (1982), both of whom have extensive discussions of the differential equation. We finish this section with a discussion of group representations and spherical functions.

If $\mathfrak{A}$ is an algebra over the complex numbers $\mathbb{C}$ a mapping of a group $\mathfrak{G}$,

$$A \to \pi(A) \in \mathfrak{A} \tag{12.11.13}$$

is a group representation provided $\mathfrak{A}$ has a multiplicative identity. $\pi(\text{identity}) = \text{identity}$ and $\pi(AB) = \pi(A)\pi(B)$. $\pi(A)$ determines an endomorphism T_A of the vector space $\mathfrak{A}$ by $T_A(B) = \pi(A)B$ satisfying $T_{AB} = T_A T_B$. We consider here the case that $\mathfrak{A}$ is completely reducible and e is an idempotent of the center of $\mathfrak{A}$ such that $\mathfrak{A}e$ is a minimal ideal. Then $\mathfrak{A}e$ is a full matrix algebra spanned by elements e_{ij} satisfying

$$e_{ik}e_{kj} = e_{ij}, \qquad e = e_{11} + \cdots + e_{pp}, \qquad 1 \le i, j \le p. \tag{12.11.14}$$

Thus we write

$$\pi(A)e = \sum_{i=1}^{p} \sum_{j=1}^{p} \bar{a}_{ij} e_{ij} \tag{12.11.15}$$

and note that

$$\pi'(A) = \pi(A)e \tag{12.11.16}$$

is a representation. Acting on $\mathfrak{A}e_{kk}$ the transformation $\pi'(A)$ may be considered to be an endomorphism whose matrix in the basis $e_{1k}, \ldots, e_{pk}$ is

$$\pi'(A)e_{qk} = \sum_{i=1}^{p} \sum_{j=1}^{p} \bar{a}_{ij} e_{ij} e_{qk} = \sum_{i=1}^{p} \bar{a}_{iq} e_{ik}, \tag{12.11.17}$$

that is, the matrix is $(\bar{a}_{ij})^t$. Therefore these representations of $\mathfrak{G}$ in End $\mathfrak{A}e_{kk}$, $1 \le k \le p$, are conjugate and each have the same matrix (12.11.17).

The *group character* is the trace of the matrix $\pi(A)$ as a linear transformation, so that the character is

$$\chi(A) = p(\bar{a}_{11} + \cdots + \bar{a}_{pp}). \tag{12.11.18}$$

In the case $\mathfrak{A}_k$ of bi-symmetric $n^k \times n^k$ matrices and the symmetric group, p is the number of distinct conjugates of a Young's diagram corresponding to a particular κ, which we denote

$$\chi_\kappa(1) = p. \tag{12.11.19}$$

We now turn to a specific interpretation.

Lemma 12.11.5. *To each zonal polynomial C_κ define*

$$D_\kappa(X) = C_\kappa(X^t X)/C_\kappa(I_n). \tag{12.11.20}$$

Then

$$\int_{O(n)} D_\kappa(XHY)\, dH = D_\kappa(X) D_\kappa(Y). \tag{12.11.21}$$

Proof. This is Theorem 12.10.5 and Definitions 12.11.2, 12.11.3. □

Functions satisfying (12.11.21) are spherical functions in the sense of Helgason (1962).

As noted in Theorem 12.11.1, there is a complex number $\phi_{2\kappa}$ such that

$$D_\kappa(X) = \phi_{2\kappa} \operatorname{tr} P_{\varepsilon_{2\kappa}} E \bigotimes_{i=1}^{2k} X, \tag{12.11.22}$$

where, if κ is $k_1 \geq \cdots \geq k_n$ then 2κ is $2k_1 \geq \cdots \geq 2k_n$.

$$\begin{aligned} D_\kappa(XA) &= \phi_{2\kappa} \operatorname{tr} P_{\varepsilon_{2\kappa}} E \left(\bigotimes_{i=1}^{2k} (XA) \right) \\ &= \phi_{2\kappa} \operatorname{tr} \left(\bigotimes_{i=1}^{2k} A \right) P_{\varepsilon_{2\kappa}} E \left(\bigotimes_{i=1}^{2k} X \right). \end{aligned} \tag{12.11.23}$$

Restricting ourselves to real matrices, from (12.11.7), the invariant subspace $\mathbf{V}_{2\kappa}$ of homogeneous polynomials of degree k in $X^t X$ is in one-to-one correspondence with $\phi_{2\kappa} \mathfrak{A}_{2k} P_{\varepsilon_{2\kappa}} E$ which is isomorphic to the space spanned by the right translates $D_\kappa(.A)$ of D_κ.

Theorem 12.11.6. *The representation $A \to \pi(A)$ of* $\mathbf{GL}(n)$ *in $\mathbf{V}_{2\kappa}$ defined by*

$$(\pi(A) f)(X) = f(XA) \tag{12.11.24}$$

is an irreducible representation and

$$D_\kappa(A) = \int_{O(n)} \operatorname{tr} \pi(HA)\, dH. \tag{12.11.25}$$

Proof (Following Helgason). As just shown, the space $\mathbf{V}_{2\kappa}$ is finite dimensional. Therefore the operator

$$\bar{E} = \int_{O(n)} \pi(H)\, dH \tag{12.11.26}$$

is well defined by the definition

$$(\bar{E}f)(X) = \int_{\mathbf{O}(n)} (\pi(H)f)(X)\,dH = \int_{\mathbf{O}(n)} f(XH)\,dH. \quad (12.11.27)$$

If f is a translate of $D_{\kappa'}$ say $f(X) = D_\kappa(XA)$, and if $H \in \mathbf{O}(n)$, then

$$f(XH) = D_\kappa(XHA), \quad (12.11.28)$$

and by (12.11.21),

$$(\bar{E}f)(X) = D_\kappa(X)D_\kappa(A). \quad (12.11.29)$$

By linearity, $\bar{E}$ maps $\mathbf{V}_{2\kappa}$ onto the one-dimensional space spanned by D_κ and if $X \in \mathbf{GL}(n)$ then

$$(\bar{E}D_\kappa)(X) = D_\kappa(X). \quad (12.11.30)$$

The fact that the right translates of D_κ span follows from (12.11.23) and the assertion (12.11.30) follows from (12.11.29). Thus $\bar{E} \neq 0$ and clearly $\bar{E}^2 = \bar{E}$, and $\bar{E}$ is a rank one transformation.

To show π is irreducible, suppose $\mathbf{V}_{2\kappa} = \bigotimes_{i=1}^{p} \mathbf{V}_i'$ with each $\mathbf{V}_i'$ an invariant subspace under π. Then each $\mathbf{V}_i'$ reduces $\bar{E}$, as follows from (12.11.26), so that since $\bar{E} \neq 0$, for some i, $\bar{E}\mathbf{V}_i' \neq 0$. Then $D_\kappa \in \mathbf{V}_i'$ and since $\pi(A)D_\kappa$ is the right translate of D_κ, it follows that $\mathbf{V}_i$ contains all the right translates of D_κ, and hence their linear span which is $\mathbf{V}_{2\kappa}$.

The linear transformation $\bar{E}\pi(A)$ maps $\mathbf{V}_{2\kappa}$ onto the one-dimensional space generated by D_κ, so that in any basis including D_κ, since

$$(\bar{E}\pi(A))D_\kappa = D_\kappa(A)D_\kappa, \quad (12.11.31)$$

the matrix of the transformation has trace $D_\kappa(A)$. Thus, since the trace is invariant under the choice of basis,

$$D_\kappa(A) = \operatorname{tr} \bar{E}\pi(A) = \operatorname{tr} \int_{\mathbf{O}(n)} \pi(HA)\,dH = \int_{\mathbf{O}(n)} \operatorname{tr} \pi(HA)\,dH, \quad (12.11.32)$$

which verifies (12.11.25). □

Remark 12.11.7. The space $\mathfrak{A}_{2k}P_{\varepsilon_{2\kappa}}E$ is defined by invariance on the left, and corresponds to the space of right translates of D_κ. The space of left translates results in $\int_{\mathbf{O}(n)} \operatorname{tr} \pi(AH)\,dH = D_\kappa(A)$.

Remark 12.11.8. The endomorphism $\bar{E}$ has rank one but the matrix $P_{\varepsilon_{2\kappa}}E$ in general does not have rank 1. Instead, as noted in (12.11.17), there will be several equivalent (conjugate) representations which result from consideration of nonbi-symmetric coefficient matrices.

Remark 12.11.9. Formula (34) of James (1964) follows from (12.11.32). For the complex zonal polynomial with respect to the unitary group discussed in Section 12.7, the zonal polynomial cannot result from integration of a group character. For, if $\operatorname{tr}\pi(A)$ is a polynomial group character then $f(A) = \int \operatorname{tr}\pi(AH)\,dH$ over the unitary group is a constant function. See Weyl (1946), page 177, Lemma (7.1.A).

Remark 12.11.10. Formulas (35), (36), (37), and (38) of James (1964) give expressions for group characters. The source of these formulas is not stated by James. The important formula (35) can be deduced from Littlewood (1950), page 188, together with formula (6.3; 1) on page 87 of Littlewood. Formula (44) of James, op. cit., is then a direct translation of Littlewood (6.2; 15), which in our notation would read

$$(\operatorname{tr} X_1^{k_1})(\operatorname{tr} X_2^{k_2}) \cdots (\operatorname{tr} X_p^{k_p}) = \sum_{\kappa'} \chi_{\kappa'}(\kappa)\chi_{\kappa'}(X), \tag{12.11.33}$$

where $\chi_{\kappa'}(X)$ is the character of the representation of weight κ', and $\chi_{\kappa'}(\kappa)$ is the character of the permutation κ in the representation κ' of the symmetric group. For us, X is $n \times n$ and $p \le k$. The case of special statistical interest is

$$(\operatorname{tr} X)^k = \sum_{\kappa} \chi_\kappa(1^k)\chi_\kappa(X). \tag{12.11.34}$$

Remark 12.11.11. Formula (41) of James (1964) states that if the partition κ is given by $k_1 \ge \cdots \ge k_n$, and if any of $k_1, \ldots, k_n$ are odd integers, then $\int_{(n)} \chi_\kappa(XH)\,dH = 0$. In these notes we do not obtain quite so strong a result. For us the representation

$$A \to \operatorname{tr} P_{\varepsilon_\gamma} E \bigotimes_{i=1}^{2k} (XA) = T_A\left(\operatorname{tr} P_{\varepsilon_\gamma} E \bigotimes_{i=1}^{2k} X\right) \tag{12.11.35}$$

has the property that $\int_{O(n)} T_{AH}\,dH = \int_{O(n)} T_{HA}\,dH = 0$ if the partition γ of $2k$ has an odd summand. See Lemma 12.10.6. This together with Theorem 12.11.6 or James (1964) formula (34) is sufficient to derive James (1964) formula (43).

These comments yield the special case

$$\begin{aligned}\int_{O(n)} (\operatorname{tr} XH)^{2k}\,dH &= \sum_{2\kappa} \chi_{2\kappa}(1^{2k}) \int_{O(n)} \chi_{2\kappa}(XH)\,dH \\ &= \sum_{2\kappa} \chi_{2\kappa}(1^{2k}) \frac{C_\kappa(X^tX)}{C_\kappa(I_n)}.\end{aligned} \tag{12.11.36}$$

See Problem 13.3.2.

12.12. Third Construction of Zonal Polynomials. The Converse Theorem

The first method of construction of the zonal polynomials was due to James (1961b) and is sketched in Section 13.5 with the details left to the reader to fill in. An entirely new second method of construction appeared in James

(1968) in which the polynomials were obtained as solutions of a differential equation. A third, and again entirely different method, has been presented in Saw (1977), and Saw's idea has been modified and difficulties eliminated in Takemura (1982). See Section 12.13 for a discussion of Takemura's development. Both developments, Saw's and Takemura's, are the outgrowth of trying to use generating functions to define zonal polynomials in an elementary teachable manner. The paper by Saw starts with a seemingly arbitrary definition from which the essential properties of zonal polynomials can be proven. After the mathematical development of Chapter 12 is made, looking back, it is clear such an arbitrary definition must be correct. In this section we use hindsight to explore the possibility of Saw's approach.

Given a partition κ of k with parts $k_1 \geq \cdots \geq k_n$ we let the monomial symmetric function $M_\kappa(S)$, using the eigenvalues of S, be

$$\sum \lambda_{i_1}^{k_1} \lambda_{i_2}^{k_2} \cdots \lambda_{i_n}^{k_n} \tag{12.12.1}$$

with the summation being taken over the distinct monomial terms. Let the ordered partitions be $\kappa_1 \geq \kappa_2 \geq \cdots \geq \kappa_r$ and define $r \times 1$ vectors

$$\begin{aligned} C^t &= (C_{\kappa_1}, \ldots, C_{\kappa_r}), \quad \text{the functions } C_\kappa \text{ as in (12.11.10)}; \\ M^t &= (M_{\kappa_1}, \ldots, M_{\kappa_r}); \\ J^t &= (J_{\kappa_1}, \ldots, J_{\kappa_r}). \end{aligned} \tag{12.12.2}$$

The functions J_κ, sometimes called the s-functions, as are the functions M_κ, are discussed in Section 12.2. See Definitions 12.2.4 and 12.2.7. The s-functions are redefined in (13.3.14). As is a consequence of Section 12.8 and as is discussed again in Section 12.13, and also in the Problems of Section 13.1, there is a lower triangular $r \times r$ matrix T_1 with rows and columns indexed by the partitions of k such that

$$C^t = M^t T_1, \tag{12.12.3}$$

where the row vectors are defined above.

If A is an $r \times r$ diagonal matrix then

$$C^t A C = M^t (T_1 A T_1^t) M, \tag{12.12.4}$$

and from the matrix of the quadratic form on the right side of (12.12.4) the matrix T_1 is recoverable except for the proportionalities introduced by A. That is, given an equation

$$C^t A C = M^t B M \tag{12.12.5}$$

with A a diagonal positive definite matrix and B an $r \times r$ positive definite matrix, one has uniquely that there exists $T \in \mathbf{T}(r)$ such that

$$TT^t = B \quad \text{and} \quad C^t = M^t T A^{-1/2}. \tag{12.12.6}$$

To show this, by (12.12.3), $C^t = M^t T_1$ and by (12.12.4) we have

$$C^t A C = M^t (T_1 A^{1/2})(T_1 A^{1/2})^t M = M^t B M. \tag{12.12.7}$$

M is a function of the eigenvalues $\lambda(S)$ of S (this variable is suppressed in these expressions) and both sides of (12.12.7) are polynomials in $\lambda_1, \ldots, \lambda_n$. The identity holds for all positive values of $\lambda_1, \ldots, \lambda_n$ and hence for all complex numbers $\lambda_1, \ldots, \lambda_n$. By Section 12.2 the values of $M(S)$ cover $\mathbb{C} \times \cdots \times \mathbb{C}$ so that it follows that the quadratic forms in (12.12.7) have the same matrix. That is,

$$B = (T_1 A^{1/2})(T_1 A^{1/2})^t. \tag{12.12.8}$$

Since B is positive definite the decomposition (12.12.8) into a product of a lower triangular matrix and its transpose is unique. Therefore $T = T_1 A^{1/2}$, which proves (12.12.6).

In order to compute the zonal polynomials as components of the vector C in principle one needs only determine a suitable equation of the form (12.12.4). The matrix product $T_1 A T_1^t$ must be known but A need not be known in order to determine C up to proportionality. In part this is what Saw, op. cit., has done, using the generating function

$$\mathbf{E} \exp(\operatorname{tr} S\mathbf{X}T\mathbf{X}^t). \tag{12.12.9}$$

We will consider the case of square matrices but the arguments can be done for S and T square, and $\mathbf{X}$ a rectangular matrix. Orthogonal invariance is introduced by assuming $\mathbf{X}$ to be a matrix of n^2 mutually independent normal $(0, 1)$ random variables, and S, T symmetric with eigenvalues $s_1 \geq \cdots \geq s_n$, $t_1 \geq \cdots \geq t_n$ respectively, so that

$$\mathbf{E}(\operatorname{tr} S\mathbf{X}T\mathbf{X}^t)^k = \mathbf{E}\left(\sum_{i=1}^{n} \sum_{j=1}^{n} s_i t_j ((\mathbf{X})_{ij})^2\right)^k. \tag{12.12.10}$$

Then (12.12.9) is the Laplace transform of the joint density function of n^2 independent χ_1^2 random variables and

$$\mathbf{E} \exp(\operatorname{tr} \theta S\mathbf{X}T\mathbf{X}^t) = \prod_{i=1}^{n} \prod_{j=1}^{n} (1 - 2\theta s_i t_j)^{-1/2}. \tag{12.12.11}$$

This Laplace transform is considered in Problems 13.3.5 and 13.3.6 where it is shown that

$$\mathbf{E} \exp(\operatorname{tr} \theta S\mathbf{X}T\mathbf{X}^t) = \sum_{k=0}^{\infty} (\theta^k/k!) \sum_{\kappa} \lambda(\kappa) J_\kappa(S) J_\kappa(T). \tag{12.12.12}$$

The coefficients $\lambda(\kappa)$ are defined in (13.3.15). On the other hand

$$\mathbf{E} \exp(\operatorname{tr} \theta S\mathbf{X}T\mathbf{X}^t) = \sum_{k=0}^{\infty} (\theta^k/k!) \, \mathbf{E}(\operatorname{tr} S\mathbf{X}T\mathbf{X}^t)^k, \tag{12.12.13}$$

so that

$$\begin{aligned} \sum_{\kappa} \lambda(\kappa) J_\kappa(S) J_\kappa(T) &= \mathbf{E} \sum_{\kappa} C_\kappa(S) C_\kappa(\mathbf{X}T\mathbf{X}^t) \\ &= \mathbf{E} \sum_{\kappa} C_\kappa(S) C_\kappa(T(\mathbf{X}^t\mathbf{X})). \end{aligned} \tag{12.12.14}$$

The random variable $\mathbf{X}^t\mathbf{X}$ has a central Wishart distribution with n degrees of freedom and the value of the expectation $\mathbf{E}\, C_\kappa(T(\mathbf{X}^t\mathbf{X}))$ is known. See Problem 13.2.6. That is,

$$\sum_\kappa \lambda(\kappa) J_\kappa(S) J_\kappa(T) = \sum_\kappa 2^k \frac{\Gamma_n(\frac{1}{2}n, \kappa)}{\Gamma_n(\frac{1}{2}n)} C_\kappa(S) C_\kappa(T). \tag{12.12.15}$$

Aside from (12.12.3), using Theorem 12.2.15, with $T_2 \in \mathbf{T}(r)$,

$$J^t = M^t T_2^t. \tag{12.12.16}$$

Hence if Λ is a diagonal matrix with diagonal entries $\lambda(\kappa)$,

$$J^t \Lambda J = M^t(T_2^t \Lambda T_2) M = (12.12.15). \tag{12.12.17}$$

Thus, if $T_3 \in \mathbf{T}(r)$ and $T_3 T_3^t = (T_2^t \Lambda T_2)$, one has that

$$M^t T_3 \text{ is proportional to } C^t. \tag{12.12.18}$$

The numbers $\lambda(\kappa)$ are computable from the formula and tables of T_2, $2 \le k \le 12$, are available in David, Kendall and Barton (1966).

The converse theory due to Saw (1977) proceeds as follows. Start with the known equation

$$\sum_\kappa \lambda(\kappa) J_\kappa(S) J_\kappa(T) = \mathbf{E}\,(\operatorname{tr} S\mathbf{X}T\mathbf{X}^t)^k. \tag{12.12.19}$$

Since the r functions which are the components of M^t are a basis of the space of homogeneous symmetric polynomials of degree k,

$$\mathbf{E}\,(\operatorname{tr} S\mathbf{X}T\mathbf{X}^t)^k = \sum_{i=1}^{r} \sum_{j=1}^{r} a_{ij} M_{\kappa_i}(S) M_{\kappa_j}(T). \tag{12.12.20}$$

That is, the left side of (12.12.20) is a symmetric function in the eigenvalues of S, which, when expressed in the basis M is a sum whose coefficients are symmetric functions in the eigenvalues of T. Linear independence of the basis functions is sufficient to show the coefficient polynomials are each homogeneous of degree k. It follows from (12.12.19), since both sides of the expression are polynomials in the eigenvalues, that the quadratic form (12.12.20) is positive definite. Thus uniquely one may write $(a_{ij}) = TT^t$ and define up to proportionality that

$$C'^t = M^t T. \tag{12.12.21}$$

By elementary calculations Saw then shows that the component functions of C' have the properties of zonal polynomials including the reproducing property. One knows from the hindsight that this must be so. The proofs starting from (12.12.20) lead to an elementary theory that is easily presentable. See Section 13.6, and Section 12.13 for Takemura's development using a different representation of (12.12.20).

We do not pursue the converse theory and instead continue with the hindsight point of view. Let $k_1 \ge k_2 \ge \cdots \ge k_p \ge 1$ be parts of the partition

κ so that i occurs π_i times. See (12.2.8). A permutation g is said to be of class κ if g is a product of $\pi_1 + \cdots + \pi_k$ cycles such that π_i of the cycles have length i, $1 \le i \le k$. There are $\lambda'(\kappa)$ permutations of class $\mathbf{k}$, where

$$(k!)\Big/\left(\prod_{i=1}^{k} i^{\pi_i}(\pi_i!)\right) = \lambda'(\kappa), \tag{12.12.22}$$

where κ is the partition of k determined by the numbers $k_1, \ldots, k_p$. To each partition κ we let I_κ be the idempotent defined as $(k_1! \cdots k_p!)^{-1}\Sigma\sigma$ over all permutations of type p for the diagram $T(\kappa)$, This idempotent determines a left ideal of the group algebra so each group element acts as an endomorphism $\pi(g)(aI_\kappa) = gaI_\kappa$. It is known that the group character ψ of this group representation is

$$\begin{aligned}&(k_1! \cdots k_p!)^{-1} \# \text{ permutations } h \text{ such that } hgh^{-1}\\ &\text{is a } p\text{, and we call this number } \psi(\kappa, g).\end{aligned} \tag{12.12.23}$$

The value of this character depends only on the class κ of g. From Weyl (1946) Chapter VII, Section 7, formula (7.6), which is similar to James (1964) formula (44), see also (12.11.34), one obtains

$$\begin{aligned} J_\kappa(S) &= (\operatorname{tr} S)^{\pi_1}(\operatorname{tr} S^2)^{\pi_2} \cdots (\operatorname{tr} S^k)^{\pi_k} \\ &= \sum_{\kappa'} \psi(\kappa', g) M_{\kappa'}(S). \end{aligned} \tag{12.12.24}$$

Replace g by its class κ''. From (12.12.22) we obtain (12.12.25)

$$J^t(S)\Lambda J(T) = \sum_{\kappa}\sum_{\kappa'}\sum_{\kappa''} \psi(\kappa, \kappa'')\psi(\kappa', \kappa'')\lambda'(\kappa'') M_\kappa(S) M_{\kappa'}(T). \tag{12.12.25}$$

In this expression $\lambda'(\kappa'')$ is the number of permutations in the class $\mathbf{k}''$, so that $\Sigma_{\kappa''}\, \lambda'(\kappa'')$ amounts to a sum over all permutations with equal weighting. Thus the matrix of the right side of (12.12.25) is

$$\sum_g \psi(\kappa, g)\psi(\kappa', g), \tag{12.12.26}$$

where $\psi(\kappa, g)$ is the character of the representation of g in the left ideal generated by I_κ. The matrix (12.12.26) is immediately computable using the orthogonality relations for group characters. An example is given below.

If $T \in \mathbf{T}(r)$ and

$$T^tT = \sum_g \psi(\kappa, g)\psi(\kappa', g) \tag{12.12.27}$$

then M^tT_t is proportional to J^t and in fact

$$M^tT^t(\Lambda')^{-1/2} = J^t. \tag{12.12.28}$$

We compare the numbers $\lambda'(\kappa)$ with the numbers $\lambda(\kappa)$, see (13.3.15) and (12.12.22), so that if κ has π_1 $1's, \ldots, \pi_k$ $k's$ then

$$\lambda(\kappa) = 2^{k - \Sigma\pi_i}\lambda'(\kappa). \tag{12.12.29}$$

If F is the diagonal matrix with diagonal entries $2^{k-\Sigma\pi_i}$ then

$$\Lambda = F\Lambda'; \qquad J^t\Lambda^{1/2} = M^tT^t(\Lambda')^{-1/2}F^{1/2}; \qquad J^t\Lambda J = M^t(T^tFT)M. \tag{12.12.30}$$

Therefore the lower triangular decomposition of T^tFT defines the zonal polynomials.

For computational purposes we may write $\psi(\kappa, \cdot) = \Sigma_{\kappa''} n(\kappa, \kappa'')e_{\kappa''}$ where the $e_{\kappa''}$ are mutually orthogonal idempotents. Then by the orthogonality relations for group characters

$$\sum_g \psi(\kappa, g)\psi(\kappa', g) = \sum_{\kappa''} n(\kappa, \kappa'')n(\kappa', \kappa''), \tag{12.12.31}$$

which is an inner product of the two characters.

For example, in the symmetric group on 3 letters, in terms of the three primative characters χ_1, χ_2, χ_3, one obtains for characters

$$\begin{array}{ll} (3) & \chi_1(g); \\ (21) & \chi_1(g) + \chi_2(g); \\ (1^3) & \chi_1(g) + 2\chi_2(g) + \chi_3(g). \end{array} \tag{12.12.32}$$

These generate the matrix

$$\begin{vmatrix} 1 & 0 & 0 \\ 1 & 1 & 0 \\ 1 & 2 & 1 \end{vmatrix} \begin{vmatrix} 1 & 1 & 1 \\ 0 & 1 & 2 \\ 0 & 0 & 1 \end{vmatrix} = \begin{vmatrix} 1 & 1 & 1 \\ 1 & 2 & 3 \\ 1 & 3 & 6 \end{vmatrix}. \tag{12.12.33}$$

We find that

$$J^t = M^t \begin{vmatrix} 1 & 1 & 1 \\ 0 & 1 & 3 \\ 0 & 0 & 6 \end{vmatrix} \quad \text{and} \quad \Lambda' = \begin{vmatrix} 2 & 0 & 0 \\ 0 & 3 & 0 \\ 0 & 0 & 1 \end{vmatrix}. \tag{12.12.34}$$

so that

$$J^t\Lambda' J = 6M^t \begin{vmatrix} 1 & 1 & 1 \\ 1 & 2 & 3 \\ 1 & 3 & 6 \end{vmatrix} M. \tag{12.12.35}$$

To obtain the zonal polynomials one needs $J^t\Lambda J$ where

$$\Lambda = \begin{vmatrix} 8 & 0 & 0 \\ 0 & 6 & 0 \\ 0 & 0 & 1 \end{vmatrix}; \qquad J^t\Lambda J = M^t \begin{vmatrix} 15 & 9 & 6 \\ 9 & 15 & 18 \\ 6 & 18 & 36 \end{vmatrix} M. \tag{12.12.36}$$

From this one obtains the lower triangular matrix

$$\begin{vmatrix} 1 & 0 & 0 \\ 3/5 & 4/5 & 0 \\ 2/5 & 6/5 & (2/5)\sqrt{5} \end{vmatrix}. \tag{12.12.37}$$

Except for the proportionality factors this gives for the zonal polynomials

$$M_3 + (3/5)M_{21} + (2/5)M_{1^3} = (1/15)(15M_3 + 9M_{21} + 6M_{1^3});$$

$$4M_{21} + 6M_{1^3}; \tag{12.12.38}$$

$$M_{1^3}.$$

These agree with the published polynomials.

12.13. Zonal Polynomials as Eigenfunctions. Takemura's Idea

Takemura (1982) has taken ideas of Saw (1977), Farrell (1980), and Farrell (1976), Section 13.6, and has produced an elegant construction of the zonal polynomials of real matrix argument as the "eigenvectors" of a linear transformation. The result of the eigenvalue problem is similar to formulas stated in Problems 13.6.10 and 13.6.11.

In revision of Farrell (1976) Section 12.2 has been changed to reflect the ideas of Farrell (1980) and Takemura (1982) about symmetric polynomials in n variables and to reflect the importance of ordered partitions in the subject. In this section, we present a brief outline of Takemura's construction. By careful workmanship Takemura has found many new forumlas for the coefficients that occur. We do not include this part of his work, which is contained in his dissertation.

Takemura observed that a development of zonal polynomials of a complex matrix argument can be done in parallel to the development of zonal polynomials of a real matrix argument and the last chapter of Takemura (1982) presents this parallel development. We do not discuss this material in this book.

Main results of the theory, the splitting formula which is Theorem 12.7.3, and the triangular character of the coefficient matrix, Theorem 12.9.3, are given treatments in this book that reflect the discussion of polynomials as functions of a matrix argument whereas the developments of Saw (1977) discussed in Sections 12.12 and 13.6 and of Takemura (1982) discussed in this section treat zonal polynomials as homogeneous symmetric functions of n variables thereby trying to bypass aspects of group representations and the algebra of *bi*-symmetric matrices.

Takemura's construction, the subject of this section, defines the vector C of zonal polynomials in terms of A, see (12.2.11) and Definition 12.2.5, that is, in terms of polynomials of the elementary symmetric functions. The equation $C^t = A^tT$, T lower triangular, is to be contrasted with Saw's development sketched in Section 12.12 and 13.6, $C^t = M^tT_1$, T_1 a lower triangular matrix. Both developments require computation of $\mathbf{E}\,C(\mathbf{W})$, $\mathbf{W}$ a Wishart random matrix. Takemura has made this computation straightforward.

In this section we write $|A|$ for the determinant of A as a usage for the section. Following Takemura we write $\lambda(A)$ for the column vector of eigenvalues of A in decreasing order, and $A(i_1, \ldots, i_h)$ the $h \times h$ submatrix of A with rows and columns indexed by $i_1, \ldots, i_h$. By computing derivatives, see 13.1.16, it follows that the elementary symmetric functions of the roots of $0 = |A - \lambda I_n|$ are

$$a_h(\lambda(A)) = \sum_{i_1 \le \cdots \le i_h} \cdots \sum |A(i_1, \ldots, i_h)|. \tag{12.13.1}$$

If the random $n \times n$ matrix $\mathbf{W}$ is Wishart, v degrees of freedom, then

$$\tau_v A_\kappa(A) = \mathbf{E}\, A_\kappa(A\mathbf{W}) \tag{12.13.2}$$

defines, with extension by linearity, a linear transformation of the invariant homogeneous symmetric functions of matrices A. See Section 12.2. If A is symmetric with eigenvalues $\lambda_1, \ldots, \lambda_n$ then orthogonal invariance of the distribution of $\mathbf{W}$ implies $\mathbf{E}\, A_\kappa(A\mathbf{W})$ is a symmetric function of $\lambda_1, \ldots, \lambda_n$. Thus we may take A to be diagonal. If A is a diagonal matrix as indicated then

$$|(A\mathbf{W})(i_1, \ldots, i_h)| = (\lambda_{i_1} \cdots \lambda_{i_h})|\mathbf{W}(i_1, \ldots, i_h)|. \tag{12.13.3}$$

It follows $\mathbf{E}\, A_\kappa(A\mathbf{W})$ is a sum of monomial terms corresponding to partitions $\le \kappa$. Thus by Theorem 12.2.13 there exists a lower triangular matrix T_v with

$$(\tau_v A_{\kappa_1}, \ldots, \tau_v A_{\kappa_r}) = (A_{\kappa_1}, \ldots, A_{\kappa_r})T_v. \tag{12.13.4}$$

Note that this is the transpose of Takemura's expression. In $\mathbf{E}\, A_\kappa(A\mathbf{W})$ the coefficient of $\lambda_1^{k_1} \cdots \lambda_n^{k_n}$ is

$$\mathbf{E}\, |\mathbf{W}(1)|^{k_1 - k_2} |\mathbf{W}(1, 2)|^{k_2 - k_3} \cdots |\mathbf{W}(1, \ldots, n)|^{k_n} = \gamma_{v\kappa}. \tag{12.13.5}$$

With the random triangular decomposition of $\mathbf{W}$ as given by the Bartlett decomposition, see Theorem 11.1.0, it follows that $|\mathbf{W}(1)| \cdots |\mathbf{W}(1, \ldots, n)| = \mathbf{t}_{11}^{k_1 - k_2}(\mathbf{t}_{11}\mathbf{t}_{22})^{k_2 - k_3} \cdots (\mathbf{t}_{11} \cdots \mathbf{t}_{nn})^{k_n} = \mathbf{t}_{11}^{k_1} \cdots \mathbf{t}_{nn}^{k_n}$. Since $\mathbf{t}_{11}, \ldots, \mathbf{t}_{nn}$ are independent chi-square random variables, the value of (12.3.5) is computable and the diagonal elements $\gamma_{v\kappa}$ of T_v are

$$\gamma_{v\kappa} = 2^n \prod_{i=1}^{n} \Gamma(k_i + \tfrac{1}{2}(v + 1 - i))/\Gamma(\tfrac{1}{2}(v + 1 - i)), \tag{12.13.6}$$

which is a polynomial in v.

Theorem 12.13.1. *The linear transformations τ_v represented in the basis $A_{\kappa_1}, \ldots, A_{\kappa_r}$ have matrices T_v which are lower triangular, and the diagonal elements of T_v are given by* (12.13.6). *Further $T_v T_\eta = T_\eta T_v$ for degrees of freedom v, η.*

PROOF. It remains to note that if $\mathbf{W}_v$, $\mathbf{W}_\eta$ are independent Wishart matrices with the indicated degrees of freedom, then, taking transposes,

$$\begin{aligned} T_v T_\eta A_\kappa(A) &= \mathbf{E}\, A_\kappa(A\mathbf{W}_v\mathbf{W}_\eta) = \mathbf{E}\, A_\kappa(\mathbf{W}_\eta \mathbf{W}_v A) \\ &= \mathbf{E}\, A_\kappa(A\mathbf{W}_\eta\mathbf{W}_v) = T_\eta T_v A_\kappa(A). \end{aligned} \tag{12.13.7}$$

If the set of vectors $A_\kappa(A)$ taken over all $A \in \mathbf{D}(n)$ form a vector space of dimension less than r, then there is an $r \times 1$ vector a such that $\Sigma_{i=1} a_i A_{\kappa_i}(A) = 0$ for all diagonal matrices A, which contradicts the linear independence of the functions $A_{\kappa_1}, \ldots, A_{\kappa_r}$. □

Remark 12.13.2. In the basis of the functions A_κ the matrices T_ν are a commutative family. If some matrix T_ν has pairwise distinct diagonal elements, and if $T_\nu x = \lambda x$, then $T_\nu(T_\eta x) = \lambda(T_\eta x)$, which, by uniqueness of the eigenvectors, implies $T_\eta x = \lambda' x$ for some λ'. Hence the set of eigenvectors of T_ν would simultaneously diagonalize T_ν for all integers $\nu \geq 1$. From (12.13.6) the r polynomials in ν can be pairwise equal for at most a finite set of ν. Hence T_ν exists with pairwise distinct diagonal elements. Similarly for T_ν^t. Using Theorem 8.1.1 it then follows that

Theorem 12.13.3 (Takemura). *There exists an $r \times r$ lower triangular matrix T_C such that for all degrees of freedom ν,*

$$T_\nu T_C = T_C \operatorname{diag}(\lambda_{\nu\kappa_1}, \ldots, \lambda_{\nu\kappa_r}). \tag{12.13.8}$$

Definition 12.13.4 (Takemura).

$$(C_{\kappa_1}, \ldots, C_{\kappa_r}) = (A_{\kappa_1}, \ldots, A_{\kappa_r}) T_C, \tag{12.13.9}$$

that is,

$$C^t = A^t T_C,$$

defines the vector of zonal polynomials.

Remark 12.13.5. This definition makes the polynomials C_κ the eigenfunctions of the operators T_ν simultaneously for all ν, which is basically the result of Constantine (1963). That is, from (12.13.8) and (12.13.9) it follows that $\mathbf{E}\, A^t(B\mathbf{W}) = A^t(B) T_\nu T_C = A^t(B) T_C \operatorname{diag}(\lambda_{\nu\kappa_1}, \ldots, \lambda_{\nu\kappa_r})$ so that $\mathbf{E}\, C_\kappa(B\mathbf{W}) = \lambda_{\nu\kappa} C_\kappa(B)$, $B \in \mathbf{S}(n)$ a positive definite matrix.

Remark 12.13.6. The discussion just given requires the extension of the definition of C_κ to all $n \times n$ matrices X diagonalizable over $\mathbb{R}$. This is done by the definition $C_\kappa(X) = C_\kappa(\operatorname{diag}(\lambda(X)))$.

Theorem 12.13.7 (Takemura Theorem 3.2). *If A and B are $n \times n$ positive definite matrices and $U \in \mathbf{O}(n)$ then*

$$\int_{\mathbf{O}(n)} C_\kappa(AUBU^t)\, dU = C_\kappa(A) C_\kappa(B) / C_\kappa(I_n). \tag{12.13.10}$$

Proof. Since the functions C_κ are a basis and since the integral is a homogeneous polynomial of degree k in the eigenvalues of A and is a homogeneous polynomial of degree k in the eigenvalues of B it follows that

$$\int_{\mathbf{O}(n)} C_\kappa(AUBU^t)\,dU = \sum_{\kappa'}\sum_{\kappa''}\alpha(\kappa',\kappa'')C_{\kappa'}(A)C_{\kappa''}(B). \tag{12.13.11}$$

Further, let $\mathbf{W}$ be Wishart with v degrees of freedom and replace A by $A^{1/2}\mathbf{W}A^{1/2}$ so that

$$\mathbf{E}\,C_{\kappa'}(A^{1/2}\mathbf{W}A^{1/2}) = \mathbf{E}\,C_{\kappa'}(A\mathbf{W}) = \lambda_{v\kappa'}C_{\kappa'}(A), \tag{12.13.12}$$

and so that

$$\mathbf{E}\,C_\kappa(A^{1/2}\mathbf{W}A^{1/2}UBU^t) = \lambda_{v\kappa}C_\kappa(AUBU^t). \tag{12.13.13}$$

Therefore (12.13.11) becomes

$$\begin{aligned}\lambda_{v\kappa}\sum_{\kappa'}\sum_{\kappa''}\alpha(\kappa',\kappa'')C_{\kappa'}(A)C_{\kappa''}(B)&\\ = \sum_{\kappa'}\sum_{\kappa''}\lambda_{v\kappa'}\alpha(\kappa',\kappa'')C_{\kappa'}(A)C_{\kappa''}(B),&\end{aligned} \tag{12.13.14}$$

valid for all v, A, B. Thus for all v and all partitions κ', κ'' of k, with κ fixed,

$$\lambda_{v\kappa'}\alpha(\kappa',\kappa'') = \lambda_{v\kappa}\alpha(\kappa',\kappa''). \tag{12.13.15}$$

As seen before we may take v so large that the numbers $\lambda_{v\kappa'}$ are all distinct. Thus if $\kappa \neq \kappa'$ it follows that $\alpha(\kappa',\kappa'') = 0$. Since

$$\int_{\mathbf{O}(n)} C_\kappa(AUBU^t)\,dU = \int_{\mathbf{O}(n)} C_\kappa(UAU^tB)\,dU = \int_{\mathbf{O}(n)} C_\kappa(AU^tBU)\,dU \tag{12.13.16}$$

it follows that the matrix $\alpha(\kappa',\kappa'')$ is symmetric. Thus the $\alpha(\kappa,\kappa)$ is the only entry that may be nonzero. Then

$$\int_{\mathbf{O}(n)} C_\kappa(AUBU^t)\,dU = \alpha C_\kappa(A)C_\kappa(B). \tag{12.13.17}$$

Since $C_\kappa(B) \neq 0$ for some B, take $A = I_n$ and obtain $\alpha = 1/C_\kappa(I_n)$. □

In (12.13.9) the vector of zonal polynomials is defined as $C^t = A^tT_C$. From (12.2.15) we have $A^t = M^tT_A$ so that $C^t = M^t(T_AT_C)$ which is in agreement with (12.12.3). This gives an independent proof that the zonal polynomials in terms of the monomials are expressed by a lower triangular matrix, with partitions as indices in decreasing order.

12.14. The Integral Formula of Kates

In his Ph. D. dissertation, Princeton (1980), Kates obtained the following formula.

Theorem 12.14.1. *Let A be an $n \times n$ symmetric positive semidefinite matrix and $\mathbf{X}$ be an $n \times n$ matrix of mutually independent normal* $(0, 1)$ *random* *variables. If the partition κ has parts $k_1 \geq \cdots \geq k_n$ then*

$$\mathbf{E}\,|(\mathbf{X}A\mathbf{X}^t)(1)|^{k_1-k_2}|(\mathbf{X}A\mathbf{X}^t)(1,2)|^{k_2-k_3}\cdots|(\mathbf{X}A\mathbf{X}^t)(1,\ldots,n)|^{k_n} \tag{12.14.1}$$

is the zonal polynomial associated with κ normalized so that evaluated at I_n the value is $\lambda_{n\kappa}$, $\lambda_{n\kappa}$ defined in (12.13.6) with $n = v$. When expressed as a sum of monomial terms in the eigenvalues of A the coefficients of the monomial terms are nonnegative integers.

The following proof is based on Takemura, op. cit.

PROOF. Let $f_n(A)$ be the polynomial in (12.14.1). If $\mathbf{W}$ is a central Wishart $n \times n$ random matrix with n degrees of freedom, so that $\mathbf{W} = \mathbf{Y}^t\mathbf{Y}$ with $\mathbf{Y}$ an $n \times n$ matrix of normal (0, 1) random variables, then

$$\begin{aligned}(\tau_n f_n)(A) &= \mathbf{E} f_n(A\mathbf{W}) = \mathbf{E} f_n(\mathbf{Y}A\mathbf{Y}^t)\\ &= \mathbf{E}\prod_{j=1}^{n}|(\mathbf{X}\mathbf{Y}A\mathbf{Y}^t\mathbf{X}^t)(1,\ldots,j)|^{k_j-k_{j+1}} \qquad (12.14.2)\\ &= \mathbf{E}\prod_{j=1}^{n}|(\mathbf{Y}\mathbf{X}A\mathbf{X}^t\mathbf{Y}^t)(1,\ldots,j)|^{k_j-k_{j+1}}.\end{aligned}$$

In this formula we take $k_{n+1} = 0$. The interchange is possible since $\mathbf{XY}$ and $\mathbf{YX}$ have the same distribution. By the Bartlett decomposition, Theorem 11.0.1, write $\mathbf{Y} = \mathbf{TU}$, $\mathbf{T}$ lower triangular and $\mathbf{U}$ orthogonal, so that, using the orthogonal invariance of $\mathbf{X}$, the continuation of (12.14.2) is

$$\begin{aligned}(\tau_n f_n)(A) &= \mathbf{E}\prod_{j=1}^{n}|(\mathbf{T}\mathbf{X}A\mathbf{X}^t\mathbf{T}^t)(1,\ldots,j)|^{k_j-k_{j+1}}\\ &= \mathbf{E}\prod_{j=1}^{n}(\mathbf{t}_{11}\cdots\mathbf{t}_{jj})^{k_j-k_{j+1}} f_n(A) \qquad (12.14.3)\\ &= \lambda_{n\kappa} f_n(A).\end{aligned}$$

Therefore f_n is an eigenfunction of τ_n for the eigenvalue $\lambda_{n\kappa}$.

If A' is $m \times m$ and $\mathbf{Z}$ is $n \times m$ so that $\mathbf{Z}^t\mathbf{Z} = \mathbf{W}$ is Wishart with n degrees of freedom, the one verifies readily that if $\mathbf{Y} = (\mathbf{Z}, \mathbf{Z}_1)$ and $\mathbf{A} = \begin{vmatrix} A' & 0 \\ 0 & 0 \end{vmatrix}$, then

$$\begin{aligned}(\tau_n f_m)(A') &= \mathbf{E} f_m(A'\mathbf{W}) = \mathbf{E} f_n(\mathbf{Z}A'\mathbf{Z}^t)\\ &= \mathbf{E} f_n(\mathbf{Y}A\mathbf{Y}^t) = \lambda_{n\kappa} f_n(A) \qquad (12.14.4)\\ &= \lambda_{n\kappa} f_m(A').\end{aligned}$$

Thus f_m is an eigenfunction of τ_v for all degrees of freedom v greater than m, which implies f_m is a zonal polynomial. See Theorem 12.13.7.

Taking $A = I_n$ and using the Bartlett decomposition for $\mathbf{XX}^t$, a repetition of the computations (12.14.2) and (12.14.3) gives the value $f_n(I_n) = \lambda_{n\kappa}$.

The Binet–Cauchy formula, see Problem 8.5.24, gives for $A = \operatorname{diag}(a_1, \ldots, a_n)$ the result that

$$|(\mathbf{X}A\mathbf{X}^t)(1, \ldots, j)| = \sum_{i_1 < \cdots < i_j} \sum a_{i_1} \cdots a_{i_j} |\mathbf{X}(\substack{1, \ldots, j \\ i_1, \ldots, i_j})|^2. \tag{12.14.5}$$

From (12.14.5) it follows that the expression for f_n as a sum of monomials has nonnegative coefficients and that the matrix taking the polynomials f_n for the ordered partitions $\kappa_1 \geq \cdots \geq \kappa_r$ to the monomials is a triangular matrix. The coefficients are expectations of sums of monomials in the independent random variables $(\mathbf{X})_{ij}$. Since the moments of normal (0, 1) random variables are integers, the last assertion of the theorem follows. □

Remark 12.14.2. The analogue of Kates's formula

$$\mathbf{E} \prod_{j=1}^{n} |(\mathbf{U}A\mathbf{U}^t)(1, \ldots, j)|^{k_j - k_{j+1}} \tag{12.14.6}$$

with the random orthogonal matrix $\mathbf{U}$ having Hear measure as distribution is discussed in Theorem 4.7.1 of Takemura, op. cit., where it is shown that this is a (unnormalized) zonal polynomial. In this vein

$$\mathbf{E} \prod_{j=1}^{n} |(\mathbf{U}A\mathbf{V})(1, \ldots, j)|^{2k_j - 2k_{j+1}} \tag{12.14.7}$$

is discussed in Theorem 4.7.2 of Takemura. The result here is the zonal polynomial in $A^t A$ associated with κ. In all cases the values of the polynomials at I_n is determined and stated by Takemura.

Remark 12.4.3. Chapter 5 of Takemura, op. cit., develops the complex analogue of the results mentioned here in Section 12.13 and 12.14. In particular the complex analogue of Kates's formula is given and by a direct computation it is shown that the Schur functions are the eigenfunctions of the operator $\bar{\tau}$ given by $(\bar{\tau} f)(A) = \mathbf{E} f(A\overline{\mathbf{W}})$, $\overline{\mathbf{W}}$ a complex Wishart matrix. For the sake of completeness we state here the analogue of Kates's formula.

$$\mathbf{E} \prod_{i=1}^{n} |(\overline{\mathbf{X}}A\overline{\mathbf{X}})(1, \ldots, j)|^{k_j - k_{j+1}} \tag{12.14.8}$$

is the complex zonal polynomial associated with κ having value at I_n given by

$$\bar{\lambda}_{n\kappa} = \prod_{i=1}^{n} \Gamma(k_i + n + 1 - i)/\Gamma(n + 1 - i). \tag{12.14.9}$$

In (12.14.8) the random matrix $\overline{\mathbf{X}}$ is an $n \times n$ matrix of independently distributed complex normal (0, 1) random variables. That is, each entry of $\overline{\mathbf{X}}$ is distributed as $\mathbf{x} + i\mathbf{y}$ where $\mathbf{x}$ and $\mathbf{y}$ are independent normal $(0, \frac{1}{2})$ random variables. Somewhat simpler expressions result if instead complex normal (0, 2) random variables are used, as explained in Section 3.5 and Section 5.2.

Remark 12.14.4. The proof of Theorem 12.14.1 presented here is basically Takemura's proof. A variant on this argument results if $\mathbf{Y}$ is a square root of a Wishart (n, k) matrix as explained in Section 5.2, see Remark 5.2.7 and Remark 5.2.11. From Theorem 5.2.13 it follows that the distribution of $\mathbf{X} = \mathbf{TU}$ and $\mathbf{Y} = \mathbf{SV}$, $\mathbf{T}$ and $\mathbf{S}$ lower triangular, is that of $\mathbf{TUSV}$ is that of $\mathbf{TSUV}$. Then as in (12.14.3), since the value depends only on the diagonal of $\mathbf{ST}$ which is the same as the diagonal of $\mathbf{TS}$, the last line of (12.14.3) then follows.

CHAPTER 13

Problems for Users of Zonal Polynomials

13.0. Introduction

In this section we present a series of results that are useful to users of zonal polynomials. Inasmuch as many of these results are relatively easy applications of the preceding theory many of these results are presented as problems. Some are original to the author. Others are based on various articles by James (1960, 1961a, 1961b, 1964, 1968) and Constantine (1963). In the use of generating functions there is overlap with the recent papers by Saw (1977) and Takemura (1982).

The problems have been grouped, roughly as follows. A section of problems related to the mathematics of zonal polynomials is first. Next follows a section on identities. This is followed by some problems on numerical coefficients in series of zonal polynomials, followed by problems on the construction of the polynomials.

The literature now has many examples of multivariate density functions that have been expressed as infinite sums of zonal polynomials. It was felt that the reader would be better off reading the examples in the existing literature and for illustration purposes only one example, the noncentral Wishart density, has been included in this chapter.

Primary sources for formulas, which for sake of accuracy should be used, are James (1964), Takemura (1982) and Muirhead (1982). Those who have tried numerical work with series of zonal polynomials report that the series converge very slowly, and resort to asymptotic expansions of hypergeometric functions is used. We do not explore this subject and the interested reader should consult the works of Muirhead as a starter for the theory of asymptotic expansions.

It is the author's view that the impact of Weyl (1946), James (1961b), Saw

(1977) and Section 12.12 together is to establish the existence of an intricate connection between the coefficients that relate the various symmetric functions and the group characters of the representations of the symmetric group on n letters in its group algebra. The work of Garsia and Remmel (1981) supports this contention. In addition, see Sections 13.4 and 13.5, which also relate to the combinatorial question.

13.1. Theory

PROBLEM 13.1.1. Let f be a two-sided invariant homogeneous polynomial of degree $2k$ in the variables X. Suppose that if $X, Y \in \mathbf{GL}(n)$ then

$$\int_{\mathbf{O}(n)} f(XHY)\, dH = f(X)f(Y). \tag{13.1.1}$$

Show that f is a (unnormalized) zonal polynomial in X, i.e., the induced polynomial of $X^t X$ is a zonal polynomial.

HINT. This can be done by writing $f(X) = \Sigma_\kappa a_\kappa \operatorname{tr} P_{\varepsilon_\kappa} E \bigotimes_{i=1}^{2k} X$. □

PROBLEM 13.1.2. If C is a zonal polynomial of degree k then

$$(\det S)^m C(S) \quad \text{and} \quad (\det S)^{-m} C(S) \tag{13.1.2}$$

are zonal polynomials of degrees $k + nm$ and $k - nm$ respectively, where $S \in \mathbf{S}(n)$, provided $k - nm \geq 0$, and $(\det S)^{-m} C(S)$ is a polynomial.

PROBLEM 13.1.3. Let I_{n-1} be the $(n-1) \times (n-1)$ identity matrix and I'_{n-1} be the $n \times n$ matrix $\begin{vmatrix} I_{n-1} & 0 \\ 0 & 0 \end{vmatrix}$. Let $Y \in \mathbf{GL}(n)$ and $X' = I'_{n-1} Y I'_{n-1}$. The polynomial

$$\operatorname{tr} P_{\varepsilon_{2\kappa}} E \bigotimes_{i=1}^{2k} X' \tag{13.1.3}$$

is a (unnormalized) zonal polynomial of X if and only if it is not the zero polynomial, where $X' = \begin{vmatrix} X & 0 \\ 0 & 0 \end{vmatrix}$.

HINT. $\bigotimes_{i=1}^{2k} I'_{n-1} \mathfrak{A}_{2k} \bigotimes_{i=1}^{2k} I'_{n-1}$ is a subalgebra of $\mathfrak{A}_{2k}$ which is an H^* algebra. The idempotents in the center determined by the symmetric group algebra are the idempotents $P_{\varepsilon_{2\kappa}} \bigotimes_{i=1}^{2k} I'_{n-1}$, as readily follows from the matrix definitions. From Section 12.10 write the matrix form of the coefficient matrix of a zonal polynomial in X. □

Remark 13.1.4. Zonal polynomials expressed as polynomials of the elementary symmetric functions of the eigenvalues have an expression that is independent of the size of the matrix. The equations of the coefficients in James (1968) again exhibit the independence from the matrix size. See also Takemura (1982) for further discussion.

Remark 13.1.5. Among possible choices of symmetric functions to work with, $\operatorname{tr} S$, $\operatorname{tr} S^2, \ldots, \operatorname{tr} S^n$ have the advantage that the matrix size enters only as the exponent n. Thus if S is $n \times n$ and $k > n$, the polynomials of degree k in $\operatorname{tr} S, \ldots, \operatorname{tr} S^n$ still span the space of homogeneous functions of degree k.

PROBLEM 13.1.6. Let f be a polynomial of n variables such that

$$C(S) = f(\operatorname{tr} S, \ldots, \operatorname{tr} S^n) \tag{13.1.4}$$

is a homogeneous polynomial of degree k in the eigenvalues of S. Suppose the coefficients of f are real but the entries of S may be complex numbers. Show

$$C(\bar{S}^t) = \overline{C(S)} \quad \text{and} \quad C(ST) = C(TS). \tag{13.1.5}$$

Further, show that the coefficient matrix of the polynomial C is in the center of $\mathfrak{A}_k$.

HINT. The matrices ST and TS, even if singular, have the same characteristic equation, hence have the same eigenvalues, multiplicities included. Therefore if L is a nonsingular matrix and $C(S) = \operatorname{tr} A \bigotimes_{i=1}^k S$, then

$$\left(\bigotimes_{i=1}^{k} L\right) A = A \left(\bigotimes_{i=1}^{k} L\right). \tag{13.1.6}$$

The matrix L may have complex number entries. □

PROBLEM 13.1.7. By the theory of orthogonal invariants for real matrices,

$$\operatorname{tr} P_{\varepsilon_{2\kappa}} E \bigotimes_{i=1}^{2k} X = \operatorname{tr} A \bigotimes_{i=1}^{k} (X^t X) \tag{13.1.7}$$

for some bi-symmetric matrix A. Does Problem 13.1.6 imply that A is in the center of $\mathfrak{A}_k$?

PROBLEM 13.1.8. In Section 12.8 the subspace $\mathbf{E}_\kappa$ was defined as follows. Let κ have parts $k_1 \geq k_2 \geq \cdots \geq k_n \geq 0$ and let the set of indices $i_1, \ldots, i_k$ belong to κ. Then the basis of $\mathbf{E}_\kappa$ is the set of all k-tuples

$$(e_{\tau(i_{\sigma(1)})}, \ldots, e_{\tau(i_{\sigma(k)})}) \tag{13.1.8}$$

where τ is a permutation of $1, \ldots, n$ and σ is a permutation of $1, \ldots, k$. It was shown that $\mathbf{E}_\kappa$ is an invariant subspace under all P_a, a in the symmetric group algebra, see (12.5.8), together with being invariant under $\bigotimes_{i=1}^k U$, $U \in \mathbf{O}(n)$ a permutation matrix.

PROBLEM 13.1.9. Let $\mathbf{K}_\kappa$ be the space of k-forms generated by the canonical basis elements

$$u_{\tau(i_{\sigma(1)})} \cdots u_{\tau(i_{\sigma(k)})} \tag{13.1.9}$$

corresponding to the basis elements (13.1.8). If a bi-symmetric matrix A commutes with the matrices $\bigotimes_{i=1}^k U$, $U \in \mathbf{O}(n)$, U a permutation matrix, then the diagonal elements of the part of A acting on $\mathbf{K}_\kappa$ are all equal.

PROBLEM 13.1.10. For each partition κ of k let $i_1^\kappa, \ldots, i_k^\kappa$ be an index set belonging to κ. Suppose A is in the center of $\mathfrak{A}_k$ and that X is an $n \times n$ diagonal matrix $X = \mathrm{diag}(x_1, \ldots, x_n)$. Then show

$$\operatorname{tr} A \bigotimes_{i=1}^{k} X = \sum_\kappa a_\kappa \sum_\tau \sum_\sigma x_{\tau(i^\kappa_{\sigma(1)})} \cdots x_{\tau(i^\kappa_{\sigma(k)})}. \tag{13.1.10}$$

PROBLEM 13.1.11. If A is in the center of $\mathfrak{A}_{2\kappa}$ and $X \in \mathbf{GL}(n)$ let $d_1, \ldots, d_n$ be the eigenvalues of $(X^t X)^{1/2}$. Then

$$\operatorname{tr} AE \bigotimes_{i=1}^{2k} X = \sum_\kappa a_\kappa \sum_\tau \sum_\sigma d^2_{\tau(i^\kappa_{\sigma(1)})} \cdots d^2_{\tau(i^\kappa_{\sigma(k)})}. \tag{13.1.11}$$

PROBLEM 13.1.12. If $U \in \mathbf{O}(n)$ show

$$\left(\bigotimes_{i=1}^{2k} U\right) P_{\varepsilon_{2\kappa}} E = P_{\varepsilon_{2\kappa}} E \left(\bigotimes_{i-1}^{2k} U\right);$$
$$P_\tau P_{\varepsilon_{2\kappa}} E = P_{\varepsilon_{2\kappa}} E P_\tau; \tag{13.1.12}$$

$P_{\varepsilon_{2\kappa}} E$ is a symmetrical matrix and is an idempotent.

See Section 12.10.

PROBLEM 13.1.13. Continue Problems 13.1.10 to 13.1.12. If $X = \mathrm{diag}(x_1^{1/2}, \ldots, x_n^{1/2})$ with nonnegative diagonal entries, then

$$\operatorname{tr} P_{\varepsilon_{2\kappa}} E \left(\bigotimes_{i=1}^{2k} X\right) = \sum_{\kappa' \le \kappa} a_{\kappa'} \sum_\tau \sum_\sigma x_{\tau(i^{\kappa'}_{\sigma(1)})} \cdots x_{\tau(i^{\kappa'}_{\sigma(k)})}. \tag{13.1.13}$$

The coefficients a_κ, are nonnegative real numbers since the diagonal elements of $P_{\varepsilon_{2\kappa'}} E$ are nonnegative (Constantine (1963), James (1968)).

HINT AND REMARK. See Section 12.8 on the nested character of the matrices for the P_{ε_κ}. The lower triangular character of the matrices is further discussed in Sections 12.12 and 12.13. □

Problem 13.1.14. To each partition κ of k into not more than n parts define a monomial symmetric function $M(X, \kappa)$ as follows. Let $i_1, \ldots, i_k$ belong to κ, see Section 12.8. Then

$$M(X, \kappa) = \sum_{\tau} \sum_{\sigma} x_{\tau(i_{\sigma(1)})} \cdots x_{\tau(i_{\sigma(k)})}, \tag{13.1.14}$$

$x_1, \ldots, x_n$ the eigenvalues of X. Let $M(X)$ be the $r \times 1$ column vector whose κ-th entry is $M(X, \kappa)$. Show there exists an upper triangular matrix A with rows and columns indexed by the partitions κ such that each entry of the vector

$$A\, M(X) = C(X) \tag{13.1.15}$$

is a (unnormalized) zonal polynomial of X^tX. That is, the κ-th entry of $C(X^tX)$ is $a_\kappa C_\kappa(X^tX)$.

Problem 13.1.15. By (13.1.13) we may write

$$\operatorname{tr} P_{\varepsilon_{2\kappa}} E\left(\bigotimes_{i=1}^{2k} X\right) = \sum_{\kappa' \leq \kappa} a_{\kappa'} M(X^tX, \kappa'). \tag{13.1.16}$$

Corresponding to a partition κ' being $k_1' \geq k_2' \geq \cdots \geq k_n'$ the monomial symmetric function is a sum of terms

$$x_1^{k_1'} x_2^{k_2'} \cdots x_n^{k_n'} = (x_1 \cdots x_n)^{k_n'} \cdots (x_1 x_2)^{k_2' - k_3'} (x_1)^{k_1' - k_2'} \tag{13.1.17}$$

taken over all permutations of the bottom indices.

Problem 13.1.16. Let the elementary symmetric functions be defined by

$$\prod_{i=1}^{n} (x - c_i) = x^n - a_1 x^{n-1} + \cdots + (-1)^n a_n. \tag{13.1.18}$$

Let κ' be given by the parts $k_1' \geq k_2' \geq \cdots \geq k_n'$ and κ be given by $k_1 \geq k_2 \geq \cdots \geq k_n$. Show that if $x_1^{k_1'} \cdots x_n^{k_n'}$ is a term of the polynomial

$$a^{k_1 - k_2} a_2^{k_2 - k_3} \cdots a_{n-1}^{k_{n-1} - k_n} a_n^{k_n} \tag{13.1.19}$$

then $\kappa' \leq \kappa$.

Problem 13.1.17 (Constantine (1963)). The zonal polynomial C_κ, except for normalization, is given by (13.1.13). Evaluate $C_\kappa(TS)$ as a function of T given that $S \in \mathbf{S}(n)$ and that $T \in \mathbf{D}(n)$. Show that

$$C_\kappa(TS) = \sum_{\kappa' \leq \kappa} \sum_{\tau} \sum_{\sigma} t_{\tau(i^{\kappa'}_{\sigma(1)})} \cdots t_{\tau(i^{\kappa'}_{\sigma(k)})} f(S, \tau, \sigma, \kappa'). \tag{13.1.20}$$

Constantine, op. cit., needed the coefficient of the term $t_1^{k_1} \cdots t_n^{k_n}$. Show this coefficient is

$$a_\kappa t_1^{k_1} \cdots t_n^{k_n} |S_1|^{k_1 - k_2} |S_2|^{k_2 - k_3} \cdots |S_n|^{k_n}, \tag{13.1.21}$$

where the determinant $|S_p|$ is the determinant of the principal $p \times p$ minor of S.

Hint and Remark. See also Section 12.13 and some different notations used by Takemura. The elementary symmetric function a_p of $\det(TS - cI_n) = 0$ is given by the sum of the $p \times p$ principal minors of TS, that is, the sum of the determinants of these minors. □

Problem 13.1.18. If a zonal polynomial is expressed as a sum of monomial symmetric functions then the coefficients are rational if rational normalization is used.

Hint. The diagonal entries of the projections $P_{\varepsilon_{2\kappa}} E$ are rational. See (13.1.13). □

13.2. Numerical Identities

Problem 13.2.1. Show that

$$\int_{\mathbf{S}(n)} \exp(-\operatorname{tr} S)(\det S)^{t-(n+1)/2} C_\kappa(ST) \prod_{j \le i} ds_{ij} = \Gamma_n(t, \kappa) C_\kappa(T). \tag{13.2.1}$$

In this problem, only show the existence of the constant $\Gamma_n(t, \kappa)$, whose value is determined in the next problem.

Hint. In (3.3.16) it is shown that $\mathbf{S}(n)$ is the factor space $\mathbf{GL}(n)/\mathbf{O}(n)$ and a measure is induced by Haar measure on $\mathbf{GL}(n)$ which is invariant under the transformations $S \to ASA^t$ of $\mathbf{S}(n)$. Except for changes in normalization the differential form for the invariant measure is given in (3.3.16), which accounts for the factor $(\det S)^{-(n+1)/2}$ in (13.2.1). Since $\operatorname{tr} USU^t = \operatorname{tr} S$ and $\det(USU^t) = \det S$, the invariance replaces $C_\kappa(ST)$ by $C_\kappa(USU^tT)$. Integrate over U. □

Problem 13.2.2 (Constantine (1963)). Show that

$$\Gamma_n(t, \kappa) = \pi^{n(n-1)/4} \prod_{i=1}^{n} \Gamma(t + k_i - \tfrac{1}{2}(i-1)). \tag{13.2.2}$$

Hint. Both sides of (13.2.1) are polynomials of the diagonal elements $t_1, \ldots, t_n$ of T. To determine the value of $\Gamma_n(t, \kappa)$ it is sufficient to find the coefficient of $t_1^{k_1} \cdots t_n^{k_n}$ on both sides of (13.2.1). Using (13.1.21) the required equation is

$$\int_{\mathbf{S}(n)} \exp(-\operatorname{tr} S)(\det S)^{t-(n+1)/2} |S_1|^{k_1-k_2} \cdots \times |S_n|^{k_n} \prod_{j \le i} ds_{ij} = \Gamma_n(t, \kappa). \tag{13.2.3}$$

Write $S = RR^t$ with $R \in \mathbf{T}(n)$ and let the elements of R be r_{ij}, $j \le i$. The Jacobian of the transformation is

$$2^n \prod_{i=1}^{n} (r_{ii})^{n-i+1}, \tag{13.2.4}$$

and the integral (13.2.3) in the new variables is

$$\begin{aligned}\Gamma_n(t, \kappa) = \int \cdots \int \exp(-\operatorname{tr} RR^t) \prod_{i=1}^{n} (r_{ii}^2)^{t+k_i-(i+1)/2} \\ \times \prod_{i=1}^{n} d(r_{ii}^2) \prod_{j<i} dr_{ij}.\end{aligned} \tag{13.2.5}$$

Evaluate this integral to obtain (13.2.2). □

Definition 13.2.3.

$$\Gamma_n(u) = \pi^{n(n-1)/4} \prod_{i=1}^{n} \Gamma(u - \tfrac{1}{2}(i-1)). \tag{13.2.6}$$

Problem 13.2.4 (Constantine (1963), Theorem 1).

$$\begin{aligned}\int_{\mathbf{S}(n)} \exp(-\operatorname{tr} RS)(\det S)^{t-(n+1)/2} C_\kappa(ST) \prod_{j\le i} ds_{ij} \\ = \Gamma_n(t, \kappa)(\det R)^{-t} C_\kappa(TR^{-1}).\end{aligned} \tag{13.2.7}$$

Remark 13.2.5 (See Problem 4.4.9). If $\mathbf{X}$ is $n \times h$ and consists of independent rows each distributed as normal $(0, \Sigma)$, then the Wishart density function for n degrees of freedom is

$$\frac{(\det S)^{(n-(h+1))/2} \exp(-\operatorname{tr}(2\Sigma)^{-1}S)}{(\det 2\Sigma)^{n/2}\Gamma_n(\frac{1}{2}n)}. \tag{13.2.8}$$

Problem 13.2.6. If $\mathbf{S}$ has as density function the Wishart density (13.2.8) then

$$\mathbf{E}\, C_\kappa(\mathbf{S}) = \frac{2^k \Gamma_h(\frac{1}{2}n, \kappa)}{\Gamma_h(\frac{1}{2}n)} C_\kappa(\Sigma). \tag{13.2.9}$$

13.3. Coefficients of Series

Remark 13.3.1. Constantine (1963) defined the zonal polynomials C_κ by requiring the normalization to satisfy

$$\text{if} \quad S \in \mathbf{S}(n) \quad \text{then} \quad (\operatorname{tr} S)^k = \sum_\kappa C_\kappa(S). \tag{13.3.1}$$

Since the polynomials C_κ are linearly independent (cf. Problem 13.1.12) this identity uniquely determines the values of $C_\kappa(I_n)$. Thus this normalization *does* depend on the size n of S. See Problem 13.1.3.

We now consider the function $\int_{\mathbf{O}(n)} (\operatorname{tr} HX)^{2k}\, dH$, with $X \in \mathbf{GL}(n)$, and write

$$\int_{\mathbf{O}(n)} (\operatorname{tr} HX)^{2k}\, dH = \sum_\kappa b_\kappa C_\kappa(X^t X). \tag{13.3.2}$$

Note that this is one term of the Laplace transform

$$\int_{\mathbf{O}(n)} \exp(HX)\, dH, \tag{13.3.3}$$

and this is the integral that arose in Chapter 4 in the discussion of the noncentral Wishart density function.

Problem 13.3.2. Determine the b_κ by letting $\mathbf{X}$ have independently and identically distributed rows each normal $(0, \Sigma)$. Obtain the answer

$$b_\kappa = ((2k)!)/(k!))2^{-2k}\Gamma_n(\tfrac{1}{2}n)/\Gamma_n(\tfrac{1}{2}n, \kappa). \tag{13.3.4}$$

Hint. If $H \in \mathbf{O}(n)$ then $\mathbf{X}$ and $H\mathbf{X}$ have the same distribution. Thus

$$\exp(\operatorname{tr} \tfrac{1}{2}s^2\Sigma) = \mathbf{E}\int_{\mathbf{O}(n)} \exp(\operatorname{tr} sH\mathbf{X})\, dH. \tag{13.3.5}$$

Expand the left side in an infinite series and the integrand of the right side in an infinite series and match the coefficients of the powers of s, first, and then the functions $C_\kappa(\Sigma)$, second. Obtain

$$\sum_\kappa C_\kappa(\tfrac{1}{2}\Sigma) = (k!)/((2k)!)\sum_\kappa \frac{b_\kappa 2^k \Gamma_n(\tfrac{1}{2}n, \kappa)}{\Gamma_n(\tfrac{1}{2}n)} C_\kappa(\Sigma), \tag{13.3.6}$$

and from (13.3.6) obtain (13.3.4). □

Definition 13.3.3 (James (1964)).

$$\begin{aligned} (a)_k &= a(a+1)\cdots(a+k-1); \\ (\tfrac{1}{2}n)_\kappa &= \Gamma_n(\tfrac{1}{2}n, \kappa)/\Gamma_n(\tfrac{1}{2}n). \end{aligned} \tag{13.3.7}$$

Problem 13.3.4. The coefficients b_κ in (13.3.4) are

$$b_\kappa = (\tfrac{1}{2})_k/(\tfrac{1}{2}n)_\kappa. \tag{13.3.8}$$

Hint. $2^{-2k}(2k)! = (k - \tfrac{1}{2})(k - (3/2)) \cdots (1/2)(k!) = (\tfrac{1}{2})_k(k!)$. □

Problem 13.3.5. If $\mathbf{X}$ is an $n \times n$ matrix consisting of n^2 mutually independent normal (0, 1) random entries and if S and T are $n \times n$ diagonal matrices (possibly with complex numbers on the diagonals) then

$$\mathbf{E}\exp(\operatorname{tr}\theta S\mathbf{X}T\mathbf{X}^t) = \prod_{i=1}^{n}\prod_{i=1}^{n}(1-2\theta s_i t_j)^{-1/2}, \tag{13.3.9}$$

where $s_1, \ldots, s_n$, $t_1, \ldots, t_n$ are the diagonal entries of S, T respectively. Show (13.3.9) and that

$$\ln \mathbf{E}\exp(\operatorname{tr}\theta S\mathbf{X}T\mathbf{X}^t) = (\tfrac{1}{2})\sum_{n=1}^{\infty} n^{-1}(2\theta)^n(\operatorname{tr} S^n)(\operatorname{tr} T^n). \tag{13.3.10}$$

Use (13.3.10) to show that

$$\mathbf{E}(\operatorname{tr} S\mathbf{X}\bar{S}\mathbf{X}^t)^n \geq 0, \tag{13.3.11}$$

where $\bar{S}$ is the conjugate of S.

Problem 13.3.6 (Saw (1977)). Compute the exponential of (13.3.10), that is, compute

$$\sum_{k=1}^{\infty}(1/k!)(\tfrac{1}{2}\sum_{n=1}^{\infty} n^{-1}(2\theta)^n(\operatorname{tr} S^n)(\operatorname{tr} T^n))^k \tag{13.3.12}$$

and show that this is the series

$$\sum_{k=1}^{\infty}(\theta^k/k!)\sum_{\kappa}\lambda(\kappa)J_\kappa(S)J_\kappa(T). \tag{13.3.13}$$

Interpret this formula as follows. Let π_j be the number of times j occurs in the partition κ, $1 \leq j \leq k$. Then

$$J_\kappa(S) = \prod_{j=1}^{k}(\operatorname{tr} S^j)^{\pi_j}. \tag{13.3.14}$$

See also Definition 12.2.7. The coefficient $\lambda(\kappa)$ may be computed, using the multinomial theorem, to be

$$\lambda(\kappa) = k!2^{k-\sum_{j=1}^{k}\pi_j}/\prod_{j=1}^{k}(j^{\pi_j}\pi_j!). \tag{13.3.15}$$

Problem 13.3.7. Formulas (13.3.10) and (13.3.13) hold if $S, T \in \mathbf{S}(n)$ or if S, T are diagonal matrices with possibly complex numbers on the diagonal.

Problem 13.3.8 (Saw (1977)). Let $S, T \in \mathbf{S}(n)$ and $\mathbf{X}$ be as in Problem 13.3.5. Show that

$$\mathbf{E}(\operatorname{tr} S\mathbf{X}T\mathbf{X}^t)^k = \sum_{\kappa} 2^k \frac{\Gamma_n(\frac{1}{2}n, \kappa)}{\Gamma_n(\frac{1}{2}n)} \frac{C_\kappa(S)C_\kappa(T)}{C_\kappa(I_n)}. \tag{13.3.16}$$

Hence, from (13.3.13),

$$\sum_{\kappa}\lambda(\kappa)J_\kappa(S)J_\kappa(T) = \sum_{\kappa} 2^k \frac{\Gamma_n(\frac{1}{2}n, \kappa)}{\Gamma_n(\frac{1}{2}n)} \frac{C_\kappa(S)C_\kappa(T)}{C_\kappa(I_n)}. \tag{13.3.17}$$

Problem 13.3.9. Derive orthogonality relations as follows. Write

$$C_\kappa = \sum_{\kappa'} k(\kappa, \kappa') J_{\kappa'} \tag{13.3.18}$$

and let K be the coefficient matrix in (13.3.18) so that the rows and columns of K are indexed by the partitions of k, ordered in the partial ordering of partitions. Substitute (13.3.18) into (13.3.17) and use the linear independence of the functions J_κ to obtain the system of equations

$$\begin{aligned} \lambda(\kappa') &= \sum_\kappa k(\kappa, \kappa')^2 2^k \frac{\Gamma_n(\frac{1}{2}n, \kappa)}{\Gamma_n(\frac{1}{2}n) C_\kappa(I_n)}; \\ 0 &= \sum_\kappa k(\kappa, \kappa') k(\kappa, \kappa'') 2^k \frac{\Gamma_n(\frac{1}{2}n, \kappa)}{\Gamma_n(\frac{1}{2}n) C_\kappa(I_n)}, \quad \kappa' \neq \kappa''. \end{aligned} \tag{13.3.19}$$

The coefficient matrix K is dependent on the number of rows in S, T only if $k > n$ and (13.3.19) holds for all matrix sizes n.

Problem 13.3.10 (Constantine (1963)). Verify the following evaluation:

$$\begin{aligned} &(\Gamma_n(a))^{-1} \int_{S>0} \exp(\operatorname{tr} - S) \exp(\operatorname{tr} SZ) (\det S)^{a-(n+1)/2}\, dS \\ &= (\det(I_n - Z))^{-a}, a > \tfrac{1}{2}(n+1). \end{aligned} \tag{13.3.20}$$

Expand $\exp(SZ)$ as an infinite sum, integrate, and obtain an infinite sum of zonal polynomials, assuming Z is symmetric. Obtain the identity

$$(\det(I_n - Z))^{-a} = \sum_{k=0}^{\infty} (k!\,\Gamma_n(a))^{-1} \sum_\kappa \Gamma_n(a, \kappa) C_\kappa(Z). \tag{13.3.21}$$

Problem 13.3.11. In the notation of Section 12.10, show that

$$\sum_\kappa \operatorname{tr} P_{\varepsilon_{2\kappa}} E \bigotimes_{i=1}^{2k} X = \int_{O(n)} (\operatorname{tr} HX)^{2k}\, dH. \tag{13.3.22}$$

Use Problem 13.3.2 to obtain

$$\sum_\kappa \operatorname{tr} P_{\varepsilon_{2\kappa}} E \bigotimes_{i=1}^{2k} X = \sum_\kappa b_\kappa C_\kappa(X^t X). \tag{13.3.23}$$

Therefore,

$$\operatorname{tr} P_{\varepsilon_{2\kappa}} E = b_\kappa C_\kappa(I_n). \tag{13.3.24}$$

Use (13.3.4) and the value (cf. James (1964), formula (21))

$$C_\kappa(I_n) = 2^{2k} k!(\tfrac{1}{2}n)_\kappa \prod_{i<j}^{p} (2k_i - 2k_j - i + j) / \prod_{i=1}^{p} (2k_i + p - i)!, \tag{13.3.25}$$

where p is the number of nonzero summands in the partition κ, to obtain

$$\operatorname{tr} P_{\varepsilon_{2\kappa}} E = (2k)! \prod_{i<j}^{p} (2k_i - 2k_j - i + j) / \prod_{i=1}^{p} (2k_i + p - i)!. \tag{13.3.26}$$

Compare this value with James (1964), formula (19).

PROBLEM 13.3.12. If **X** is $n \times h$ with independently distributed rows each normally distributed with covariance matrix Σ, and if $\mathbf{E}\mathbf{X} = M$, then the joint density function is

$$\begin{aligned} f(X) &= (2\pi)^{-nh/2} (\det \Sigma)^{-n/2} \exp(-\tfrac{1}{2} \operatorname{tr} \Sigma^{-1} X^t X) \\ &\quad \times \exp(-\tfrac{1}{2} \operatorname{tr} \Sigma^{-1} M^t M) \exp(\operatorname{tr} \Sigma^{-1} X^t M). \end{aligned} \tag{13.3.27}$$

By the results of Chapter 4, if a random variable **Y** has density function

$$\int_{\mathbf{O}(n)} f(H^t Y)\, dH, \tag{13.3.28}$$

then $\mathbf{X}^t\mathbf{X}$ and $\mathbf{Y}^t\mathbf{Y}$ have the same distribution, namely the noncentral Wishart density function. The evaluation of (13.3.28) requires evaluation of the integral

$$\int_{\mathbf{O}(n)} \exp(\Sigma^{-1} X^t H M)\, dH, \tag{13.3.29}$$

which may be expressed as an infinite sum of zonal polynomials. Using the idea of Chapter 4, the noncentral Wishart density function is then (13.3.27) multiplied by a normalization. Hence, write down the density function of $\mathbf{X}^t\mathbf{X}$.

PROBLEM 13.3.13. Refer to (12.10.27), (12.11.36), (13.3.2) and (13.3.4). Infer that

$$\chi_{2\kappa}(1^{2k}) = b_\kappa C_\kappa(I_n) = \operatorname{tr} P_{\varepsilon_{2\kappa}} E. \tag{13.3.30}$$

See also (13.3.24).

13.4. On Group Representations

Part of Theorem 12.11.6 may be given a more algebraic proof along the lines of the proof to Theorem 12.10.4. As shown in Section 12.11 the coefficient matrices of the homogeneous polynomials $f(X^t X)$ of degree $2k$ are AE, $A \in \mathfrak{A}_{2k}$. Thus the representation is $X \to \pi(X) \in \operatorname{End}(\mathfrak{A}_{2k} E)$ defined by

$$\pi(X)(AE) = \Big(\bigotimes_{i=1}^{2k} X\Big)(AE). \tag{13.4.1}$$

By linearity the transformations $\pi(X)$ extend to transformations T_A, $A \in \mathfrak{A}_{2k}$, and the T_A form an H^* algebra, $T_A(XE) = A(XE)$, A, $X \in \mathfrak{A}_{2k}$.

PROBLEM 13.4.1. If $S \in \operatorname{End}(\mathfrak{A}_{2k}E)$ and $ST_A = T_A S$ for all $A \in \mathfrak{A}_{2k}$ then there exists $B \in \mathfrak{A}_{2k}$ such that if $A \in \mathfrak{A}_{2k}$ then

$$S(AE) = (AE)(EBE) = AEBE. \tag{13.4.2}$$

This problem simply expresses the well-known result that left multiplication commutes with right multiplication when considered as endomorphisms. Write S_{EBE} for the transformation defined in (13.4.2).

PROBLEM 13.4.2. The map $EBE \to S_{EBE}$ is a one-to-one algebra homomorphism.

PROBLEM 13.4.3. The algebras $\{T_A | A \in \mathfrak{A}_{2k}\}$ and $\{S_{EBE} | B \in \mathfrak{A}_{2k}\}$ have the same center.

HINT. See Lemma 12.6.2. □

That is, if $A \in \mathfrak{A}_{2k}$ and T_A is in the center then there exists $B \in \mathfrak{A}_{2k}$ such that

$$A(XE) = (XE)(EBE) = XEBE \quad \text{for all} \quad X \in \mathfrak{A}_{2k}. \tag{13.4.3}$$

PROBLEM 13.4.4. Determine the $A \in \mathfrak{A}_{2k}$ satisfying (13.4.3).

HINT. Either $T_A = 0$ for all $A \in \mathfrak{A}_{2k} P_{\varepsilon_{2\kappa}}$ or the map $A \to T_A$ is one-to-one on this set. If not zero, then $T_A T_B = T_B T_A$ if and only if $AB = BA$. The solutions to (13.4.3) must satisfy, therefore, $AB = BA$ for all $B \in \mathfrak{A}_{2k}$. Therefore

$$A = \sum_{\{2\kappa | P_{\varepsilon_{2\kappa}} E \neq 0\}} a_{2\kappa} P_{\varepsilon_{2\kappa}} E. \tag{13.4.4}$$

□

PROBLEM 13.4.5. The representation $X \to \pi(X) \in \operatorname{End}(\mathfrak{A}_{2k}E)$ is either zero or is irreducible.

HINT. The commutator algebra is Abelian. □

13.5. First Construction of Zonal Polynomials

This section presents in outline a method by which, in principle, zonal polynomials may be constructed. The method is based on James (1961b) and is essentially the method by which James first constructed the polynomials. However we do not make the detailed analysis of the equivalences of various group representations that is to be found in James (1961b). The sketch of this section is a problem in that the reader may try to fill in some or all the missing arguments.

As in Section 12.10, if A is the *bi*-symmetric coefficient matrix of the trace function, that is,

$$\operatorname{tr} A \bigotimes_{i=1}^{2k} X = (\operatorname{tr} X^t X)^k, \tag{13.5.1}$$

then

$$\phi^{-1} P_c A = \phi^{-1} b_{2\kappa}^{-1} P_c E \tag{13.5.2}$$

where c is the Young's symmetrizer for the diagram $T(2\kappa)$ and the number $\phi^{-1} = \chi_{2\kappa}(1)/(2k)!$ as shown in Lemma 12.4.11.

From Section 12.10 the $((i_1, \ldots, i_{2k}), (j_1, \ldots, j_{2k}))$-entry of A is

$$((2k)!)^{-1} \sum_g \delta(i_{g(1)}, i_{g(2)}) \cdots \delta(j_{g(2k-1)}, j_{g(2k)}), \tag{13.5.3}$$

the summation being over all permutations g of $1, 2, \ldots, 2k$. For brevity we write

$$\begin{aligned} i = (i_1, \ldots, i_{2k}) \quad &\text{and} \quad j = (j_1, \ldots, j_{2k}); \quad \text{and,} \\ g(i) = (i_{g(1)}, \ldots, i_{g(2k)}), \quad &\text{similarly for } g(j). \end{aligned} \tag{13.5.4}$$

For a basis element $u_{i_1} \cdots u_{i_{2k}}$ we write for short u_i with the transformed basis element $P_g u_i = u_{g^{-1}(i)}$. Then $P_c A$ applied to the basis element u_i yields

$$\begin{aligned} \Sigma_j (A)_{i_1 \cdots i_{2k} j_1 \cdots j_{2k}} P_c(u_j) &= \Sigma_j (A)_{ij} P_c(u_j) \\ &= \Sigma_j \sum_p \sum_q \varepsilon(q) (A)_{ij} u_{(pq)^{-1}(j)} \\ &= \Sigma_j \sum_p \sum_q \varepsilon(q) (A)_{i(pq)((pq)^{-1}(j))} u_{(pq)^{-1}(j)} \\ &= \Sigma_j \sum_p \sum_q \varepsilon(q) (A)_{i(pq)(j)} u_j. \end{aligned} \tag{13.5.5}$$

In James's construction one notes that the matrix entry $(A)_{i\,pq(j)}$ is

$$((2k)!)^{-1} \sum_g \delta(i_{g(1)}, i_{g(2)}) \cdots \delta(j_{g(pq(1))}, j_{g(pq(2))}) \cdots. \tag{13.5.6}$$

One then writes the pairs

$$(1, 2), (3, 4), \ldots, (2k - 1, 2k) \tag{13.5.7}$$

and

$$(pq(1), pq(2)), \ldots, (pq(2k - 1), pq(2k)).$$

From these $2k$ pairs one constructs cycles as follows. We explain by example. In James's example $2k = 6$ and the pairs are $(1, 3)$, $(2, 4)$, $(5, 6)$ and $(1, 2)$, $(3, 6)$, $(4, 5)$. Then $1 \to 3 \to 6 \to 5 \to 4 \to 2 \to 1$, a cycle of length 6. Call the set of $2k$ pairs in (13.5.7) $\# pq$, and let $2k'_1, \ldots, 2k'_n = 2\kappa'$ be the partition determined by the cycle lengths. Then, with $S = X^t X$, the term

$$b_\kappa \phi^{-1} \varepsilon(q) (\operatorname{tr} S^{k'_1})(\operatorname{tr} S^{k'_2}) \cdots (\operatorname{tr} S^{k'_n}) \tag{13.5.8}$$

occurs in the polynomial $P_{\varepsilon_{2\kappa}} E \bigotimes_{i=1}^{2k} X$ and this polynomial is the sum of the terms (13.5.8) taken over all permutations pq such that p is of type p and q is of type q for the diagram $T(2\kappa)$.

In this construction of the zonal polynomials the factor ϕ^{-1} is common to all terms and can be ignored so long as proportionality is the only consideration. In the polynomial the term $(\operatorname{tr} S)^k$ results from those permutations pq which map the pairs $(1, 2), (3, 4), \ldots, (2k - 1, 2k)$ among themselves. There are

$$2^k(k_1)! \cdots (k_n)!(1!)^{p_1}(2!)^{p_2} \cdots (n!)^{p_n} \tag{13.5.9}$$

such permutations pq where the $p_1, \ldots, p_n$ are the number of times $1, \ldots, n$ occur in the diagram $T(\kappa)$ as a column length.

For example, if $k = 3$ and the diagram $T(2\kappa)$ has a single row $1, 2, \ldots, 6$ then $2^3(3!) = 48$ and the symmetric group of 6 letters factored by this subgroup has 15 cosets corresponding to the 15 isomorphic cycles produced. These 15 classes then break down into

$$\begin{aligned} &1 \text{ class produces three cycles of length } 2; \\ &6 \text{ classes produce two cycles, one of length 2, one of length } 4; \\ &8 \text{ classes produce a single cycle of length } 6. \end{aligned} \tag{13.5.10}$$

In this Young's diagram there is only one permutation of type q so $\varepsilon(q) = +1$. Thus the desired polynomial, up to proportionality, is

$$(\operatorname{tr} S)^3 + 6(\operatorname{tr} S)(\operatorname{tr} S^2) + 8(\operatorname{tr} S^3). \tag{13.5.11}$$

The polynomial in (13.5.11) is the zonal polynomial $Z_{(3)}(S)$, where (3) is the partition of 3 into 3, 0, 0.

PROBLEM 13.5.1. In the zonal polynomials $Z_{\kappa'}$ when expressed as a weighted sum of terms $(\operatorname{tr} S^{k_1'})(\operatorname{tr} S^{k_2'}) \cdots (\operatorname{tr} S^{k_n'})$, the coefficients are integers. Show this.

PROBLEM 13.5.2. Look in a table of symmetric functions and determine number of partitions of $k = 12$, hence the number of zonal polynomials of degree 12 in the variables of S.

13.6. A Teaching Version

The outline given here is based on Saw (1977) and has now been superseded by Takemura's development, see Section 12.13. We set out in a series of problems a series of steps whereby zonal polynomials may be defined and some of the basis properties proven. The computations required are relatively easy thus leading to a presentation that is teachable, i.e., not requiring undue mathematical training. The sequel does not follow Saw, op. cit., exactly.

Notations are defined in Section 12.12. It is important to note that the following argument works because certain definitions can be made independently of the size n of the matrices involved. Thus an immediate consequence of this method of definition is that the zonal polynomials have a definition not depending on the size of the matrices. See Problem 13.1.4.

PROBLEM 13.6.1. If $0 = \Sigma_{i=1}^{r} a_i M_{\kappa_i}(S)$ for all positive definite matrices S then $a_i = 0, 1 \le i \le r$.

PROBLEM 13.6.2. Let $T \in \mathbf{T}(r)$ and $C^t = M^t T$. Let $C^t = (C_{\kappa_1}, \ldots, C_{\kappa_r})$. If a is an $r \times 1$ vector such that $a^t C(S) = 0$ for all positive definite matrices S than $a = 0$.

PROBLEM 13.6.3. There exists an $r \times r$ matrix (a_{ij}) such that if $n \ge 1$ and if S, T, $\mathbf{X}$ are $n \times n$ matrices, if the entries of $\mathbf{X}$ are independent normal $(0, 1)$ random variables, and if S, T are symmetric, then

$$\mathbf{E}(\operatorname{tr} S\mathbf{X}T\mathbf{X}^t)^k = \sum_{i=1}^{r} \sum_{j=1}^{r} a_{ij} M_{\kappa_i}(S) M_{\kappa_j}(T). \tag{13.6.1}$$

Show further that, since $\mathbf{X}$ and $\mathbf{X}^t$ have the same distribution,

$$\mathbf{E}(\operatorname{tr} S\mathbf{X}T\mathbf{X}^t)^k = \mathbf{E}(\operatorname{tr} T\mathbf{X}S\mathbf{X}^t)^k. \tag{13.6.2}$$

Therefore, by Problem 13.6.1, the matrix (a_{ij}) is symmetric.

PROBLEM 13.6.4. The matrix (a_{ij}) of (13.6.1) is positive definite.

HINT. By (13.3.13),

$$\mathbf{E}(\operatorname{tr} S\mathbf{X}T\mathbf{X}^t)^k = \sum_{\kappa} \lambda(\kappa) J_\kappa(S) J_\kappa(T). \tag{13.6.3}$$

Further by (12.12.16),

$$\sum_{\kappa} \lambda(\kappa) J_\kappa(S) J_\kappa(T) = M^t(S)(T_2^t \Lambda T_2) M(T). \tag{13.6.4}$$

The matrix $T_2^t \Lambda T_2$ is positive definite. Use Problem 13.6.1. □

Definition 13.6.5. Let $T_3 \in \mathbf{T}(r)$ and $(a_{ij}) = T_3 T_3^t$. Then

$$C^t \underset{\text{def}}{=} M^t T_3 A^{-1/2}, \tag{13.6.5}$$

where the diagonal matrix $A \in \mathbf{D}(r)$ is to be specified.

Corollary 13.6.6. *Except possibly for the choice of A, the definition* (13.6.5) *is independent of the matrix size n, and as functions*

$$C^t A C = M^t T_3 T_3^t M. \tag{13.6.6}$$

Evaluated, if $n \geq 1$ and S and T are in $\mathbf{S}(n)$ *then*

$$C^t(S)AC(T) = \mathbf{E}(\operatorname{tr} S\mathbf{X}T\mathbf{X}^t)^k. \tag{13.6.7}$$

PROBLEM 13.6.7. If T is a diagonal $n \times n$ matrix and if $\mathbf{W}$ is an $n \times n$ matrix which has a Wishart density function then $\mathbf{E}\, C_{\kappa_i}(T\mathbf{W})$ is a symmetric homogeneous polynomial in the diagonal entries of T. There exists $D^t \in \mathbf{T}(r)$ such that $D = (d_{ij})$ and

$$\sum_{j=1}^{r} d_{ij} C_\kappa(T) = \mathbf{E}\, C_\kappa(T\mathbf{W}). \tag{13.6.8}$$

HINT. This result is similar to a lemma in Constantine (1963) and has a similar proof. □

PROBLEM 13.6.8. There exists a matrix $B = (b_{ij})$ such that

$$\int_{\mathbf{O}(n)} (\operatorname{tr} SHTH^t)^k \, dH = \sum_{i=1}^{r} \sum_{j=1}^{r} b_{ij} C_{\kappa_i}(S) C_{\kappa_j}(T). \tag{13.6.9}$$

The matrix B is symmetric. The b_{ij} depend on the matrix size n of S and T.

PROBLEM 13.6.9. If S, $T \in \mathbf{S}(n)$ and $\mathbf{X}$ is $n \times n$ with mutually independent normal (0, 1) entries then

$$\begin{aligned} \mathbf{E}(\operatorname{tr} S\mathbf{X}T\mathbf{X}^t)^k &= \mathbf{E} \int_{\mathbf{O}(n)} (\operatorname{tr} SH(\mathbf{X}T\mathbf{X}^t)H^t)^k \, dH \\ &= \sum_{i=1}^{r} \sum_{j=1}^{r} b_{ij} C_{\kappa_i}(S) \mathbf{E} C_{\kappa_j}(T\mathbf{X}^t\mathbf{X}) \qquad (13.6.10) \\ &= \sum_{i=1}^{r} \sum_{j=1}^{r} \sum_{k=j}^{r} b_{ij} d_{jk} C_{\kappa_i}(S) C_{\kappa_j}(T). \end{aligned}$$

That is, $A = BD$. Show that the matrices B and D are diagonal matrices.

PROBLEM 13.6.10. If $\mathbf{W}$ is an $n \times n$ symmetric matrix which has a Wishart distribution with covariance matrix Σ then

$$d_{ii} C_{\kappa_i}(T\Sigma) = \mathbf{E}\, C_{\kappa_i}(T\mathbf{W}). \tag{13.6.11}$$

Remark 13.6.11. If $\mathbf{W}$ is an $n \times n$ symmetric matrix that has a Wishart distribution with covariance matrix Σ and h degrees of freedom then

$$\mathbf{E}\, C_\kappa(\mathbf{W}) = C_\kappa(\Sigma)(2^k \Gamma_n(\tfrac{1}{2}h, \kappa)/\Gamma_n(\tfrac{1}{2}h)). \tag{13.6.12}$$

See (13.2.9). This must be proven by an argument similar to that of Constantine (1963). The proof of (13.6.12) is one vague point in Saw's presentation that has been rendered precise by Takemura.

PROBLEM 13.6.12. Given two partitions $\kappa_i \neq \kappa_j$ of k there exists an h such that $\Gamma_n(\frac{1}{2}h, \kappa_i)$ and $\Gamma_n(\frac{1}{2}h, \kappa_j)$ are distinct.

HINT. Form the ratio. The ratio may be irrational due to $\Gamma(\frac{1}{2})$ and hence not equal 1, or else the ratio is the ratio of two integers. Choose n in the later case so one of the numerator or denominator has a prime factor not in the other.

PROBLEM 13.6.13. The equation (13.6.6) is universal. That is, T_3 is chosen independent of the matrix size and the normalization A may be chosen so as not to depend on n, for example $A = I_r$.

PROBLEM 13.6.14. Show that

$$\int_{\mathbf{O}(n)} C_{\kappa_i}(SHTH^t)\,dH = \sum_{j=1}^{r}\sum_{h=1}^{r} e_{jh} C_{\kappa_j}(S) C_{\kappa_h}(T) \tag{13.6.13}$$

where the constants e_{jh} do depend on the size of the matrices S, T.

PROBLEM 13.6.15. Show that

$$\int_{\mathbf{O}(n)} C_{\kappa_i}(SHTH^t)\,dH = C_{\kappa_i}(S)C_{\kappa_i}(T)/C_{\kappa_i}(I_n). \tag{13.6.14}$$

HINT. By definition of the functions C_κ, $C_\kappa(AB) = C_\kappa(BA)$. Therefore, if $\mathbf{T}$ has a Wishart distribution with covariance Σ then

$$\begin{aligned} d_{ii}\int_{\mathbf{O}(n)} C_{\kappa_i}(H^tSH\Sigma)\,dH &= \int_{\mathbf{O}(n)} \mathbf{E}\, C_{\kappa_i}(H^tSH\mathbf{T})\,dH \\ &= \mathbf{E}\int_{\mathbf{O}(n)} C_{\kappa_i}(SH\mathbf{T}H^t)\,dH = \sum_{j=1}^{r}\sum_{h=1}^{r} e_{jh} C_{\kappa_j}(S)\mathbf{E}C_{\kappa_h}(\mathbf{T}) \\ &= \sum_{j=1}^{r}\sum_{h=1}^{r} e_{jh} d_{hh} C_{\kappa_j}(S) C_{\kappa_h}(\Sigma). \end{aligned} \tag{13.6.15}$$

Thus

$$d_{ii} e_{jh} = e_{jh} d_{hh}. \tag{13.6.16}$$

The coefficients e_{jh} are dependent on n while the d_{hh} are given by (13.6.12). Use Problem 13.6.12, and let the number of degrees of freedom tend over all sufficiently large positive integers. □

CHAPTER 14
Multivariate Inequalities

14.0. Introduction

There are now some very fine sources on inequalities, of which one of the comprehensive treatments is Marshall and Olkin (1979), which contains a very extensive bibliography. The more recent paper by Eaton (1982) lists some of the recent literature. It is the intent of this chapter to be somewhat selective in presenting fewer results but with more complete and accessible arguments than will be found in other sources.

The first part of the chapter originated with a 1975 course on optimal design taught by Jack C. Kiefer at Cornell University. In his lectures Kiefer applied majorization to the development of inequalities on matrices relevant to optimal design problems, and discussed Loewner's results on monotone and convex matrix functions. This author has done a lot of filling in around the material of Kiefer's 1975 lectures and has felt it worthwhile to preserve this treatment, which constitutes the material of Sections 14.2, 14.3, 14.4, and 14.5 of this chapter. It will be seen that much of the included material is about the eigenvalue functions $\lambda_1(A) \geq \lambda_2(A) \geq \cdots \geq \lambda_n(A)$, which is especially relevant in view of the continued use of eigenvalues in inference. In particular one obtains insight into measurability, continuity, monotonicity and convexity of these functions, and this is necessary mathematical background for the mathematics of inference.

The earlier and easier optimality results said that, if available, one should use designs with lots of symmetry. Hence BIBD, Latin squares, Youden squares, Hadamard matrices, orthogonal arrays, to begin a list, have been studied. Proofs become harder when complete symmetry is not possible and this is reflected in the papers of Kiefer on Youden squares, Kiefer and Galil on D-optimum weighing designs, and in recent papers by Kiefer's former student Cheng. Several recent inequalities, due to Cheng, that give insight into the asymmetric problem, are included in Section 14.4.

Once the chapter on inequalities was started, it was decided to include material on concave and convex measures, and on the *FKG*-inequality. The author felt that since he had coherent treatments of some of the essentials, not easily available in the case of convex and concave measures, and since the results have importance in the study of inference, this material should be included.

Although an effort has been made to be lucid in the statement of results and arguments, the chapter is otherwise very condensed. There is little in the way of interconnecting discussion and few examples of use, except that frequently one result builds upon another. Even so this has produced a long chapter. Eaton's paper, op. cit., is 33 pages in length.

14.1. Lattice Ordering of the Positive Definite Matrices

Theorem 14.1.1. *The positive definite $n \times n$ matrices are a lattice in the ordering of positive definite matrices, and*

$$\max(A, B) + \min(A, B) = A + B.$$

Proof. Let P be nonsingular with P^tAP and P^tBP diagonal matrices. See Problem 8.5.13. Define max and min by

$$(P^t \min(A, B)P)_{ij} = 0 \quad \text{if} \quad i \neq j,$$

and

$$= \min((P^tAP)_{ii}, (P^tBP)_{ii}),$$

for

$$1 \le i \le n.$$

Similarly for max. For notation let $M = \max(A, B)$, and for this proof as elsewhere in this book e_i is the $n \times 1$ unit vector with 1 in the i-th coordinate. Then for every i, one of

$$P^t(M - A)Pe_i = 0 \quad \text{or} \quad P^t(M - B)Pe_i = 0,$$

hence that

$$(M - A)Pe_i = 0 \quad \text{or} \quad (M - B)Pe_i = 0.$$

If $M \geq N \geq A, B$ so that $M - N \geq 0$ and $N - A \geq 0$, $N - B \geq 0$, then either

$$0 = e_i^tP^t(M - A)Pe_i = e_i^tP^t(M - N + N - A)Pe_i,$$

or

$$0 = e_i^tP^t(M - N + N - B)Pe_i, \quad \text{for} \quad 1 \le i \le n.$$

It follows that $0 = e_i^t P^t(M - N)Pe_i$ for $1 \le i \le n$, so that $0 = P^t(M - N)P$ and since P is nonsingular, $M = N$. Similarly for the minimum.

For numbers a, and b, $a + b = \min(a, b) + \max(a, b)$. So $\min(A, B) + \max(A, B) = P^{t^{-1}}(P^tAP + P^tBP)P^{-1} = A + B$. □

Definition 14.1.2. Suppose $\mathfrak{C}$ is a set of $n \times n$ symmetric matrices. If $A \in \mathfrak{C}$ and $B \le A$ implies $B \in \mathfrak{C}$ then we will say $\mathfrak{C}$ is a *decreasing* set of symmetric matrices. If $B \ge A$ implies $B \in \mathfrak{C}$ then we will say $\mathfrak{C}$ is an *increasing* set of symmetric matrices.

Theorem 14.1.3. *If $\mathfrak{C}_0$ and $\mathfrak{C}_1$ are increasing convex sets of symmetric matrices and $0 < \alpha < 1$ then $\alpha\mathfrak{C}_0 + (1 - \alpha)\mathfrak{C}_1$ is an increasing and convex set of matrices.*

Proof. We consider containing half-spaces for the form $\{A | \operatorname{tr} AB \ge c\}$, B a symmetric matrix and c a real number. Let $\{A | \operatorname{tr} AB \ge c\} \supset \alpha\mathfrak{C}_0 + (1 - \alpha)\mathfrak{C}_1$. If $A_0 \in \mathfrak{C}_0$ and $A_1 \in \mathfrak{C}_1$ then $\max(A_0, A_1) \in \mathfrak{C}_0 \cap \mathfrak{C}_1$ and if $x \in \mathbb{R}^n$ then for the $n \times n$ matrix xx^t, $nxx^t + \max(A_0, A_1) \in \mathfrak{C}_0 \cap \mathfrak{C}_1$. Thus $n^{-1}c \le \operatorname{tr}(xx^t)B + n^{-1} \operatorname{tr} B \max(A_0, A_1)$, and, letting $n \to \infty$, $0 \le x^tBx$ for all $x \in \mathbb{R}^n$. Consequently, if $A \ge \alpha A_0 + (1 - \alpha)A_1$ it follows from B being positive semidefinite that $\operatorname{tr} AB \ge \operatorname{tr}(\alpha A_0 + (1 - \alpha)A_1)B \ge c$. Since $\alpha\mathfrak{C}_0 + (1 - \alpha)\mathfrak{C}_1$ is convex it is the intersection of all half-spaces $\{A | \operatorname{tr} AB \ge c\}$. Thus $\alpha\mathfrak{C}_0 + (1 - \alpha)\mathfrak{C}_1$ is an increasing convex set of positive semidefinite matrices. □

Remark 14.1.4. The Hotelling-T^2 statistic has the form $x^tS^{-1}x$ and the sets $\{(x, S) | x^tS^{-1}x \le c\}$ are convex and increasing sets. See Lemma 14.3.8. In the discussion of log-concave measures in Section 14.6, particularly the examples of that section, it will be of interest to take convex mixtures of certain sets in a discussion of power of tests.

14.2. Majorization

The study of majorization depends on the classical Birkhoff result that every $n \times n$ doubly stochastic matrix is a convex mixture of $n \times n$ permutation matrices. We prove this result below and then apply it to inequalities. The aim of this section is to present complete arguments for some of the basic results. A comprehensive treatment can be found in Marshall and Olkin (1979).

One word of caution about the results stated in this section. Many results are correct for vectors with positive or negative entries, for symmetric matrices generally. Results are stated with a nonnegativity assumption only as needed.

It will be seen on reading this section that the main mathematical tool being used is Abel's identity. It will be helpful to review Remark 14.2.12 in advance of reading this section. Applications of the inequalities are given in Sections 14.3 and 14.4.

Definition 14.2.1. If $n \times 1$ vectors x and y satisfy $\Sigma_{i=1}^{j} e_i^t x \geq \Sigma_{i=1}^{j} e_i^t y, 1 \leq j \leq n$, with equality if $j = n$, then x is said to *majorize y*.

Definition 14.2.2. An $n \times 1$ vector x is *monotone* if and only if the entries of x are in decreasing order.

The set of monotone y majorized by x is contained in the set of vectors Px, P an $n \times n$ doubly stochastic matrix. This result is formally stated and proven below.

Definition 14.2.3. The diagonals of an $n \times n$ matrix A are the sets of n-tuples $((A)_{1,\sigma(1)}, \ldots, (A)_{n,\sigma(n)})$ taken over all permutations σ of $1, \ldots, n$.

Remark 14.2.4. The permutation matrix P_σ with entries $\delta_{i,\sigma(j)}$ yields products $AP_\sigma = ((A)_{i,\sigma(j)})$ which has a diagonal that corresponds to the n-tuple mentioned in Definition 14.2.3.

Lemma 14.2.5 (Frobenius–König Theorem). Let $n \geq 2$ and suppose every diagonal of the $n \times n$ matrix A contains a zero. Then A contains an $h \times k$ submatrix of zeros with $h + k \geq n + 1$.

Conversely, if A has an $h \times k$ submatrix of zeros and A has a diagonal with no zeros then $h + k \leq n$.

Proof. We prove the last statement first. By permutation of rows and columns arrange A to appear as $\begin{vmatrix} A_{11} & 0 \\ A_{21} & A_{22} \end{vmatrix}$ with 0 an $h \times k$ matrix. Then a nonzero diagonal $(A)_{i,\sigma(i)}$ with $1 \leq i \leq h$ must have $1 \leq \sigma(i) \leq n - k$ so that $\sigma(1), \ldots, \sigma(h)$ are h distinct numbers in the set $\{1, \ldots, n - k\}$ so that $h \leq n - k$ or $h + k \leq n$.

The proof of the first statement is by induction on n. If $n = 1$ the theorem follows trivially. If $n = 2$ and A is not the zero matrix then A must contain a 1×2 or 2×1 submatrix of zeros if A has no nonzero diagonal. In general, permute the rows and columns of A to obtain the new matrix $B = \begin{vmatrix} B_{11} & B_{12} \\ 0 & B_{22} \end{vmatrix}$, where B_{11} is $q \times q$ and B_{22} is $(n - q) \times (n - q)$. The block of zeros is $n - q \times q$. Deny the result. If B_{11} has a nonzero diagonal then every diagonal of B_{22} contains a zero. By induction B_{22} contains an $r \times s$ submatrix of zeros and $r + s \geq n - q + 1$. After permutation of the rows and columns of B_{22} we may suppose the resulting matrix looks like $\begin{vmatrix} C_{11} & C_{12} \\ 0 & C_{22} \end{vmatrix}$.

Then the matrix that results for these permutations on B has an $r \times (q + s)$ block of zeros and $r + q + s \geq n + 1$.

If we suppose B_{22} has a nonzero diagonal then denial implies that B_{11} has no nonzero diagonals. Hence a similar argument gives a similar construction of a block of zeros. □

Lemma 14.2.6. *Let A be a doubly stochastic matrix and* $A = \begin{vmatrix} A_{11} & A_{12} \\ A_{21} & A_{22} \end{vmatrix}$ *be such that A_{22} is a $p \times q$ matrix of zeros. Then $p + q \leq n$.*

PROOF. Let a_{ij} be the sum of the entries in A_{ij}. Then $a_{12} = q$, $a_{21} = p$, $a_{11} + a_{12} = n - p$, and $a_{11} + a_{21} = n - q$. Thus $0 \leq a_{11} = n - p - q$ or $p + q \leq n$. □

Corollary 14.2.7. *A doubly stochastic matrix has a nonzero diagonal.*

Theorem 14.2.8 (Birkhoff). *The doubly stochastic matrices are the convex hull of the permutation matrices.*

PROOF. Let A be an $n \times n$ doubly stochastic matrix and P_1 a nonzero diagonal of A. Then there is an $\varepsilon_1 > 0$ such that the entries of $A - \varepsilon_1 P_1$ are nonnegative but has more zero entries than A. $(A - \varepsilon_1 P_1)/(1 - \varepsilon_1)$ is doubly stochastic and the argument continues until only zeros are remaining. Thus inductively $A_{m+1} = (A_m - \varepsilon_m P_m)/(1 - \varepsilon_m)$ and $A = A_1$. At the last step A_{m+1} is a permutation matrix P_{m+1} and it is clear by successive back substitutions that each A_j is a convex mixture of permutation matrices. □

Lemma 14.2.9. *Let $P = (p_{ij})$ be an $n \times n$ doubly stochastic matrix and $x_1 \geq x_2 \geq \cdots \geq x_n$. Then if $1 \leq k \leq n$, $\Sigma_{i=1}^k \Sigma_{j=1}^n p_{ij} x_j \leq \Sigma_{i=1}^k x_i$.*

PROOF. Use the identity

$$\sum_{i=1}^{k} \sum_{j=k+1}^{n} p_{ij} = \sum_{i=k+1}^{n} \sum_{j=1}^{k} p_{ij}. \tag{14.2.1}$$

Then, since the $p_{ij} \geq 0$ and the $x_k - x_j \geq 0$ for $j \geq k$,

$$\begin{aligned} \sum_{i=1}^{k} \sum_{j=1}^{n} p_{ij} x_j &\leq \sum_{i=1}^{k} \sum_{j=1}^{k} p_{ij} x_j + \left(\sum_{i=1}^{k} \sum_{j=k+1}^{n} p_{ij} \right) x_k \\ &= \sum_{i=1}^{k} \sum_{j=1}^{k} p_{ij} x_j + \left(\sum_{i=k+1}^{n} \sum_{j=1}^{k} p_{ij} \right) x_k \\ &\leq \sum_{j=1}^{k} \sum_{i=1}^{k} p_{ij} x_j + \sum_{j=1}^{k} \sum_{i=k+1}^{n} p_{ij} x_j \\ &= \sum_{j=1}^{k} \left(\sum_{i=1}^{n} p_{ij} \right) x_j = \sum_{j=1}^{k} x_j. \end{aligned} \tag{14.2.2}$$

□

Theorem 14.2.10. *Let $x^t = (x_1, \ldots, x_n)$ and $y^t = (y_1, \ldots, y_n)$ be monotone vectors. Let $\mathbf{P}$ be the set of $n \times n$ doubly stochastic matrices. Then $\sup_{P \in \mathbf{P}} x^t P y = x^t y$. If x and y are strictly monotone then $x^t P y = x^t y$ if and only if $P = I_n$. More generally, if the entries of x and y are nonnegative then $\sup_{P \in \mathbf{P}} \Sigma_{i=1}^k \Sigma_{j=1}^n p_{ij} x_i y_j = \Sigma_{i=1}^k x_i y_i$.*

PROOF. We use Abel's identity, see Remark 14.2.12. Then, with $y_{n+1} = 0$, using (14.2.1),

$$\begin{aligned} x_1 y_1 + \cdots + x_n y_n &\geq \sum_{k=1}^n (y_k - y_{k+1}) \sum_{i=1}^k \sum_{j=1}^n p_{ij} x_j \\ &= \sum_{i=1}^n \sum_{k=i}^n \sum_{j=1}^n (y_k - y_{k+1}) p_{ij} x_j \\ &= \sum_{i=1}^n \sum_{j=1}^n \sum_{k=i}^n (y_k - y_{k+1}) p_{ij} x_j \\ &= \sum_{i=1}^n \sum_{j=1}^n y_i p_{ij} x_j . \end{aligned} \tag{14.2.3}$$

Since equality holds for $P = I_n$, the first statement of the theorem follows. To show the second statement, use Theorem 14.2.8, and write $P = \Sigma_\sigma \alpha_\sigma P_\sigma$ where P_σ is the $n \times n$ permutation matrix determined by σ. Then $x^t P y = \Sigma_\sigma \alpha_\sigma (x^t P_\sigma y)$ and as is easily shown, unless $P_\sigma = I_n$, in the strictly monotone case, $x^t P_\sigma y < x^t y$.

In the nonnegative case, with $P = (p_{ij})$,

$$\begin{aligned} \sum_{i=1}^k \sum_{j=1}^n p_{ij} x_i y_j &\leq \sum_{i=1}^k \left(\sum_{j=1}^k p_{ij} y_j + \sum_{j=k}^n p_{ij} y_k \right) x_i \\ &= \sum_{i=1}^k \sum_{j=1}^k p'_{ij} x_i y_j \leq x_1 y_1 + \cdots + x_k y_k \end{aligned}$$

for the new doubly stochastic matrix $P' = (p'_{ij})$. □

Corollary 14.2.11. *If P is an $n \times n$ doubly stochastic matrix and x is a monotone $n \times 1$ vector then Px is majorized by x. The vector Px need not be monotone.*

PROOF. Let $f_j = e_1 + \cdots + e_j$. Then $\Sigma_{i=1}^j (Px)_i = f_j^t (Px) \leq f_j^t x = x_1 + \cdots + x_j$, holding for $1 \leq j \leq n$, where f_j is defined as above. If $j = n$ then $\Sigma_{i=1}^n \Sigma_{j=1}^n (P)_{ij} x_j = \Sigma_{j=1}^n x_j \Sigma_{i=1}^n (P)_{ij} = \Sigma_{j=1}^n x_j$. The example $\begin{vmatrix} 1/3 & 2/3 \\ 2/3 & 1/3 \end{vmatrix} \begin{vmatrix} 3 \\ -3 \end{vmatrix} = \begin{vmatrix} -1 \\ 1 \end{vmatrix}$ shows Px need not be monotone. □

Remark 14.2.12. Abel's identity says $a_1 b_1 + \cdots + a_n b_n = (a_1 - a_2) b_1 + (a_2 - a_3)(b_1 + b_2) + \cdots + (a_{n-1} - a_n)(b_1 + \cdots + b_{n-1}) + a_n (b_1 + \cdots + b_n)$.

Theorem 14.2.13 (The Majorization Theorem). *Let x be a nonzero vector and the convex set $\mathbf{V} = \{z \mid \text{exists } P \in \mathbf{P} \text{ with } z = Px\}$. If x majorizes y and if y is a monotone vector, then $y \in \mathbf{V}$. Alternatively, if x is monotone, then $y \in \mathbf{V}$ if and only if x majorizes $P_\sigma y$ for every permutation σ.*

Remark 14.2.14. The identity $\begin{vmatrix} a & 1-a \\ 1-a & a \end{vmatrix} \begin{vmatrix} 1 \\ -1 \end{vmatrix} = \begin{vmatrix} -2 \\ 2 \end{vmatrix}$ has no solution with $0 \le a \le 1$. Thus the theorem need not hold if y is not monotone.

PROOF. Let $x^t = (x_1, \ldots, x_n)$ and $y^t = (y_1, \ldots, y_n)$. Assume x majorizes y. We suppose $y \notin V$ and obtain a contradiction. By the separation theorems for finite dimensional convex sets, there exists a vector $a^t = (a_1, \ldots, a_n)$ and a real number c such that if $P \in \mathbf{P}$ then $a^t P^t x \le c < a^t y$. We use here the fact that $\mathbf{P}$ is compact, hence $\mathbf{V}$ is a compact convex set of vectors.

If σ is a permutation of $1, \ldots, n$ with $P = P_\sigma$ we obtain $\Sigma_{i=1}^n a_i x_{\sigma(i)} \le c < \Sigma_{i=1}^n a_i y_i$. Let τ be a permutation of $1, \ldots, n$ such that $a_{\tau(1)} \ge \cdots \ge a_{\tau(n)}$. Choose $\sigma = \tau^{-1}$ so that $\Sigma_{i=1}^n a_{\tau(i)} x_i \le c < \Sigma_{i=1}^n a_{\tau(i)} y_{\tau(i)}$. By Abel's identity

$$\sum_{i=1}^{n} a_{\tau(i)} x_i = \sum_{i-1}^{n-1} (a_{\tau(i)} - a_{\tau(i+1)})(x_1 + \cdots + x_i) + a_{\tau(n)}(x_1 + \cdots + x_n)$$

$$\ge \sum_{i=1}^{n-1} (a_{\tau(i)} - a_{\tau(i+1)})(y_1 + \cdots + y_i) + a_{\tau(n)}(y_1 + \cdots + y_n)$$

$$= \sum_{i=1}^{n} a_{\tau(i)} y_i \ge \sum_{i=1}^{n} a_{\tau(i)} y_{\tau(i)} = \sum_{i=1}^{n} a_i y_i.$$

The last inequality follows since $a_{\tau(1)} \ge a_{\tau(2)} \ge \cdots \ge a_{\tau(n)}$ and y monotone imply that $\Sigma_{i-1}^n a_{\tau(i)} y_i = \sup_\pi \Sigma_{i-1}^n a_{\tau(i)} y_{\pi(i)}$ taken over all permutations π of $1, \ldots, n$.

This contradiction shows the assumption $y \notin V$ is contradictory.

To see that the last statement of the theorem is correct, if $y \in \mathbf{V}$ then $y = Px$ and $P_\sigma y \in \mathbf{V}$ so x majorizes $P_\sigma y$. Conversely, if for all σ, x majorizes $P_\sigma y$, then we may assume y is monotone and hence by the first part of the theorem, $y \in \mathbf{V}$. □

Theorem 14.2.15. *Let f be a convex real valued function defined on a convex set $\mathbf{D} \subset \mathbb{R}^n$ such that if P_σ is a permutation matrix and $x \in \mathbf{D}$ then $f(x) = f(P_\sigma x)$. If y is monotone and is majorized by x then $f(y) \le f(x)$.*

PROOF. By Theorem 14.2.13 there exists a doubly stochastic matrix P with $y = Px = \Sigma_\sigma a_\sigma P_\sigma$ a convex mixture of permutation matrices so that $f(y) = f(\Sigma_\sigma a_\sigma P_\sigma x) \le \Sigma_\sigma a_\sigma f(P_\sigma x) = f(x)$. □

EXAMPLE 14.2.16. $f(x) = \Sigma_{i=1}^n (e_i^t x)^{-r}$ where $r > 0$ and the components of x, $e_i^t x > 0$, $1 \le i \le n$.

EXAMPLE 14.2.17. More generally, if $g\colon [0, \infty) \to \mathbb{R}$ is a convex function then $f(x) = \Sigma_{i=1}^n g(e_i^t x)$ satisfies the above condition.

These examples are extended in the following discussion.

Theorem 14.2.18. *Let x be a strictly monotone $n \times 1$ vector and y be a monotone $n \times 1$ vector. If for all increasing convex functions $f\colon \mathbb{R} \to \mathbb{R}$, $\Sigma_{i=1}^n f(e_i^t x) \geq \Sigma_{i=1}^n f(e_i^t y)$, then $\Sigma_{i=1}^h (e_i^t x) \geq \Sigma_{i=1}^h (e_i^t y)$ for $1 \leq h \leq n$.*

PROOF. Let $f(t) = (t - x_1)^+$. By hypothesis $0 = \Sigma_{i=1}^n f(e_i^t x) \geq \Sigma_{i=1}^n f(e_i^t y) \geq 0$ so that $f(e_1^t y) = 0$ and $e_1^t y \leq e_1^t x$.

To abbreviate with $e_i^t x = x_i$ and $e_i^t y = y_i$, suppose it is shown that $y_1 + \cdots + y_j \leq x_1 + \cdots + x_j$ for $j \leq i - 1$. Take $f(t) = (t - x_i)^+$. Then $\Sigma_{j=1}^n f(x_j) = x_1 + \cdots + x_{i-1} - (i-1)x_i$ which follows since x is a strictly monotone vector.

Case I. If $y_i > x_i$ then $\Sigma_{j=1}^n f(y_j) \geq f(y_1) + \cdots + f(y_i) = y_1 + \cdots + y_i - ix_i$ and $x_1 + \cdots + x_i - ix_i = \Sigma_{j=1}^n f(x_j) \geq y_1 + \cdots + y_i - ix_i$.

Case II. If $x_i \geq y_i$ then since $x_1 + \cdots + x_{i-1} \geq y_1 + \cdots + y_{i-1}$, the result follows by adding the two inequalities. □

Theorem 14.2.19. *Let x be an $n \times 1$ monotone vector and f be a nondecreasing nonnegative convex function of a real variable. Let P be an $n \times n$ doubly stochastic matrix such that $y = Px$. Then if $1 \leq h \leq n$, $f(e_1^t x) + \cdots + f(e_h^t x) \geq f(e_1^t y) + \cdots + f(e_h^t y)$.*

PROOF. $e_i^t y = \Sigma_{j=1}^n (P)_{ij}(e_j^t x)$, a convex mixture. Therefore $f(e_i^t y) \leq \Sigma_{j=1}^n (P)_{ij} f(e_j^t x)$. Since x is a monotone vector and f is a nondecreasing nonnegative function, by Lemma 14.2.9, $\Sigma_{i=1}^h \Sigma_{j=1}^n (P)_{ij} f(e_j^t x) \leq \Sigma_{j=1}^h f(e_j^t x)$. □

The following two results are due to Ostrowski (1952).

Theorem 14.2.20. *Let $\mathbf{M}^n$ be the set of strictly monotone $n \times 1$ vectors. Let $f\colon \mathbf{M}^n \to \mathbb{R}$ be such that if x majorizes y then $f(x) \geq f(y)$. Suppose f has first partial derivatives $\partial_i f$, $1 \leq i \leq n$. Then the partial derivatives of f satisfy*

$$(\partial_i f)(x) \geq (\partial_{i+1} f)(x), \qquad x \in \mathbf{M}^n, \qquad 1 \leq i \leq n-1. \qquad (14.2.4)$$

Conversely, if the partial derivatives satisfy (14.2.4) *then f is Schur-convex, i.e., x majorizes y implies $f(x) \geq f(y)$.*

PROOF. Let $g(s, t) = f(x_1, \ldots, x_{i-1}, s, t, x_{i+2}, \ldots, x_n)$. Then on $\mathbf{M}^n$, with $s = x_i$ and $t = x_{i+1}$, the hypothesis implies $g(s - x, t + x)$ is, for x near zero, a decreasing function of x. The derivative with respect to x evaluated at $x = 0$ is $g_1(s, t) - g_2(s, t) \leq 0$. That is, $\partial_i f \geq \partial_{i+1} f$.

Conversely, if x majorizes y and the sequences are monotone, consider the convex mixture $(1-t)x+ty$. By Abel's identity, see Remark 14.2.12, $(d/dt)f((1-t)x+ty) = \Sigma_{i=1}^n (\partial_i f)((1-t)x+ty)(y_i - x_i) = \Sigma_{i=1}^{n-1}(\Sigma_{j=1}^i (y_j - x_j))(\partial_i f - \partial_{i+1} f) \leq 0$. We use here the fact that $\Sigma_{i=1}^n (y_i - x_i) = 0$. Therefore $f((1-t)x+ty)$ is a decreasing function of t and $f(x) \geq f(y)$ follows. □

Theorem 14.2.21. *Suppose f is a function of n variables with partial derivatives*

$$0 \leq \partial_1 f \leq \partial_2 f \leq \cdots \leq \partial_n f. \tag{14.2.5}$$

If x and y are $n \times 1$ vectors such that

$$\sum_{i=j}^n (e_i^t x) \geq \sum_{i=j}^n (e_i^t y), \qquad 1 \leq j \leq n, \tag{14.2.6}$$

then $f(x) \geq f(y)$.

Proof. By Abel's identity, and with $e_i^t x = x_i$ and $e_i^t y = y_i$,

$$\begin{aligned}(d/dt)f((1-t)x+ty) &= \sum_{i=1}^n (y_i - x_i)\partial_i f \\ &= \left(\sum_{i=1}^n (y_i - x_i)\right)\partial_1 f \\ &\quad + \sum_{j=2}^n \left(\sum_{i=j}^n (y_i - x_i)\right)(\partial_i f - \partial_{i-1} f) \leq 0.\end{aligned} \tag{14.2.7}$$

Thus the minimum is at $t = 1$ which implies $f(y) \leq f(x)$. □

14.3. Eigenvalues and Singular Values

Definition 14.3.1. If A is an $n \times n$ symmetric matrix with real entries then $\lambda_i(A)$ is the i-th largest eigenvalue of A. The $n \times 1$ column vector $\lambda(A)$ is the vector of eigenvalues in decreasing order.

Theorem 14.3.2. *Let A and B be $n \times n$ symmetric matrices. Then* $\sup_{U \in \mathbf{O}(n)} \operatorname{tr} UAU^tB = \Sigma_{i=1}^n \lambda_i(A)\lambda_i(B)$.

Proof. The supremum is unchanged if A is replaced by VAV^t and B is replaced by WBW^t, $V, W \in \mathbf{O}(n)$. We may thus assume A and B are diagonal matrices with $A = \operatorname{diag}(\lambda_1(A), \ldots, \lambda_n(A))$ and $B = \operatorname{diag}(\lambda_1(B), \ldots, \lambda_n(B))$. Then $\operatorname{tr} UAU^tB = \Sigma_{i=1}^n \Sigma_{j=1}^n (U)_{ij}^2 \lambda_j(A)\lambda_i(B)$. The matrix with ij-entry $(U)_{ij}^2$ is a doubly stochastic matrix and the supremum is assumed at $U = I_n$. See Theorem 14.2.9. □

In order to state the next theorem we let $\mathbf{V}$ be the set of all p-dimensional subspaces of $\mathbb{R}^n$.

Theorem 14.3.3. *If* $A \in \mathbf{S}(n)$ *then*

$$\lambda_p(A) = \sup_{V \in \mathbf{V}} \inf_{|x|=1, x \in V} x^t A x. \tag{14.3.1}$$

PROOF. We abbreviate to $\lambda_i = \lambda_i(A)$, with corresponding orthonormal eigenvectors $x_1, \ldots, x_n$. In the p-dimensional subspace spanned by $x_1, \ldots, x_p$, using the same norm, we have $|x_p| = 1$ and $x_j^t A x_j = \lambda_j$. If x is any unit vector in the subspace with $x = \alpha_1 x_1 + \cdots + \alpha_p x_p$, then $x^t A x = \alpha_1^2 \lambda_1 + \cdots + \alpha_p^2 \lambda_p \geq (\alpha_1^2 + \cdots + \alpha_p^2)\lambda_p = \lambda_p$. Hence sup inf $\geq \lambda_p(A)$ taken over this subspace.

To obtain the converse, let V be a p-dimensional subspace. The vectors $x_p, x_{p+1}, \ldots, x_n$ span a space W of dimension $(n - p) + 1$ so there is a nonzero vector $x \in V \cap W$, and by a scale change we may suppose $|x| = 1$ and that $x = \beta_p x_p + \cdots + \beta_n x_n$. Then $x^t A x = \Sigma_{j=p}^n \beta_j^2 \lambda_j(A) \leq \lambda_p(A) \Sigma_{j=p}^n \beta_j^2 = \lambda_p(A)$. Thus for every p-dimensional subspace V, $\inf_{|x|=1, x \in V} x^t A x \leq \lambda_p(A)$. □

Corollary 14.3.4. *If A and B are in* $\mathbf{S}(n)$ *and* $A \geq B$ *in the ordering of symmetric matrices then* $\lambda_j(A) \geq \lambda_j(B)$ *for* $1 \leq j \leq n$.

Theorem 14.3.5. $\lambda_1(A)$, $\lambda_1(A) + \lambda_2(A), \ldots, \lambda_1(A) + \cdots + \lambda_i(A), \ldots$ *are convex functions of* $A \in \mathbf{S}(n)$, *and are continuous, hence Borel measurable functions of* A.

PROOF. By Theorem 14.3.2, $\lambda_1(A) + \cdots + \lambda_i(A)$ is the supremum of a set of linear functions hence is a convex function. Since within the space of symmetric matrices a given matrix is an interior point of a bounded open set, and since $\lambda_1, \ldots, \lambda_i$ are bounded functions on bounded sets, continuity follows by the classical argument of Hardy, Littlewood and Polya (1952). By successive subtractions continuity of the individual functions $\lambda_1, \ldots, \lambda_n$ follows. Hence measurability. □

Corollary 14.3.6. *If* $A_1, A_2 \in \mathbf{S}(n)$ *and* $A_1 - A_2$ *is positive semidefinite, then* $\lambda_1(A_1) + \cdots + \lambda_i(A_1) \geq \lambda_1(A_2) + \cdots + \lambda_i(A_2)$ *for* $1 \leq i \leq n$.

PROOF. Use $B = \begin{vmatrix} I_i & 0 \\ 0 & 0 \end{vmatrix}$ so that $B^2 = B = B^t$ and thus $\operatorname{tr} U(A_1 - A_2)U^t B = \operatorname{tr}(BU)(A_1 - A_2)(BU)^t \geq 0$. The result now follows from Theorem 14.3.2. □

Some of the basic matrix inequalities are derived from the following observations. Since $\Sigma_{i=1}^h \lambda_i(\alpha A + (1 - \alpha)B) \leq \alpha \Sigma_{i=1}^h \lambda_i(A) + (1 - \alpha)\Sigma_{i=1}^h \lambda_i(B)$

and since $\Sigma_{i=1}^n \lambda_i(\alpha A + (1-\alpha)B) = \operatorname{tr}(\alpha A + (1-\alpha)B) = \alpha \Sigma_{i=1}^n \lambda_i(A) + (1-\alpha)\Sigma_{i=1}^n \lambda_i(B)$, so that majorization and monotonicity hold, it follows that if f is a symmetric convex function of n variables then using Theorem 14.2.15, the following result holds.

Theorem 14.3.7. *If f is a symmetric convex function of n variables then $f(\lambda_1(A), \ldots, \lambda_n(A))$ is a convex function of symmetric matrices.*

PROOF. By majorization, $f(\lambda_1(\alpha A + (1-\alpha)B), \ldots, \lambda_n(\alpha A + (1-\alpha)B)) \le f(\alpha\lambda_1(A) + (1-\alpha)\lambda_1(B), \ldots, \alpha\lambda_n(A) + (1-\alpha)\lambda_n(B))$, and by convexity this is $\le \alpha f(\lambda_1(A), \ldots, \lambda_n(A)) + (1-\alpha)f(\lambda_1(B), \ldots, \lambda_n(B))$. □

EXAMPLE here include $f(x_1, \ldots, x_n) = \Sigma_{i=1}^n x_i^r$ where $x_i > 0$ and $r < 0$ or $r \ge 1$, applied to positive definite matrices. In each case f is symmetric and convex so that for positive definite matrices A, $\Sigma_{i=1}^n \lambda_i(A)^r$ is a convex function of matrix argument.

The negative exponent case can be treated somewhat differently as follows.

Lemma 14.3.8. *In the ordering of positive definite matrices, if $X \in \mathbf{S}(n)$ is positive definite, if Y is $p \times n$, then for all $p \times 1$ vectors x, $x^t(YX^{-1}Y^t)x$ is a convex function of the variables X and Y.*

PROOF. $x^t(AX + Y)X^{-1}(AX + Y)^t x \ge 0$ and equals zero if $A = -YX^{-1}$. Expansion shows that $-x^t(AXA^t + AY^t + YA^t)x \le x^t(YX^{-1}Y^t)x$. The left side is linear in the variables X and Y and the right side is the supremum over all $p \times n$ matrices A. □

Theorem 14.3.9. *If $f\colon (0, \infty)x \cdots x(0, \infty) \to \mathbb{R}$ is convex, symmetric, and an increasing function of each variable, then $f(\lambda_1(A^{-1}), \ldots \lambda_n(A^{-1}))$ is a convex function of positive definite matrices $A \in \mathbf{S}(n)$.*

PROOF. By the preceding lemma, $(\alpha A + (1-\alpha)B)^{-1} \le \alpha A^{-1} + (1-\alpha)B^{-1}$ so that by Theorem 14.2.2, $\lambda_i((\alpha A + (1-\alpha)B)^{-1}) \le \lambda_i(\alpha A^{-1} + (1-\alpha)B^{-1})$. Substitution into f and use of Theorem 14.3.7 now completes the proof. □

Corollary 14.3.10. *The function $(\Sigma_{i=1}^n \lambda_i(A)^{-r})^{1/r}$, $r > 0$, is a convex function of positive definite matrices A.*

PROOF. The function $f(x) = (\Sigma_{i=1}^n x_i^{-r})^{1/r}$ is a convex function of vectors $x^t = (x_1, \ldots, x_n)$. To see this, if $r \ge 1$ then $f(\alpha x + (1-\alpha)y) = (\Sigma_{i=1}^n (\alpha x_i + (1-\alpha)y_i)^{-r})^{1/r} \le (\Sigma_{i=1}^n (\alpha x_i^{-1} + (1-\alpha)y_i^{-1})^r)^{1/r} \le \alpha f(x) + (1-\alpha)f(y)$. If $0 < r \le 1$ then $f(\alpha x + (1-\alpha)y) \le (\alpha \Sigma_{i=1}^n x_i^{-r} + (1-\alpha)\Sigma_{i=1}^n y_i^{-r})^{1/r} \le \alpha f(x) + (1-\alpha)f(y)$ since $t^{1/r}$ is a convex function if $1/r \ge 1$. Therefore Theorem 14.3.7 applies. □

As is well known, as a function of $r > 0$, $(\Sigma_{i=1}^n \lambda_i(A^{-1})^r)^{1/r}$ is an increasing function of $r > 0$. Hence the limit as $r \downarrow 0$ exists and is of special interest in optimality theory, corresponding to D-optimality.

Lemma 14.3.11. $\lim_{r \downarrow 0} (x_1^r + \cdots + x_n^r)^{1/r} = (x_1 \cdots x_n)^{1/n}$.

PROOF. Define a function $g(r) = (x_1^r + \cdots + x_n^r)^{1/r}$. Then $\ln g(r) = [\ln (x_1^r + \cdots + x_n^r)]/r$ is an indeterminant form $0/0$ as $r \downarrow 0$. By L'Hospital's rule, the limit is the same as $\lim_{r \downarrow 0} d/dr \ln (x_1^r + \cdots + x_n^r) = \lim_{r \downarrow 0} (\Sigma_{i=1}^n x_i^r \ln x_i)/(x_1^r + \cdots + x_n^r) = \ln (x_1 \cdots x_n)^{1/n}$. Hence the result follows by taking exponentials. □

In optimality theory one then has $(\det A^{-1})^{1/n} = \lim_{r \downarrow 0} (\Sigma_{i=1}^n \lambda_i(A)^{-r})^{1/r}$. As we now show, the function $\det A$ is a log-concave function.

Lemma 14.3.12. $\ln(\det A)$ *is a concave function of* $n \times n$ *positive definite matrices* $A \in \mathbf{S}(n)$, *and* $(\det A)^{-1/n}$ *is a convex function of positive definite matrices.*

PROOF. Let A and B be positive definite $n \times n$ matrices. Choose C nonsingular so that $CAC^t = I_n$ and $CBC^t = D$ is the diagonal matrix $D = \operatorname{diag}(d_1, \ldots, d_n)$. See Problem 8.5.13. Then

$$\begin{aligned} \ln \det (\alpha A + (1-\alpha)B) &= -\ln \det CC^t + \ln \det C(\alpha A + (1-\alpha)B)C^t \\ &= \ln \det A + \ln \det (\alpha I_n + (1-\alpha)D) \\ &= \ln \det A + \sum_{i=1}^n \ln (\alpha 1 + (1-\alpha)d_i) \\ &\geq \ln \det A + \sum_{i=1}^n (\alpha \ln 1 + (1-\alpha) \ln d_i) \\ &= \alpha \ln \det A + (1-\alpha)[\ln \det (CC^t)^{-1} + \ln \det D] \\ &= \alpha \ln \det A + (1-\alpha) \ln \det B. \end{aligned} \tag{14.3.2}$$

This proves the first assertion. Since $(\det A)^{-1/n}$ is a pointwise limit of convex functions, see Lemma 14.3.11 and Corollary 14.3.10, the limit function is convex. □

The inequality of Weyl, Problem 8.5.29, says that if, over the complex numbers, A has eigenvalues $\lambda_1, \ldots, \lambda_n$, and $(A^*A)^{1/2}$ has eigenvalues $\nu_1, \ldots, \nu_n$ such that $|\lambda_1| \geq \cdots \geq |\lambda_n|$ and $\nu_1 \geq \cdots \geq \nu_n$, then for $1 \leq h \leq n$, $\Sigma_{i=1}^h \ln |\lambda_i| \leq \Sigma_{i=1}^h \ln \nu_i$, and equality holds for $h = n$. An immediate consequence of the majorization theorem is then the following result.

Theorem 14.3.13 (Weyl (1949)). *If f is a symmetric function of n variables and if f is convex, then*

$$f(\ln|\lambda_1|, \ldots, \ln|\lambda_n|) \leq f(\ln \nu_1, \ldots, \ln \nu_n). \tag{14.3.3}$$

And the corollary,

Corollary 14.3.14. *If $s > 0$ then $\Sigma_{i=1}^h |\lambda_i|^s \leq \Sigma_{i=1}^h \nu_i^s$, $1 \leq h \leq n$.*

PROOF. The function $f(x) = \Sigma_{i=1}^n \exp(sx_i)$ is a convex symmetric function, increasing in each variable. By Theorem 14.2.19 the result follows. □

An important special case of Corollary 14.3.14 is given a separate statement and proof here. The theorem gives more information about when majorization can hold.

Theorem 14.3.15 (Cohen (1966) uses this result). *Let M be an $n \times n$ matrix having real entries. Then* $\operatorname{tr} M \leq \operatorname{tr}(M^tM)^{1/2}$. *Equality holds if and only if $M^t = M$.*

PROOF. By Theorem 8.1.11, M has singular value decomposition $M = U\operatorname{diag}(\lambda_1^{1/2}, \ldots, \lambda_n^{1/2})V$ with U, V in $\mathbf{O}(n)$. Thus if U has column vectors $u_1, \ldots, u_n$, and V has row vectors $v_1^t, \ldots, v_n^t$, then

$$\begin{aligned}\operatorname{tr} M &= \sum_{i=1}^n u_i^t M u_i = \sum_{i=1}^n e_i^t \operatorname{diag}(\lambda_1^{1/2}, \ldots, \lambda_n^{1/2}) V u_i \\ &= \sum_{i=1}^n \lambda_i^{1/2} e_i^t V u_i = \sum_{i=1}^n \lambda_i^{1/2} v_i^t u_i \leq \sum_{i=1}^n \lambda_i^{1/2} \|v_i\| \, \|u_i\| \qquad (14.3.4)\\ &= \operatorname{tr}(M^tM)^{1/2}.\end{aligned}$$

In using the Cauchy–Schwarz inequality, $v_i^t u_i < \|v_i\| \, \|u_i\|$ unless u_i and v_i are proportional. Proportionality of unit vectors implies they are equal or differ by a factor of -1. A sign change gives $-\lambda_i^{1/2}$ and strict inequality. Hence $M = U\operatorname{diag}(\lambda_1^{1/2}, \ldots, \lambda_n^{1/2})U^t$ and if equality holds then $M = M^t$. □

In the subject of *Universal Optimality*, optimality functionals of matrix argument $\Phi(A) = f(\lambda(A))$ are considered, $\lambda(A)$ the vector of eigenvalues in decreasing order, and $A^t = A$, A an $n \times n$ matrix. This definition makes Φ orthogonally invariant, i.e., if $U \in \mathbf{O}(n)$ then $\Phi(A) = \Phi(UAU^t)$. The following result appears in Bondar (1983).

Theorem 14.3.16. *Φ is a convex function of symmetric matrices if and only if*

(1) *$f: \mathbf{M}^n \to \mathbb{R}$ is convex (see Theorem 14.2.20);*
(2) *(Schur-convex) if λ_1 majorizes λ_2 and if λ_1 and λ_2 are monotone vectors then $f(\lambda_1) \geq f(\lambda_2)$.*

PROOF. If A is a diagonal matrix, $A = \operatorname{diag}(\lambda(A))$, then it is clear that convexity of Φ implies that of f. If λ_1 majorizes λ_2 and the vectors are monotone, then by Theorem 14.2.8 and 14.2.13, there exists a doubly stochastic matrix P with $\lambda_2 = P\lambda_1$, and permutation matrices P_σ so that $P = \Sigma_\sigma \alpha_\sigma P_\sigma$, a convex mixture. Then $\lambda_2 = (\Sigma_\sigma \alpha_\sigma P_\sigma)\lambda_1$ and $\operatorname{diag}(\lambda_2) = \Sigma_\sigma \alpha_\sigma P_\sigma \operatorname{diag}(\lambda_1) P_\sigma^t$. Then, since Φ is convex, $f(\lambda_2) = \Phi(\operatorname{diag}(\lambda_2)) \le \Sigma_\sigma \alpha_\sigma \Phi(P_\sigma \operatorname{diag}(\lambda_1) P_\sigma^t) = \Phi(\operatorname{diag}(\lambda_1)) = f(\lambda_1)$, and f is Schur-convex.

Conversely, from Theorem 14.3.5, Schur-convexity of f implies $f(\lambda(\alpha A + (1-\alpha)B)) \le f(\alpha\lambda(A) + (1-\alpha)\lambda(B))$ so that by convexity of f, $\Phi(\alpha A + (1-\alpha)B) \le \alpha\Phi(A) + (1-\alpha)\Phi(B)$ follows. □

Remark 14.3.17. Most results stated in Sections 14.2 and 14.3 are stated for and are correct for symmetric matrices which may have negative eigenvalues. Optimality theory as considered by Bondar (1983) as well as by Cheng, Kiefer and others earlier, consider only functions Φ of positive definite matrices defined on monotone vectors with positive entries.

14.4. Results Related to Optimality Considerations

Lemma 14.4.1. *If $A \in \mathbf{S}(n)$ is positive definite and $r > 0$ then $(\Sigma_{i=1}^n \lambda_i(A)^{-r})^{1/r}$ is an increasing function of r and $\lim_{r\downarrow 0} (\Sigma_{i=1}^n \lambda_i(A)^{-r})^{1/r} = (\det A^{-1})^{1/n}$. Also, $(1/n)\operatorname{tr} A \ge (\det A)^{1/n}$ and equality holds if and only if A is a multiple of the identity.*

Proof. As is well known $(\Sigma_{i=1}^n x_i^r)^{1/r}$ is an increasing function of r. Since $(\Sigma_{i=1}^n \lambda_i(A^{-1})^r)^{1/r} = (\Sigma_{i=1}^n \lambda_{n-i}(A)^{-r})^{1/r} = (\Sigma_{i=1}^n \lambda_i(A)^{-r})^{1/r}$, the first assertion follows. By Lemma 14.3.11 the second assertion follows. To obtain the last assertion, $(1/n)\operatorname{tr} A = (1/n)\Sigma_{i=1}^n \lambda_i(A) = (1/n)\Sigma_{i=1}^n \exp\ln\lambda_i(A) \ge \exp((1/n)\Sigma_{i=1}^n \ln\lambda_i(A)) = (\det A)^{1/n}$. □

The following theorem is useful in proving some simple optimality results for balanced designs such as Latin squares. Let f_k be the $k \times 1$ vector having all entries equal one.

Theorem 14.4.2 (Farrell (1973)). *Let A be an $n \times h$ matrix of 0,1 entries such that $Af_h = f_n$. Let $D = A^tA - B$ be a symmetric positive semidefinite matrix such that $Df_h = 0$. Assume $B \ge 0$. Then $\operatorname{tr} D \le (h-1)n/h$.*

Proof. Let $x_1, \ldots, x_{h-1}$ be an orthonormal set of eigenvectors for the nonzero eigenvalues of D and $y_1, \ldots, y_{h-1}$ an orthonormal set of eigenvectors for the nonzero eigenvalues of $(1/h!)\Sigma_\sigma P_\sigma D P_\sigma^t$, the P_σ the $h \times h$ permutation matrices. Then A^tA is a diagonal matrix and

$$\begin{aligned}
\operatorname{tr} D &= \sum_{i=1}^{h-1} x_i^t D x_i = \operatorname{tr} D \sum_{i=1}^{h-1} x_i x_i^t \\
&= \operatorname{tr} D(I_h - f_h f_h^t/h) = \operatorname{tr} DP_\sigma^t(I_h - f_h f_h^t/h)P_\sigma \\
&= (1/h!)\Sigma_\sigma \operatorname{tr}(P_\sigma D P_\sigma^t)(I_h - f_h f_h^t/h) \\
&= (1/h!)(y_1 + \cdots + y_{h-1})^t \Sigma_\sigma P_\sigma D P_\sigma^t (y_1 + \cdots + y_{h-1}) \quad (14.4.1) \\
&= (h-1)(1/h!) \sup_{\|y\|=1} y^t(\Sigma_\sigma P_\sigma D P_\sigma^t) y \\
&\le (h-1)(1/h!) \sup_{\|y\|=1} y^t(\Sigma_\sigma P_\sigma A^t A P_\sigma^t) y \\
&= ((h-1)/h) \operatorname{tr} A^t A = (h-1)n/h. \qquad \square
\end{aligned}$$

The following lemma is needed in order to prove Cheng's inequalities, see Theorems 14.4.4 and 14.4.6.

Lemma 14.4.3. *Let f and g be strictly convex functions of a real variable with everywhere defined second derivatives. Subject to $x + y + ht$ constant, $g(x) + g(y) + hg(t)$ constant, the function $z(t) = f(x(t)) + f(y(t)) + hf(t)$ is a strictly decreasing function if f''/g'' is strictly decreasing, and is a strictly increasing function if f''/g'' is strictly increasing.*

Proof. Taking derivatives with respect to t gives

$$\begin{aligned}
dx/dt &= h(g'(y) - g'(t))/(g'(x) - g'(y)); \\
dy/dt &= -h(g'(x) - g'(t))/(g'(x) - g'(y)).
\end{aligned} \quad (14.4.2)$$

Substitution into dz/dt and use of the generalized mean value theorem gives

$$dz/dt = h(g'(y) - g'(t))((f''(\theta_1)/g''(\theta_1)) - (f''(\theta_2)/g''(\theta_2))) \quad (14.4.3)$$

where $\theta_1 \in (x, y)$ and $\theta_2 \in (y, t)$. We assume $x < y < t$. Since g is strictly convex, $dx/dt > 0$ and $dy/dt < 0$. Therefore the sign of dz/dt is negative if f''/g'' is a decreasing function, positive if f''/g'' is an increasing function. The derivative dz/dt cannot vanish if f''/g'' is strictly monotone. □

The following inequalities, developed by Cheng, were presented at the Kiefer–Wolfowitz Memorial Conference, Cornell University, 1983, in a more general form than reported here. The following proofs are this author's. The most important thing here is the method of proof, which may be needed for treatment of special cases.

Theorem 14.4.4 (Cheng (1983)). *Let $f: [0, \infty) \to \mathbb{R}$ and $g: [0, \infty) \to \mathbb{R}$ be strictly convex functions with second derivatives on $(0, \infty)$ such that g is increasing and f''/g'' is a strictly decreasing function. Then subject to $x_1 \ge \cdots \ge x_n$, $x_1 + \cdots + x_n$ a constant, and $g(x_1) + \cdots + g(x_n)$ a constant, the value $f(x_1) + \cdots + f(x_n)$ is minimized by $x_1 \ge x_2 = x_3 = \cdots = x_n$.*

PROOF. By induction. Suppose $x_{n-h} > x_{n-h+1} = \cdots = x_n$. Let $x = x_1$, $y = x_{n-h}$, and $t = x_{n-h+1}$. Then subject to $x + y + ht$ constant and $g(x) + g(y) + hg(t)$ constant we want to decrease $f(x) + f(y) + hf(t)$. Since g is convex and increasing $\lim_{t\to\infty} g(t) = \infty$. By Lemma 14.4.3, $y(t)$ decreases so there is a value t_0 at which $y(t_0) = t_0$. By the lemma, $f(x(t_0)) + f(y(t_0)) + hf(t_0) < f(x) + f(y) + hf(t)$. We now have new values $x(t_0) \geq x_1 \geq \cdots \geq x_{n-h} \geq y(t_0) = t_0$. Thus as long as there is a $j \geq 2$ with $x_j > x_{j+1}$ it is possible to make an adjustment, until $x_2 = x_3 = \cdots = x_n$ is obtained. □

Remark 14.4.5. In the case that f''/g'' is strictly increasing the alternate form of Lemma 14.4.3 is $t > x > y$ with $ht + x + y$ constant, $hg(t) + g(x) + g(y)$ constant. Then $dx/dt < 0$ and $dy/dt > 0$. In minimization of $f(x_1) + \cdots + f(x_n)$ one now decreases x_1 which increases x_2 and decreases x_n. If f''/g'' is now increasing as a function of t, decreasing the value of t decreases $hf(t) + f(x) + f(y)$. An inductive argument similar to that used to prove Theorem 14.4.4 now establishes the following.

Theorem 14.4.6 (Cheng (1983)). *Let f and g be strictly convex functions defined on $[0, \infty)$ to $\mathbb{R}$ such that f''/g'' is a strictly increasing function. Subject to $x_1 \geq \cdots \geq x_n$, $x_1 + \cdots + x_n$ constant and $g(x_1) + \cdots + g(x_n)$ constant, the sum $f(x_1) + \cdots + f(x_n)$ is minimized by $x_1 = x_2 = \cdots = x_{n-1} \geq x_n$.*

Remark 14.4.7. A somewhat different set of conditions result if one applies the majorization inequalities, Theorems 14.2.19 and 14.2.20. For example, if $g: \mathbb{R} \to \mathbb{R}$ is convex, nondecreasing, and if $f(g^{-1}(\ ))$ is convex and nondecreasing, then x majorizes y implies

$$\sum_{i=1}^{h} g(x_i) \geq \sum_{i=1}^{h} g(y_i), \qquad h = 1, \ldots, n. \tag{14.4.4}$$

See Theorem 14.2.19. Hence

$$\begin{aligned}\sum_{i=1}^{h} f(x_i) &= \sum_{i=1}^{h} f(g^{-1}(g(x_i))) \\ &\geq \sum_{i=1}^{h} f(g^{-1}(g(y_i))) = \sum_{i=1}^{h} f(y_i),\end{aligned} \tag{14.4.5}$$

by Theorem 14.2.20 slightly modified. That is, $f(x) \geq f(y)$.

This proves the following result.

Theorem 14.4.8. *If g is a nondecreasing convex function and if $f(g^{-1}(\ ))$ is a nondecreasing convex function then x majorizes y implies $g(x) \geq g(y)$ and $f(x) \geq f(y)$.*

14.5. Loewner Ordering

If A is an $n \times n$ symmetric matrix and $U \in \mathbf{O}(n)$ such that UAU^t is a diagonal matrix with diagonal entries $\lambda_i(A) = (UAU^t)_{ii}$, then $f(A)$ is defined to be the symmetric matrix such that $Uf(A)U^t = \operatorname{diag}(f(\lambda_1(A)), \ldots, f(\lambda_n(A)))$.

The most common examples of such functions f are $f(x) = x^2, f(x) = x^{1/2}$, and $f(x) = 1/x$, respectively $f(A) = A^2$, $f(A) = A^{1/2}$, and $f(A) = A^{-1}$ being the values obtained. $A > 0$ has 2^n square roots of which $f(A)$ is the positive definite square root. For certain f such as $x^{1/2}$, the domain of f on matrices may be a proper subset of the symmetric matrices.

Definition 14.5.1. f is nondecreasing of order n means, if A and B are $n \times n$ symmetric matrices and $A \le B$ then $f(A) \le f(B)$.

Definition 14.5.2. f is nondecreasing of all orders means f is defined for $n \times n$ symmetric matrices for all n and f is nondecreasing of order n for all $n \ge 1$.

Definition 14.5.3. f is convex of all orders if $f(\alpha A + (1-\alpha)B) \le \alpha f(A) + (1-\alpha)f(B)$ for all $n \times n$ symmetric matrices and all $n \ge 1$ such that the eigenvalues of A and B are in the domain of f.

EXAMPLE 14.5.4. See Lemma 14.3.8. $f(A) = A^{-1}$ is convex and nonincreasing of all orders. Similarly by the same result $f(A) = A^2$ is a convex function. As is shown next, A^2 is not monotone. Take $A = \begin{vmatrix} 0 & 1 \\ 1 & 0 \end{vmatrix}$ and $B = \begin{vmatrix} \frac{1}{2} & \frac{1}{2} \\ \frac{1}{2} & \frac{1}{2} \end{vmatrix}$. Then $B \ge A$. But A^2 is the identity and $B^2 = B$ and $B^2 - A^2$ is negative definite.

Theorem 14.5.5. *$A^{1/2}$ is a nondecreasing matrix function.*

PROOF. We give a simple inductive argument to show that $A^{1/2}$ is nondecreasing. Define the square root iteratively, $A_{n+1} = \frac{1}{2}(A_n + A_n^{-1}A)$, with $A_0 = I_n$. All elements of this sequence commute with A and the limit is $A^{1/2}$. Here if $B \le A$ similarly define B_n for $n \ge 1$ and show by induction that $B_n \le A_n$, $n \ge 1$. In particular $B_1 \le A_1$ is clear and $B_1 B^{-1} \ge A_1 A^{-1}$ holds. If $B_n \le A_n$ and $B_n B^{-1} \ge A_n A^{-1}$ then $B_{n+1} = \frac{1}{2}(B_n + B_n^{-1}B) \le \frac{1}{2}(A_n + A_n^{-1}A) = A_{n+1}$. Also $A_n^{-1}A = AA_n^{-1}$ and $B_n^{-1}B = BB_n^{-1}$. Then $A_{n+1}A^{-1} = \frac{1}{2}(A_n A^{-1} + A_n^{-1}) \le \frac{1}{2}(B_n B^{-1} + B_n^{-1}) = B_{n+1}B^{-1}$. Hence passing to the limit, $B^{1/2} \le A^{1/2}$. □

See Section 8.4 for the corresponding result on Hilbert spaces.

Since $A \to A^2$ is a convex function, $\alpha A + (1-\alpha)B = \alpha(A^{1/2})^2 + (1-\alpha)(B^{1/2})^2 \ge (\alpha A^{1/2} + (1-\alpha)B^{1/2})^2$. Since the map $A \to A^{1/2}$ is nondecreasing, $(\alpha A + (1-\alpha)B)^{1/2} \ge \alpha A^{1/2} + (1-\alpha)B^{1/2}$.

Theorem 14.5.6.

$A \to A^{-1}$ *is convex, nonincreasing of all orders.*

$A \to A^{1/2}$ *is concave, nondecreasing of all orders.*

$A \to A^{2}$ *is convex of all orders but is not monotone.*

Some of the main results are stated wthout proofs which may be found in Bendat and Sherman (1955).

Theorem 14.5.7. *A function f defined on (a, b) is nondecreasing of all orders if and only if it is analytic in (a, b), can be analytically continued into the entire upper half plane, and there represents an analytic function whose imaginary part is positive. For nonincreasing functions the imaginary part is negative.*

Corollary 14.5.8. *$f(x) = x^p$ on $(0, \infty)$. If $0 \le p \le 1$ then f is nondecreasing. If $-1 \le p \le 0$ f is nonincreasing.*

Proof. Write $z = R(\cos\theta + i\sin\theta)$ and $z^p = R^p(\cos p\theta + i\sin p\theta)$. On the upper half plane $0 \le \theta \le \pi$ and we must have $0 \le p\theta \le \pi$ for a nondecreasing function, and $0 \ge p\theta \ge -\pi$ for a nonincreasing function. □

Example 14.5.9. The function $f(x) = x/(1 - tx)$ is defined if $x < 1/t$ and we consider the interval $(-1/t, 1/t)$. The function is analytic on the upper half plane and has positive imaginary part there. Thus if μ is a nonnegative countably additive measure on $(-1/R, 1/R)$ we may define $f(x) = \int_{-1/R}^{1/R} x/(1 - tx)\mu(dt)$ which is defined if $x \in (-R, R)$. Since the imaginary part will be positive, $f(A)$ is a nondecreasing function.

Theorem 14.5.10. *If f is nondecreasing of all orders, if $f(x) = \Sigma_{n=0}^{\infty} a_n x^n$ with radius of convergence R, then f is uniquely representable as $f(x) = \int_{-1/R}^{1/R} x/(1 - tx)\mu(dt)$.*

Theorem 14.5.11. *f is a convex operator function defined on (a, b) if and only if for each $x_0 \in (a, b)$, the analytic function $(f(z) - f(x_0))/(z - x_0)$ is a monotonic operator function of all orders.*

Example 14.5.12. $f(x) = (1 + x)^{-1/2}$ defines a nonincreasing convex operator function.

Direct Proof. As previously shown, $A \to A^{1/2}$ is nondecreasing and concave. Then $(I_n + \alpha A + (1 - \alpha)B)^{1/2} \ge \alpha(I_n + A)^{1/2} + (1 - \alpha)(I_n + B)^{1/2}$. Then $(I_n + \alpha A + (1 - \alpha)B)^{-1/2} \le (\alpha(I_n + A)^{1/2} + (1 - \alpha)(I_n + B)^{1/2})^{-1} \le \alpha(I_n + A)^{-1/2} + (1 - \alpha)(I_n + B)^{-1/2}$. That proves convexity. To prove monotonicity, let

$B \leq A$. Then $(I_n + B)^{-1} \geq (I_n + A)^{-1}$ and since $A \to A^{1/2}$ is nondecreasing, $(I_n + B)^{-1/2} \geq (I_n + A)^{-1/2}$. □

EXAMPLE 14.5.13. The preceding example is of interest in MANOVA. The nonzero eigenvalues of $\mathbf{Y}(\mathbf{Y}^t\mathbf{Y} + \mathbf{Z}^t\mathbf{Z})^{-1}\mathbf{Y}^t$ are the same as those of $(\mathbf{Y}^t\mathbf{Y} + \mathbf{Z}^t\mathbf{Z})^{-1}(\mathbf{Y}^t\mathbf{Y}) = (I_n + (\mathbf{Y}^t\mathbf{Y})^{-1}(\mathbf{Z}^t\mathbf{Z}))^{-1}$, which are therefore the same as those of $(I_n + \mathbf{Z}(\mathbf{Y}^t\mathbf{Y})^{-1}\mathbf{Z}^t)^{-1}$. See Problem 8.5.22. Then the square roots have the same nonzero eigenvalues. For any convex acceptance region $\mathbf{C}$ which is monotone decreasing, let $\mathbf{D} = \{S \mid (I_n + S)^{-1/2} \in \mathbf{C}\}$. If $S_1 \in \mathbf{D}$ and $S_2 \in \mathbf{D}$ then $(I_n + \alpha S_1 + (1-\alpha)S_2)^{-1/2} \leq \alpha(I_n + S_1)^{-1/2} + (1-\alpha)(I_n + S_2)^{-1/2}$ so since the right side is in $\mathbf{C}$ so is the left side of the inequality, so that $\alpha S_1 + (1-\alpha)S_2 \in \mathbf{D}$. Therefore $\mathbf{D}$ is convex and nondecreasing.

Theorem 14.5.14. *If f is defined on $(-\infty, \infty)$, if $f(0) = f'(0) = 0$, and if $f(x) = \Sigma_{n=2}^{\infty} a_n x^n$, then f is a convex matrix function if and only if $f(x)/x$ is nondecreasing of all orders.*

Corollary 14.5.15. *Applied to the function $f(x) = x^p, p > 1$, the function $f(x)/x$ is (see Corollary 14.5.7) monotone if and only if $0 \leq p - 1 \leq 1$ or $1 \leq p \leq 2$. We examine the other cases, i.e., $p < 1$, below.*

Theorem 14.5.16 (Proof from Jack Kiefer (1975)). *The function $f(x) = x^p$ defined on $(0, \infty)$ is a convex matrix function if and only if $-1 \leq p < 0$ or $1 \leq p \leq 2$.*

PROOF. To apply Theorem 14.5.10, we look at $(f(z) - f(x))/(z - x) = x^{p-1}[(z/x)^p - 1]/[(z/x) - 1]$ with $x > 0$ a real number. To discuss monotonicity, by Theorem 14.5.6, it then suffices to look at the imaginary part of $w^p - 1/w - 1$, w complex. This function is analytic at $w = 1$, hence for all $w > 0$ and real. To see this, the line integral $(2\pi i)^{-1} \int_{\mathbf{L}} (w^p - 1)/(w - 1)\, dw$ is the value of the analytic function $w^p - 1$ at $w = 1$, i.e., 0, for all simple closed curves $\mathbf{L}$ about 1. Hence the function is analytic at $w = 1$. Write $w = R(\cos\theta + i \sin\theta)$ and substitute. After finding the imaginary part it is sufficient to look at

$$\begin{aligned}\mathrm{Im} &= R^{p+1}\sin(p-1)\theta + R\sin\theta - R^p \sin p\theta \\ &= R\sin\theta\left[R^p\frac{\sin(p-1)\theta}{\sin\theta} - R^{p-1}\frac{\sin p\theta}{\sin\theta} + 1\right].\end{aligned} \tag{14.5.1}$$

Case. $0 < p < 1$. Not convex. As a harmonic function on the half circle bounded by $[-R, R]$ together with $R(\cos\theta + i\sin\theta)$, $0 \leq \theta \leq \pi$, the minimum and maximum values are assumed on the boundary. On the real axis, $\theta = 0$ and $\mathrm{Im} = 0$. On the circular part, R large, $0 < \theta < \pi$, and $p - 1 < 0$, so that $\sin(p-1)\theta/\sin\theta < 0$. Thus at $\theta = \pi/2$, Im will be negative for R large and the imaginary part is not everywhere positive.

Case $p = 0$. The map $f(A) = I_n$ for all A is convex.

Case $-1 < p < 0$. Convex. Again, if $\theta = 0$ then Im $= 0$. On any sequence $R_n \to \infty$ and $\theta_n \to \theta$, $0 \le \theta < \pi$, the quantity in the brackets of (14.5.1) converges to 1, since $p < 0$, while $R_n \sin\theta_n > 0$. In the case $R_n \to \infty$ and $\theta_n \to \pi$ then $-\pi < p\theta_n < 0$ and $-2\pi < (p-1)\theta_n < -\pi$ so that for n large the terms of Im are positive.

Case $p = 1$. The identity map is convex.

The cases $p = -1$, $p < -1$, and $p > 1$ may be treated in a similar way. □

Theorem 14.5.17 (Proof from Jack Kiefer (1975)). *Let* $z = |z|(\cos\theta + i\sin\theta)$ *and* $\ln z = \ln|z| + i\theta$, $0 \le \theta < 2\pi$, *define the branch of the logarithm. The function* $-\ln z$ *is a convex operator function.*

Proof. We need $(-\ln z + \ln x)/(z - x) = x^{-1}(-\ln(z/x))/((z/x) - 1)$. Thus it is sufficient to look at the imaginary part of $-\ln z/(z-1) = [-\ln R - i\theta]/[(R\cos\theta - 1) + i\sin\theta]$ with

$$\begin{aligned} \mathrm{Im} &= -\theta(R\cos\theta - 1) + R\sin\theta\ln R. \\ &= R_n \ln R_n[\sin\theta_n - \theta_n(R_n\cos\theta_n - 1)/R_n\ln R_n]. \end{aligned} \tag{14.5.2}$$

Consider sequences $R_n \to \infty$ and $\theta_n \to \theta$ where $0 \le \theta \le \pi$.

Case $0 < \theta < \pi$. For R large the part in brackets is nearly $\sin\theta > 0$.

Case $\theta = \pi$. The part in brackets becomes and remains positive.

Case $\theta = 0$. $\sin\theta/\theta$ converges to one so that $(\theta_n R_n \ln R_n)^{-1}$ Im converges to one. □

14.6. Concave and Convex Measures

The following material is based on a paper by Rinott (1976). Considered are measures ν on the Borel subsets of $\mathbf{C} \subset \mathbb{R}^n$, where $\mathbf{C}$ is an open convex subset, which satisfy an inequality of the form, if $0 \le \alpha \le 1$ and $\mathbf{A}_0 \subset \mathbf{C}$, $\mathbf{A}_1 \subset \mathbf{C}$, and s is given, then

$$\nu(\alpha\mathbf{A}_0 + (1-\alpha)\mathbf{A}_1) \ge (\alpha\nu(\mathbf{A}_0)^s + (1-\alpha)\nu(\mathbf{A}_1)^s)^{1/s}. \tag{14.6.1}$$

Such measure are called *concave*. A famous example is the Brunn–Minkowsky inequality for n-dimensional Lebesgue measure μ_n with $\mathbf{C} = \mathbb{R}^n$, which satisfies the inequality

$$\mu_n(\alpha\mathbf{A}_0 + (1-\alpha)\mathbf{A}_1)^{1/n} \ge \alpha\mu_n(\mathbf{A}_0)^{1/n} + (1-\alpha)\mu_n(\mathbf{A}_1)^{1/n}. \tag{14.6.2}$$

This inequality is proven by induction in Lemma 14.6.2 and Lemma 14.6.3, and provides the basis for proof of the following theorem.

Theorem 14.6.1. *Let P be a nonnegative σ-finite countably additive Borel measure on the Borel subsets of the open convex set $\mathbf{C} \subset \mathbb{R}^n$. Assume P is absolutely continuous relative to Lebesgue measure μ_n, with g a version of $dP/d\mu_n$.*

(1) *If $\ln g$ is a concave function then $\ln P$ is a concave set function. Conversely, if g is continuous and $\ln P$ is a concave set function then $\ln g$ is a concave function (the domain of g is $\mathbf{C}$).*
(2) *If $-\infty < s < 0$ then $g^{s/(1-sn)}$ is a convex function on $\mathbf{C} \to \mathbb{R}$ if and only if $P = v$ satisfies (14.6.1) for all Borel subsets $\mathbf{A}_0$ and $\mathbf{A}_1$ of $\mathbf{C}$.*
(3) *If $0 < s < 1/n$ then $g^{s/(1-sn)}$ is a concave function on $\mathbf{C} \to \mathbb{R}$ if and only if $P = v$ satisfies (14.6.1) for all Borel subsets $\mathbf{A}_0$ and $\mathbf{A}_1$ of $\mathbf{C}$.*

Remark 14.6.2. *C.* Borel is cited by Rinott as showing that any concave measure must be absolutely continuous with respect to Lebesgue measure.

Keep in mind that nonconstant concave and convex functions defined on $\mathbf{C} = \mathbb{R}^n$ must be unbounded. If v satisfies (4.6.1) for all Borel subsets of $\mathbb{R}^n$ then for a fixed convex set $\mathbf{A}$ and vectors a_0 and a_1,

$$\alpha(\mathbf{A} + a_0) + (1 - \alpha)(\mathbf{A} + a_1) = \mathbf{A} + (\alpha a_0 + (1 - \alpha)a_1). \qquad (14.6.3)$$

By (14.6.1) it follows that if $s > 0$ then $v(\mathbf{A} + a)^s$ is a concave function, if $s < 0$ a convex function, of vectors a, hence must be the constant function or unbounded. If $\mathbf{A}$ is the set of vectors having positive coordinates and $c^t = (1, \ldots, 1)$ then $\lim_{t\to\infty} v(\mathbf{A} + tc) = 0$ since v is countably additive. Hence v is nonconstant and must be unbounded. Then, probability measures which satisfy (2) or (3) of the theorem must have restricted sets of support $\mathbf{C}$. In case $s = 0$, $\ln P(\mathbf{A} + a)$ is a concave function and since $\lim_{t\downarrow 0} \ln t = -\infty$, the boundedness problem does not arise. Probability problems with convex measure will, then, usually involve log-concave measures and density functions.

Note that

$$\lim_{s\downarrow 0} (\alpha P(A_0)^s + (1 - \alpha)P(A_1)^s)^{1/s} = P(A_0)^\alpha P(A_1)^{1-\alpha}. \qquad (14.6.4)$$

See Lemma 14.3.12 of this chapter.

Lemma 14.6.3. *If $p < 1, p \neq 0$, and $(1/p) + (1/q) = 1$ then for all real numbers $a > 0$ and $b > 0$, $ab \geq p^{-1}a^p + q^{-1}b^q$. Equality holds if and only if $a = b^{1/(p-1)}$.*

If $p > 1$ and $(1/p) + (1/q) = 1$ then for all real numbers $a > 0$ and $b > 0$, $ab \leq p^{-1}a^p + q^{-1}b^q$. Equality holds if and only if $a = b^{1/(p-1)}$.

Proof. Fix b and set $f(a) = ab - p^{-1}a^p - q^{-1}b^q$. The first derivative is $f'(a) = b - a^{p-1}$ and the root is $a_0 = b^{1/(p-1)}$. Also $f(a_0) = 0$. If $p < 1$ then

a^{p-1} is decreasing (strictly) and the sign change of f' is $-$ to $+$ and f has a unique minimum at a_0. If $p > 1$ then a^{p-1} is increasing (strictly) and the sign change of f' is $+$ to $-$. Thus f has a unique maximum at a_0. □

Lemma 14.6.4 (Hölder Inequality). *If $p < 1, p \neq 0$, and $p^{-1} + q^{-1} = 1$, then*

$$\int |fg|\, d\mu \geq \left(\int |f|^p\, d\mu\right)^{1/p} \left(\int |g|^q\, d\mu\right)^{1/q}. \tag{14.6.5}$$

If $p > 1$ and $p^{-1} + q^{-1} = 1$ then

$$\int |fg|\, d\mu \leq \left(\int |f|^p\, d\mu\right)^{1/p} \left(\int |g|^q\, d\mu\right)^{1/q}. \tag{14.6.6}$$

In each case strict inequality holds except in the case $\mu(\{x | (\int |f|^p\, d\mu)^{-1} |f(x)|^p \neq (\int |g|^q\, d\mu)^{-1} |g(x)|^q\}) = 0$.

Proof. Write $|f|_r = (\int |f|^r\, d\mu)^{1/r}$. Set $a = |f|/|f|_p$ and $b = |g|/|g|_q$ and use Lemma 14.6.3. Then in the case $p < 1$ $|fg|/|f|_p|g|_q \geq |f|^p/p(|f|_p)^p + |g|^q/q(|g|_q)^q$. Integration with respect to μ and use of $p^{-1} + q^{-1} = 1$ gives the conclusion. In case $p > 1$ the inequality reverses and the result follows.

The condition for strict inequality follows from the observation that if $a = b^{1/(p-1)}$ then $a^p = b^q$. □

Following in three lemmas is a proof of the Brunn–Minkowski inequality.

Lemma 14.6.5. *If μ is Lebesgue measure on $\mathbb{R}$ then for all Borel sets* **A** *and* **B**, *$\mu(\mathbf{A} + \mathbf{B}) \geq \mu(\mathbf{A}) + \mu(\mathbf{B})$. The inequality is not strict.*

Proof. We assume first that **A** and **B** are compact and thus have finite measure. Let $x = \sup\{a | a \in \mathbf{A}\}$ and $y = \inf\{b | b \in \mathbf{B}\}$. Then $x \in \mathbf{A}$ and $y \in \mathbf{B}$. Consider the sets $x + \mathbf{B}$ and $y + \mathbf{A}$. If $z \in x + \mathbf{B}$ then $z \geq x + y$ and if $z \in y + \mathbf{A}$ then $z \leq x + y$. Therefore $(x + \mathbf{B}) \cap (y + \mathbf{A}) = \{x + y\}$ and $\mu(\mathbf{A} + \mathbf{B}) \geq \mu((x + \mathbf{B}) \cup (y + \mathbf{A})) = \mu(x + \mathbf{B}) + \mu(y + \mathbf{A}) = \mu(\mathbf{A}) + \mu(\mathbf{B})$. Next, if **A** and **B** are Borel subsets of finite measure then choose $\varepsilon > 0$ and compact sets $\mathbf{C} \subset \mathbf{A}$ and $\mathbf{D} \subset \mathbf{B}$ such that $\mu(\mathbf{A} - \mathbf{C}) \leq \varepsilon/2$ and $\mu(\mathbf{B} - \mathbf{D}) \leq \varepsilon/2$. Then $\mu(\mathbf{A}) + \mu(\mathbf{B}) \leq \mu(\mathbf{C}) + \mu(\mathbf{D}) + \varepsilon \leq \mu(\mathbf{C} + \mathbf{D}) + \varepsilon \leq \mu(\mathbf{A} + \mathbf{B}) + \varepsilon$. Since $\varepsilon > 0$ is arbitrary the general result follows by an inner regularity argument.

If **A** and **B** are intervals then $\mathbf{A} + \mathbf{B}$ is convex and hence is an interval. Let $\mathbf{A} = (a_1, a_2)$ and $\mathbf{B} = (b_1, b_2)$. Then $\mathbf{A} + \mathbf{B} = (a_1 + b_1, a_2 + b_2)$ so that $\mu(\mathbf{A} + \mathbf{B}) = (a_2 + b_2) - (a_1 + b_1) = \mu(\mathbf{A}) + \mu(\mathbf{B})$. Therefore the subadditivity is not a strict inequality. □

Lemma 14.6.6. *Let μ_n be n-dimensional Lebesgue measure and $\mathbf{A} \subset \mathbb{R}^{n+1}$ be a Borel subset. Write $\mathbf{A}_x$ for the section of* **A** *on the $(n + 1)$st-coordinate. If* **A** *is a finite union of measurable convex sets then $\mu_n(\mathbf{A}_x)$ is a piecewise continuous function of x.*

PROOF. This lemma is proved simultaneously with Lemma 14.6.7 by induction on n. If $\mathbf{A}$ is a measurable convex set and (14.6.7) holds and if the sections $\mathbf{A}_x$ and $\mathbf{A}_y$ are nonempty then $\mathbf{A}_{\alpha x+(1-\alpha)y} \supset \alpha A_x + (1-\alpha)\mathbf{A}_y$. Thus by (14.6.7), $\mu_n(\mathbf{A}_x)^{1/n}$ is a concave function of those x such that $\mathbf{A}_x$ is nonempty. This implies that $\mu_n(\mathbf{A}_x)$ has at most two points of discontinuity. The indicator function of $\mathbf{A}_1 \cup \cdots \cup \mathbf{A}_k$ is $1 - \prod_{i=1}^k (1 - 1_{\mathbf{A}_i}(y, x))$, $y \in \mathbb{R}^n$, $x \in \mathbb{R}$. Expand the product and integrate by $\mu_n(dy)$. This shows that if $\mathbf{A}_1, \ldots, \mathbf{A}_k$ are measurable convex sets then $\mu_n(\mathbf{A}_1 \cup \cdots \cup \mathbf{A}_k)$ is a sum of piecewise continuous functions. □

Lemma 14.6.7 (Inductive Step, Brunn–Minkowski). *If n-dimensional Lebesgue measure μ_n satisfies*

$$\mu_n(\alpha\mathbf{A} + (1-\alpha)\mathbf{B})^{1/n} \geq \alpha\mu_n(\mathbf{A})^{1/n} + (1-\alpha)\mu_n(\mathbf{B})^{1/n} \tag{14.6.7}$$

then the corresponding result holds with n replaced by $n + 1$.

PROOF. The collection of sets which are finite unions of measurable convex sets is measure dense in the Borel subsets since μ_n is a regular measure. Thus it is sufficient to prove (14.6.7) for sets $\mathbf{A}$ and $\mathbf{B}$ which are finite unions of convex sets. Let $\mathbf{A}$ and $\mathbf{B}$ be Borel subsets of $\mathbb{R}^{n+1}$, $\mathbf{C} = \alpha\mathbf{A} + (1-\alpha)\mathbf{B}$ and $\mathbf{C}_x$ be the section of $\mathbf{C}$ on the $(n+1)$st-coordinate. By Fubini's theorem $\mu_{n+1}(\mathbf{C}) = \int \mu_n(\mathbf{C}_x)\,dx$. Using sections $\mathbf{A}_x$ and $\mathbf{B}_x$ define numbers $m_{\mathbf{A}} = \int \mu_n(\mathbf{A}_x)\,dx$ and $m_{\mathbf{B}} = \int \mu_n(\mathbf{B}_x)\,dx$ and functions $z_{\mathbf{A}}(\tau)$ and $z_{\mathbf{B}}(\tau)$ by the equation $\tau m_{\mathbf{A}} = \int_{-\infty}^{z_{\mathbf{A}}(\tau)} \mu_n(\mathbf{A}_x)\,dx$, $\tau m_{\mathbf{B}} = \int_{-\infty}^{z_{\mathbf{B}}(\tau)} \mu_n(\mathbf{B}_x)\,dx$. Then $z_{\mathbf{A}}$ and $z_{\mathbf{B}}$ are nondecreasing functions and for example $d(\tau m_{\mathbf{A}})/d\tau = m_{\mathbf{B}} = z'_{\mathbf{A}}(\tau)\mu_n(\mathbf{A}_{z_{\mathbf{A}}(\tau)})$. Set $x(\tau) = \alpha z_{\mathbf{A}}(\tau) + (1-\alpha)z_{\mathbf{B}}(\tau)$ and change variable in

$$\begin{aligned}\mu_{n+1}(\mathbf{C}) &= \int_{-\infty}^{\infty} \mu_n(\mathbf{C}_x)\,dx = \int_0^1 \mu_n(\mathbf{C}_{\alpha z_{\mathbf{A}}(t)+(1-\alpha)z_{\mathbf{B}}(t)})\,dx/dt\,dt \\ &= \int_0^1 \mu_n(\mathbf{C}_{x(t)})\left[(\alpha m_{\mathbf{A}}/\mu_n(\mathbf{A}_{z_{\mathbf{A}}(t)})) + ((1-\alpha)m_{\mathbf{B}}/\mu_n(\mathbf{B}_{z_{\mathbf{B}}(t)}))\right]dt.\end{aligned} \tag{14.6.8}$$

With $p = -1/n$ and $q = 1/(n+1)$, using Lemma 14.6.4,

$$\begin{aligned}&\frac{\alpha m_{\mathbf{A}}}{\mu_n(\mathbf{A}_{z_{\mathbf{A}}(t)})} + \frac{(1-\alpha)m_{\mathbf{B}}}{\mu_n(\mathbf{B}_{z_{\mathbf{B}}(t)})} \\ &\qquad \geq (\alpha m_{\mathbf{A}}^{1/(n+1)} + (1-\alpha)m_{\mathbf{B}}^{1/(n+1)})^{n+1}(\alpha\mu_n(\mathbf{A}_{z_{\mathbf{A}}(t)})^{1/n} + (1-\alpha)\mu_n(\mathbf{B}_{z_{\mathbf{B}}(t)})^{1/n})^{-n}.\end{aligned} \tag{14.6.9}$$

From the definition of $x(\tau)$ it follows that $(\alpha\mathbf{A} + (1-\alpha)\mathbf{B})_{x(t)} \supset \alpha\mathbf{A}_{z_{\mathbf{A}}(t)} + (1-\alpha)\mathbf{B}_{z_{\mathbf{B}}(t)}$. From the n-dimensional case

$$\mu_n(\mathbf{C}_{x(t)}) \geq \mu_n(\alpha\mathbf{A}_{z_{\mathbf{A}}(t)} + (1-\alpha)\mathbf{B}_{z_{\mathbf{B}}(t)}) \geq (\alpha\mu_n(\mathbf{A}_{z_{\mathbf{A}}(t)})^{1/n} + (1-\alpha)\mu_n(\mathbf{B}_{z_{\mathbf{B}}(t)})^{1/n})^n.$$

Substitute this and (14.6.9) into the integral (14.6.8) to obtain

$$\mu_{n+1}(\mathbf{C}) \ge (\alpha m_{\mathbf{A}}^{1/(n+1)} + (1-\alpha)m_{\mathbf{B}}^{1/(n+1)})^{n+1} \int_0^1 dx. \qquad \square$$

Remark 14.6.8. The set function $\mu_n^{1/n}$ is not strictly concave. For ε-spheres $\alpha\mathbf{S}_\varepsilon(x) + (1-\alpha)\mathbf{S}_\varepsilon(y) \subset \mathbf{S}_\varepsilon(\alpha x + (1-\alpha)y)$. Lemma 14.6.7 gives the result that

$$\begin{aligned}\mu_n(\mathbf{S}_\varepsilon(0))^{1/n} &= \mu_n(\mathbf{S}_\varepsilon(\alpha x + (1-\alpha)y))^{1/n}\\ &\ge \mu_n(\alpha\mathbf{S}_\varepsilon(x) + (1-\alpha)\mathbf{S}_\varepsilon(y))^{1/n} \qquad (14.6.10)\\ &\ge \alpha\mu_n(\mathbf{S}_\varepsilon(x))^{1/n} + (1-\alpha)\mu_n(\mathbf{S}_\varepsilon(y))^{1/n} = \mu_n(\mathbf{S}_\varepsilon(0))^{1/n}.\end{aligned}$$

Lemma 14.6.9 (Rinott). *Let μ be a Borel measure on $\mathbb{R}^{n+1}$ with density function $f(x_1, \ldots, x_{n+1}) = \exp(-x_{n+1})$ relative to Lebesgue measure. If $\mathbf{A}_0$ and $\mathbf{A}_1$ are Borel subsets, if $0 \le \alpha \le 1$, then*

$$\mu(\alpha\mathbf{A}_0 + (1-\alpha)\mathbf{A}_1) \ge \mu(\mathbf{A}_0)^\alpha \mu(\mathbf{A}_1)^{1-\alpha}. \qquad (14.6.11)$$

Proof. The proof uses the geometric mean, namely for positive numbers $a \ne b$, $\alpha a + (1-\alpha)b = \alpha e^{\ln a} + (1-\alpha)e^{\ln b} > \exp(\ln a^\alpha + \ln b^{1-\alpha}) = a^\alpha b^{1-\alpha}$. In the following we use the notation that $T_{\mathbf{A}}(x) = \{y \mid (y, x) \in \mathbf{A}\}$. Let $\mu(\mathbf{A}_i) = m_i$ and define $z_i(\tau)$ by

$$\tau m_i = \int_{-\infty}^{z_i(\tau)} \mu_n(T_{\mathbf{A}_i}(x))e^{-x}\,dx. \qquad (14.6.12)$$

Note that $\tau = 1$ gives $\int_{-\infty}^{\infty} \mu_n(T_{\mathbf{A}_i}(x))e^{-x}\,dx = m_i = \mu(\mathbf{A}_i)$. Make the change of variable $x(t) = \alpha z_0(t) + (1-\alpha)z_1(t)$ and proceed as in Lemma 14.6.7. Then $z_i'(t) = m_i e^{z_i(t)}/\mu_n(T_{\mathbf{A}_i}(z_i(t)))$ and if $\mathbf{B} = \alpha\mathbf{A}_0 + (1-\alpha)\mathbf{A}_1$ then

$$\begin{aligned}\mu(\mathbf{B}) &= \int_0^1 \mu_n(T_{\mathbf{B}}(\alpha z_0(t) + (1-\alpha)z_1(t)))\exp -(\alpha z_0(t) + (1-\alpha)z_1(t))\\ &\quad\times\left[\frac{\alpha m_0 \exp(z_0(t))}{\mu_n(T_{\mathbf{A}_0}(z_0(t)))} + \frac{(1-\alpha)m_1\exp(z_1(t))}{\mu_n(T_{\mathbf{A}_1}(z_1(t)))}\right]dt \qquad (14.6.13)\\ &\ge m_0^\alpha m_1^{1-\alpha}\int_0^1 \mu_n(T_{\mathbf{B}}(x(t)))\mu_n(T_{\mathbf{A}_0}(z_0(t)))^{-\alpha}\mu_n(T_{\mathbf{A}_1}(z_1(t)))^{-1+\alpha}\,dt.\end{aligned}$$

This step uses the inequality stated at the start of the proof. For sections of sets $T_{\alpha\mathbf{A}_0+(1-\alpha)\mathbf{A}_1}(\alpha z_0 + (1-\alpha)z_1) \supset \alpha T_{\mathbf{A}_0}(z_0) + (1-\alpha)T_{\mathbf{A}_1}(z_1)$. Hence relative to Lebesgue measure, using Lemma 14.6.7,

$$\begin{aligned}\mu_n(T_{\mathbf{B}}(x(t))) &\ge \mu_n(\alpha T_{\mathbf{A}_0}(z_0(t)) + (1-\alpha)T_{\mathbf{A}_1}(z_1(t)))\\ &\ge (\alpha\mu_n(T_{\mathbf{A}_0}(z_0(t)))^{1/n} + (1-\alpha)\mu_n(T_{\mathbf{A}_1}(z_1(t)))^{1/n})^n \qquad (14.6.14)\\ &\ge (\mu_n(T_{\mathbf{A}_0}(z_0(t)))^{\alpha/n}\mu_n(T_{\mathbf{A}_1}(z_1(t)))^{(1-\alpha)/n})^n.\end{aligned}$$

Substitution into (14.6.13) then gives the result (14.6.11). $\square$

Remark 14.6.10. Strict inequality results at (14.6.11) when $0 < \alpha < 1$ unless $z_0'(t) = z_1'(t)$ almost surely, so that $z_0(t) = z_1(t)$ for all t. This implies $\mu_n(T_{\mathbf{A}_0}(x)) = \mu_n(T_{\mathbf{A}_1}(x))$ almost surely μ_1.

Lemma 14.6.11. *Let μ on $\mathbb{R}^{n+1}$ have the density function $f(x_1, \ldots, x_n, x_{n+1}) = |(1 - sn)/s| x_{n+1}^{-1+(1-sn)/s} = g(x_{n+1})$ relative to Lebesgue measure. If $0 \le \alpha \le 1$ and $s \ne 0$, $s < 1/(n+1)$, then*

$$\mu(\alpha\mathbf{A}_0 + (1-\alpha)\mathbf{A}_1) \ge (\alpha\mu(\mathbf{A}_0)^s + (1-\alpha)\mu(\mathbf{A}_1)^s)^{1/s}.$$

Proof. Copy the proofs of the preceding lemmas. Taking a section, $m_i = \mu(\mathbf{A}_i) = \int_{+\infty}^{\infty} \mu_n(T_{\mathbf{A}_i}(x))g(x)\,dx$, and $\tau m_i = \int_{-\infty}^{z_i(\tau)} \mu_n(T_{\mathbf{A}_i}(x))g(x)\,dx$. With $\mathbf{B} = \alpha\mathbf{A}_0 k + (1-\alpha)\mathbf{A}_1$, the change of variable then gives

$$\mu(\mathbf{B}) = \int_0^1 \mu_n(T_{\mathbf{B}}(x(t)))g(x(t))\,dx/dt\,dt.$$

Application of the Brunn–Minkowski inequality then gives

$$\begin{aligned}\mu(\mathbf{B}) \ge \int_0^1 &(\alpha\mu_n(T_{\mathbf{A}_0}(z_0(t))))^{1/n} + (1-\alpha)\mu_n(T_{\mathbf{A}_1}(z_1(t)))^{1/n})^n \\ &\times g(x(t))\left[\frac{\alpha m_0}{g(z_0)\mu_n(T_{\mathbf{A}_0})} + \frac{(1-\alpha)m_1}{g(z_1)\mu_n(T_{\mathbf{A}_1})}\right] dt.\end{aligned} \tag{14.6.15}$$

Apply the Hölder inequality, Lemma 14.6.4, with $s < 1/(n+1)$, so that with $\gamma = s/(s-1)$, $s^{-1} + \gamma^{-1} = 1$. Then the term in square brackets above is bounded by

$$(\alpha m_0^s + (1-\alpha)m_1^s)^{1/s} \times (\alpha(g(z_0)\mu_n(T_{\mathbf{A}_0}))^{-\gamma} + (1-\alpha)(g(z_1)\mu_n(T_{\mathbf{A}_1}))^{-\gamma})^{1/\gamma}. \tag{14.6.16}$$

The desired result will then follow by showing, after substitution of (14.6.16) and omission of $(\alpha m_0^s + (1-\alpha)m_1^s)^{1/s}$, that the resulting integral is ≥ 1.

Set $\beta = ((1 - sn)/s) - 1$ so that

$$n + \beta + \gamma^{-1} = 0. \tag{14.6.17}$$

Then

$$g(x) = |1 + \beta| x^\beta. \tag{14.6.18}$$

Case I. $s > 0$. Then $\beta > 0$ and $\gamma < 0$. Because of homogeneity the normalization $|1 + \beta|$ on g cancels in the integrals, hence is not written. The integrand is

$$\begin{aligned}&(\alpha\mu_n(T_{\mathbf{A}_0})^{1/n} + (1-\alpha)\mu_n(T_{\mathbf{A}_1})^{1/n})^n x(t)^\beta \\ &\quad\times (\alpha(z_0^\beta\mu(T_{\mathbf{A}_0}))^{-\gamma} + (1-\alpha)(z_1^\beta\mu(T_{\mathbf{A}_1}))^{-\gamma})^\gamma.\end{aligned} \tag{14.6.19}$$

The first two terms of (14.6.19) can be written, after taking the $-\gamma$ power, note that $-\gamma < 0$, and use the fact that $-n\gamma - \beta\gamma = 1$,

$$\begin{aligned}(\alpha\mu_n(T_{\mathbf{A}_0})^{1/n} &+ (1-\alpha)\mu_n(T_{\mathbf{A}_1})^{1/n})^{-n\gamma}x^{-\beta\gamma}\\ &= (\alpha\mu_n(T_{\mathbf{A}_0})^{-\gamma/(-n\gamma)} + (1-\alpha)\mu_n(T_{\mathbf{A}_1})^{-\gamma/(-n\gamma)})^{-ng}\\ &\times(\alpha z_0^{-\beta\gamma/(-\beta\gamma)} + (1-\alpha)z_1^{-\beta\gamma/(-\beta\gamma)})^{-\beta\gamma}\\ &\geq \alpha z_0^{-\beta\gamma}\mu_n(T_{\mathbf{A}_0})^{-\gamma} + (1-\alpha)z_1^{-\beta\gamma}\mu_n(T_{\mathbf{A}_1})^{-\gamma}.\end{aligned} \tag{14.6.20}$$

Substitution into (14.6.19) then shows that the integral is ≥ 1.

Case II. $s < 0$. Then $\beta < 0$ and $\gamma > 0$. A similar application of the Hölder inequality to the first two terms of (14.6.20) gives the desired result. □

Lemma 14.6.12. *Suppose ν is a Borel measure on $\mathbb{R}^n$ absolutely continuous with respect to Lebesgue measure μ_n. Suppose $d\nu/d\mu_n$ has a continuous representation f. If the measure ν is log-concave then $\ln f$ is a concave function.*

Proof. We suppose $\nu(\alpha\mathbf{A} + (1-\alpha)\mathbf{B}) \geq \nu(\mathbf{A})^\alpha\nu(\mathbf{B})^{1-\alpha}$. Let $\mathbf{S}_\varepsilon(x)$ be the sphere of radius ε about x. Then

$$\mathbf{S}_\varepsilon(\alpha x + (1-\alpha)y) \supset \alpha\mathbf{S}_\varepsilon(x) + (1-\alpha)\mathbf{S}_\varepsilon(y). \tag{14.6.21}$$

Then by the translation invariance of Lebesgue measure, $\mu_n(\mathbf{S}_\varepsilon(x)) = \mu_n(\mathbf{S}_\varepsilon(y)) = \mu_n(\mathbf{S}_\varepsilon(\alpha x + (1-\alpha)y))$ so that

$$\begin{aligned}(\nu(\mathbf{S}_\varepsilon(x))/\mu_n(\mathbf{S}_\varepsilon(x)))^\alpha(\nu(\mathbf{S}_\varepsilon(y))/\mu_n(\mathbf{S}_\varepsilon(y)))^{1-\alpha}&\\ \leq \nu(\mathbf{S}_\varepsilon(\alpha x + (1-\alpha)y))/\mu_n(\mathbf{S}_\varepsilon(\alpha x + (1-\alpha)y)).&\end{aligned} \tag{14.6.22}$$

As $\varepsilon \downarrow 0$ the limit exists and we obtain

$$f(x)^\alpha f(y)^{1-\alpha} \leq f(\alpha x + (1-\alpha)y). \tag{14.6.23}$$

□

Proof of Theorem 14.6.1. The method of proof of Lemma 14.6.12 shows how, if g is continuous, the results for measure imply the results for density functions. Consequently we only give proofs that density functions satisfying the stated condition imply the corresponding result for measures.

Given $\mathbf{A} \subset \mathbb{R}^n$ let $\mathbf{G}_\mathbf{A}(f) = \{(x, y) \mid x \in \mathbf{A},\ y \in \mathbb{R},\ f(x) \leq y\}$. As is easy to see, if f is a convex function then the set $\mathbf{G}_\mathbf{A}(f)$ convex, and conversely, provided $\mathbf{A}$ is a convex set.

Case $s = 0$. Then $-\ln g = f$ is a convex function. For any Borel set $\mathbf{A}$ let $\mathbf{A}^* = \mathbf{G}_\mathbf{A}(-\ln g)$. Then relative to the measure μ of Lemma 14.6.9

$$\begin{aligned}P(\mathbf{A}) &= \int\cdots\int_\mathbf{A} g(x_1, \ldots, x_n)\,dx_1\cdots dx_n\\ &= \int\cdots\int_\mathbf{A} dx_1\cdots dx_n \int_{-\ln g}^{\infty} e^{-x}\,dx = \mu(\mathbf{A}^*).\end{aligned} \tag{14.6.24}$$

The convexity of $-\ln g$ implies

$$(\alpha\mathbf{A}_0 + (1-\alpha)\mathbf{A}_1)^* \supset \alpha\mathbf{A}_0^* + (1-\alpha)\mathbf{A}_1^*. \tag{14.6.25}$$

Then

$$\begin{aligned} P(\alpha\mathbf{A}_0 + (1-\alpha)\mathbf{A}_1) &= \mu((\alpha\mathbf{A}_0 + (1-\alpha)\mathbf{A}_1)^*) \geq \mu(\alpha\mathbf{A}_0^* + (1-\alpha)\mathbf{A}_1^*) \\ &\geq \mu(\mathbf{A}_0^*)^\alpha \mu(\mathbf{A}_1^*)^{1-\alpha} = P(\mathbf{A}_0)^\alpha P(\mathbf{A}_1)^{1-\alpha}. \end{aligned} \tag{14.6.26}$$

The converse is proven below.

Case $s > 0$. Let $\mathbf{H}_{\mathbf{A}}(g) = \{(x, y) | x \in \mathbf{A}, \quad y \leq g(x)\} = \mathbf{A}^*$. The exponents $s/(1-sn) > 0$ and $(1-sn)/s > 1$. Further

$$(\alpha\mathbf{A}_0 + (1-\alpha)\mathbf{A}_1)^* \supset \alpha\mathbf{A}_0^* + (1-\alpha)\mathbf{A}_1^* \tag{14.6.27}$$

where $\mathbf{A}_i^* = \mathbf{H}_{\mathbf{A}_i}(g)$. Therefore

$$\begin{aligned} \mu(\mathbf{A}^*) &= \int \cdots \int_{\mathbf{A}} dx_1 \cdots dx_n \int_0^{g^{s/(1-sn)}} |(1-sn)/s| x^{-1+(1-sn)/s}\, dx \\ &= \int \cdots \int_{\mathbf{A}} f(x_1, \ldots, x_n)\, dx_1 \cdots dx_n = P(\mathbf{A}). \end{aligned} \tag{14.6.28}$$

Then using Lemma 14.6.11,

$$\begin{aligned} P(\alpha\mathbf{A}_0 + (1-\alpha)\mathbf{A}_1) &= \mu((\alpha\mathbf{A}_0 + (1-\alpha)\mathbf{A}_1)^*) \geq \mu(\alpha\mathbf{A}_0^* + (1-\alpha)\mathbf{A}_1^*) \\ &\geq (\alpha\mu(\mathbf{A}_0^*)^s + (1-\alpha)\mu(\mathbf{A}_1^*)^s)^{1/s} \\ &= (\alpha P(\mathbf{A}_0)^s + (1-\alpha)P(\mathbf{A}_1)^s)^{1/s}. \end{aligned} \tag{14.6.29}$$

Case $s < 0$. With $\mathbf{G}_{\mathbf{A}}(g)$ as before, if $s < 0$ then $(1-sn)/s < 0$ and (14.6.27) holds but (14.6.28) is replaced by

$$\begin{aligned} \mu(\mathbf{A}^*) &= \int \cdots \int dx_1 \cdots dx_n \int_{g^{s/(1-sn)}}^{\infty} |(1-sn)/s| x^{-1+(1-sn)/s}\, dx \\ &= P(\mathbf{A}). \end{aligned} \tag{14.6.30}$$

Then the final step is the same as in (14.6.29).

A proof of the converse in the cases $s > 0$ and $s < 0$ is as follows. About x and y take spheres $\mathbf{S}_{\varepsilon r}(x)$ and $\mathbf{S}_{\varepsilon t}(y)$. Then $\mathbf{S} = \mathbf{S}_{\varepsilon(\alpha r + (1-\alpha)t)}(\alpha x + (1-\alpha)y) \supset \alpha\mathbf{S}_{\varepsilon r}(x) + (1-\alpha)\mathbf{S}_{\varepsilon t}(y)$ so $\mu(\mathbf{S}) \geq (\alpha\mu(\mathbf{S}_{\varepsilon r}(x))^s + (1-\alpha)\mu(\mathbf{S}_{\varepsilon t}(y))^s)^{1/s}$. For ε small the integrals are approximately (constant) (radius)^n $g(\cdot)^s$, so letting $\varepsilon \downarrow 0$,

$$g(\alpha x + (1-\alpha)y)(\alpha r + (1-\alpha)t)^n \geq (\alpha r^{sn} g(x)^s + (1-\alpha)t^{sn} g(y)^s)^{1/s}. \tag{14.6.31}$$

The values r and t are arbitrary so we take

$$r = g(x)^{s/(1-sn)} \quad \text{and} \quad t = g(y)^{s/(1-sn)}. \tag{14.6.32}$$

The right side of (14.6.31) becomes, using $s + ns^2/(1-sn) = s/(1-sn)$,

$$= (\alpha g(x)^{s/(1-sn)} + (1-\alpha)g(y)^{s/(1-sn)})^{1/s}. \tag{14.6.33}$$

The resulting inequality is then

$$g(\alpha x + (1-\alpha)y) \geq (\alpha g(x)^{s/(1-sn)} + (1-\alpha)g(y)^{s/(1-sn)})^{(1-sn)/s}, \tag{14.6.34}$$

which proves the converse.

EXAMPLE 14.6.13. Let $\mathbf{X}(T)$ be a Gaussian process. If $0 = t_0 < t_1 < \cdots < t_n < 1$ the joint density of $\mathbf{X}(t_1), \ldots, \mathbf{X}(t_n)$ is a joint normal density, hence is log-concave, the case $s = 0$. Consider $\Sigma_{j=2}^n \mathbf{X}(t_j)^2(t_j - t_{j-1})$. If

$$\mathbf{A}(b) = \left\{(x_2, \ldots, x_n) \,\middle|\, \sum_{j=2}^n x_j^2(t_j - t_{j-1}) \leq b\right\} \tag{14.6.35}$$

then $\mathbf{A}(\alpha b_0 + (1-\alpha)b_1) \supset \alpha\mathbf{A}(b_0) + (1-\alpha)\mathbf{A}(b_1)$. Therefore

$$\begin{aligned} P&\left(\sum_{j=2}^n \mathbf{X}(t_j)^2(t_j - t_{j-1}) \leq \alpha b_0 + (1-\alpha)b_1\right) \\ &= P(\mathbf{X}(t_2), \ldots, \mathbf{X}(t_n) \in \mathbf{A}(\alpha b_0 + (1-\alpha)b_1)) \\ &\geq P(\mathbf{X}(t_2), \ldots, \mathbf{X}(t_n) \in \alpha\mathbf{A}(b_0) + (1-\alpha)\mathbf{A}(b_1)) \\ &\geq P\left(\sum_{j=2}^n \mathbf{X}(t_j)^2(t_j - t_{j-1}) \leq b_0\right)^{\alpha} P\left(\sum_{j=2}^n \mathbf{X}(t_j)^2(t_j - t_{j-1}) \leq b_1\right)^{1-\alpha}. \end{aligned} \tag{14.6.36}$$

Therefore the distribution of $\Sigma_{j=2}^n \mathbf{X}(t_j)^2(t_j - t_{j-1})$ is log-concave. As the maximum interval of the partition converges to zero obtain $\int_0^1 \mathbf{X}(t)^2\,dt$ has a distribution function that is log-concave.

Theorem 14.6.14 (Prékopka). *The convolution of two* log-*concave probability measures on* $\mathbb{R}^n$ *is* log-*concave.*

PROOF. Given log-concave density functions f on $\mathbb{R}^n$ and g on $\mathbb{R}^m$ the product density fg on $\mathbb{R}^{m+n}$ is log-concave. We want $m = n$ for convolutions and suppose random variables $\mathbf{X}$ and $\mathbf{Y}$ are independently distributed with density functions f and g respectively. Then

$$P(\mathbf{X} + \mathbf{Y} \in \mathbf{A}) = \int 1_{\mathbf{A}}(x + y) f(x) g(y)\, dx\, dy. \tag{14.6.37}$$

Given Borel subsets $\mathbf{A}$ and $\mathbf{B}$ define $\mathbf{C}$ and $\mathbf{D}$ by

$$\mathbf{C} = \{(x, y) \mid x + y \in \mathbf{A}\}; \qquad \mathbf{D} = \{(x, y) \mid x + y \in \mathbf{B}\}. \tag{14.6.38}$$

If $(x, y) \in \alpha\mathbf{C} + (1-\alpha)\mathbf{D}$ then $(x, y) = \alpha(x_0, y_0) + (1-\alpha)(x_1, y_1)$ such that $x_0 + y_0 \in \mathbf{A}$ and $x_1 + y_1 \in \mathbf{B}$. Then $x + y \in \alpha\mathbf{A} + (1-\alpha)\mathbf{B}$ and

$$1_{\alpha\mathbf{C}+(1-\alpha)\mathbf{D}}(x, y) \leq 1_{\alpha\mathbf{A}+(1-\alpha)\mathbf{B}}(x + y). \tag{14.6.39}$$

Therefore using the log-concave property of the product measure

$$\begin{aligned} P(\mathbf{X}+\mathbf{Y}\in\alpha\mathbf{A}+(1-\alpha)\mathbf{B}) &\geq P((\mathbf{X},\mathbf{Y})\in\alpha\mathbf{C}+(1-\alpha)\mathbf{D}) \\ &\geq P((\mathbf{X},\mathbf{Y})\in\mathbf{C})^{\alpha}P((\mathbf{X},\mathbf{Y})\in\mathbf{D})^{1-\alpha} \qquad (14.6.40) \\ &= P(\mathbf{X}+\mathbf{Y}\in\mathbf{A})^{\alpha}P(\mathbf{X}+\mathbf{Y}\in\mathbf{B})^{1-\alpha}. \end{aligned}$$ □

Theorem 14.6.15 (T. W. Anderson (1955)). *Let* $\mathbf{E}$ *be a convex set in* $\mathbb{R}^n$ *such that* $\mathbf{E}=-\mathbf{E}$. *Let* $f\colon \mathbb{R}^n\to[0,\infty)$ *satisfy* $f(x)=f(-x)$ *and* $\{x\,|\,f(x)\geq u\}=\mathbf{K}_u$ *is convex,* $u\geq 0$. *If* $\int_{\mathbf{E}} f(x)\mu_n(dx)<\infty$ *and* $0\leq\alpha\leq 1$ *then* $\int_{\mathbf{E}} f(x+\alpha y)\mu_n(dx) \geq \int_{\mathbf{E}} f(x+y)\mu_n(dx)$, $y\in\mathbb{R}^n$.

Proof. For nonvoid sets $\alpha[(\mathbf{E}+x_0)\cap\mathbf{K}_u]+(1-\alpha)[(\mathbf{E}+x_1)\cap\mathbf{K}_u]\subset(\mathbf{E}+(\alpha x_0+(1-\alpha)x_1))\cap\mathbf{K}_u$. By the Brunn–Minkowski inequality

$$\begin{aligned} &\mu_n((\mathbf{E}+(\alpha x_0+(1-\alpha)x_1))\cap\mathbf{K}_u)^{1/n} \\ &\quad\geq \alpha\mu_n((\mathbf{E}+x_0)\cap\mathbf{K}_u)^{1/n}+(1-\alpha)\mu_n((\mathbf{E}+x_1)\cap\mathbf{K}_u)^{1/n}. \end{aligned} \qquad (14.6.41)$$

By hypothesis $\mathbf{E}$ and $\mathbf{K}_u$ are symmetric sets so the function, which is concave if $\mu_n((\mathbf{E}+x)\cap\mathbf{K}_u)^{1/n}>0$, has its maximum at $x=0$. For $(\mathbf{E}-x)\cap\mathbf{K}_u = (-\mathbf{E}-x)\cap(-\mathbf{K}_u) = -(\mathbf{E}+x)\cap\mathbf{K}_u$ shows it to be symmetric. Write

$$\begin{aligned} \int_{\mathbf{E}} f(x+\alpha y)\mu_n(dx) &= \int_{\mathbf{E}+\alpha y} f(x)\mu_n(dx) \\ &= \int 1_{\mathbf{E}+\alpha y}(x)f(x)\mu_n(dx) \\ &= \int_0^\infty \mu_n((\mathbf{E}+\alpha y)\cap\mathbf{K}_u)\,du \qquad (14.6.42) \\ &\geq \int_0^\infty \mu_n((\mathbf{E}+y)\cap\mathbf{K}_u)\,du \\ &= \int f(x+y)\mu_n(dx). \end{aligned}$$ □

Remark 14.6.16. Under the hypotheses of Theorem 14.6.15, if $\mathbf{E}=D\mathbf{E}$ and $\mathbf{K}_u=D\mathbf{K}_u$, $u\geq 0$, for all diagonal orthogonal matrices D, then

$$\int f(x+Dy)\mu_n(dx) \geq \int f(x+D'y)\mu_n(dx) \qquad (14.6.43)$$

for all D, $D'\in\mathbf{D}(n)$ with positive entries such that $D\leq D'$.

Corollary 14.6.17 (See Sherman (1955)). *If* $f\colon\mathbb{R}^n\to[0,\infty)$ *and* $g\colon\mathbb{R}^n\to[0,\infty)$ *satisfy* $f(x)=f(-x)$, $g(x)=g(-x)$, *and* $\mathbf{K}_u=\{x\,|\,g(x)\geq u\}$, $\mathbf{H}_u=\{x\,|\,g(x)\geq u\}$ *are convex sets, then the convolution* $(f*g)(tx)$ *is a nonincreasing function of* $t\geq 0$.

PROOF. Note that $f(x) = \int_0^\infty 1_{\mathbf{K}_u}(x)\,du$ and $g(x) = \int_0^\infty 1_{\mathbf{H}_v}(x)\,dv$. Then,

$$\begin{aligned}(f * g)(tx) &= \int_0^\infty \int_0^\infty \int 1_{\mathbf{K}_u}(tx - y) 1_{\mathbf{H}_v}(y) \mu_n(dy)\,du\,dv \\ &= \int_0^\infty \int_0^\infty \mu_n(\mathbf{H}_v \cap (\mathbf{K}_u + tx))\,du\,dv.\end{aligned} \tag{14.6.44}$$

By Theorem 14.6.15 and the remark contained in its proof, $\mu_n(\mathbf{H}_v \cap (\mathbf{K}_u + tx))^{1/n}$ is a symmetric function of t that is a concave function of those t for which the values are positive. Hence this is a nonincreasing function of $t \geq 0$. □

Remark 14.6.18. The density function of infinitely divisible random variables are convolutions, hence have a monotonicity described by Corollary 14.6.17.

Lemma 14.6.19 (Fefferman, Jodeit, Perlman (1972)). *Assume f is $\mathbf{C}_\infty$, $f(x) = f(-x)$, $\{x | f(x) \geq u\}$ is convex, $u \geq 0$. Let $D_\lambda = \operatorname{diag}(\lambda, 1, \ldots, 1)$ and μ be the uniform measure on the unit shell $\mathbf{S} = \{y | y^t y = 1\}$. Then the surface integral*

$$\int_{\mathbf{S}} f(D_\lambda x) \mu(dx) \tag{14.6.45}$$

is a nonincreasing function of $\lambda \geq 0$.

PROOF. Write $x^t = (x_1, x')$, and $a(x') = (1 - |x'|^2)^{1/2}$. Going under the integral sign the derivative $d/d\lambda$ is

$$\int_{\mathbf{S}} x_1 f(D_\lambda x) \mu(dx). \tag{14.6.46}$$

Let $\mathbf{B}$ be the unit ball, $\mathbf{B} = \{y | y^t y \leq 1\}$. Let $g(x) = f(D_\lambda x)$. The convolution $g * 1_{\mathbf{B}}$ evaluated, $g * 1_{\mathbf{B}}(tx)$, is by Theorem 14.6.17, decreasing in t. Since all derivatives exist and an absolute maximum occurs at $t = 0$,

$$\begin{aligned}0 \geq (d^2/dt^2)(g * 1_{\mathbf{B}})(te_1)\big|_{t=0} &= \int \partial_1^2 g(y) 1_{\mathbf{B}}(y)\,dy \\ &= \int_{|x'| \leq 1} \int_{-a(x')}^{a(x')} f(\lambda x_1, x')\,dx_1\,dx, \\ &= 2 \int_{|x'| \leq 1} \lambda^{-1} (\partial_1 f)(\lambda x_1, x')\,dx' \\ &= \lambda^{-1} \int_{\mathbf{S}} x_1 (\partial_1 f)(D_\lambda y) \mu(dy).\end{aligned} \tag{14.6.47}$$

The transformation to a surface integral at the last step is discussed in Section 7.7. See in particular (7.7.4). Since $\lambda > 0$, the result now follows. □

Theorem 14.6.20 (Fefferman, Jodeit, Perlman (1972)). *If μ is uniform measure of unit mass on the unit shell* $\mathbf{S}$ *of* $\mathbb{R}^n$ *then for all $n \times n$ matrices A of norm ≤ 1 and convex sets* $\mathbf{C}$ *such that* $\mathbf{C} = -\mathbf{C}$,

$$\mu((A\mathbf{C}) \cap \mathbf{S}) \le \mu(\mathbf{C} \cap \mathbf{S}). \tag{14.6.48}$$

Proof. If (14.6.48) holds for all convex sets $\mathbf{C}$ such that $\mathbf{C} = -\mathbf{C}$, then for a matrix B of norm not greater than one,

$$\mu(((AB)\mathbf{C}) \cap \mathbf{S}) = \mu((A(B\mathbf{C})) \cap \mathbf{S}) \le \mu((B\mathbf{C}) \cap \mathbf{S}) \le \mu(\mathbf{C} \cap \mathbf{S}). \tag{14.6.49}$$

Further, there exists $U \in \mathbf{O}(n)$ with $UA = (A^tA)^{1/2}$. See Section 8.1. Thus

$$\mu((A\mathbf{C}) \cap \mathbf{S}) = \mu((A^tA)^{1/2}\mathbf{C}) \cap \mathbf{S}) = \mu(V(A^tA)^{1/2}V^t(V\mathbf{C}) \cap \mathbf{S}) \tag{14.6.50}$$

and since μ is orthogonally invariant it is sufficient to discuss diagonal matrices A and convex symmetric sets $\mathbf{C}$ such that A has a single nonunit on the diagonal.

The distance function $d(x) = \inf_{y \in \mathbf{C}} |x - y|$ is a convex function of $x \in \mathbb{R}^n$ if $\mathbf{C}$ is convex and compact. For if $d(x_1) = |x_1 - y_1|$ and $d(x_2) = |x_2 - y_2|$ then

$$\begin{aligned} \alpha d(x_1) + (1-\alpha)d(x_2) &\ge \inf_{y \in \mathbf{C}} |(\alpha x_1 + (1-\alpha)x_2) - y| \\ &= d(\alpha x_1 + (1-\alpha)x_2). \end{aligned} \tag{14.6.51}$$

Thus $\exp - \varepsilon d(x)$ is a log-concave function which is equal one on $\mathbf{C}$ and is less than one if $x \notin \mathbf{C}$ (closed convex sets). Then $\lim_{\varepsilon \uparrow \infty} \exp - \varepsilon d(x) = 1_{\mathbf{C}}(x)$.

If f is a log-concave function then by Theorem 14.6.14 $f * (\exp - \varepsilon d(x))$ is log-concave so that letting $\varepsilon \to \infty$, $f * 1_{\mathbf{C}}$ is log-concave. Thus $\{x \mid f * 1_{\mathbf{C}}(x) \ge u\}$ is convex.

If f_ε is the mean zero multivariate normal density function with covariance matrix εI_n then we may smooth the function $1_{\mathbf{C}}$ by taking the convolution $f = f_\varepsilon * 1_{\mathbf{C}}$ which will be a $\mathbf{C}_\infty$-function. By Lemma 14.6.19 $\int f_\varepsilon(y) 1_{\mathbf{C}}(D_\lambda x - y)$ is nondecreasing in λ. Let $\varepsilon \downarrow 0$. The limit is, by the bounded convergence theorem, $1_{\mathbf{C}}(D_\lambda x)$ at all points interior and exterior to $\mathbf{C}$. Thus for convex sets with polygonal boundary the μ-measure of the boundary of $\mathbf{C}$ is zero, and the bounded convergence theorem implies

$$\lim_{\varepsilon \downarrow 0} \int_{\mathbf{S}} f(D_\lambda x)\mu(dx) = \int_{\mathbf{S}} 1_{\mathbf{C}}(D_\lambda x)\mu(dx) = \mu(D_\lambda^{-1}\mathbf{C} \cap \mathbf{S}). \tag{14.6.52}$$

This is a nondecreasing function of λ. Since the general closed convex set $\mathbf{C}$ is a monotone limit of a decreasing sequence of convex sets with polygonal boundaries, the result now follows from countable additivity of μ. □

A limited version of the following corollary was proven by Anderson (1955). Fefferman, Jodeit and Perlman in their paper cite several other references containing special cases of the corollary.

Corollary 14.6.21 (Fefferman, Jodeit, Perlman (1972)). *If ν is a Borel measure on $\mathbb{R}^n$ which is orthogonally invariant, if Σ_1 and Σ_2 are positive definite symmetric matrices such that $\Sigma_1 \geq \Sigma_2$, then*

$$\nu(\Sigma_1^{-1/2}\mathbf{C}) \leq \nu(\Sigma_2^{-1/2}\mathbf{C}) \tag{14.6.53}$$

for all closed convex sets $\mathbf{C}$.

PROOF. The theory of Chapter 10 implies that the measure ν factors into the product of a measure ν' on $[0, \infty)$ and the measure μ on the unit shell. Then Theorem 14.6.20 implies

$$\mu(\Sigma_1^{-1/2}\mathbf{C} \cap (r\mathbf{S})) \leq \mu(\Sigma_2^{-1/2}\mathbf{C} \cap (r\mathbf{S})) \tag{14.6.54}$$

and hence

$$\nu(\Sigma_1^{-1/2}\mathbf{C}) = \int \mu(\Sigma_1^{-1/2}\mathbf{C} \cap (r\mathbf{S}))\nu'(dr) \leq \nu(\Sigma_2^{-1/2}\mathbf{C}). \tag{14.6.55}$$

Note that $\Sigma_1^{-1/2}\Sigma_2^{-1/2}$ has norm ≤ 1. □

Remark 14.6.22. The authors Fefferman, Jodeit, and Perlman (1972) obtain somewhat stronger results than are stated here by considering separately the interior and closure of a convex set.

14.7. The FKG-Inequality

In 1977 Kemperman presented an elegant proof of the FKG-inequality for measures on partially ordered spaces. This inequality which arose from the study of phase shifts in statistical mechanics has statistical applications in the study of unbiasedness of tests of hypotheses. See for example the paper by Olkin and Perlman (1980) on the eigenvalues of a noncentral Wishart matrix. Much of the paper by Kemperman is devoted to a discussion of which lattices allow such results. We do not discuss the pathology and only give Kemperman's basic proofs. Then a random variable proof due to this author is given since random variable constructions sometimes allow better insights.

A set $\mathfrak{X}$ with a partial ordering making $\mathfrak{X}$ a lattice is assumed. Max and min are designated by $\vee$ and $\wedge$. $\mathfrak{X}$, $\mathfrak{F}$, λ is a countably additive measure space with a σ-algebra of measurable sets $\mathfrak{F}$ and a σ-finite measure λ. Given measures λ_1 and λ_2 absolutely continuous relative to λ with $f_1 = d\lambda_1/d\lambda$ and $f_2 = d\lambda_2/d\lambda$ the condition of interest, the FKG-*condition*, is

$$f_1(x)f_2(y) \leq f_1(x \wedge y)f_2(x \vee y), \quad \text{all} \quad x, y \in \mathfrak{X}. \tag{14.7.1}$$

In case $\mathfrak{X}$ has a total ordering then (14.7.1) reduces to

$$f_1(y)f_2(x) - f_1(x)f_2(y) \geq 0 \tag{14.7.2}$$

when $x \geq y$, otherwise known as having a monotone likelihood ratio.

Possible consequences of (14.7.1) to be examined are, λ_2 is a *dilation* of λ_1, meaning, if g is nondecreasing and nonnegative then

$$\int g\,d\lambda_1 \le \int g\,d\lambda_2; \tag{14.7.3}$$

λ_2 is a *strong dilation* of λ_1, meaning, for pairs (g_1, g_2) such that $x_1 \le x_2$ implies $g_1(x_1) \le g_2(x_2)$ it follows that

$$\int g_1\,d\lambda_1 \le \int g_2\,d\lambda_2. \tag{14.7.4}$$

A space in which (14.7.1) implies λ_2 is a strong dilation of λ_1 is called an FKG-*space*.

Following are three results proven by Kemperman, the last of which is the result of principal interest here.

Theorem 14.7.1 (Kemperman (1977), Theorem 3). *Let $\mathfrak{X}$ be totally ordered and $\mathfrak{X}$, $\mathfrak{F}$, λ be a measure space. Let λ_1 and λ_2 be absolutely continuous relative to λ with density functions $f_1 = d\lambda_1/d\lambda$ such that* (14.7.1) *holds. Then λ_2 is a strong dilation of λ_1.*

PROOF. Define sets $\mathbf{A}_i(u) = \{x | g_i(x) > u\}$. Suppose $g_i \ge 0$ so that by Fubini's theorem

$$\int g_i\,d\lambda_i = \int \left(\int_0^{g_i(x)} du\right) \lambda_i(dx) = \int_0^\infty \lambda_i(\mathbf{A}_i(u))\,du. \tag{14.7.5}$$

Hence it is sufficient to show that $\lambda_1(\mathbf{A}_1(u)) \le \lambda_2(\mathbf{A}_2(u))$. By hypothesis $g_1(x) \le g_2(x)$ for all x so that $\mathbf{A}_1(u) \subset \mathbf{A}_2(u)$ for all u follows, and thus $\mathbf{A}_1^c(u) \supset \mathbf{A}_2^c(u)$ for the set complements. Suppose $x_1 \in \mathbf{A}_2^c(u)$ and $x_2 \in \mathbf{A}_1^c(u)$, so that $g_2(x_1) \le u < g_1(x_2)$. Since the space is totally ordered $x_1 > x_2$ is not possible, hence $x_1 \le x_2$. From (14.7.1) it follows that $f_1(x_2)f_2(x_1) \le f_1(x_1)f_2(x_2)$ and

$$\begin{aligned} &\iint_{\mathbf{A}_2^c x \mathbf{A}_1} f_1(x_2)f_2(x_1)\lambda(dx_1)\lambda(dx_2) \\ &\qquad \le \iint_{\mathbf{A}_2^c x \mathbf{A}_1} f_1(x_1)f_2(x_2)\lambda(dx_1)\lambda(dx_2) \end{aligned} \tag{14.7.6}$$

so that

$$\lambda_2(\mathbf{A}_2^c)\lambda_1(\mathbf{A}_1) \le \lambda_1(\mathbf{A}_2^c)\lambda_2(\mathbf{A}_1) \tag{14.7.7}$$

and

$$(1 - \lambda_2(\mathbf{A}_2))\lambda_1(\mathbf{A}_1) \le (1 - \lambda_1(\mathbf{A}_2))\lambda_2(\mathbf{A}_1).$$

From $\mathbf{A}_1(u) \subset \mathbf{A}_2(u)$ and monotonicity of the measures it follows that $(1 - \lambda_2(\mathbf{A}_2))\lambda_1(\mathbf{A}_1) \le (1 - \lambda_1(\mathbf{A}_1))\lambda_2(\mathbf{A}_2)$ or that $\lambda_1(\mathbf{A}_1) \le \lambda_2(\mathbf{A}_2)$, as was to be shown. □

Lemma 14.7.2. *Let $\mathfrak{X} = \mathfrak{X}_1 \times \mathfrak{X}_2$ and μ be a probability measure on $\mathfrak{X}$. Let Q be a Markov kernel on $\mathfrak{X}$, that is,*

$$\begin{aligned}&Q(x, \cdot) \text{ is a probability measure on } \mathfrak{X}_2 \text{ for all}\\ &x \in \mathfrak{X}_1 \text{ and } Q(\cdot, \mathbf{B}) \text{ is } \mathfrak{X}_1 \text{ measurable for all}\\ &\text{measurable subsets } \mathbf{B} \text{ of } \mathfrak{X}_2.\end{aligned} \tag{14.7.8}$$

Let ν be a σ-finite measure on $\mathfrak{X}_1$ such that $\mu(\mathbf{A} \times \mathbf{B}) = \int_{\mathbf{A}} Q(x, \mathbf{B})\nu(dx)$. If f is $\mathfrak{X}$ measurable and $g(x) = \int f(x, y)Q(x, dy)$ for all x then $\int f(x, y)\mu(d(x, y)) = \int g(x)\nu(dx)$.

Proof. If $f(x, y) = 1_{\mathbf{A}}(x)1_{\mathbf{B}}(y)$, a product of indicator functions, then $g(x) = 1_{\mathbf{A}}(x)Q(x, \mathbf{B})$ and $\int g(x)\nu(dx) = \int_{\mathbf{A}} Q(x, \mathbf{B})\nu(dx) = \int f(x, y)\mu(d(x, y))$. By Fubini's theorem, linear combinations of simple functions of the form $1_{\mathbf{A}}(x)1_{\mathbf{B}}(y)$ are dense in the $\mathbf{L}_1$-space of $\mathfrak{X}$. The result then follows by a standard limiting argument. □

Theorem 14.7.3 (Theorem 4, Kemperman, op. cit.). *Let $\mathfrak{X}_1$ and $\mathfrak{X}_2$ be lattices and $\mathfrak{X} = \mathfrak{X}_1 \times \mathfrak{X}_2$, Q_1 and Q_2 be Markov kernels as described in Lemma 14.7.2, ν_1 and ν_2 be probability measures on $\mathfrak{X}$, ν_2 a strong dilation of ν_1. Let $\mathbf{N}_i$ be sets such that $\nu_i(\mathbf{N}_i) = 0$ and if $x_i \in \mathfrak{X}_i \cap \mathbf{N}_i^c$ then $Q_2(x_2, \cdot)$ is a strong dilation of $Q_1(x_1, \cdot)$. Define $\mu_i(\mathbf{A} \times \mathbf{B}) = \int_{\mathbf{A}} Q_i(x, \mathbf{B})\nu_i(dx)$. Then μ_2 is a strong dilation of μ_1.*

Proof. $(x, y) \le (x', y')$ is to mean $x \le x'$ and $y \le y'$. Let $f_i: [0, \infty]$ satisfy, if $(x, y) \le (x', y')$ then $f_1(x, y) \le f_2(x', y')$. If $x \in \mathfrak{X}_1$ and $x \in (\mathbf{N}_1 \cup \mathbf{N}_2)^c$ then define g_i by $g_i(x) = \int f_i(x, y)Q_i(x, dy)$. If $x \in \mathbf{N}_1$ define $g_1(x) = 0$ while if $x \in \mathbf{N}_2$ define $g_2(x) = \infty$.

We show that if $x \le x'$ and $x \in (\mathbf{N}_1 \cup \mathbf{N}_2)^c$ then $g_1(x) \le g_2(x')$. Then use of Lemma 14.7.2 and integration by ν_i gives the desired result. If $y \le y'$ are elements of $\mathfrak{X}_2$ then $(x, y) \le (x', y')$ and by hypothesis $f_1(x, y) \le f_2(x', y')$. Further, if $x, x' \in (\mathbf{N}_1 \cup \mathbf{N}_2)^c$ then by hypothesis $Q_2(x', \cdot)$ is a strong dilation of $Q_1(x, \cdot)$ so integration gives $g_1(x) \le g_2(x')$. □

Lemma 14.7.4. *Suppose $\mathfrak{X}_1$ is a lattice and $\mathfrak{X}_2$ is totally ordered, $(\mathfrak{X}_1, \mathfrak{F}_1, \lambda_1)$ and $(\mathfrak{X}_2, \mathfrak{F}_2, \lambda_2)$ are σ-finite measure spaces, $\lambda = \lambda_1 \times \lambda_2$, and f_1, f_2 are probability density functions relative to λ which satisfy the* FKG-*condition that*

$$f_1(x_1, y_1)f_2(x_2, y_2) \le f_1(x_1 \wedge x_2, y_1 \wedge y_2)f_2(x_1 \vee x_2, y_1 \vee y_2). \tag{14.7.9}$$

Then the marginal density functions $g_i(x) = \int f_i(x, y)\lambda_2(dy)$ satisfy the FKG-*condition.*

Proof (Due to Preston). Write the products as double integrals

$$g_1(x_1)g_2(x_2) = \iint f_1(x_1, y_1)f_2(x_2, y_2)\lambda_2(dy_1)\lambda_2(dy_2) \tag{14.7.10}$$

and

$$g_1(x_1 \wedge x_2)g_2(x_1 \vee x_2) = \iint f_1(x_1 \wedge x_2, y_1)f_2(x_1 \vee x_2, y_2)\lambda_2(dy_1)\lambda_2(dy_2).$$

Hence (following Kemperman) it is sufficient to show

$$a + b \leq c + d, \tag{14.7.11}$$

where

$$\begin{aligned} a &= f_1(x_1, y_1)f_2(x_2, y_2); \\ b &= f_1(x_1, y_2)f_2(x_2, y_1); \\ c &= f_1(x_1 \wedge x_2, y_1)f_2(x_1 \vee x_2, y_2); \\ d &= f_1(x_1 \wedge x_2, y_2)f_2(x_1 \vee x_2, y_1). \end{aligned}$$

Think of x_1 and x_2 as fixed and a, b, c, and d as functions of y_1 and y_2. Then $a + b$ and $c + d$ are symmetric functions of y_1 and $y_2 \in \mathfrak{X}_2$. Since $\mathfrak{X}_2$ is linearly ordered it is sufficient to consider $y_1 \leq y_2$. Then $y_1 \wedge y_2 = y_1$ and $y_1 \vee y_2 = y_2$ and (14.7.9) then becomes

$$f_1(x_1, y_1)f_2(x_2, y_2) \leq f_1(x_1 \wedge x_2, y_1)f_2(x_1 \vee x_2, y_2), \quad \text{or} \quad a \leq c. \tag{14.7.12}$$

Similarly $b \leq c$. Hence if $c = 0$ then $a = b = 0$ and (14.7.11) holds. Suppose then $c > 0$. Then again, using (14.7.9),

$$\begin{aligned} ab &= f_1(x_1, y_1)f_2(x_2, y_2)f_1(x_1, y_2)f_2(x_2, y_1) \\ &\leq f_1(x_1 \wedge x_2, y_1)f_2(x_1 \vee x_2, y_1)f_1(x_1 \wedge x_2, y_2)f_2(x_1 \vee x_2, y_2) \\ &= cd. \end{aligned} \tag{14.7.13}$$

Since $c \neq 0$, $(c + d) - (a + b) = (1/c)[(c - a)(c - b) + (cd - ab)] \geq 0$. That completes the proof of the lemma. □

Theorem 14.7.5 (Theorem 5, Kemperman, op. cit.). *Suppose* $(\mathfrak{X}_1, \mathfrak{F}_1, \lambda_1)$ *is a* FKG-*space and* $\mathfrak{X}_2$ *is totally ordered,* $(\mathfrak{X}_2, \mathfrak{F}_2, \lambda_2)$ *a* σ-*finite measure space. Then the product space in the product ordering is an* FKG-*space.*

PROOF. Let $\lambda = \lambda_1 \times \lambda_2$ and μ_1, μ_2 be probability measures on $(\mathfrak{X}_1 \times \mathfrak{X}_2, \mathfrak{F}_1 \times \mathfrak{F}_2)$ absolutely continuous with respect to λ such that $f_i = d\mu_i/d\lambda$ satisfy the FKG-condition (14.7.1). Define marginal measures $\eta_i(\mathbf{A}) = \int_{\mathbf{A}} \int f_i(x, y)\lambda_1(dx)\lambda_2(dy)$ with density functions $g_i(x) = \int f_i(x, y)\lambda_2(dy)$. By Lemma 14.7.4, $g_1(x_1)g_2(x_2) \leq g_1(x_1 \wedge x_2)g_2(x_1 \vee x_2)$. We construct Markov kernels and apply Theorem 14.7.3. Let $\mathbf{N}_i = \{x | g_i(x) = 0 \text{ or } = \infty\}$. Then $\eta_i(\mathbf{N}_i) = \int_{\mathbf{N}_i} g_i(x)\, d\lambda_1 = \int_{\{x|g_i(x)=\infty\}} g_i(x)\, d\lambda_1 = 0$. If $x \in \mathfrak{X}_1$ and $x \notin \mathbf{N}_1 \cup \mathbf{N}_2$ then define

$$Q_i(x, \mathbf{B}) = \int_{\mathbf{B}} f_i(x, y)\lambda_2(dy)/g_i(x). \tag{14.7.14}$$

This defines Markov kernels Q_i which relative to λ_2 have density functions $f_i(x, y)/g_i(x)$. By hypothesis

$$\begin{aligned} &f_1(x, y_1)f_2(x, y_2)/g_1(x)g_2(x) \\ &\quad \leq f_1(x, y_1 \wedge y_2)f_2(x, y_1 \vee y_2)/g_1(x)g_2(x). \end{aligned} \tag{14.7.15}$$

So, since $\mathfrak{X}_1$ is an FKG-space, the measure $Q_2(x, \cdot)$ is a strong dilation of the measure $Q_1(x, \cdot)$.

By the preceding part of the proof, η_2 is a strong dilation of η_1. By Theorem 14.7.3, μ_2 is a strong dilation of μ_1 where

$$\mu_i(\mathbf{A} \times \mathbf{B}) = \int_{\mathbf{A}} \int_{\mathbf{B}} f_i(x, y)\lambda_2(dy), \tag{14.7.16}$$

as was to be shown

Theorem 14.7.6 (See Theorem 2, Kemperman, op. cit.). *Let* $(\mathfrak{X}, \mathfrak{F}, \lambda)$ *be an* FKG-*space and let* λ_2 *be a strong dilation of* λ_1. *Assume* λ_1 *and* λ_2 *are absolutely continuous relative to* λ. *Let* $d\lambda_i/d\lambda = f_i$ *such that* f_1 *and* f_2 *satisfy the* FKG-*condition. Let* f, g *and* h *be nonnegative* $\mathfrak{F}$ *measurable functions such that* f *and* h *are nondecreasing and* g *is nonincreasing. Then*

$$\int fg\, d\lambda_1 \int h\, d\lambda_2 \leq \int g\, d\lambda_1 \int fh\, d\lambda_2 \tag{14.7.17}$$

provided $0 \cdot \infty = 0$ *is the interpretation made.*

PROOF. If $\int h\, d\lambda_2 = 0$ the result follows. If $\int g\, d\lambda_1 = 0$ then $\int fg\, d\lambda_1 = 0$ and the result holds. Thus we assume $\int g\, d\lambda_1 > 0$ and $\int h\, d\lambda_2 > 0$. By homogeneity we may assume $\int g\, d\lambda_1 = 1$ and $\int h\, d\lambda_2 = 1$ and are to show that

$$\int fgf_1\, d\lambda \leq \int fhf_2\, d\lambda. \tag{14.7.18}$$

Here gf_1 and hf_2 are probability density functions relative to λ and since g is nonincreasing and h nondecreasing,

$$g(x)f_1(x)h(y)f_2(y) \leq g(x \wedge y)h(x \vee y)f_1(x \wedge y)f_2(x \vee y). \tag{14.7.19}$$

Since $(\mathfrak{X}, \mathfrak{F}, \lambda)$ is an FKG-space and since f is nondecreasing, the result now follows. □

Corollary 14.7.7. *Under the hypotheses of Theorem* 14.7.6,

$$\begin{aligned} &\int fg\, d\lambda_1 \leq \int f d\lambda_2 \int g\, d\lambda_1 \qquad (g \text{ is nonincreasing}); \\ &\int f d\lambda_1 \int h\, d\lambda_2 \leq \int fh\, d\lambda_2 \qquad (h \text{ is nondecreasing}). \end{aligned} \tag{14.7.20}$$

With $g = h$ *and* $\lambda_1 = \lambda_2$ *this is called the* FKG-*inequality.*

Proof. The problem is to extend Theorem 14.7.6 to functions taking negative values. Consider the second inequality with

$$\int (f + c_1)\, d\lambda_1 \int (h + c_2)\, d\lambda_2 \le \int (f + c_1)(h + c_2)\, d\lambda_2 \quad (14.7.21)$$

equivalent to the desired result. Since truncation below does not change monotonicity the corollary follows by trunction, translation, then use of monotone convergence. □

The following lemmas and theorem give a random variable construction for the real valued case of the FKG-inequality.

Lemma 14.7.8. *Let* $\mathbf{X} \in \mathbb{R}^n$ *and* $\mathbf{Y} \in \mathbb{R}$ *be random variables and* $f_1, f_2 \colon \mathbb{R}^n \times \mathbb{R}$ *be density functions for* $(\mathbf{X}, \mathbf{Y})$ *relative to* $\nu \times \lambda$. *Assume the* FKG-*condition* (14.7.1) *holds. Then the conditional distribution functions*

$$\begin{aligned} F_i(t|x) &= \int_{-\infty}^{t} f_i(x,y)\lambda(dy) \bigg/ \int_{-\infty}^{\infty} f_i(x,y)\lambda(dy), \\ &= 0 \text{ if the divisor is zero,} \end{aligned} \quad (14.7.22)$$

satisfy, if $x_1 \le x_2$ *then* $F_2(t|x_2) \le F_1(t|x_1)$.

Proof. $F_2(t|x_2) \le F_1(t|x_1)$ means

$$\begin{aligned} &\int_{-\infty}^{t} f_2(x_2,y)\lambda(dy) \int_{-\infty}^{\infty} f_1(x_1,y)\lambda\, dy \\ &\quad \le \int_{-\infty}^{t} f_1(x_1,y)\lambda(dy) \int_{-\infty}^{\infty} f_2(x_2,y)\lambda(dy), \end{aligned} \quad (14.7.23)$$

which is equivalent to

$$\int_{-\infty}^{t} \lambda(dy_1) \int_{t}^{\infty} \lambda(dy_2)[f_2(x_2,y_1)f_1(x_1,y_2) - f_1(x_1,y_1)f_2(x_2,y_2)] \le 0. \quad (14.7.24)$$

Here $y_1 \le t \le y_2$ and thus $y_1 = y_1 \wedge y_2$ and $y_2 = y_1 \vee y_2$. By hypothesis $x_1 \le x_2$ so $x_1 = x_1 \wedge x_2$ and $x_2 = x_1 \vee x_2$. By assumption of the FKG-condition (14.7.1), the integrand is nonpositive and the result follows. □

Lemma 14.7.9. *Under the assumptions of Lemma* 14.7.8, *let* $\mathbf{U}$ *be a uniform* $[0, 1]$ *random variable independent of* $\mathbf{X}$. *Then* $P(F_i^{-1}(\mathbf{U}|\mathbf{X}) \le t$ *and* $\mathbf{X} \in \mathbf{A}) = P(\mathbf{X} \in \mathbf{A}, \mathbf{Y} \le t)$.

Proof. Since $\mathbf{U}$ is independent of $\mathbf{X}$, $P(F_i^{-1}(\mathbf{U}|\mathbf{X}) \le t, \mathbf{X} \in \mathbf{A}) = P(\mathbf{U} \le F_i(t|\mathbf{X})$ and $\mathbf{X} \in \mathbf{A}) = \int_{\mathbf{A}} F_i(t|x)\eta(dx) = P(\mathbf{X} \in \mathbf{A}$ and $\mathbf{Y} \le t)$. □

Theorem 14.7.10. *If* $(\mathbf{X}_1, \ldots, \mathbf{X}_n) \in \mathbb{R}^n$ *have density functions* f_i, $i = 1, 2$, *relative to a product measure* $\prod_{i=1}^n \lambda_i(dx_i)$ *which satisfy the* FKG-*condition* (14.7.1) *then there exist functions* h_{ij}, $i = 1, 2$, $1 \le j \le n$, *and independently uniformly distributed random variables* $\mathbf{U}_1, \ldots, \mathbf{U}_n$ *such that the joint distribution of* $h_{ij}(U_1, \ldots, U_j)$, $j = 1, \ldots, n$, *is that of* $\mathbf{X}_1, \ldots, \mathbf{X}_n$ *under* f_i. *Further*

$$h_{1j}(\mathbf{U}_1, \ldots, \mathbf{U}_j) \le h_{2j}(\mathbf{U}_1, \ldots, \mathbf{U}_j), \qquad 1 \le j \le n. \tag{14.7.25}$$

Proof. By Lemma 14.7.4 the marginal densities of $\mathbf{X}_1, \ldots, \mathbf{X}_j$ relative to $\prod_{i=1}^j \lambda_i(dx_i)$ satisfy the FKG-condition. Define inductively

$$\begin{aligned} \mathbf{Y}_i &= F_1^{-1}(\mathbf{U}_i | \mathbf{Y}_1, \ldots, \mathbf{Y}_{i-1}), \qquad i \ge 2, \\ \mathbf{Z}_i &= F_2^{-1}(\mathbf{U}_i | \mathbf{Y}_1, \ldots, \mathbf{Y}_{i-1}), \qquad i \ge 2, \end{aligned} \tag{14.7.26}$$

and

$$\mathbf{Y}_1 = F_1^{-1}(\mathbf{U}_1) \le F_2^{-1}(\mathbf{U}_1) = \mathbf{Z}_1.$$

Suppose it has been shown that $\mathbf{Y}_2 \le \mathbf{Z}_2, \ldots, \mathbf{Y}_{i-1} < \mathbf{Z}_{i-1}$. By Lemma 14.7.8 if $\bar{x} \le x$ then $F_2(t|x) \le F_1(t|\bar{x})$ and $F_1^{-1}(t|\bar{x}) \le F_2^{-1}(t|x)$. Thus $\mathbf{Y}_i = F_1^{-1}(\mathbf{U}_i | \mathbf{Y}_1, \ldots, \mathbf{Y}_{i-1}) \le F_2^{-1}(\mathbf{U}_i | \mathbf{Z}_1, \ldots, \mathbf{Z}_{i-1}) = \mathbf{Z}_i$. Thus by induction

$$(\mathbf{Y}_1, \ldots, \mathbf{Y}_n) \le (\mathbf{Z}_1, \ldots, \mathbf{Z}_n). \tag{14.7.27}$$

As is well known the distributions of $\mathbf{Y}_1$ and $\mathbf{Z}_1$ are those of $\mathbf{X}_1$ under f_1 and f_2. By Lemma 14.7.9 if $\mathbf{X}_1, \ldots, \mathbf{X}_{j-1}$ and $\mathbf{Y}_1, \ldots, \mathbf{Y}_{j-1}$ have the same distribution then so do $\mathbf{Y}_1, \ldots, \mathbf{Y}_{j-1}, \mathbf{Y}_j = F_1^{-1}(\mathbf{U}_j | \mathbf{Y}_1, \ldots, \mathbf{Y}_{j-1})$ and $\mathbf{X}_1, \ldots, \mathbf{X}_j$. Similarly for $\mathbf{Z}_1, \ldots, \mathbf{Z}_j$. Hence by induction the theorem follows. □

Lemma 14.7.11. *A sufficient condition that the* FKG-*condition hold for* f_1 *and* f_2 *defined on* $\mathbb{R}^n$ *is that if* $\ln f_1$ *and* $\ln f_2$ *have piecewise continuous partial derivatives then* $\partial_i \ln f_2(x) \ge \partial_i \ln f_1(y)$ *for all* $x \ge y$ *and* $1 \le i \le n$.

Proof. Let $f = \ln f_2$ and $g = \ln f_1$. Given that $x \ge y$ and that $\partial_i f(x) \ge \partial_i g(y)$ let $z: [0, 1] \to \mathbb{R}^n$ be a path having nonnegative nondecreasing coordinates and with $z(0) = 0$. Then $z(t) + x \ge z(t) + y$ so that

$$(\partial z)^t \Delta f(x + z(t)) \ge (\partial z)^t \Delta g(y + z(t)), \tag{14.7.28}$$

and integration with respect to t gives

$$f(x + z(1)) - f(x) \ge g(y + z(1)) - g(y). \tag{14.7.29}$$

Thus

$$f_2(x + z(1)) f_1(y) \ge f_2(x) f_1(y + z(1)). \tag{14.7.30}$$

□

Remark 14.7.12. There is a converse result that is discussed by Olkin and Perlman (1980).

EXAMPLE 14.7.13. (Olkin and Perlman (1980)). The function f defined by $f(x) = \prod_{i \le j}(x_i - x_j)$ for $x^t = (x_1, \ldots, x_n)$ and $x_1 \ge \cdots \ge x_n \ge 0$ satisfies $f(x \vee y)f(x \wedge y) \ge f(x)f(y)$. Olkin and Perlman verify this by a direct computation on the two sides.

EXAMPLE 14.7.14. If $\mathbf{X}$ is $n \times k$ with independently normally distributed rows having covariance I_k and if $\mathbf{E}\mathbf{X} = M$ then by the theory of Section 9.3, see (9.3.14), and Chapter 10, the noncentral density function of the singular values of $\mathbf{X}$ is of the form

$$f_M(D) = \text{const}\, e^{-\operatorname{tr}(X^tX + M^tM)/2} \int_{\mathbf{O}(n)} \int_{\mathbf{O}(k)} e^{\operatorname{tr} AX\mathbf{G}^tM^t}\, dA\, dG \quad (14.7.31)$$

with integrating measure on the eigenvalues given by the density

$$\left(\prod_{i=1}^{k} c_i\right)^{n-k} \prod_{i \le j} (c_i^2 - c_j^2) \quad (14.7.32)$$

relative to Lebesgue measure on $\mathbb{R}^k$. The function

$$g_M(D) = \int_{\mathbf{O}(n)} \int_{\mathbf{O}(k)} e^{\operatorname{tr} AX\mathbf{G}^tM^t} \quad (14.7.33)$$

with $D = \operatorname{diag} \lambda((X^tX)^{1/2}) = \operatorname{diag}(c_1, \ldots, c_k)$ is a convex function of $c_1, \ldots, c_k$ with is invariant under all sign changes of $c_1, \ldots, c_k$. It is therefore a nondecreasing function of each c_j and hence $g_M(D) \ge 1$ in value. If $M = 0$ then $g_0(D) \equiv 1$. It follows that given diagonal matrices D_1 and D_2 with nonnegative diagonal entries

$$g_M(D_1 \vee D_2)g_0(D_1 \wedge D_2) > g_M(D_1)g_0(D_2). \quad (14.7.34)$$

Using Example 14.7.13 it follows that the pair f_M and f_0 satisfy the FKG-condition. By Theorem 14.7.6, if h is a nondecreasing function on $\mathbb{R}^k \to (-\infty, \infty)$ then

$$E_M h(\lambda((\mathbf{X}^t\mathbf{X})^{1/2})) \ge \mathbf{E}_0 h(\lambda((\mathbf{X}^t\mathbf{X})^{1/2})). \quad (14.7.35)$$

EXAMPLE 14.7.15. The invariant in MANOVA is

$$\lambda((\mathbf{X}^t\mathbf{X} + \mathbf{Y}^t\mathbf{Y})^{-1/2}\mathbf{X}^t\mathbf{X}(\mathbf{X}^t\mathbf{X} + \mathbf{Y}^t\mathbf{Y})^{-1/2}). \quad (14.7.36)$$

The distribution of this invariant depends on the parameter $M\Sigma^{-1/2}$ and we suppose $\Sigma = I_k$. If $h: \mathbb{R}^k \to (-\infty, \infty)$ is measurable and $\mathbf{E}_Y$ means integration on the variables Y, then orthogonal invariance of the eigenvalue function and the joint normal density function ($\Sigma = I_k$) implies that if $\Lambda(X) = \operatorname{diag}(\lambda(X^tX)^{1/2})$ then (14.7.36) integrated equals

$$\mathbf{E}_Y h((\Lambda(\mathbf{X}) + \mathbf{Y}^t\mathbf{Y})^{-1/2}\Lambda(\mathbf{X})(\Lambda(\mathbf{X}) + \mathbf{Y}^t\mathbf{Y})^{-1/2}) \quad (14.7.37)$$

and this is a function of the eigenvalues of $\mathbf{X}^t\mathbf{X}$. If h is a nondecreasing function, then, since

$$\lambda((\Lambda + \mathbf{Y}^t\mathbf{Y})^{-1/2}\Lambda(\Lambda + \mathbf{Y}^t\mathbf{Y})^{-1/2}) \tag{14.7.38}$$

is a nondecreasing function of Λ, by Example 14.7.14 and (14.7.35) it follows that

$$\mathbf{E}_M h \geq \mathbf{E}_0 h. \tag{14.7.39}$$

14.8. Problems

PROBLEM 14.8.1. If A is an $n \times n$ positive definite matrix then $\operatorname{tr} A \leq (\det A)^{1/n}$. Equality holds if and only if $A = I_n$.

PROBLEM 14.8.2. Let A be an $n \times n$ positive semidefinite matrix of rank r. Show that $\operatorname{tr} A \leq \sup_P (\det (P^tAP))^{1/r}$ taken over all $n \times r$ matrices P such that $P^tP = I_r$. Equality holds if and only if the nonzero eigenvalues of A are equal. [Suggestion: diagonalize A by UAU^t. Then $(UP)^t(UP) = I_r$. Write $P^t = (P_1^t, P_2^t)$ so that $P_1^tP_1 \leq I_r$.]

PROBLEM 14.8.3. Let A be an $n \times n$ positive definite matrix. Then $\Sigma_{i=j+1}^n \lambda_i(A^{1/2})$ is a nondecreasing concave function of A. [Hint: $\lambda_i(\cdot)$ is nondecreasing, $\Sigma_{i=j+1}^n \lambda_i(\cdot)$ is concave, and $A^{1/2}$ a concave nondecreasing matrix function.]

PROBLEM 14.8.4. Let $A \in \mathbf{S}(n)$. Then $\{X | X \in \mathbf{S}(n)$ and $X \leq A\} = \bigcap_{B \in \mathbf{S}(n)} \{X | \operatorname{tr} AB \geq \operatorname{tr} XB$ and $X \in \mathbf{S}(n)\}$. [If X is not less than A then there exists x with $x^t(A - X)x < 0$].

PROBLEM 14.8.5. The matrix $X^tX = \sup_A(-A^tA - A^tX - X^tA)$ is thus a convex function of X. Certain linear combinations of eigenvalues of X^tX can be expressed as (see Theorem 14.3.2) $\sup_{U \in \mathbf{O}(n)} U(X^tX)U^tB$. If B is positive definite, show that this is a convex function of X. Then, by Problem 14.8.8, if the density function of $\mathbf{X}$ is the log-concave function f, $\ln P(\{\sup_{U \in \mathbf{O}(n)} U(\mathbf{X}^t\mathbf{X})U^tB \leq a\})$ is a concave function of a.

PROBLEM 14.8.6. In particular, continue Problem 14.8.5. The functions $\lambda_1(X^tX) + \cdots + \lambda_i(X^tX)$ are convex functions on X and $\ln P(\{\lambda_1(\mathbf{X}^t\mathbf{X}) + \cdots + \lambda_i(\mathbf{X}^t\mathbf{X}) \leq a\})$ is a concave function of a.

PROBLEM 14.8.7. If X is $n \times p$ then relative to np-dimensional Lebesgue measure μ_{np}, $(\mu_{np}(\{\lambda_1(\mathbf{X}^t\mathbf{X}) + \cdots + \lambda_i(\mathbf{X}^t\mathbf{X})\}) \leq a)^{1/np}$ is a concave function of a. Show that the values are finite by $\lambda_1(X^tX) + \cdots + \lambda_i(X^tX) \leq \operatorname{tr} X^tX = \Sigma_{i=1}^n \Sigma_{j=1}^p (X)_{ij}^2 \leq a$, which is a compact subset of $\mathbb{R}^{np}$.

PROBLEM 14.8.8. Let $f: \mathbb{R}^n \to \mathbb{R}$ be a probability density function relative to Lebesgue measure μ_n such that if $P(\mathbf{A}) = \int_{\mathbf{A}} f(x)\mu_n(dx)$ and $0 \leq \alpha \leq 1$ then

$P(\alpha\mathbf{A} + (1-\alpha)\mathbf{B}) \geq P(\mathbf{A})^\alpha P(\mathbf{B})^{1-\alpha}$. Let $\mathbf{A}$ be a convex set and show that $\ln P(\mathbf{A} + a)$ is a concave function of $n \times 1$ vectors a.

Problem 14.8.9. Under the conditions of Problem 14.8.8, let $\mathbf{A}$ be the acceptance region of a size α test. If $\mathbf{H}_0 \subset \mathbb{R}^n$ and the parameter η is a location parameter, then $P(\mathbf{A} + \eta)$ is the probability of acceptance when η is the parameter. Suppose $\mathbf{H}_0$ is convex with nonvoid interior and that if η is a boundary point of $\mathbf{H}_0$ then $P(\mathbf{A} + \eta) = 1 - \alpha$. Show that the test is unbiased.

Problem 14.8.10. Under the conditions of Problems 14.8.8 and 14.8.9, if the probability measure P is strictly log-concave, if the test is unbiased, then the boundary of $\mathbf{H}_0$ cannot contain any line segments.

Problem 14.8.11. If P is log-concave and $\mathbf{A}$, $\mathbf{B}$ are convex sets such that $\mathbf{B} \subset \mathbf{A}$ then $\ln P((\mathbf{A} - \mathbf{B}) + a)$ is a concave function of a. [$P((\mathbf{A} - \mathbf{B}) + a)$ can be interpreted as the difference of two power functions.

Problem 14.8.12. Let $\mathbf{X}$ be $m \times p$, $N(\mu, \Sigma)$, and $\mathbf{Y}$ be $n \times q$, $N(0, \Sigma)$, with $\mathbf{X}$ and $\mathbf{Y}$ independent. Let

$$\mathbf{H}_0 : \mu = 0, \quad \Sigma \text{ arbitrary};$$

$$\mathbf{H}_1 : \mu \neq 0, \quad \Sigma \text{ arbitrary}.$$

Suppose the acceptance region $\mathbf{A}$ is convex and symmetric. Use the Anderson inequality, see Theorem 14.6.15, to show the test with acceptance region $\mathbf{A}$ cannot have constant power on $\mathbf{H}_0$.

Problem 14.8.13. Let P be a convex measure and define η by $\alpha = P(x_1 \geq \mathbf{X}_1, \ldots, x_{n-1} \geq \mathbf{X}_{n-1}, x_n \geq \eta(x_1, \ldots, x_{n-1}))$. Show η to be a concave function.

Problem 14.8.14. Let $f: \mathbb{R}^n \to \mathbb{R}$ be log-concave and the parameter be a location parameter. If $\mathbf{A}$ is the convex acceptance region of an unbiased test and if $\mathbf{H}_0$ is a closed set, then $\mathbf{H}_0$ is convex.

Problem 14.8.15. Let $\phi: \mathbb{R}^n \to \mathbb{R}$ be a convex function and let critical regions of tests be sets $\{x | \phi(x) \leq a\}$. Then show $\alpha\{x|\phi(x) \leq a\} + (1-\alpha)\{x|\phi(x) \leq b\} \subset \{x|\phi(x) \leq \alpha a + (1-\alpha)b\}$. If the density function f is log-concave then show $\ln P(\{\phi(\mathbf{X}) \leq a\})$ is a concave function of a. For example, the Hotelling T^2-statistic and the central Wishart density.

Problem 14.8.16. If $f_1, \ldots, f_n, g_1, \ldots, g_n$ are real valued functions of a real variable such that g_i/f_i is nondecreasing, $1 \leq i \leq n$, then $\prod_{i=1}^n f_i(x_i)$ and $\prod_{i=1}^n g_i(x_i)$ satisfy the FKG-condition (14.7.1).

PROBLEM 14.8.17. If $f(x)f(y) \le f(x \wedge y)f(x \vee y)$ for all x, y, and if g is a nonnegative nondecreasing function, then f and fg satisfy the FKG-condition (14.7.1).

PROBLEM 14.8.18. Let X, $M \in \mathbb{R}^n$. If $M_1 \le M_2$ then the pair $\exp -\frac{1}{2}\operatorname{tr}(X - M_i)(X - M_i)^t$ satisfies the FKG-condition. Hence the M_2-measure is a strong dilation of the M_1-measure. [Remark: $X_1 + X_2 = (X_1 \wedge X_2) + (X_1 \vee X_2)$.]

PROBLEM 14.8.19. If $\mathbf{X}$ is $n \times h$ normal $(0, \Sigma)$ then the $\Sigma_1 < \Sigma_2$ densities of $\mathbf{S} = \mathbf{X}^t\mathbf{X}$ satisfy the FKG-condition (14.7.1) relative to the lattice ordering of symmetric matrices. [See Section 14.1. Although the integrating measure is $h(h+1)/2$-dimensional product measure, the partial ordering is not the product order. Hence the theory of Section 14.7 does not apply.]

PROBLEM 14.8.20. When $\Sigma = I_h$ the density of the eigenvalues $\mathbf{t}_1 > \mathbf{t}_2 > \cdots > \mathbf{t}_h$ of $\mathbf{X}^t\mathbf{X} = \mathbf{S}$ for the noncentral Wishart matrix $\mathbf{S}$ has the form $f(t_1, \ldots, t_h)g(t_1, \ldots, t_h; \lambda_1, \ldots, \lambda_h)$. Here

$$f(t_1, \ldots, t_h) = \prod_{i=1}^{h} (t_i^{(n-h-1)} \exp(-\tfrac{1}{2}t_i)) \prod_{i<j} (t_i - t_j).$$

Show that the function g is increasing in $t' = (t_1, \ldots, t_n)$. Show that the function f satisfies the FKG-condition (14.7.1).

For short write $g(t, \lambda)$. It is unknown to this author whether $g(\cdot\,; \lambda_1)$, $g(\cdot\,; \lambda_2)$ satisfy the FKG-condition when $\lambda_1 \le \lambda_2$. This question was left unanswered by Olkin and Perlman (1980).

PROBLEM 14.8.21. Let $\mathbf{X}$ be $n \times k$, independently distributed rows, each normal with covariance matrix I_k. Let $\mathbf{E}\mathbf{X} = M$. Show that any admissible test of $H_0: M = 0$ vs $H_1: M \ne 0$ based on the eigenvalues of $\mathbf{X}^t\mathbf{X}$ must have acceptance region $\mathbf{A} \subset \mathbf{R}^{nk}$ which is convex. Show that the class of these tests is a complete class and that all such admissible tests are unbiased.

HINT. Using the methods of Section 10.1 the significant part of the density function is

$$f(E) = \int_{\mathbf{O}(k)} \int_{\mathbf{O}(n)} \exp(\operatorname{tr} V(E, 0)UM)\, dV\, dU \tag{14.8.1}$$

where $E = \operatorname{diag}(\lambda((X^tX)^{1/2}))$. The function f is a convex function of E and $f(W_1EW_2) = f(E)$ for all $k \times k$ orthogonal matrices W_1, W_2. Thus f is a symmetric function of k variables and is sign change invariant in each variable. Using Lemma 14.3.8, Corollary 14.3.4, Theorem 14.3.5, and Theorem 14.3.7, if follows that

$$f(\operatorname{diag}(\lambda(S^{1/2}))) \tag{14.8.2}$$

is a convex function of X since (14.8.1) is, and is convex in $S^{1/2}$. Hence the tests which are Bayes within the class of tests based on λ have acceptance regions convex in X. By Stein (1956b) all such tests are admissible. Since the multivariate normal density function is log-concave, and since $\mathbf{A}$ is convex, with

$$\mathbf{A} = \{X | \lambda((X^t X)^{1/2}) \in \mathbf{B}\}, \tag{14.8.3}$$

it follows that $\lambda(((X + M)^t(X + M))^{1/2}) \in \mathbf{B}$ if and only if $X \in (-M) + \mathbf{A}$. Consequently the power function is log-concave in M with maximum at $M = 0$. □

Bibliography

Aitkin, M. A. (1969). Some tests for correlation matrices. *Biometrika*, **56**, 443–446. (Use of the "delta" method for the computation of asymptotic series.)

Albert, A. A. (1939). *Structure of Algebra*. American Mathematical Society Collequium Publication **xxiv**.

Anderson, G. A. (1965). An asymptotic expansion for the distribution of the latent roots of the estimated covariance matrix. *Ann. Math. Statist.*, **36**, 1153–1173.

Anderson, G. A. (1970). An asymptotic expansion for the noncentral Wishart distribution. *Ann. Math. Statist.*, **41**, 1700–1707. (Uses zonal polynomials as the roots go to infinity.)

Anderson, T. W. (1946). The noncentral Wishart distribution and certain problems of multivariate statistics. *Ann. Math. Statist.*, **17**, 409–431.

Anderson, T. W. (1955). The integral of a symmetric unimodel function over a symmetric convex set and some probability inequalities. *Proc. Amer. Math. Soc.*, **6**, 170–176. (For showing multivariate tests are unbiased.)

Anderson, T. W. (1958). *An Introduction to Multivariate Statistical Analysis*. John Wiley & Sons, Inc., New York.

Anderson, T. W. (1963). Asymptotic theory for principal component analysis. *Ann. Math. Statist.*, **34**, 122–148.

Anderson, T. W. and Styan, G. P. H. (1980). Cochran's theorem, rank additivity and tripotent matrices. Technical Report 43, Department of Statistics, Stanford University.

Anderson, T. W., Gupta, S. S., and Styan, G. P. H. (1972). A Bibliography of Multivariate Statistical Analysis. John Wiley & Sons, Inc., New York.

Andersson, S. A. (1975). Invariant normal models. *Ann. Statist.*, **3**, 132–154.

Andersson, S. A. (1978). Invariant measures. Technical Report 129. Stanford University, Department of Statistics, Stanford University.

Andersson, S. A. (1981). Distributions of maximal invariants using quotient measures. Technical Report No. 7., Department of Statistics, University of Washington, Seattle, Washington.

Andersson, S. A. (1982). Distributions of maximal invariants using quotient measures. *Ann. Statist.*, **10**, 955–961.

Andersson, S. A., Brøns, H. K., and Jensen, S. T. (1983). Distribution of eigenvalues in multivariate statistical analysis. *Ann. Statist.*, **11**, 392–415.

Bartlett, M. S. (1933). On the theory of statistical regression. *Proc. Royal Soc. of Edinburgh*, **53**, 260–283. (Bartlett decomposition.)

Basu, D. (1955). On statistics independent of a complete sufficient statistic. *Sankhyā*, **15**, 377–380.

Bechhofer, R. E., Kiefer, J., and Sobel, M. (1968). *Sequential Identification and Ranking Procedures*. The University of Chicago Press.

Bechenbach, E. F., and Bellman, R. (1965). *Inequalities*. Springer-Verlag, Berlin.

Bendat, J. and Sherman, S. (1955). Monotone and convex operator functions. *Trans. Amer. Math. Soc.*, **79**, 58–71.

Berger, J. and Wolpert, R. L. (1981). Estimating the mean function of a Gaussian process and the Stein effect. Mimeograph Series #81-17. Statistics Department, Purdue University.

Berger, J. and Wolpert, R. (1982). Incorporating prior information in minimax estimation of the mean of a Gaussian process. *Statistical Decision Theory and Related Topics III*. (S. S. Gupta and J. O. Berger, ed.) Academic Press, New York.

Bhattacharya, P. K. (1966). Estimating the mean of a multivariate normal population with general quadratic loss function. *Ann. Math. Statist.*, **37**, 1819–1824.

Bhattacharya, R. N. (1971). Rates of weak convergence and asymptotic expansions for classical central limit theorems. *Ann. Math. Statist.*, **42**, 241–259. (This paper contains a list of useful references.)

Bingham, C. (1972). An asymptotic expansion for the distribution of the eigenvalues of a 3 by 3 Wishart matrix. *Ann. Math. Statist.*, **43**, 1498–1506.

Birnbaum, A. (1955). Characterization of complete classes of tests of some multivariate hypotheses, with applications to likelihood ratio tests. *Ann. Math. Statist.*, **26**, 21–36.

Bondar, J. V. (1976). Borel cross-sections and maximal invariants. *Ann. Statist.*, **4**, 866–877.

Bondar, J. V. (1983). On universal optimality of designs—Definitions and a simple necessary and sufficient condition. *Can. Jour. Statist*, 11, 325–331.

Bondar, J. V., and Milnes, P. (1981). Amenability: A survey for statistical applications of Hunt–Stein and related conditions on groups. *Z. Wahr.*, **57**, 103–128.

Box, G. E. P. (1949). A general distribution theory for a class of likelihood criteria. *Biometrika*, **36**, 317–346. (Obtains asymptotic series by use of Fourier transforms.)

Braaksma, L. J. (1964). Asymptotic expansions and analytic continuations for a class of Barnes-integrals. *Compositio Mathematica*, **15**, 239–341. (Defines *H*-functions.)

Brillinger, D. R. (1963). Necessary and sufficient conditions for a statistical problem to be invariant under a Lie group. *Ann. Math. Statist.*, **34**, 492–500.

Brillinger, D. R. (1975). *Time series: data analysis and theory*. Holden-Day, Inc., San Francisco.

Brown, L. D. (1971). Admissible estimators, recurrent diffusions and insoluble boundary value problems. *Ann. Math. Statist.*, **42**, 855–903.

Brown, L. D. (1980). A Hunt–Stein theorem. Preprint.

Brown, L. D. (1982). Comments on "Invariance, minimax sequential estimation, and continuous time processes" (by J. Kiefer (1959) *Ann. Math. Statist.*, **30**, 573–601.) Preprint.

Brown, L. D., and Farrell, R. H. (1983). All admissible linear estimators of a multivariate Poisson mean. To Appear. *Ann. Statist.*

Chambers, J. M. (1967). On methods of asymptotic approximation for multivariate distributions. *Biometrika*, **54**, 367–383. (Edgeworth expansions, perturbation approximations.)

Cheng, C.-S. (1978). Optimality of certain asymmetrical experimental designs. *Ann. Statist.*, **6**, 1239–1261.

Cheng, C.-S. (1980). Optimality of some weighing and 2^n fractional factorial designs. *Ann. Statist.*, **8**, 436–446.

Cohen, A., and Strawderman, W. E. (1971). Unbiasedness of tests for homogeneity of variances. *Ann. Math. Statist.*, **42**, 355–360.

Constantine, A. G. (1960). Multivariate Distributions. CSIRO Report.

Constantine, A. G. (1963). Noncentral distribution problems in multivariate analysis. *Ann. Math. Statist.*, **34**, 1270–1285.

Constantine, A. G. (1966). The distribution of Hotelling's generalized T_0^2. *Ann. Math. Statist.*, **37**, 215–225.

Constantine, A. G., and James, A. T. (1958). On the general canonical correlation distribution. *Ann. Math. Statist.*, **29**, 1146–1166.

Consul, P. C. (1969). The exact distributions of likelihood criteria for different hypotheses. 171–181, *Multivariate Analysis II.* (P. R. Krishnaiah, ed.) Academic Press, New York. (Meijer's functions, Mellin inversion.)

Das Gupta, S. (1969). Properties of power functions of some tests concerning dispersion of matrices of multivariate normal distributions. *Ann. Math. Statist.*, **40**, 697–701. (Unbiasedness proven.)

Das Gupta, S., Anderson, T. W., and Mudholkar, G. S. (1964). Monotonicity of the power functions of some tests of the multivariate linear hypothesis. *Ann. Math. Statist.*, **35**, 200–205.

David, F. N., Kendall, M. G., and Barton, D. E. (1966). *Symmetric Functions and Allied Tables.* Cambridge University Press.

Davis, A. W. (1972). On the marginal distributions of the latent roots of the multivariate beta matrix. *Ann. Math. Statist.*, **43**, 1664–1670.

Davis, A. W. (1980). Invariant polynomials with two matrix arguments, extending zonal polynomials. *Multivariate Analysis V*, (P. R. Krishnaiah, ed.) North-Holland, Amsterdam.

Davis, A. W. (1981). On the construction of a class of invariant polynomials in several matrices, extending the zonal polynomials. *Ann. Inst. Statist. Math.*, **33**, 297–313.

Davis, A. W., and Hill, G. W. (1968). Generalized asymptotic expansions of Cornish–Fisher type. *Ann. Math. Statist.*, **39**, 1264–1273.

De Waal, D. J. (1972). On the expected values of the elementary symmetric functions of a noncentral Wishart matrix. *Ann. Math. Statist.*, **43**, 344–347.

Deemer, W. L., and Olkin, I. (1951). The Jacobians of certain matrix transformations useful in multivariate analysis. *Biometrika*, **38**, 345–367.

Dempster, A. P. (1969). *Elements of Continuous Multivariate Analysis.* Addison Wesley, Reading, Mass.

Dieudonné, J. (1972). *Treatise on Analysis, Vol. III.* Academic Press, New York.

Doob, J. L. (1958). Lecture notes on probability theory. Personal notes of R. H. Farrell, unpublished.

Dunford, N., and Schwartz, J. T. (1958). *Linear Operators Part I: General Theory.* Interscience Publishers, New York.

Dunkl, C. F., and Ramirez, D. E. (1971). *Topics in Harmonic Analysis.* Appleton-Century-Croft, New York.

Dykstra, R. L. (1970). Establishing the positive definiteness of the sample covariance matrix. *Ann. Math. Statist.*, **41**, 2153–2154.

Dynkin, E. B., and Mandelbaum, A. (1983). Symmetric statistics, Poisson point processes, and multiple Wiener integrals. *Ann. Statist.*, **11**, 739–745.

Eaton, M. L. (1970). Gauss–Markov estimation for multivariate linear models: A coordinate free approach. *Ann. Math. Statist.*, **41**, 528–538.

Eaton, M. L. (1972). *Multivariate Statistical Analysis.* Institute of Mathematical Statistics University of Copenhagen, August 1972.

Eaton, M. L. (1982). A review of selected topics in multivariate probability inequalities. *Ann. Statist.*, **10**, 11–43.

Eaton, M. L. (1983). *Multivariate Statistics. A Vector Space Approach.* John Wiley & Sons Inc., New York.

Eaton, M. L. and Perlman, M. D. (1973). The non-singularity of generalized covariance matrices. *Ann. Statist.*, **1**, 710–717.

Eaton, M. L. and Perlman, M. D. (1974). A monotonicity property of the power function of some invariant tests of MANOVA. *Ann. Statist.*, **2**, 1022–1028.

Erdelyi, A. (1956). *Asymptotic Expansions*. Dover Publications, Inc, New York.

Erdelyi, A., Magnus, W., Oberhettinger, and Tricomi, F. G. (1953). *Higher Transcendental Functions*. McGraw Hill, New York.

Fan, Ky (1949). On a theorem of Weyl concerning eigenvalues of linear transformations. *Proc. Nat. Acad. Sci.*, **35**, 652–655.

Farrell, R. H. (1962). Representations of invariant measures. *Ill. J. Math.*, **6**, 447–467.

Farrell, R. H. (1968). Towards a theory of generalized Bayes tests. *Ann. Math. Statistic.*, **39**, 1–22.

Farrell, R. H. (1973). Max-min designs in the analysis of variance. *Multivariate Analysis III* (P. R. Krishnaiah, ed.). Academic Press, New York.

Farrell, R. H. (1976). *Techniques of Multivariate Calculation*. Lecture Notes in Mathematics 520. Springer-Verlag, Berlin.

Farrell, R. H. (1980). Calculation of complex zonal polynomials. *Multivariate Analysis V* (P. R. Krishnaiah, ed.). North Holland, Amsterdam.

Fefferman, C., Jodeit, Jr., M. and Perlman, M. D. (1972). A spherical surface measure inequality for convex sets. *Proc. Am. Math. Soc.*, **33**, 114–119.

Feller, W. (1966). *An Introduction to Probability Theory and its Applications, Vol. II*. John Wiley & Sons, Inc., New York.

Frankel, L. R., and Hotelling, H. (1938). The transformation of statistics to simplify their distributions. *Ann. Math. Statist.*, **9**, 87–96.

Garsia, A. M. and Remmel, J. (1981). Symmetric functions and raising operators. *Linear and Multilinear Algebra*, **10**, 15–43.

Gelfond, A. O. (1971). *Residues and their Applications*. Moscow: MIR Publishers.

Giri, N., and Kiefer, J. (1964). Local and asymptotic minimax properties of multivariate tests. *Ann. Math. Statist.*, **35**, 21–35.

Giri, N. (1968a). Locally and asymptotically minimax tests of a multivariate problem. *Ann. Math. Statist.*, **39**, 171–178.

Giri, N. (1968b). On tests of the equality of two covariance matrices. *Ann. Math. Statist.*, **39**, 275–277.

Giri, N., Kiefer, J. and Stein, C. (1963). Minimax character of Hotelling's T^2 test in the simplest case. *Ann. Math. Statist.*, **34**, 1524–1535.

Gleser, L. J. (1966). A note on the sphericity test. *Ann. Math. Statist.*, **37**, 464–467. (Proves unbiasedness.).

Gleser, L. J. (1968). On testing a set of correlation coefficients for equality: Some asymptotic results. *Biometrika*, **55**, 513–517.

Gleser, L. J., and Olkin, I. (1970). Linear models in multivariate analysis. *Essays in Probability and Statistics* (R. C. Bose et. al. eds.) University of North Carolina Press.

Good, I. J. (1969). Conditions for a quadratic form to have a chi-squared distribution. *Biometrika*, **56**, 215–216.

Graybill, F. A., and Milliken, G. (1969). Quadratic forms and idempotent matrices with random elements. *Ann. Math. Statist.*, **40**, 1430–1438.

Gupta, S. S. (1963). Bibliography on the multivariate normal integrals and related topics. *Ann. Math. Statist.*, **34**, 829–838.

Halmos, P. R. (1950). *Measure Theory*. Van Nostrand, New York.

Halmos, P. R. (1951). *Introduction to Hilbert Space and the Theory of Spectral Multiplicity*. Chelsea Publishing Company, New York.

Halmos, P. R. (1974). *Measure Theory*. Springer-Verlag, New York.

Hardy, G. H., Littlewood, J. E., and Polya, G. (1952). *Inequalities*. Cambridge University Press.

Helgason, S. (1962). *Differential Geometry and Symmetric Spaces.* Academic Press, New York.

Herz, C. S. (1955). Bessel functions of a matrix argument. *Annals of Mathematics*, **61**, 474–523.

Hodges, J. L. (1955). On the noncentral beta distribution. *Ann. Math. Statist.*, **26**, 648–653.

Hsu, P. L. (1938). Notes on Hotelling's generalized T. *Ann. Math. Statist.*, **9**, 231–243.

Itô, K. (1956). Asymptotic formula for the distribution of Hotelling's generalized T_0^2 statistic. *Ann. Math. Statist.*, **27**, 1091–1105.

Itô, K. (1960). Asymptotic formula for the distribution of Hotelling's generalized T_0^2 statistic II. *Ann. Math. Statist.*, **31**, 1148–1153.

Itô, K. (1970). The topological support of Gauss measure on Hilbert Space. *Nagoya Math. J.*, **38**, 181–183.

James, A. T. (1954). Normal multivariate analysis and the orthogonal group. *Ann. Math. Statist.*, **25**, 40–75.

James, A. T. (1955a). The noncentral Wishart distribution. *Proc. Royal Society Series*, **A 229**, 364–366.

James, A. T. (1955b). A generating function for averages over the orthogonal group. *Proc. Royal Society Series*, **A 229**, 367–375.

James, A. T. (1957). The relationship algebra of an experimental design. *Ann. Math. Statist.*, **28**, 993–1002.

James, A. T. (1960). The distribution of the latent roots of the covariance matrix. *Ann. Math. Statist.*, **31**, 151–158.

James, A. T. (1961a). The distribution of noncentral means with known covarance. *Ann. Math. Statist.*, **32**, 874–882.

James, A. T. (1961b). Zonal polynomials of the real positive definite symmetric matrices. *Annals of Mathematics*, **74**, 456–469. (This is the theory paper.)

James, A. T. (1964). Distributions of matrix variates and latent roots derived from normal samples. *Ann. Math. Statist.*, **35**, 475–501. (Survey paper of all known results. Almost no proofs.)

James, A. T. (1966). Inference on latent roots by calculation of hypergeometric functions of matrix argument. *Multivariate Analysis* (P. R. Krishnaiah, ed.) Academic Press, New York.

James, A. T. (1968). Calculation of zonal polynomial coefficients by use of the Laplace–Beltrami operator. *Ann. Math. Statist.*, **39**, 1711–1718. (Expression in monomials showing coefficients are positive.)

James, A. T. (1969). Tests of equality of latent roots of the covariance matrix. *Multivariate Analysis II* (P. R. Krishnaiah, ed.) Academic Press, New York.

James, A. T., and Wilkinson, G. N. (1971). Factorization of the residual operator and canonical decomposition of nonorthogonal factors in the analysis of variance. *Biometrika*, **58**, 279–294.

James, G. and Kerber, A. (1981). *The Representation Theory of the Symmetric Group.* Addison-Wesley, Reading, Mass.

John, S. (1971). Some optimal multivariate tests. *Biometrika*, **58**, 123–127.

John, S. (1972). The distribution of a statistic used for testing sphericity of a normal distribution. *Biometrika*, **59**, 169–173.

Johnstone, I. M. (1981). *Admissible Estimation of Poisson Means, Birth-Death Processes, and Discrete Dirichlet Problems.* Ph. D. Dissertation, Cornell University.

Kabe, D. G. (1965). Generalization of Sverdrup's lemma and its applications to multivariate distribution theory. *Ann. Math. Statist.*, **36**, 671–676.

Karlin, S. (1960). Notes on Multivariate Analysis. Preprint, distributed to statistics students at Stanford University, Spring 1960.

Kates, L. K. (1980). *Zonal Polynomials.* Ph. D. Dissertation, Princeton University.

Kemperman, J. H. B. (1977). On the FKG-inequality for measures on a partially

ordered space. *Indag. Math.*, **39**, 313–331.
Kettenring, J. R. (1971). Canonical analysis of several sets of variables. *Biometrika*, **58**, 433–451.
Khatri, C. G. (1967). Some distribution problems connected with the characteristic roots of $S_1 S_2^{-1}$. *Ann. Math. Statist.*, **38**, 944–948.
Khatri, C. G., and Pillai, K. C. S. (1968). On the noncentral distributions of two test criteria in multivariate analysis of variance. *Ann. Math. Statist.*, **39**, 215–226.
Kiefer, J. (1957). Invariance, minimax sequential estimation, and continuous time processes. *Ann. Math. Statist.*, **28**, 573–601.
Kiefer, J. (1958). On the nonrandomized optimality and randomized nonoptimality of symmetrical designs. *Ann. Math. Statist.*, **29**, 675–699.
Kiefer, J. (1966). Multivariate optimality results. *Multivariate Analysis*, (P. R. Krishnaiah, ed.) Academic Press, New York.
Kiefer, J. (1974). Equivalence theory for optimal designs. (Approximate theory.) *Ann. Statist.*, **2**, 849–879.
Kiefer, J., and Schwartz, R. (1965). Admissible Bayes character of T^2-, R^2-, and other fully invariant tests for classical multivariate normal problems. *Ann. Math. Statist.*, **36**, 747–770.
Knutson, D. (1973). *λ-Rings and the Representation Theory of the Symmetric Group*. Lecture Notes in Mathematics 308, Springer-Verlag, Berlin.
Koehn, U. (1970). Global cross sections and the densities of maximal invariants. *Ann. Math. Statist.*, **41**, 2045–2056.
Koehn, U., and Thomas, D. L. (1975). On statistics independent of a sufficient statistic. *American Statistician*, **29**, 40–41.
Korin, B. P. (1968). On the distribution of a statistic used for testing a covariance matrix. *Biometrika*, **55**, 171–178. (Obtains asymptotic series and checks accuracy of previous results.)
Kruskal, W. (1954). The monotonicity of the ratio of two noncentral t-density functions. *Ann. Math. Statist.*, **25**, 162–165.
Kruskal, W. (1968). When are Gauss–Markov and least squares estimators identical? A coordinate free approach. *Ann. Math. Statist.*, **39**, 70–75.
Kruskal, W. K. (1961). The coordinate-free approach to Gauss–Markov estimation, and its application to missing and extra observations. *Fourth Berkeley Symposium on Mathematical Statistics and Probability, Vol. 1*, University of California Press.
Kshirsagar, A. M. (1961). The noncentral multivariate beta distribution. *Ann. Math. Statist.*, **32**, 104–111. (Special functions and random variable methods.)
Kullback, S. (1934). An application of characteristic functions to the distribution problem of statistics. *Ann. Math. Statist.*, **5**, 263–307. (Obtains the distribution of a product of three or more chi-squares.)
Kullback, S. (1959). *Information Theory and Statistics*. John Wiley & Sons, Inc., New York.
Kuo, H. (1975). *Gaussian Measures in Banach Sapces*. Lecture Notes in Mathematics 403. Springer-Verlag, Berlin.
Kushner, H. B., Lebow, A., and Meisner, M. (1981). Eigenfunctions of expected value operators in the Wishart distribution II. *J. Multivariate Anal.*, **11**, 418–433.
Kushner, H. B. and Meisner, M. (1980). Eigenfunctions of expected value operators in the Wishart distribution. *Ann. Statist.*, **8**, 977–988.
Lee, Y.-S. (1971). Asymptotic formulae for the distribution of a multivariate test statistic: Power comparisons of certain multivariate tests. *Biometrika*, **58**, 647–651.
Lehmann, E. L. (1959). *Testing Statistical Hypotheses*. John Wiley & Sons, Inc., New York.
Lehmann, E. L. (1983). *Theory of Point Estimation*. John Wiley & Sons, Inc., New York.
Lipster, R. S. and Shiryaev, A. N. (1977). *Statistics of Random Processes I. General Theory*. Springer-Verlag, Berlin.

Littlewood, D. E. (1940). *The Theory of Group Characters and Matrix Representations of Groups*. Oxford, The Clarendon Press.
Littlewood, D. E. (1950). *The Theory of Group Characters and Matrix Representations of Groups*. Oxford at the Clarendon Press.
Loewner, C. (1950). Some classes of functions defined by difference or differential inequalities. *Bull. Amer. Math. Soc.*, **56**, 308–319.
Loomis, L. H. (1953). *An Introduction to Abstract Harmonic Analysis*. Van Nostrand, New York.
Macdonald, I. G. (1979). *Symmetric Functions and Hall Polynomials*. Oxford Mathematical Monographs, Clarendon Press, Oxford.
Magnus, J. R., and Neudecker, H. (1979). The commutation matrix: Some properties and applications. *Ann. Statist.*, **7**, 381–394. (Calculus of vec and applications to covariances.)
Malley, J. D. (1983). Statistical and algebraic independence. *Ann. Statist.*, **11**, 341–345.
Mandelbaum, A. (1983). Linear estimators and measurable linear transformations on a Hilbert Space. Cornell University Preprint.
Mandelbaum, A. (1983). *Linear estimators of the mean of a Gaussian distribution on a Hilbert Space*. Ph.D. Dissertation, Cornell University.
Marden, J. I. (1982). Minimal complete classes of tests of hypotheses with multivariate one-sided alternatives. *Ann. Statist.*, **10**, 962–970.
Marden, J. I. (1983). Admissibility of invariant tests in the general multivariate analysis of variance problem. *Ann. Statist.*, **11**, 1086–1099.
Marshall, A. W., and Olkin, I. (1979). *Inequalities: Theory of Majorization and its Applications*. Academic Press, New York.
Mathai, A. M., and Rathie, P. N. (1971). The exact distributions of Wilk's criterion. *Ann. Math. Statist.*, **42**, 1010–1019. (An example of the inversion of Mellin transforms.)
Mathai, A. M., and Saxena, R. K. (1969). Distribution of a product and the structural set up of densities. *Ann. Math. Statist.*, **40**, 1439–1448. (Use of H-functions to obtain distribution of a product of noncentral chi-squares.)
Mathai, A. M., and Saxena, R. K. (1978). *The H-Function with Applications in Statistics and Other Disciplines*. John Wiley & Sons, Inc., New York.
Mauldon, J. G. (1955). Pivotal quantities for Wishart's and related distributions. *Jour. Royal Statistical Society Series*, **B 17**, 79–85. (Also obtained the lower triangular decomposition of the covariance matrix.)
MacRae, E. C. (1974). Matrix derivatives with an application to an adaptive linear decision problem. *Ann. Statist.*, **2**, 337–346.
Miller, K. S. (1964). *Multidimensional Gaussian Distributions*. John Wiley & Sons, Inc., New York.
Moran, P. A. P. (1970). On asymptotically optimal tests of composite hypotheses. *Biometrika*, **57**, 47–55.
Mudholkar, G. S. (1965). A class of tests with monotone power functions for two problems in multivariate statistical analysis. *Ann. Math. Statist.*, **36**, 1794–1801.
Mudholkar, G. S. (1966a). On confidence bounds associated with multivariate analysis of variance and non-independence between two sets of variates. *Ann. Math. Statist.*, **37**, 1736–1746. (Matrix inequalities.)
Mudholkar, G. S. (1966b). The integral of an invariant unimodal function over an invariant convex set—an inequality and applications. *Proc. Amer. Math. Soc.*, **17**, 1327–1333. (Unbiasedness.)
Mudholkar, G. S., and Rao, P. S. R. S. (1967). Some sharp multivariate Tchebycheff inequalities. *Ann. Math. Statist.*, **38**, 393–400.
Muirhead, R. (1970a). Systems of partial differential equations for hypergeometric functions of a matrix argument. *Ann. Math. Statist.*, **41**, 991–1001.
Muirhead, R. (1970b). Asymptotic distributions of some multivariate tests. *Ann. Math. Statist.*, **42**, 1002–1010.

Muirhead, R. (1972a). On the test of independence between two sets of variates. *Ann. Math. Statist.*, **43**, 1491–1497.

Muirhead, R. (1972b). The asymptotic noncentral distribution of Hotelling's generalized T_0^2. *Ann. Math. Statist.*, **43**, 1671–1677.

Muirhead, R. J. (1982). *Aspects of Multivariate Statistical Theory*. John Wiley & Sons, Inc., New York.

Nachbin, L. (1965). *The Haar Integral*. Van Nostrand, New York.

Naimark, M. A. (1970). *Normed Rings*. Groningen: Walters-Noordhoff.

Neudecker, H. (1969). Some theorems on matrix differentiation with special reference to Kronecker products. *J. Am. Statist. Assoc.*, **64**, 953–963. (Calculus of vec, ψ, and Jacobians.)

Okamoto, M. (1963). An asymptotic expansion for the distribution of the linear discriminant function. *Ann. Math. Statist.*, **34**, 1286–1301.

Okamoto, M. (1973). Distinctness of the eigenvalues of a quadratic form in a multivariate sample. *Ann. Statist.*, **1**, 763–765.

Olkin, I. (1951). On distribution problems in multivariate analysis. Institute of Statistics Mimeograph Series, No. 43, pp. 1–126.

Olkin, I. (1952). Note on 'The Jacobians of certain matrix transformations useful in multivariate analysis.' *Biometrika*, **40**, 43–46.

Olkin, I. (1959). A class of integral identities with matrix argument. *Duke Math. J.*, **26**, 207–213.

Olkin, I. (1966). Special topics in matrix theory and inequalities. Mimeographed notes recorded by M. L. Eaton, Department of Statistics, Stanford University.

Olkin, I., and Perlman, M. D. (1980). Unbiasedness of invariant tests for MANOVA and other multivariate problems. *Ann. Statist.*, **8**, 1326–1341.

Olkin, I., and Roy, S. N. (1954). On multivariate distribution theory. *Ann. Math. Statist.*, **25**, 329–339.

Olkin, I., and Rubin, H. (1964). Multivariate beta distributions and independence properties of Wishart distributions. *Ann. Math. Statist.*, **35**, 261–269.

Olkin, I., and Siotani, M. (1964). Asymptotic distribution of functions of a correlation matrix. Technical Report No. 6, Stanford University.

Ord, J. K. (1968). Approximations to distribution functions which are hypergeometric series. *Biometrika*, **55**, 243–248.

Ostrowski, A. (1952). Sur quelques applications des fonctions convexes et concaves au sens de I. Schur. *J. Math. Pures Appl.*, **31**, 253–292.

Parkhurst, A. M., and James, A. T. (1974). Zonal polynomials of order 1 through 12. *Selected Tables in Mathematical Statistics, Vol. 2*, American Mathematical Society, Providence.

Pearson, E. S. (1968). Studies in the history of probability and statistics xx. Some early correspondence between W. S. Gosset, R. A. Fisher, and Karl Pearson, with notes and comments. *Biometrika*, **55**, 445–457.

Penrose, R. A. (1955). A generalized inverse for matrices. *Proc. Camb. Phil. Soc.*, **51**, 406–413.

Perlman, M. D. (1969). One-sided testing problems in multivariate analysis. *Ann. Math. Statist.*, **40**, 549–567.

Perlman, M. D. (1980). Unbiasedness of the likelihood ratio test for equality of several covariance matrices and equality of several multivariate normal populations. *Ann. Statist.*, **8**, 247–263.

Pillai, K. C. S., and Jouris, G. M. (1969). On the moments of elementary symmetric functions of the roots of two matrices. *Ann. Inst. Statist. Math.*, **21**, 309–320. (Functions of I-S as series of polynomials of S.)

Pillai, K. C. S., and Sugiyama, T. (1969). Noncentral distributions of the largest latent roots of three matrices in multivariate analysis. *Ann. Inst. Statist. Math.*, **21**, 321–327. (More constants.)

Pillai, K. C. S. (1966). Noncentral multivariate beta distribution and the moments of traces of some matrices. *Multivariate Analysis* (P. R. Krishnaiah, ed.) Academic Press, New York.

Pillai, K. C. S. (1977). Distributions of characteristic roots in multivariate analysis. Part II. Non-null distributions. *Can. J. Statist.*, **5**, 1–62.

Pitman, E. J. G. (1939). Tests of hypotheses concerning location and scale parameters. *Biometrika*, **31**, 200–215. (Unbiasedness of tests.)

Press, S. J. (1966). Linear combinations of noncentral chi-square variates. *Ann. Math. Statist.*, **37**, 480–487.

Rao, C. R. (1976). Estimation of parameters in a linear model. *Ann. Statist.*, **4**, 1023–1037. (The 1975 Wald Memorial Lecture.)

Remmel, J. B. and Whitney, R. (1983). Multiplying Schur functions. Preprint, Department of Mathematics, University of California, San Diego.

Riesz, F. and Sz.-Nagy, B. (1955). *Functional Analysis.* Frederic Ungar Publishing Co., New York.

Rinott, Y. (1976). On convexity of measures. *Ann. Prob.*, **4**, 1020–1026.

Rozanov, J. A. (1971). Infinite-dimensional Gaussian distribution. *American Mathematical Society Translations of the Proceedings of the Steklov Institute of Mathematics 108.*

Salaevskii, Y. (1971). Essay in Investigations in Classical Problems of Probability Theory and Mathematical Statistics by V. M. Kalinin and O. V. Shalaevskii. *V. A. Steklov Inst. Seminars Vol. 13.*

Saw, J. (1970). The multivariate linear hypothesis with nonnormal errors and a classical setting for the structure of inference in a special case. *Biometrika*, **57**, 531–535.

Saw, J. G. (1977). Zonal polynomials: An alternative approach. *J. Multivariate Anal.*, **7**, 461–467.

Scheffé, H. (1959). *The Analysis of Variance.* John Wiley & Sons, Inc., New York.

Schwartz, R. E. (1966a). *Properties of invariant multivariate tests.* Cornell University, Ph. D. Dissertation.

Schwartz, R. E. (1966b). Fully invariant proper Bayes tests. *Multivariate Analysis*, (P. R. Krishnaiah, ed.) Academic Press, New York.

Schwartz, R. E. (1967a). Admissible tests in multivariate analysis of variance. *Ann. Math. Statist.*, **38**, 698–710.

Schwartz, R. E. (1967b). Locally minimax tests. *Ann. Math. Statist.*, **38**, 340–359.

Schwartz, R. E. (1969). Invariant proper Bayes tests for exponential families. *Ann. Math. Statist.*, **40**, 270–283.

Serre, J. P. (1977). *Linear Representations of Finite Groups.* Springer-Verlag, New York.

Shah, B. K. (1970). Distribution theory of positive definite quadratic form with matrix argument. *Ann. Math. Statist.*, **41**, 692–697.

Shanbhag, D. N. (1970). On the distribution of a quadratic form. *Biometrika*, **57**, 222–223.

Shepp, L. A. (1965). Distinguishing a sequence of random variables from a translate of itself. *Ann. Math. Statist.*, **36**, 1107–1112.

Shepp, L. A. (1966). Radon–Nikodym derivatives of Gaussian measures. *Ann. Math. Statist.*, **37**, 321–354.

Sherman, S. (1955). A theorem on convex sets with applications. *Ann. Math. Statist.*, **26**, 763–767.

Shorrock, R. W., and Zidek, J. V. (1976). An improved estimator of the generalized variance. *Ann. Statist.*, **4**, 629–638.

Skorohod, A. V. (1974). *Integration in Hilbert Space.* Springer-Verlag, Berlin.

Srivastava, M. S. (1968). On the distribution of a multiple correlation matrix: noncentral multivariate beta distributions. *Ann. Math. Statist.*, **39**, 227–232.

Stein, C. (1956a). Inadmissibility of the usual estimator for the mean of a multivariate normal distribution. *Proceedings of the Third Berkeley Symposium on Mathematical Statistics and Probability, Vol. 1.* University of California Press.

Stein, C. (1956b). The admissibility of Hotelling's T^2-test. *Ann. Math. Statist.*, **27**, 616–623.
Stein, C. (1956c). Some problems of multivariate analysis I. Technical Report No. 6, Department of Statistics, Stanford University. (Referenced by U. Koehn for the first source of the maximal invariant theory, see Chapter 10.)
Stein, C. (1959). Notes on Multivariate Analysis. Preprint distributed to statistics students at Stanford University around 1959.
Stein, C. M. (1966). Multivariate Analysis. Mimeograph notes recorded by M. L. Eaton, Stanford University, Department of Statistics.
Sugiura, N. (1973). Derivatives of the characteristic root of a symmetric or a Hermitian matrix with two applications in multivariate analysis. *Communications in Statistics*, **1**, 393–417.
Sugiura, N. and Fujikoshi, Y (1969). Asymptotic expansions of the non-null distributions of the likelihood ratio criteria for multivariate linear hypothesis and independence. *Ann. Math. Statist.*, **40**, 942–952.
Sugiura, N., and Nagao, H. (1968). Unbiasedness of some test criteria for the equality of one or two covariance matrices. *Ann. Math. Statist.*, **39**, 1686–1692.
Sugiura, N. (1969). Asymptotic expansions of the distributions of the likelihood ratio criteria for covariance matrix. *Ann. Math. Statist.*, **40**, 2051–2063.
Takemura, A. (1982). *A statistical approach to zonal polynomials*. Ph. D. Dissertation, Stanford University.
Takemura, A. (1984). *Zonal polynomials*. Institute of Mathematical Statistics Lecture Notes—Monograph Series, Vol. 4.
Towber, J. (1981). Personal communications while visiting Cornell University.
Wasow, W. (1956). On the asymptotic transformations of certain distributions into the normal distribution. *Proceedings of the 6th Symposium Applied Mathematics of Amer. Math. Soc. Vol. iv*, 251–259, New York: McGraw Hill.
Weyl, H. (1946). *The Classical Groups*. Princeton University Press, Princeton, N.J.
Weyl, H. (1949). Inequalities between two kinds of eigenvalues of a linear transformation. *Proc. Nat. Acad. Sci.*, **35**, 408–411.
Widder, D. V. (1941). *The Laplace Transform*. Princeton University Press, Princeton, N.J.
Wiener, N. (1933). *The Fourier Integral and Certain of its Applications*. Dover Publications, Inc., New York.
Wijsman, R. A. (1957). Random orthogonal transformations and their use in some classical distribution problems in multivariate analysis. *Ann. Math. Statist.*, **28**, 415–423.
Wijsman, R. A. (1958). Lectures on Multivariate Analysis. Notes recorded by R. H. Farrell.
Wijsman, R. A. (1966). Cross-sections of orbits and their application to densities of maximal invariants. *Proceedings of the Fifth Berkeley Symposium on Mathematical Statistics and Probability, Vol. 1*, University of California Press.
Wijsman, R. A. (1983). Monotonicity in the noncentrality parameter of the ratio of two noncentral t-densities. *Ann. Statist.*, **11**, 1008–1010.
Wijsman, R. A. (1984). Global cross sections as a tool for factorization of measures and distribution of maximal invariants. Preprint, Department of Mathematics, University of Illinois at Urbana.
Wilks, S. S. (1962). *Mathematical Statistics*. John Wiley & Sons, Inc., New York.
Wishart, J. (1928). The generalized product moment distribution in samples from a normal multivariate population. *Biometrika*, **20A**, 32–52.
Wong, E. (1971). *Stochastic Processes in Information and Dynamical Systems*. McGraw Hill, New York.
Zidek, J. (1978). Deriving unbiased risk estimators of multinormal mean and regression coefficient estimators using zonal polynomials. *Ann. Statist.*, **6**, 769–782.

Index